METHODS IN MOLECULAR BIOLOGY

Series Editor
John M. Walker
School of Life and Medical Sciences
University of Hertfordshire
Hatfield, Hertfordshire, AL10 9AB, UK

For further volumes:
http://www.springer.com/series/7651

c-di-GMP Signaling

Methods and Protocols

Edited by

Karin Sauer

Department of Biological Sciences
Binghamton University
Binghamton, NY, USA

Humana Press

Editor
Karin Sauer
Department of Biological Sciences
Binghamton University
Binghamton, NY, USA

ISSN 1064-3745 ISSN 1940-6029 (electronic)
Methods in Molecular Biology
ISBN 978-1-4939-7239-5 ISBN 978-1-4939-7240-1 (eBook)
DOI 10.1007/978-1-4939-7240-1

Library of Congress Control Number: 2017943962

Printed on acid-free paper

This Humana Press imprint is published by Springer Nature
The registered company is Springer Science+Business Media LLC
The registered company address is: 233 Spring Street, New York, NY 10013, U.S.A.

Preface

From the relative obscurity of an allosteric activator of a bacterial cellulose synthase, dimeric ($3' \rightarrow 5'$) GMP (cyclic di-GMP or c-di-GMP) has emerged as one of the most common, important, and truly universal bacterial second messengers. Cyclic di-GMP plays key roles in lifestyle changes of many bacteria, including transition from the planktonic to the sessile lifestyle, which aids in the establishment of multicellular biofilm communities, and from the virulent state in acute infections to the less virulent but more resilient state characteristic of chronic infectious diseases. C-di-GMP has also been shown to regulate motility, the cell cycle, and differentiation and to participate in interkingdom signaling, with c-di-GMP being recognized by mammalian immune systems as a uniquely bacterial molecule. Modulating c-di-GMP signaling pathways is based on c-di-GMP levels, with the second messenger being enzymatically modulated by diguanylate cyclases (DCG), proteins containing a GGDEF domain, and phosphodiesterases (PDE) containing either an EAL or HD-GYP domain. Additionally, riboswitches regulate gene expression in response to cyclic di-GMP concentrations in many but not all bacteria.

This volume of the *Methods in Molecular Biology* series provides a collection of protocols for many of the common experimental approaches used in the burgeoning field of c-di-GMP-dependent signaling to synthesize, detect, quantitate, and modulate the levels of c-di-GMP present in cells. Additionally, procedures to detect and evaluate the interaction of c-di-GMP with proteins and bacterial response to varying c-di-GMP levels including virulence, swarming, and matrix production are included. Additionally, some less common but up-and-coming approaches focusing on the inhibition of c-di-GMP signaling are included.

This book is divided into eight major parts, reflecting the breath of techniques used in the field of c-di-GMP. The chapters are as follows: synthesis of c-di-GMP, detection and quantitation of c-di-GMP, visualizing c-di-GMP levels using biosensors, indirect detection of c-di-GMP levels, modulation of c-di-GMP levels and bacterial responses, measuring c-di-GMP modulating activities, c-di-GMP binding proteins, and targeting c-di-GMP signaling. The methods chapters are preceded by a review on the discovery of the intracellular signaling molecule c-di-GMP. Presented methods are diverse and range from thin layer chromatography (TLC) and mass spectrometry to fluorescence-activated cell sorting (FACS), footprinting, pulldown assays, and isothermal titration calorimetry to methods aiming at inhibiting c-di-GMP-dependent signaling and virulence models.

All chapters are written in the same format as that used in the *Methods in Molecular Biology*™ series. Each chapter opens with a description of the basic theory behind the method being described. The Materials section lists all the chemicals, reagents, buffers, and other materials necessary for carrying out the protocol. Since the principal goal of the book is to provide experimentalists with a full account of the practical steps necessary for carrying out each protocol successfully, the Methods section contains detailed step-by-step descriptions of every protocol that should result in the successful completion of each method. The Notes section complements the Methods section by indicating how best to deal with any problem or difficulty that might arise when using a given technique. Considering the contribution of c-di-GMP to biofilm formation, with the human pathogen *Pseudomonas aeruginosa* being a paradigm organism for the study of biofilm communities, the book is most detailed for *P. aeruginosa* but includes also protocols for other model species such as *Escherichia coli*, *Salmonella enterica* serovar Typhimurium, *Xanthomonas campestris,* and *Myxococcus xanthus.*

Together, I hope that this volume will be an essential part of many laboratory libraries. However, I hope that this book is more often on the bench top than in the book shelf and inspire researchers to step out of their comfort zone and try their hands on new approaches.

Binghamton, NY, USA ***Karin Sauer***

Contents

Part IV Indirect Detection of c-di-GMP Levels

Part V Modulation of c-di-GMP Levels and Bacterial Responses

Part VI Measuring c-di-GMP Modulating Activities

Part VII c-di-GMP Binding Proteins

Part VIII Targeting c-di-GMP Signaling

Contributors

GLADYS ALEXANDRE • *Department of Biochemistry, Cellular and Molecular Biology, The University of Tennessee, Knoxville, TN, USA*

HENRIK ALMBLAD • *Department of Biological Sciences, University of Calgary, Calgary, AB, Canada*

SHI-QI AN • *Division of Molecular Microbiology, College of Life Sciences, University of Dundee, Dundee, UK*

JENS BO ANDERSEN • *Department of Immunology and Microbiology, Faculty of Health and Medical Sciences, University of Copenhagen, Copenhagen, Denmark*

VERONIKA ANGERER • *Institute of Biology III, Albert-Ludwigs-University Freiburg, Freiburg, Germany*

HEIKE BÄHRE • *Research Core Unit Metabolomics, Institute of Pharmacology, Hannover Medical School, Hannover, Germany*

EHUD BANIN • *The Mina and Everard Goodman Faculty of Life Sciences, Center for Advanced Materials and Nanotechnology, Bar-Ilan University, Ramat-Gan, Israel*

CLAUDINE BARAQUET • *Department of Microbiology, University of Washington, Seattle, WA, USA; Université de Toulon, MAPIEM, EA4323, La Garde, France*

JENS BERTHELSEN • *Department of Immunology and Microbiology, Faculty of Health and Medical Sciences, University of Copenhagen, Copenhagen, Denmark*

BRADLEY R. BORLEE • *Department of Microbiology, Immunology and Pathology, Infectious Disease Research Center, Colorado State University, Fort Collins, CO, USA*

GRACE I. BORLEE • *Department of Microbiology, Immunology and Pathology, Infectious Disease Research Center, Colorado State University, Fort Collins, CO, USA*

JACOB R. CHAMBERS • *Department of Biological Sciences, Binghamton Biofilm Research Center, Binghamton University, Binghamton, NY, USA*

SONG LIN CHUA • *Singapore Center on Environmental Life Sciences Engineering, Nanyang Technological University, Singapore, Singapore*

ANNIKA CIMDINS • *Department of Microbiology, Tumor and Cell Biology, Karolinska Institutet, Stockholm, Sweden; Institute of Hygiene, University of Münster, Münster, Germany*

RICHARD B. COOLEY • *Department of Molecular Medicine, College of Veterinary Medicine, Cornell University, Ithaca, NY, USA; Department of Biochemistry and Biophysics, Oregon State University, Corvallis, OR, USA*

FRANCESCA CUTRUZZOLÀ • *Department of Biochemical Sciences, Istituto Pasteur Italia-Fondazione Cenci Bolognetti, Sapienza University of Rome, Rome, Italy*

XIAO-XIA DU • *Biodynamic Optical Imaging Center (BIOPIC), School of Life Sciences, Peking University, Beijing, China; State Key Laboratory of Protein and Plant Gene Research, School of Life Sciences, Peking University, Beijing, China*

DORIT ELI • *The Mina and Everard Goodman Faculty of Life Sciences, Center for Advanced Materials and Nanotechnology, Bar-Ilan University, Ramat-Gan, Israel*

LARS-OLIVER ESSEN • *Department of Chemistry, Philipps-University Marburg, Marburg, Germany*

ALAIN FILLOUX • *Department of Life Sciences, MRC Centre for Molecular Bacteriology and Infection, Imperial College London, London, UK*

ANASTASIA FOMICHEVA • *Department of Molecular Biology, University of Wyoming, Laramie, WY, USA*
MICHAEL Y. GALPERIN • *National Center for Biotechnology Information, National Library of Medicine, National Institutes of Health, Bethesda, MD, USA*
G. GIARDINA • *Department of Biochemical Sciences, Istituto Pasteur Italia-Fondazione Cenci Bolognetti, Sapienza University of Rome, Rome, Italy*
MICHAEL GIVSKOV • *Department of Immunology and Microbiology, Faculty of Health and Medical Sciences, Costerton Biofilm Center, University of Copenhagen, Copenhagen, Denmark; Singapore Center on Environmental Life Sciences Engineering, Nanyang Technological University, Singapore, Singapore*
TIMO GLATTER • *Proteomics Core Facility, Biozentrum, University of Basel, Basel, Switzerland; Facility for Mass Spectrometry and Proteomics, Max-Planck Institute for Terrestrial Microbiology, Marburg, Germany*
MARK GOMELSKY • *Department of Molecular Biology, University of Wyoming, Laramie, WY, USA*
JULIE GROIZELEAU • *Department of Immunology and Microbiology, Faculty of Health and Medical Sciences, University of Copenhagen, Copenhagen, Denmark*
JAN GUNDLACH • *Department of General Microbiology, Georg-August-University Göttingen, Göttingen, Germany*
MING C. HAMMOND • *Department of Chemistry, University of California, Berkeley, USA; Department of Molecular and Cell Biology, University of California, Berkeley, USA*
JOE J. HARRISON • *Department of Biological Sciences, University of Calgary, Calgary, AB, Canada*
CAROLINE S. HARWOOD • *Department of Microbiology, University of Washington, Seattle, WA, USA*
REGINE HENGGE • *Institut für Biologie/Mikrobiologie, Humboldt-Universität zu Berlin, Berlin, Germany*
CHRISTINA HERZBERG • *Department of General Microbiology, Georg-August-University Göttingen, Göttingen, Germany*
YASUHIKO IRIE • *Department of Biology, University of Dayton, Dayton, OH, USA*
URS JENAL • *Infection Biology, Biozentrum, University of Basel, Basel, Switzerland*
CHRISTOPHER J. JONES • *Department of Microbial Infection and Immunity, Center for Microbial Interface Biology, Ohio State University, Columbus, OH, USA; Division of Pulmonary, Critical Care, and Sleep Medicine, Department of Internal Medicine, Ohio State University, Columbus, OH, USA*
VOLKHARD KAEVER • *Research Core Unit Metabolomics, Institute of Pharmacology, Hannover Medical School, Hannover, Germany*
JAN KAMPF • *Department of General Microbiology, Georg-August-University Göttingen, Göttingen, Germany*
BARBARA I. KAZMIERCZAK • *Department of Medicine, Yale University, New Haven, CT, USA; Department of Microbial Pathogenesis, Yale University, New Haven, CT, USA*
PETYA V. KRASTEVA • *Institute for Integrative Biology of the Cell (I2BC), Université Paris-Saclay, CEA, CNRS, Université Paris Sud, Gif-sur-Yvette, France*
BENOÎT-JOSEPH LAVENTIE • *Infection Biology, Biozentrum, University of Basel, Basel, Switzerland*
LIVIA LEONI • *Department of Science, University Roma Tre, Rome, Italy*
ZHAO-XUN LIANG • *Division of Chemical Biology and Biotechnology, School of Biological Sciences, Nanyang Technological University, Singapore, Singapore*

F. MANTONI • *Department of Biochemical Sciences, Istituto Pasteur Italia-Fondazione Cenci Bolognetti, Sapienza University of Rome, Rome, Italy*
KEVIN H. MARTIN • *Department of Microbiology, Immunology and Pathology, Infectious Disease Research Center, Colorado State University, Fort Collins, CO, USA*
BRUNO Y. MATSUYAMA • *Department of Physics and Interdisciplinary Science, Institute of Physics of São Carlos, University of São Paulo, São Carlos, SP, Brazil*
J. MAXWELL DOW • *School of Microbiology, University College Cork, Cork, Ireland*
RONAN R. MCCARTHY • *Department of Life Sciences, MRC Centre for Molecular Bacteriology and Infection, Imperial College London, London, UK*
MARCOS V.A.S. NAVARRO • *Department of Physics and Interdisciplinary Science, Institute of Physics of São Carlos, University of São Paulo, São Carlos, SP, Brazil*
LINDSEY O'NEAL • *Department of Biochemistry, Cellular and Molecular Biology, The University of Tennessee, Knoxville, TN, USA*
CLEMENT OPOKU-TEMENG • *Purdue Institute for Drug Discovery, Purdue University, West Lafayette, IN, USA; Department of Chemistry, Center for Drug Discovery, Purdue University, West Lafayette, IN, USA; Biochemistry Graduate Program, University of Maryland, College Park, MD, USA*
A. PAIARDINI • *Department of Biology and Biotechnology "Charles Darwin", Sapienza University of Rome, Rome, Italy*
ALESSIO PAONE • *Department of Biochemical Sciences, Istituto Pasteur Italia-Fondazione Cenci Bolognetti, Sapienza University of Rome, Rome, Italy*
SARIKA VISHNU PAWAR • *Microbial Diversity Research Centre, D. Y. Patil Biotechnology and Bioinformatics Institute, Pune, India*
JOHANN PELTIER • *Laboratoire Pathogenèse des Bactéries Anaérobies, Institut Pasteur, Paris Cedex 15, France; Université Paris Diderot, Sorbonne Paris Cité, Paris Cedex 15, France*
OLGA E. PETROVA • *Department of Biological Sciences, Binghamton Biofilm Research Center (BBRC), Binghamton University, Binghamton, NY, USA*
UTE RÖMLING • *Department of Microbiology, Tumor and Cell Biology, Karolinska Institutet, Stockholm, Sweden*
GIORDANO RAMPIONI • *Department of Science, University Roma Tre, Rome, Italy*
TREVOR E. RANDALL • *Department of Biological Sciences, University of Calgary, Calgary, AB, Canada*
S. RINALDO • *Department of Biochemical Sciences, Istituto Pasteur Italia-Fondazione Cenci Bolognetti, Sapienza University of Rome, Rome, Italy*
MORTEN RYBTKE • *Department of Immunology and Microbiology, Faculty of Health and Medical Sciences, Costerton Biofilm Center, University of Copenhagen, Copenhagen, Denmark*
MIN-HYUNG RYU • *Department of Molecular Biology, University of Wyoming, Laramie, WY, USA; Department of Biological Engineering, Massachusetts Institute of Technology, Cambridge, MA, USA*
LOTTE SØGAARD-ANDERSEN • *Department of Ecophysiology, Max Planck Institute for Terrestrial Microbiology, Marburg, Germany*
KARIN SAUER • *Department of Biological Sciences, Binghamton Biofilm Research Center (BBRC), Binghamton University, Binghamton, NY, USA*
DIEGO O. SERRA • *Institut für Biologie/Mikrobiologie, Humboldt-Universität zu Berlin, Berlin, Germany*
GEOFFREY B. SEVERIN • *Department of Biochemistry and Molecular Biology, Michigan State University, East Lansing, MI, USA*

ROGER SIMM • *Norwegian Veterinary Institute, Oslo, Norway; Department of Oral Biology, Faculty of Dentistry, University of Oslo, Oslo, Norway*
HERMAN O. SINTIM • *Purdue Institute for Drug Discovery, Purdue University, West Lafayette, IN, USA; Department of Chemistry, Purdue University, West Lafayette, IN, USA; Purdue Institute of Inflammation, Immunology and Infectious Disease, West Lafayette, IN, USA*
DOROTA SKOTNICKA • *Department of Ecophysiology, Max Planck Institute for Terrestrial Microbiology, Marburg, Germany*
HOLGER SONDERMANN • *Department of Molecular Medicine, College of Veterinary Medicine, Cornell University, Ithaca, NY, USA*
OLGA SOUTOURINA • *Laboratoire Pathogenèse des Bactéries Anaérobies, Institut Pasteur, Paris Cedex 15, France; Université Paris Diderot, Sorbonne Paris Cité, Paris Cedex 15, France; Institute for Integrative Biology of the Cell (I2BC), CEA, CNRS, Univ. Paris-Sud, Université Paris-Saclay, Gif-sur-Yvette Cedex, France*
JÖRG STÜLKE • *Department of General Microbiology, Georg-August-University Göttingen, Göttingen, Germany*
XIAO-DONG SU • *Biodynamic Optical Imaging Center (BIOPIC), School of Life Sciences, Peking University, Beijing, China; State Key Laboratory of Protein and Plant Gene Research, School of Life Sciences, Peking University, Beijing, China*
JI-LIANG TANG • *State Key Laboratory for Conservation and Utilization of Subtropical Agro-bioresources, College of Life Science and Technology, Guangxi University, Nanning, People's Republic of China*
TIM TOLKER-NIELSEN • *Department of Immunology and Microbiology, Faculty of Health and Medical Sciences, Costerton Biofilm Center, University of Copenhagen, Copenhagen, Denmark*
KATRIN TREFFON • *Department of General Microbiology, Georg-August-University Göttingen, Göttingen, Germany*
MARTINA VALENTINI • *Department of Life Sciences, MRC Centre for Molecular Bacteriology and Infection, Imperial College London, London, UK*
PRABHADEVI VENKATARAMANI • *Division of Chemical Biology and Biotechnology, School of Biological Sciences, Nanyang Technology University, Singapore, Singapore*
XIN C. WANG • *Department of Molecular and Cell Biology, University of California, Berkeley, USA*
CHRISTOPHER M. WATERS • *Department of Microbiology and Molecular Genetics, Michigan State University, East Lansing, MI, USA*
ANNEGRET WILDE • *Institute of Biology III, Albert-Ludwigs-University Freiburg, Freiburg, Germany*
DANIEL J. WOZNIAK • *Department of Microbial Infection and Immunity, Center for Microbial Interface Biology, Ohio State University, Columbus, OH, USA; Department of Microbiology, Ohio State University, Columbus, OH, USA*
JOEY KUOK HOONG YAM • *Singapore Center on Environmental Life Sciences Engineering, Nanyang Technological University, Singapore, Singapore*
LIANG YANG • *Singapore Center on Environmental Life Sciences Engineering, Nanyang Technological University, Singapore, Singapore*
JONGCHAN YEO • *Department of Chemistry, University of California, Berkeley, USA*
JIE ZHOU • *Purdue Institute for Drug Discovery, Purdue University, West Lafayette, IN, USA; Department of Chemistry, Purdue University, West Lafayette, IN, USA*

Chapter 1

Discovery of the Second Messenger Cyclic di-GMP

Ute Römling and Michael Y. Galperin

Abstract

The nearly ubiquitous bacterial second messenger cyclic di-GMP is involved in a multitude of fundamental physiological processes such as sessility/motility transition and the switch between the acute and chronic infection status, combined with cell cycle control. The discovery of cyclic di-GMP, though, has been an example par excellence of scientific serendipity. We recapitulate here its years-long discovery process as an activator of the cellulose synthase of the environmental bacterium *Komagataeibacter xylinus* and its consequences for follow-up research. Indeed, the discovery of cyclic di-GMP as a ubiquitous second messenger contributed to the change in perception of bacteria as simple unicellular organisms just randomly building-up multicellular communities. Subsequently, cyclic di-GMP also paved the way to the identification of other pro- and eukaryotic cyclic dinucleotide second messengers.

Key words Biofilm, Cellulose biosynthesis, Cyclic dinucleotide, Moshe Benziman

1 Main Article

It has now been 30 years since Moshe Benziman and his group published their seminal *Nature* paper on the identification of the long-sought activator of the bacterial cellulose synthase as the cyclic dimeric (3′→5′) guanosine monophosphate (cyclic di-GMP) [1]. That paper signified the end of the long search for a low-molecular weight activator of in vitro cellulose biosynthesis, and, at the same time, marked the beginning of an entirely new area of research into the cellular role(s) of a novel bacterial second messenger, cyclic di-GMP, and mechanisms of its action. The history of the discovery of cyclic di-GMP has been described in detail [2–4]. Here, we present just a brief synopsis of this remarkable story that puts the current studies of cyclic di-GMP in a context.

Cellulose, poly-β-(1→4)-D-glucose, is probably the most abundant biopolymer on this planet. It is the key component of plant cell walls and has found numerous uses as firewood, lumber, paper, and sewing material [5]. In 1886, British scientist Adrian Brown showed that cellulose could be synthesized by certain bacteria [6].

Karin Sauer (ed.), *c-di-GMP Signaling: Methods and Protocols*, Methods in Molecular Biology, vol. 1657,
DOI 10.1007/978-1-4939-7240-1_1, © Springer Science+Business Media LLC 2017

In fact, plant cellulose synthase most likely has bacterial origin and was acquired by ancient plants from the cyanobacterial ancestors of their chloroplasts [7–9]. The process of cellulose biosynthesis by plants has long remained enigmatic and is still not fully understood. Accordingly, bacterial biosynthesis of cellulose, more amenable to experimental research, has been intensively studied throughout the 20th century. Some of such studies have been conducted in the 1930s and the 1940s at the Hebrew University of Jerusalem by Manfred Aschner (1901–1989) and Shlomo Hestrin (1914–1962) using the α-proteobacterium *Acetobacter xylinum* (current name, *Komagataeibacter xylinus*), which is an effective producer of pure microcrystalline cellulose fibers [10, 11]. These studies were subsequently continued by Moshe Benziman (1928–2003) and his group [2].

The basic biochemistry of cellulose biosynthesis in algal, plant, and bacterial cells has been resolved by mid-1970s. It had been shown that the whole process begins with the glycolytic intermediate glucose-6-phosphate, which becomes isomerized to glucose-1-phosphate. Glucose-1-phosphate then reacts with UTP, forming uridine-5′-diphosphate-α-D-glucose (UDP-glucose). UDP-glucose serves as a substrate for the membrane-bound cellulose synthase, which produces cellulose by transferring glucosyl residues from UDP-glucose to the growing β-D-1,4-glucan chain [12]. However, while whole cells of *K. xylinus* demonstrated robust production of cellulose, all attempts to purify active cellulose synthase were unsuccessful [3]. Even partly purified membrane fractions retained only about 0.2% of the cellulose synthase activity of the whole cells [12]. It subsequently became clear that the enzyme required membrane-bound and soluble component(s) to be active.

Benziman's group embarked on a long search for the conditions that would enable purification of an active enzyme. One milestone in this quest was the discovery of a specific activation of the enzyme fraction by micromolar amounts of GTP ($K_a = 34\ \mu M$). To their surprise, a GTP analog guanosine 5′-[gamma-thio]triphosphate (GTPγS, which cannot be hydrolyzed to GDP and P_i) proved to be an even more effective stimulator than GTP with an even lower activation constant $K_a = 17\ \mu M$. With the exception of these two, no other nucleotide or nucleotide derivative could serve as an effective activator of the cellulose synthase [13]. Importantly, GDP, GMP, cGMP, guanosine 5′-[γ-thio]diphosphate and guanosine 5′-[(β,γ-imino]triphosphate were completely inactive. These observations suggested that the actual activator could be a GTP derivative.

Further on, activation by GTP could only be seen in the membrane fraction obtained in the presence of 20% polyethylene glycol (PEG-4000). It became clear that GTP interacted with some additional protein factors that were associated with the membrane-bound cellulose synthase only in the presence of PEG-4000. The presence of this protein factor and GTP activated cellulose synthesis almost 200-fold and achieved the synthesis rates as high as 40% of those obtained with whole cells [13]. Armed with this understanding, Aloni and colleagues succeeded in solubilizing the active cellulose synthase complex. The digitonin-solubilized enzyme still contained the GTP-interacting protein, still retained its capability to respond to GTP, and had essentially the same catalytic and regulatory properties as the membrane-bound form [14].

The next step was to characterize the GTP-binding protein and figure out whether it was an enzyme. This protein was found to bind to an agarose-hexane-GTP column and could be eluted by GTP. It was shown that this protein indeed acted on GTP, converting it to some guanine-containing activating factor. This factor was a low molecular mass, heat-stable compound that could be radioactively labeled when derived from [8-^{3}H]GTP and [α-^{32}P] GTP but not from [γ-^{32}P]GTP. In the presence or absence of this compound, GTP, GDP, GMP, cGMP, ppGppp, GppppG, GpppppG, and guanosine 3′-diphosphate-S-diphosphate were all checked for their ability to stimulate cellulose synthase activity. Neither of them showed any effect, indicating that the activating factor was a previously unknown guanylate derivative [15]. Using chemical analysis, the relative ratios of guanine, ribose, and phosphate in this molecule were shown to be 1:1:1, whereas enzymatic analysis suggested the presence of 2′–5′ or 3′–5′ phosphodiester bonds. So, while its precise structure remained to be determined, the activating factor emerged as a cyclic nucleotide composed of GMP residues with 2′–5′ or 3′–5′ phosphodiester linkages [15].

In their final effort to characterize the activator molecule, Benziman and colleagues used DEAE-Sephadex chromatography to show that the analyzed compound consisted of not more than two GMP moieties, whereas its sensitivity to ribonuclease T1 indicated that these two GMP moieties were linked by a 3′–5′ phosphodiester bond. Mass-spectroscopic measurements estimated the molecular weight of this compound to be 690, which corresponded to the molecular weight of a cyclic diguanylic acid. Finally, the chemically synthesized cyclic bis(3′→5′) diguanylic acid was shown to stimulate cellulose synthase activity and have the same properties as the native activator in a variety of chemical and enzymatic tests [1].

Curiously, despite its importance and novelty, the *Nature* paper by Ross and colleagues reporting the identification of cyclic di-GMP as the activator of cellulose synthase [1] has not attracted much attention. In the 12 years, from 1987 till 2000, cyclic di-GMP has been mentioned in only 10 papers, all but one of which came from Benziman's laboratory. Intriguingly though, these papers addressed several of the fundamental questions in cyclic dinucleotide second messenger signaling, equally important and still unresolved today. Benziman and his close collaborators touched upon the generality of cyclic di-GMP signaling in bacteria by demonstrating activation of a cellulose synthase in an α-proteobacterium different than *K. xylinus* [16, 17], identification of diguanylate cyclases and phosphodiesterases [5, 18–20], the presence of a second unrelated phosphodiesterase to break down pGpG [1, 21, 22] and took up the quest for the determination of the molecular basis of the enzymatic activity of cyclic di-GMP turnover proteins [23]. Equally important, determination of the cyclic di-GMP concentration in the bacterial cell [24, 25] and the nonlinear correlation between cyclic di-GMP concentrations and physiological output [18, 26] came up in the course of the analysis of regulation of cellulose biosynthesis by three different diguanylate cyclases in *K. xylinus*. Furthermore, the biological impact of chemically synthesized cyclic di-GMP analogues was tested [27–29] and screens for inhibitors of cyclic di-GMP diguanylate cyclases were initiated [30, 31], Last, but not least, interkingdom crosstalk of cyclic di-GMP was addressed [32–34] among other issues.

In this century, however, the situation with cyclic di-GMP has changed substantially (Fig. 1). Based on independent observations around the beginning of the century [35–38], Benziman's legacy left the identification of GGDEF and EAL domains as diguanylate cyclases and phosphodiesterases [36, 39–41], signals that regulate cyclic di-GMP turnover proteins [42], cyclic di-GMP receptors [43, 44], and the widespread physiological impact of the ubiquitous second messenger cyclic di-GMP as a major sessility/motility life style, infection style and cell cycle regulator in Bacteria [45–47]

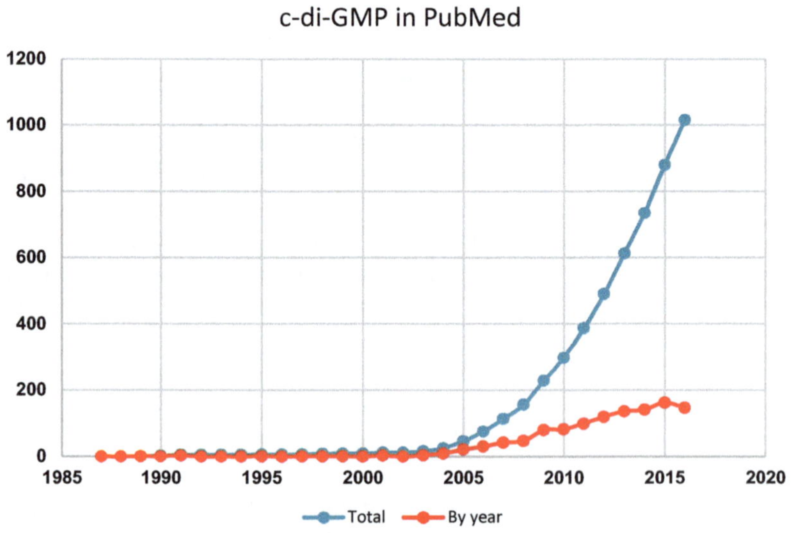

Fig. 1 The history of the c-di-GMP field in publications. The number of papers containing the word "di-GMP" or "diGMP" or "cyclic diguanylate" in PubMed is plotted over time, from 1987 to 2016 (incomplete data for 2016)

to be discovered by others. Unfortunately, and perhaps to his personal disappointment, Benziman's work could not achieve his other goal—to find a way to regulate and enhance cellulose biosynthesis in plants. As we know today, plant cellulose synthases do not require allosteric regulation by cyclic di-GMP [9].

The GGDEF and EAL domains that had been discovered by Benziman and coworkers in the diguanylate cyclases and phosphodiesterases involved in cyclic di-GMP turnover [18] have been found in multiple copies in a variety of diverse Gram-positive and Gram-negative bacteria linearly corrrelated with genome size in different phylogenetic groups [48–50]. Accordingly, cellulose biosynthesis has been detected in various bacterial species throughout the phylogenetic tree, including model organisms *Escherichia coli* and *Salmonella typhimurium* [51, 52]. It soon also became clear that, along with curli fimbriae, cyclic di-GMP-regulated cellulose production plays a key role in biofilm formation in *Escherichia*, *Salmonella*, *Citrobacter*, *Enterobacter*, and *Klebsiella* genera [53]. These findings opened the flood gates, with hundreds of works on cyclic di-GMP published every year (Fig. 1).

Although several cyclic di-GMP binding mechanisms had already been detected [54], it has taken much longer to uncover the exact mechanism of cellulose synthase activation by cyclic di-GMP, even after the identification of the cyclic di-GMP-binding PilZ domain at the C-terminus of the membrane-bound cellulose synthase subunit BcsA [43, 55]. Only after the elucidation of the crystal structure of the bacterial cellulose synthase complex [56], it became clear that it contains a conserved gating loop that blocks access of UDP-glucose to the active site. Upon cyclic di-GMP binding to the PilZ domain, the gating loop moves away from the active site cleft and allows the proper functioning of the enzyme [17].

The discovery of cyclic di-GMP was subsequently followed by the serendipitous detection of prokaryotic cyclic di-AMP and cyclic GAMP, two more cyclic dinucleotides with distinct physiological roles and phylogenetic distribution [57, 58]. Intriguingly, although synthesized by distinct enzyme families, these cyclic dinucleotides seem to be connected through enzyme promiscuity as variants of cyclic di-GMP synthesizing GGDEF domain proteins have recently been shown to produce cyclic GAMP [59]. Of note, the eukaryotic version of cyclic GAMP synthase seems to be a central component of the innate immune surveillance system [60–62]. Finally, to close the circle, cyclic-di-GMP has even reached the eukaryotic world to be involved in cell differentiation in the social amoeba *Dictyostelium discoideum* [47]. We will curiously await the next surprises that this signaling molecule will provide for us.

Acknowledgments

We thank Dr. Dorit Amikam for helpful comments. UR is supported by the Swedish Research Council Natural Sciences and Engineering, the Karolinska Institutet and Petrus and Augusta Hedlund Foundation; MYG is supported by the NIH Intramural Research Program at the U.S. National Library of Medicine.

References

1. Ross P, Weinhouse H, Aloni Y, Michaeli D, Weinberger-Ohana P, Mayer R, Braun S, de Vroom E, van der Marel GA, van Boom JH, Benziman M (1987) Regulation of cellulose synthesis in *Acetobacter xylinum* by cyclic diguanylic acid. Nature 325:279–281
2. Ross P, Mayer R, Benziman M (1991) Cellulose biosynthesis and function in bacteria. Microbiol Rev 55(1):35–58
3. Delmer DP (2000) Structure and biosynthesis of cellulose. Part II: biosynthesis. In: Kung S-D, Yang S-F (eds) Discoveries in plant biology, vol Volume 3. World Scientific Publishing Co., Singapore; Hackensack, NJ; London, pp 199–216
4. Amikam D, Weinhouse H, Galperin MY (2010) Moshe Benziman and the discovery of cyclic di-GMP. In: Wolfe AJ, Visick KL (eds) The second messenger cyclic di-GMP. ASM Press, Washington, DC, pp 11–23
5. Römling U, Galperin MY (2015) Bacterial cellulose biosynthesis: diversity of operons, subunits, products and functions. Trends Microbiol 23(9):545–557. doi:10.1016/j.tim.2015.05.005
6. Brown AJ (1886) On acetic ferment which forms cellulose. J Chem Soc Trans (London) 49:432–439
7. Nobles DR, Romanovicz DK, Brown RM Jr (2001) Cellulose in cyanobacteria. Origin of vascular plant cellulose synthase? Plant Physiol 127(2):529–542
8. Nobles DR, Brown RM Jr (2004) The pivotal role of cyanobacteria in the evolution of cellulose synthases and cellulose synthase-like proteins. Cellulose 11:437–448
9. Pear JR, Kawagoe Y, Schreckengost WE, Delmer DP, Stalker DM (1996) Higher plants contain homologs of the bacterial *celA* genes encoding the catalytic subunit of cellulose synthase. Proc Natl Acad Sci U S A 93 (22):12637–12642
10. Aschner M, Hestrin S (1946) Fibrillar structure of cellulose of bacterial and animal origin. Nature 157:659
11. Hestrin S, Aschner M, Mager J (1947) Synthesis of cellulose by resting cells of *Acetobacter xylinum*. Nature 159:64–65
12. Swissa M, Aloni Y, Weinhouse H, Benizman M (1980) Intermediary steps in *Acetobacter xylinum* cellulose synthesis: studies with whole cells and cell-free preparations of the wild type and a celluloseless mutant. J Bacteriol 143 (3):1142–1150
13. Aloni Y, Delmer DP, Benziman M (1982) Achievement of high rates of *in vitro* synthesis of 1,4-β-D-glucan: activation by cooperative interaction of the *Acetobacter xylinum* enzyme system with GTP, polyethylene glycol, and a protein factor. Proc Natl Acad Sci U S A 79 (21):6448–6452
14. Aloni Y, Cohen R, Benziman M, Delmer D (1983) Solubilization of the UDP-glucose:1,4-β-D-glucan 4-β-D-glucosyltransferase (cellulose synthase) from *Acetobacter xylinum*. A comparison of regulatory properties with those of the membrane-bound form of the enzyme. J Biol Chem 258(7):4419–4423
15. Ross P, Aloni Y, Weinhouse C, Michaeli D, Weinberger-Ohana P, Mayer R, Benziman M (1985) An unusual guanyl oligonucleotide regulates cellulose synthesis in *Acetobacter xylinum*. FEBS Lett 186(2):191–196
16. Amikam D, Benziman M (1989) Cyclic diguanylic acid and cellulose synthesis in *Agrobacterium tumefaciens*. J Bacteriol 171 (12):6649–6655
17. Morgan JL, McNamara JT, Zimmer J (2014) Mechanism of activation of bacterial cellulose synthase by cyclic di-GMP. Nat Struct Mol Biol 21(5):489–496. doi: 10.1038/nsmb.2803 [pii]
18. Tal R, Wong HC, Calhoon R, Gelfand DH, Fear AL, Volman G, Mayer R, Ross P, Amikam D, Weinhouse H, Cohen A, Sapir S, Ohana P, Benziman M (1998) Three *cdg* operons control cellular turnover of cyclic di-GMP in *Acetobacter xylinum*: genetic organization and occurrence of conserved domains in isoenzymes. J Bacteriol 180(17):4416–4425

19. Schirmer T (2016) C-di-GMP synthesis: structural aspects of evolution, catalysis and regulation. J Mol Biol 428(19):3683–3701. doi:10.1016/j.jmb.2016.07.023
20. Schmidt AJ, Ryjenkov DA, Gomelsky M (2005) The ubiquitous protein domain EAL is a cyclic diguanylate-specific phosphodiesterase: enzymatically active and inactive EAL domains. J Bacteriol 187(14):4774–4781. doi:10.1128/JB.187.14.4774-4781.2005
21. Orr MW, Donaldson GP, Severin GB, Wang J, Sintim HO, Waters CM, Lee VT (2015) Oligoribonuclease is the primary degradative enzyme for pGpG in *Pseudomonas aeruginosa* that is required for cyclic-di-GMP turnover. Proc Natl Acad Sci U S A 112(36):E5048–E5057. doi: 10.1073/pnas.1507245112
22. Cohen D, Mechold U, Nevenzal H, Yarmiyhu Y, Randall TE, Bay DC, Rich JD, Parsek MR, Kaever V, Harrison JJ, Banin E (2015) Oligoribonuclease is a central feature of cyclic diguanylate signaling in *Pseudomonas aeruginosa*. Proc Natl Acad Sci U S A 112(36):11359–11364. doi: 10.1073/pnas.1421450112
23. Rao F, Qi Y, Chong HS, Kotaka M, Li B, Li J, Lescar J, Tang K, Liang ZX (2009) The functional role of a conserved loop in EAL domain-based cyclic di-GMP-specific phosphodiesterase. J Bacteriol 191(15):4722–4731. doi:10.1128/JB.00327-09
24. Weinhouse H, Sapir S, Amikam D, Shilo Y, Volman G, Ohana P, Benziman M (1997) c-di-GMP-binding protein, a new factor regulating cellulose synthesis in *Acetobacter xylinum*. FEBS Lett 416(2):207–211
25. Simm R, Morr M, Remminghorst U, Andersson M, Römling U (2009) Quantitative determination of cyclic diguanosine mono phosphate concentrations in nucleotide extracts of bacteria by matrix-assisted laser desorption/ionization-time-of-flight mass spectrometry. Anal Biochem 386(1):53–58. doi: 10.1016/j.ab.2008.12.013
26. Massie JP, Reynolds EL, Koestler BJ, Cong JP, Agostoni M, Waters CM (2012) Quantification of high-specificity cyclic diguanylate signaling. Proc Natl Acad Sci U S A 109 (31):12746–12751. doi:10.1073/pnas.1115663109
27. Blommers MJ, Haasnoot CA, Walters JA, van der Marel GA, van Boom JH, Hilbers CW (1988) Solution structure of the 3′-5′ cyclic dinucleotide d(pApA). A combined NMR, UV melting, and molecular mechanics study. Biochemistry 27(22):8361–8369
28. Hsu C-Y, Dennis D, Jones RA (1985) Synthesis and physical characterization of bis 3′→5′ cyclic dinucleotides (Np-Np):RNA polymerase inhibitors. Nucleosides Nucleotides 4:377–389
29. Witte G, Hartung S, Buttner K, Hopfner KP (2008) Structural biochemistry of a bacterial checkpoint protein reveals diadenylate cyclase activity regulated by DNA recombination intermediates. Mol Cell 30(2):167–178. doi:10.1016/j.molcel.2008.02.020
30. Ohana P, Delmer DP, Carlson RW, Glushka J, Azadi P, Bacic T, Benziman M (1998) Identification of a novel triterpenoid saponin from *Pisum sativum* as a specific inhibitor of the diguanylate cyclase of *Acetobacter xylinum*. Plant Cell Physiol 39(2):144–152
31. Lieberman OJ, Orr MW, Wang Y, Lee VT (2014) High-throughput screening using the differential radial capillary action of ligand assay identifies ebselen as an inhibitor of diguanylate cyclases. ACS Chem Biol 9(1):183–192. doi:10.1021/cb400485k
32. Amikam D, Steinberger O, Shkolnik T,Ben-Ishai Z (1995) The novel cyclic dinucleotide 3′-5′ cyclic diguanylic acid binds to p21^{ras} and enhances DNA synthesis but not cell replication in the Molt 4 cell line. Biochem J 311(Pt 3):921–927
33. Karaolis DK, Means TK, Yang D, Takahashi M, Yoshimura T, Muraille E, Philpott D, Schroeder JT, Hyodo M, Hayakawa Y, Talbot BG, Brouillette E, Malouin F (2007) Bacterial c-di-GMP is an immunostimulatory molecule. J Immunol 178(4):2171–2181
34. Danilchanka O, Mekalanos JJ (2013) Cyclic dinucleotides and the innate immune response. Cell 154(5):962–970. doi:10.1016/j.cell.2013.08.014
35. Jones HA, Lillard JW Jr, Perry RD (1999) HmsT, a protein essential for expression of the haemin storage (Hms+) phenotype of *Yersinia pestis*. Microbiology 145(Pt 8):2117–2128. doi: 10.1099/13500872-145-8-2117
36. Römling U, Rohde M, Olsen A, Normark S, Reinköster J (2000) AgfD, the checkpoint of multicellular and aggregative behaviour in *Salmonella typhimurium* regulates at least two independent pathways. Mol Microbiol 36 (1):10–23
37. Hecht GB, Newton A (1995) Identification of a novel response regulator required for the swarmer-to-stalked-cell transition in *Caulobacter crescentus*. J Bacteriol 177(21):6223–6229
38. Tischler AD, Camilli A (2004) Cyclic diguanylate (c-di-GMP) regulates *Vibrio cholerae* biofilm formation. Mol Microbiol 53(3):857–869. doi: 10.1111/j.1365-2958.2004.04155.x
39. Barends TR, Hartmann E, Griese JJ, Beitlich T, Kirienko NV, Ryjenkov DA, Reinstein J, Shoeman RL, Gomelsky M, Schlichting I (2009)

Structure and mechanism of a bacterial light-regulated cyclic nucleotide phosphodiesterase. Nature 459(7249):1015–1018. doi:10.1038/nature07966

40. Chan C, Paul R, Samoray D, Amiot NC, Giese B, Jenal U, Schirmer T (2004) Structural basis of activity and allosteric control of diguanylate cyclase. Proc Natl Acad Sci U S A 101 (49):17084–17089. doi:10.1073/pnas.0406134101
41. Ausmees N, Mayer R, Weinhouse H, Volman G, Amikam D, Benziman M, Lindberg M (2001) Genetic data indicate that proteins containing the GGDEF domain possess diguanylate cyclase activity. FEMS Microbiol Lett 204 (1):163–167
42. Chang AL, Tuckerman JR, Gonzalez G, Mayer R, Weinhouse H, Volman G, Amikam D, Benziman M, Gilles-Gonzalez MA (2001) Phosphodiesterase A1, a regulator of cellulose synthesis in *Acetobacter xylinum*, is a heme-based sensor. Biochemistry 40(12):3420–3426
43. Amikam D, Galperin MY (2006) PilZ domain is part of the bacterial c-di-GMP binding protein. Bioinformatics 22(1):3–6
44. Sudarsan N, Lee ER, Weinberg Z, Moy RH, Kim JN, Link KH, Breaker RR (2008) Riboswitches in eubacteria sense the second messenger cyclic di-GMP. Science 321(5887):411–413. doi:10.1126/science.1159519
45. Simm R, Morr M, Kader A, Nimtz M, Romling U (2004) GGDEF and EAL domains inversely regulate cyclic di-GMP levels and transition from sessility to motility. Mol Microbiol 53 (4):1123–1134. doi:10.1111/j.1365-2958.2004.04206.x
46. Paul R, Weiser S, Amiot NC, Chan C, Schirmer T, Giese B, Jenal U (2004) Cell cycle-dependent dynamic localization of a bacterial response regulator with a novel di-guanylate cyclase output domain. Genes Dev 18 (6):715–727. doi:10.1101/gad.289504
47. Tischler AD, Camilli A (2005) Cyclic diguanylate regulates Vibrio cholerae virulence gene expression. Infect Immun 73(9):5873–5882
48. Galperin MY, Natale DA, Aravind L, Koonin EV (1999) A specialized version of the HD hydrolase domain implicated in signal transduction. J Mol Microbiol Biotechnol 1 (2):303–305
49. Galperin MY (2001) Conserved 'hypothetical' proteins: new hints and new puzzles. Comp Funct Genomics 2(1):14–18
50. Galperin MY, Nikolskaya AN, Koonin EV (2001) Novel domains of the prokaryotic two-component signal transduction systems. FEMS Microbiol Lett 203(1):11–21
51. Zogaj X, Nimtz M, Rohde M, Bokranz W, Römling U (2001) The multicellular morphotypes of *Salmonella typhimurium* and *Escherichia coli* produce cellulose as the second component of the extracellular matrix. Mol Microbiol 39(6):1452–1463
52. Römling U (2002) Molecular biology of cellulose production in bacteria. Res Microbiol 153 (4):205–212
53. Zogaj X, Bokranz W, Nimtz M, Römling U (2003) Production of cellulose and curli fimbriae by members of the family *Enterobacteriaceae* isolated from the human gastrointestinal tract. Infect Immun 71(7):4151–4158
54. Chou SH, Galperin MY (2016) Diversity of cyclic di-GMP-binding proteins and mechanisms. J Bacteriol 198(1):32–46
55. Ryjenkov DA, Simm R, Romling U, Gomelsky M (2006) The PilZ domain is a receptor for the second messenger c-di-GMP: the PilZ domain protein YcgR controls motility in enterobacteria. J Biol Chem 281(41):30310–30314
56. Morgan JL, Strumillo J, Zimmer J (2013) Crystallographic snapshot of cellulose synthesis and membrane translocation. Nature 493 (7431):181–186
57. Davies BW, Bogard RW, Young TS, Mekalanos JJ (2012) Coordinated regulation of accessory genetic elements produces cyclic di-nucleotides for *V. cholerae* virulence. Cell 149(2):358–370
58. Hallberg ZF, Wang XC, Wright TA, Nan B, Ad O, Yeo J, Hammond MC (2016) Hybrid promiscuous (Hypr) GGDEF enzymes produce cyclic AMP-GMP (3', 3'-cGAMP). Proc Natl Acad Sci USA 113(7):1790–1795
59. Sun L, Wu J, Du F, Chen X, Chen ZJ (2013) Cyclic GMP-AMP Synthase is a cytosolic DNA sensor that activates the type I interferon pathway. Science 339 (6121):786–791
60. Ablasser A, Goldeck M, Cavlar T, Deimling T, Witte G, Röhl I, Hopfner KP, Ludwig J, Hornung V (2013) cGAS produces a 2'-5'-linked cyclic dinucleotide second messenger that activates STING. Nature 498(7454):380–384
61. Gao P, Ascano M, Wu Y, Barchet W, Gaffney BL, Zillinger T, Serganov AA, Liu Y, Jones RA, Hartmann G, Tuschl T, Patel DJ (2013) Cyclic [G(2',5')pA(3',5')p] is the metazoan second messenger produced by DNA-activated cyclic GMP-AMP synthase. Cell 153(5):1094–1107
62. Schaap P, (2013) Cyclic di-nucleotide signaling enters the eukaryote domain. IUBMB Life 65 (11):897–903

Part I

Synthesis of c-di-GMP

Chapter 2

Enzymatic Production of c-di-GMP Using a Thermophilic Diguanylate Cyclase

Prabhadevi Venkataramani and Zhao-Xun Liang

Abstract

C-di-GMP has emerged as a prevalent bacterial messenger that controls a multitude of bacterial behaviors. Having access to milligram or gram quantities of c-di-GMP is essential for the biochemical and structural characterization of enzymes and effectors involved in c-di-GMP signaling. Although c-di-GMP can be synthesized using chemical methods, diguanylate cyclases (DGC)-based enzymatic synthesis is the most efficient method of preparing c-di-GMP today. Many DGCs are not suitable for c-di-GMP production because of poor protein stability and the presence of a c-di-GMP-binding inhibitory site (I-site) in most DGCs. We have identified and engineered a thermophilic DGC for efficient production of c-di-GMP for characterizing c-di-GMP signaling proteins and riboswitches. Importantly, residue replacement in the inhibitory I-site of the thermophilic DGC drastically relieved product inhibition to enable the production of hundreds of milligrams of c-di-GMP using 5–10 mg of this robust biocatalyst.

Key words c-di-GMP, Diguanylate cyclase, Thermophilic enzyme, *Thermotoga maritima*

1 Introduction

C-di-GMP (3′, 5′-cyclic diguanylate) is an intracellular messenger that can be found in many environmental and pathogenic bacteria [1–3]. Accumulating evidence suggests that c-di-GMP plays an active role in the chronic and acute infections caused by many human and plant pathogenic bacteria. There is strong interest in the microbiology community to identify and characterize the c-di-GMP signaling enzymes and effectors involved in virulence expression and biofilm formation. C-di-GMP has also attracted much attention in the aftermath of the findings that c-di-GMP can bind to specific receptors in human cell and modulate host cellular response by inhibiting basal and growth factor-induced proliferation of human carcinoma cells [4–6]. Hence, the use of c-di-GMP and its structural analogs as vaccine adjuvant or immunomodulatory molecule with immunoprophylactic properties is being actively explored [4].

Karin Sauer (ed.), *c-di-GMP Signaling: Methods and Protocols*, Methods in Molecular Biology, vol. 1657,
DOI 10.1007/978-1-4939-7240-1_2, © Springer Science+Business Media LLC 2017

Having access to milligrams or gram quantities of c-di-GMP is essential for biochemical and structural characterization of the proteins and riboswitches involved in c-di-GMP signaling. C-di-GMP can be synthesized chemically or enzymatically. While the chemical synthesis approach is time-consuming and environmentally unsustainable because of the multistep nature of the synthetic process [7–12], enzymatic production of c-di-GMP by using diguanylate cyclases (DGC) only involves a single step that condenses two GTP molecules into c-di-GMP [13]. However, the synthesis of c-di-GMP by using WspR and many other mesophilic DGC proteins suffers from poor yield as a result of poor thermostability and strong product inhibition [14]. In the search for a more robust and efficient DGC for c-di-GMP production, we identified a thermophilic DGC (TM1788) from *Thermotoga maritima* after assessing the possibility of producing soluble protein. The 241 residue-containing TM1788 is predicted to contain a single GGDEF domain (88–241 aa) and a membrane-embedded N-terminal segment (1–86). A gene construct (referred to as tDGC) that encodes the stand-alone GGDEF domain (82–241 amino acids) was synthesized with codon-optimized for *E. coli* overexpression. The recombinant tDGC protein was soluble and enzymatically active, but the yield of c-di-GMP production using this enzyme is low due to strong product inhibition. The inhibition by c-di-GMP was substantially alleviated by replacing a single residue (Arg158) in the I-site with Ala [15, 16]. This highly efficient $tDGC_{R158A}$ mutant (tDGCm) has been used in our lab for the production of c-di-GMP or ^{32}P-labelled c-di-GMP for biochemical and structural studies [14, 17–22]. Solution and X-ray crystallography studies revealed that while tDGC exists as a dimer in solution with c-di-GMP bound in the I-site at the dimerization interface (Fig. 1a), tDGCm exists as a c-di-GMP-free monomer in solution (Fig. 1b) [23]. In contrast to mesophilic enzymes that usually exhibit poor thermostability under in vitro conditions, tDGCm retained 90% of its enzymatic activity after 24 h when incubated at 45 °C and 10 h at 55 °C (Fig. 1c). The tDGCm protein can be immobilized in sol-gel blocks or particles to further extend its shelf-life at room temperature [16]. The sol-gel immobilized tDGCm can be stored at room temperature for up to 6 months and used as a convenient chemical catalyst for c-di-GMP production. In this protocol, we describe the procedures used in our lab for the expression and purification of tDGCm and enzymatic preparation of c-di-GMP in solution. Although we consider tDGCm as the most efficient enzyme currently available for the large-scale production of c-di-GMP given its thermostability and high enzymatic activity, the readers should be aware that several other mesophilic DGCs were assessed and optimized for the synthesis of c-di-GMP in other research labs [24–26].

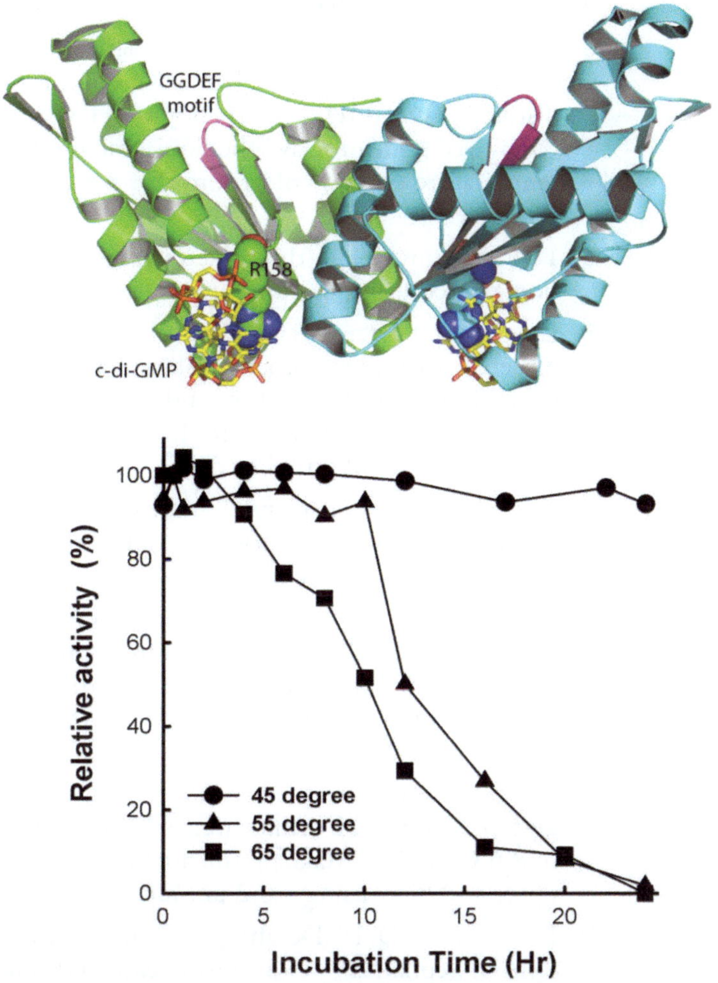

Fig. 1 Structure and thermostability of tDGC. (**a**) Crystal structure of the dimeric tDGC [23]. The conserved GGDEF motifs are highlighted in *pink*. The dimeric c-di-GMP molecules bound at the I-sites are shown in *stick* representation. The residue R158 from the RXXD motif for binding c-di-GMP in the I-site is shown in *sphere* representation. Replacement of R158 is crucial for reliving product inhibition and increasing c-di-GMP production. (**b**) Crystal structure of the monomeric $tDGC_{R158A}$ mutant (tDGCm). (**c**) Thermostability test suggests that tDGCm maintains its enzymatic activity for over 24 h at temperatures lower than 45 °C in our reaction buffer

2 Materials

2.1 Production and Purification of tDGCm

1. Overexpression plasmid pET-tDGCm. The overexpression plasmid is derived from the pET28b(+) vector and harbors a codon-optimized *tDGCm* gene for *E. coli* overexpression. The *tDGCm* gene was cloned into the pET28b(+) vector using the NdeI and XhoI sites to produce N-terminal $(His)_6$-tagged tDGCm. The overexpression plasmid can be obtained from the Liang lab at Nanyang Technological University (E-mail: zxliang@ntu.edu.sg).

2. BL21(DE3) *E. coli* strain.
3. Luria–Bertani medium: 10 g of bacto tryptone, 5 g of yeast extract, and 10 g of NaCl in a 2 l conical flask. Add double distilled H_2O (ddH_2O) or nanopure H_2O to make up the final volume of 1 l. Autoclave the liquid medium and store at room temperature or in a cold room (for storage more than 1 day) until use.
4. 1 M Isopropyl β-D-1-thiogalactopyranoside (IPTG) stock solution filtered using 0.2 μm filter disc.
5. Lysis buffer: 50 mM Tris–HCl (ultrapure), 300 mM NaCl, 5% glycerol, 1% β-mercaptoethanol, and 1 mM phenylmethyl sulfonyl fluoride (PMSF). Mix and adjust pH to 8.0 with HCl. Make up the final volume to 1 l using ddH_2O and filter the buffer using a 0.2 μm filter disc.
6. Wash buffers: W1: 50 ml Lysis buffer that contains 20 mM imidazole (Molecular biology grade >99%) and W2: 20 ml Lysis buffer that contains 50 mM imidazole.
7. Elution buffer: 20 mM Tris–HCl, 200 mM NaCl, 5% glycerol with 200 mM, 300 mM, or 500 mM of imidazole, respectively. Adjust the pH of the three buffers to 8.0 with HCl. Make up the final volume to 1 l using distilled water and filter using a 0.2 μm filter disc.
8. Gel filtration buffer: 50 mM Tris–HCl, 300 mM NaCl, 5% glycerol, and 1 M dithiothreitol (DTT). Mix and adjust pH to 8.0 with HCl. Make up the final volume to 1 l using distilled water and filter using a 0.2 μm filter disc.
9. Disposable PD-10 Desalting Column (14.5 × 50 mm), with 8 ml of Sephadex™ G-25 resin beads (GE Healthcare).
10. 15% sodium dodecyl sulfate–polyacrylamide gel (SDS-PAGE).
11. Bradford dye reagent or other assay kits for protein concentration measurement.
12. Quartz or disposable cuvettes for UV-Vis spectrophotometer.
13. UV-Vis spectrophotometer (SHIMADZU UV-1600).
14. ÄKTA Fast protein purification system (GE Healthcare) (Optional).
15. HiLoad 16/600 Superdex 75 pg gel-filtration size-exclusion column (Optional).
16. Amicon concentrator (10 kDa, 15 ml) from Millipore.
17. Sonicator. SonicsVibra cell VCX500—Tip diameter 1/2″ (13 mm) with threaded end and replaceable tip (volume 10–250 ml).
18. Beckman centrifuge and rotors JA10 rotor (Max speed: 17,700 × *g* or 10,000 rpm); JA25.50 rotor (75,600 × *g* or 25,000 rpm). Refrigeration is needed for both rotors.

19. Shaking incubator.
20. 0.2 and 0.45 μm Minisart syringe filters.
21. Ni^{2+}-nitrilotriacetic acid (NTA) resin (5 ml, Qiagen).
22. Liquid nitrogen.

2.2 Enzymatic Synthesis of c-di-GMP

1. Reaction buffer: 50 mM Tris–HCl, 300 mM NaCl, and 20 mM $MgCl_2$. Dissolve the buffer salts in 190 ml ddH_2O, adjust pH to 8.0 using HCl, make up the final volume to 200 ml with ddH_2O and filter the buffer using a 0.2 μm filter disc.
2. Analytical high performance liquid chromatography (HPLC) system (LC1200, Agilent Technologies) equipped with a 150 × 4.6 mm reverse phase C18 column.
3. HPLC solvent: mix 20 ml of triethyl ammonium bicarbonate (20 mM) buffer and 90 ml methanol (HPLC grade) in 890 ml distilled water to make a 1 l solution. Adjust pH to 7.0 using acetic acid and filter using a 0.2 μm filter disc.
4. GTP stock prepared in situ: Dissolve 1 g GTP powder (sealed and kept in −20 °C freezer) in 1.9 ml distilled water to obtain 1 M stock solution.
5. Water bath set to 45 and 95 °C.
6. Timer.
7. Stirring plate.

2.3 Purification and Quantification of c-di-GMP

1. Preparative HPLC LC-8A equipped with a fraction collector.
2. Preparative HPLC column (Phenomenex, Jupiter RP-C18 300A, 250 × 21.2 mm, 5 μm).
3. Rotary vacuum evaporator (EYELA Rotary evaporator-N-1000, with KIF LAB-vacuum controller and EYELA Digital Waterbath SB-1000).
4. Lyophilizer and lyophilizer flask (Labconco).
5. Methanol (HPLC grade).
6. HPLC buffer: Mix 20 ml of triethyl ammonium bicarbonate (20 mM) buffer and 90 ml methanol in 890 ml distilled water to make 1 l solution. Adjust pH to 7.0 using acetic acid and filter the solvent using a 0.2 μm filter disc.
7. Tris–HCl buffer, 5 mM Tris, pH 7.0.
8. UV-Vis spectrophotometer (Shimadzu UV-1600).

3 Methods

3.1 Expression and Purification of tDGCm

1. The $tDGC_{R158A}$ mutant (tDGCm) gene was cloned into the expression vector pET28b(+) to yield the plasmid pET-

tDGCm for the expression of N-terminal $(His)_6$-tagged recombinant protein. The plasmid pET-tDGCm can be transformed into *E. coli* BL21 (DE3) or other *E. coli* expression strains by standard electroporation methods. Store the *E. coli* cells as 20% glycerol stocks in a −80 °C freezer.

2. For protein expression, inoculate 2 ml of autoclaved LB medium (Kanamycin 50 μg/ml) with the frozen BL21(DE3) cells with an inoculation loop or toothpick. Grow the cells overnight at 37 °C in a shaking incubator.
3. Inoculate 1 l autoclaved LB medium (Kanamycin 50 μg/ml) with 2 ml of the overnight culture in a 2 l conical flask and grow the cells at 37 °C (180 rpm) till they reach an optical density (OD_{600nm}) of 0.8 (3.5–4 h). Cool down the culture broth to 16 °C by adjusting the temperature of the incubator. After the temperature reaches 16 °C in about 40 min, induce protein expression using 0.8 mM IPTG and continue the cultivation at 16 °C (180 rpm) overnight (*see* **Note 1**).
4. Harvest the cells by centrifugation using a JA10 rotor at 11,300 × *g* (8000 rpm) for 8 min. Resuspend the cells in 20 ml of lysis buffer and lyse the cells using a sonicator (settings—02 s ON and 01 s OFF at 40% amp for 15 min) (*see* **Note 2**). Separate the supernatant from cell debris by centrifugation using a JA25.50 rotor at 48,200 × *g* (20,000 rpm) for 30 min and filter the supernatant using a 0.45 μm minisart® syringe filter (cellulose actetate).
5. Wash 5 ml of Ni^{+2}-NTA resin slurry with the lysis buffer in a 10 ml syringe without the plunger. We use a porous filter-disc recycled from a PD-10 column to prevent the resin from falling out. Wash the resin with the lysis buffer a few times to remove ethanol from the resin storage buffer. Allow 5 min during each wash cycle for the resin to settle down. Mix the filtrate obtained above with the resin in the syringe and incubate the mixture at 4 °C for 30 min.
6. Wash the resin with 50 ml of buffer W1 and then with 20 ml of buffer W2. Elute the recombinant protein using a step-gradient with a series of elution buffers containing 200 mM (10 ml), 300 mM (5 ml) and 500 mM (5 ml) imidazole, respectively. Check the content of the eluted fractions using PAGE gel electrophoresis (15% SDS-PAGE gel). Pool the fractions that contain more than 90% recombinant protein.
7. Remove the excess imidazole from the pooled protein fractions using either a PD-10 desalting column (see the instruction from GE Healthcare about the use of the PD-10 column) or a Superdex column (*see* **Note 3**). After desalting, concentrate the protein using a 10 kDa Amicon concentrator at 4000 × *g* to a final concentration of 5–10 mg/ml.

8. To determine the final protein concentration, add 200 μl of Bradford reagent to 800 μl of ddH_2O in a 1 ml cuvette and mix well. Add 1 μl of the protein solution, mix and incubate for 5 min. Measure the absorbance at 595 nm. Calculate the concentration using a BSA (bovine serum albumin) standard curve.
9. Divide the protein solvent into 200 μl aliquots and freeze the protein samples using liquid nitrogen and store the samples in a −80 °C freezer (*see* **Note 4**).

3.2 Enzymatic Production of c-di-GMP

1. Before setting up the enzymatic reaction, prepare the analytical HPLC system by connecting and equilibrating the HPLC column (*see* **Note 5**). Set the wavelength of the UV detector of the HPLC system to 254 nm.
2. Thaw 5 mg of the frozen tDGCm protein sample and transfer the protein solution into 30 ml reaction buffer in a 50 ml Falcon tube. Add 22 μl of freshly prepared 1 M GTP stock solution to the solution to a final concentration of 0.75–0.8 mM GTP (*see* **Note 6**). Mix the solution gently using a spatula and close the lid and incubate at 45 °C in a water bath (*see* **Note 7**).
3. It normally takes 40–50 min for more than 90% of GTP to be converted to c-di-GMP at 45 °C (*see* **Note 8**). The conversion of GTP to c-di-GMP can be conveniently monitored by using an analytical HPLC. Monitoring the progress of the enzymatic reaction is strongly recommended if the users have access to an analytical HPLC. We recommend checking the turnover of GTP 45 min after adding GTP by analyzing 2 μl reaction mixture (isocratic gradient). The turnover of GTP can be estimated by comparing the area of the GTP peak at time zero (before adding enzyme) and any time afterward.
4. When the turnover of GTP reaches 90%, add the next batch of GTP (22 μl of 1 M GTP solution) to the same reaction mixture and continue to incubate the reaction mixture at 45 °C for another 50 min. Repeat the addition of GTP and enzymatic reaction till the turnover of GTP becomes low (<50% after 45 min incubation). Typically, 8–12 batches of GTP can be added in a period of ~10 h before the turnover at 45 min drops below 50%.
5. Following the enzymatic reactions, heat the reaction mixture at 95 °C in a water both for 5 min to denature and precipitate the enzyme. Centrifuge at 6000 × *g* and filter the supernatant using a 0.45 μm filter disc. Keep the filtrate in a −20 °C freezer before the purification process.

3.3 Purification of c-di-GMP Using Preparative HPLC

1. Set up the preparative HPLC system by preparing the solvent and equilibrating the HPLC column (isocratic gradient) (*see* **Notes 9** and **10**). Set the wavelength of the detector to 254 nm and set up the fraction collector according to the size of the collecting tubes.
2. Take out the frozen supernatant from the freezer. Inject 5 ml of the supernatant through the manual sample loading loop (*see* **Note 11**). Start the run and collect the fractions that contain c-di-GMP. A typical chromatogram from one of our HPLC runs is shown in Fig. 2. The peak of the remaining GTP appears at 7 min and the peak of c-di-GMP appears at 18 min. A small peak that represents the linear dinucleotide intermediate can also be seen on the chromatogram.
3. Collect and pool the fractions containing c-di-GMP. The volume of each fraction is 2 ml and usually a total of 7–9 fractions are to be collected for each run. Repeat the HPLC separation for the rest of the supernatant by loading 5 ml each time. At the

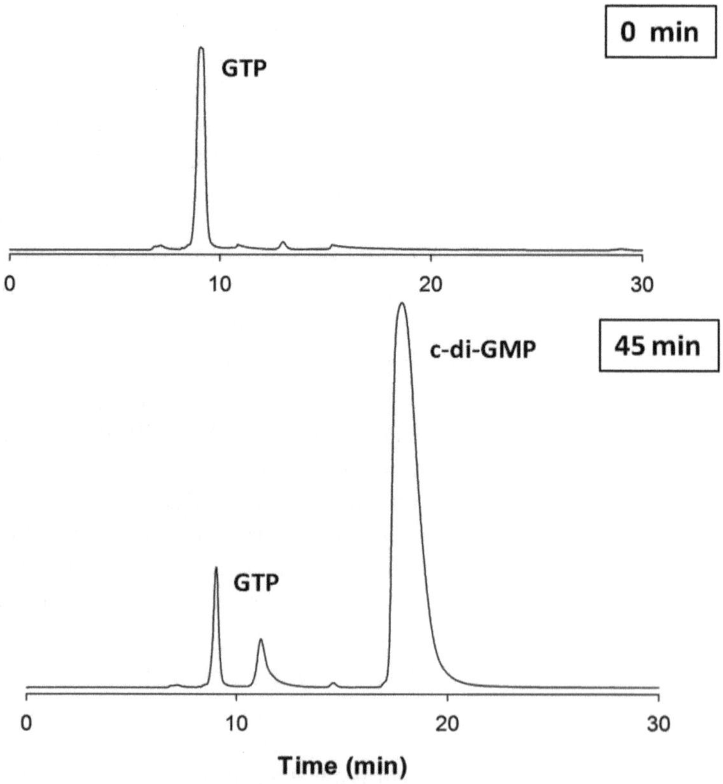

Fig. 2 Preparative HPLC chromatogram showing the turnover of GTP to c-di-GMP. The small peak at 11 min represents a linear dinucleotide intermediate that is often observed during c-di-GMP production

end of the last run, wash the chromatographic column with methanol for 10 min before disconnecting it.

4. We use a two-step procedure to remove the organic solvent and water from the HPLC fractions that contain c-di-GMP. First remove the organic solvent from the pooled fractions using a rotary evaporator with the temperature of the water bath set at 35 °C. The rotary evaporator should be monitored constantly to avoid bumping (*see* **Note 12**). After the removal or organic solvent, transfer the solution to 50 ml falcon tubes. Use a lyophilizer or freeze-dryer to remove H_2O to yield white powder. For researchers who have access to a GeneVac evaporator, the organic solvent and H_2O can be conveniently removed in one step.
5. After solvent removal, dissolve the powder in either a small volume of ddH_2O or 5 mM Tris buffer (pH 7.0) for quantification and storage.
6. Measure the concentration of the c-di-GMP solution using a UV-Vis spectrophotometer according to Beer's law. Dilution of the stock c-di-GMP is required for accurate measurement of the absorbance. The concentration of c-di-GMP can be calculated using the absorbance at 254 nm and extinction coefficient (26,100 M^{-1} cm^{-1}). For example, if the diluted c-di-GMP solution (dilution factor = 500) has an Abs_{254} of 0.85 Au, then the concentration of the stock c-di-GMP solution will be $(0.85/(26{,}100\ M^{-1}\ cm^{-1} \times 1\ cm)) \times 500 = 16.3$ mM. The c-di-GMP solution can be stored in −20 °C or −80 °C freezer (*see* **Note 13**).

4 Notes

1. Under our expression conditions, tDGCm is expressed as soluble protein in reasonably high yield (12–15 mg per liter culture). We usually only observe a small fraction (<5%) of tDGCm in inclusion body.
2. If only smaller sonication probes are available, the culture can be split into smaller volumes and sonicated separately. Alternatively, a French press or microfluidizer can be used for lysing the *E. coli* cells. Our lab uses all three methods and we do not observe significant difference in the activity of tDGCm using different methods of cell lysis.
3. For protein produced by 1 l culture broth, we found that the use of a prepacked desalting PD10 column is the fastest way of removing imidazole. However, for the preparation of more than 10 mg protein, we use an AKTA FPLC system and a

size-exclusion Superdex 75 HR 16/60 column (Amersham Biosciences) for desalting and removing imidazole. The buffer we use for size-exclusion chromatography contains 50 mM Tris–HCl, pH 8.0, 300 mM NaCl, 5% glycerol, and 1 mM DTT.

4. We found that the tDGCm protein stored (at −80 °C) for a long period of time (>3 months) is not as efficient as freshly prepared protein in c-di-GMP production. For best result, we recommend the use of freshly prepared enzyme or enzyme that has been stored for a short period of time (<2 weeks).

5. For analytical HPLC column equilibration, connect the buffer bottles (Buffer A: 20 mM TEAB + 9% methanol and Buffer B: 100% methanol) to the two pumps—A and B respectively. Connect the column to the adapters of the HPLC machine and wash the column with 100% methanol using a flow-rate of 1 ml/min for 10 min. Then equilibrate the column with Buffer A using a flow-rate of 1 ml/min for 20 min.

6. GTP should be kept dry in a sealed container in a −20 °C freezer. Due to hydrolysis of GTP, it is always a good practice to prepare fresh GTP solution in situ for the enzymatic reaction. The production yield decreases after GTP reached ca. 0.8 mM, suggesting inhibition of enzymatic activity at high substrate concentration. Hence, it is important to keep the GTP concentration below 0.8 mM.

7. When setting up the enzymatic reaction, it is important to gently stir the solution with a spatula for 5–10 s to disperse the enzyme molecules.

8. Conditions that include pH, salt and temperature were screened to for the enzymatic reaction. The enzyme tDGCm is most efficient at pH 7.5 and at relatively high salt conditions (250–300 mM NaCl). Thermostability analysis showed that the activity of the enzyme drops slowly with time at 45 and 55 °C (Fig. 1c).

9. It is recommended to use a guard column to protect the HPLC column.

10. For preparative HPLC column equilibration, follow the same steps and use the same buffers as detailed for the analytical HPLC in **Note 5**, but change the flow-rate of the column to 10 ml/min.

11. As determined by the sample-loading loop, the maximum volume of the concentrated supernatant that can be loaded onto our preparative HPLC column is 5 ml. The users can load more sample with a longer sample-loading loop as long as the separation is not compromised. Alternatively, if the users want to reduce the number of HPLC runs, they can reduce the

total volume of the reaction mixture by lyophilization and then redissolve the powder in a small volume of ddH_2O.

12. Bumping occurs when liquid is heated very rapidly and large vapor bubbles are formed to push the liquid out of the container. This rapid expulsion of boiling liquid could pose a serious hazard to researchers.

13. The procedures described here allow us to readily synthesize hundreds of milligrams of c-di-GMP with standard equipment from biochemistry laboratories. For example, with 10 mg of enzyme, we were able to obtain more than 800 mg of c-di-GMP by adding multiple batches of substrate within a period of several hours. There should not be any difficulty to produce c-di-GMP on the scales of multi-grams with more enzyme and longer reaction time.

Acknowledgment

The work on c-di-GMP in the lab of ZXL is supported by a Tier II ARC grant from the Ministry of Education of Singapore and a Tier I ARC grant from Nanyang Technological University (Singapore).

References

1. Galperin MY (2004) Bacterial signal transduction network in a genomic perspective. Environ Microbiol 6:552–567
2. Romling U, Gomelsky M, Galperin MY (2005) C-di-GMP: the dawning of a novel bacterial signalling system. Mol Microbiol 57 (3):629–639
3. Römling U, Galperin MY, Gomelsky M (2013) Cyclic di-GMP: the first 25 years of a universal bacterial second messenger. Microbiol Mol Biol Rev 77(1):1–52
4. Chandra D, Quispe-Tintaya W, Jahangir A, Asafu-Adjei D, Ramos I, Sintim HO, Zhou J, Hayakawa Y, Karaolis DK, Gravekamp C (2014) STING ligand c-di-GMP improves cancer vaccination against metastatic breast cancer. Cancer Immunol Res 2(9):901–910
5. Karaolisa DKR, Chengb K, Lipskyc M, Elnabawia A, Catalanod J, Hyodoe M, Hayakawae Y, Raufman J-P (2005) 3′,5′-Cyclic diguanylic acid (c-di-GMP) inhibits basal and growth factor-stimulated human colon cancer cell proliferation. Biochem Biophys Res Commun 329:40–45
6. Karaolis DKR, Newstead MW, Zeng X, Hyodo M, Hayakawa Y, Bhan U, Liang H, Standiford TJ (2007) Cyclic Di-GMP stimulates protective innate immunity in bacterial *Pneumonia*. Infect Immun 75(10):4942–4950
7. Hayakawa Y, Nagata R, Hirata A, Hyodo M, Kawai R (2003) A facile synthesis of cyclic bis (3′→5′)diguanylic acid. Tetrahedron 59:6465–6471
8. Ross P, Mayer R, Weinhouse H, Amikam D, Huggirat Y, Benziman M, de Vroom E, Fidder A, de Paus P, Sliedregt LA (1990) The cyclic diguanylic acid regulatory system of cellulose synthesis in *Acetobacter xylinum*. Chemical synthesis and biological activity of cyclic nucleotide dimer, trimer, and phosphothioate derivatives. J Biol Chem 265:18933–18943
9. Zhang Z, Gaffney BL, Jones RA (2004) c-di-GMP displays a monovalent metal ion-dependent polymorphism. J Am Chem Soc 126(51):16700–16701
10. Yan H, Aguilar AL (2007) Synthesis of 3′,5′-cyclic diguanylic acid (cdiGMP) using 1-(4-chlorophenyl)-4-ethoxypiperidin-4-yl as a protecting group for 2′-hydroxy functions of ribonucleosides nucleosides. Nucleosides Nucleotides Nucleic Acids 26:189–204
11. Kiburu I, Shurer A, Yan L, Sintim HO (2008) A simple solid-phase synthesis of the ubiquitous bacterial signaling molecule, c-di-GMP and analogues. Mol Biosyst 4:518–520

12. Clivio P, Coantic-Castex S, Guillaume D (2013) (3′-5′)-Cyclic dinucleotides: synthetic strategies and biological potential. Chem Rev 113(10):7354–7401
13. Chan C, Paul R, Samoray D, Amiot NC, Giese B, Jenal U, Schirmer T (2004) Structural basis of activity and allosteric control of diguanylate cyclase. Proc Natl Acad Sci U S A 101 (49):17084–17089
14. Rao F, Yang Y, Qi Y, Liang Z-X (2008) Catalytic mechanism of c-di-GMP specific phosphodiesterase: a study of the EAL domain-containing RocR from *Pseudomonas aeruginosa*. J Bacteriol 190:3622–3631
15. Rao F, Pasunooti S, Ng Y, Zhuo W, Lim L, Liu W, Liang Z-X (2009) Enzymatic synthesis of c-di-GMP using a thermophilic diguanylate cyclase. Anal Biochem 389:138–142
16. Pasunooti S, Surya W, Tan SN, Liang ZX (2010) Sol-gel immobilization of a thermophilic diguanylate cyclase for enzymatic production of cyclic-di-GMP. J Mol Catal B: Enzym 67(1-2):98–103
17. Chen MW, Kotaka M, Vonrhein C, Bricogne G, Rao F, Chuah ML, Svergun D, Schneider G, Liang ZX, Lescar J (2012) Structural insights into the regulatory mechanism of the response regulator RocR from *Pseudomonas aeruginosa* in cyclic Di-GMP signaling. J Bacteriol 194 (18):4837–4846
18. Qi Y, Chuah MLC, Dong X, Xie K, Luo Z, Tang K, Liang Z-X (2011) Binding of C-di-GMP in the non-catalytic EAL domain of FimX induces a long-range conformational change. J Biol Chem 286:2910–2917
19. Qi Y, Rao F, Liang Z-X (2009) A flavin-binding PAS domain regulates c-di-GMP synthesis in *Ax*DGC2 from *Acetobacter xylinum*. Biochemistry 48:10275–10285
20. Qi Y, Xu L, Dong X, Yau YH, Ho CL, Koh SL, Shochat SG, Chou SH, Tang K, Liang ZX (2012) Functional divergence of FimX in PilZ binding and type IV pilus regulation. J Bacteriol 194(21):5922–5931
21. Rao F, See RY, Zhang DW, Toh DC, Ji Q, Liang ZX (2010) YybT Is a signaling protein that contains a cyclic dinucleotide phosphodiesterase domain and a GGDEF domain with ATPase activity. J Biol Chem 285(1):473–482
22. Xu L, Venkataramani P, Ding Y, Liu Y, Deng Y, Yong GL, Xin L, Ye R, Zhang L, Yang L, Liang Z-X (2016) A Cyclic di-GMP-binding adaptor protein interacts with histidine kinase to regulate two-component signaling. J Biol Chem 291(31):16112–16123
23. Deepthi A, Liew CW, Liang ZX, Swaminathan K, Lescar J (2014) Structure of a diguanylate cyclase from thermotoga maritima: insights into activation, feedback inhibition and thermostability. PLoS One 9(10):e110912
24. Spehr V, Warrass R, Höcherl K, Ilg T (2011) Large-scale production of the immunomodulator c-di-GMP from GMP and ATP by an enzymatic cascade. Appl Biochem Biotechnol 165 (3):761–775
25. Zahringer F, Massa C, Schirmer T (2011) Efficient enzymatic production of the bacterial second messenger c-di-GMP by the diguanylate cyclase YdeH from *E. coli*. Appl Biochem Biotechnol 163(1):71–79
26. Christen M, Christen B, Folcher M, Schauerte A, Jenal U (2005) Identification and characterization of a cyclic di-GMP-specific phosphodiesterase and its allosteric control by GTP. J Biol Chem 280(35):30829–30837

Chapter 3

Synthesis of [^{32}P]-c-di-GMP for Diguanylate Cyclase and Phosphodiesterase Activity Determinations

Barbara I. Kazmierczak

Abstract

Diguanylate cyclases that synthesize and phosphodiesterases that hydrolyze the second messenger cyclic-di-GMP (c-di-GMP) are at the center of bacterial signaling pathways that control behaviors relevant to all aspects of microbial physiology and pathogenesis (Romling et al., Microbiol Mol Biol Rev 77(1):1–52, 2013). Bioinformatics tools can easily predict the presence of the diguanylate cyclase GGDEF domain, or the EAL and HD-GYP domains associated with phosphodiesterase activity. However, experimental confirmation of enzymatic activity is still necessary, as many proteins contain degenerate domains that lack catalytic activity but nonetheless function as c-di-GMP receptors.

Key words Diguanylate cyclase, Phosphodiesterase, Cyclic-di-GMP

1 Introduction

Diguanylate cyclase and phosphodiesterase domains are often coupled to regulatory domains within a protein [1], and their enzymatic activity highly regulated, which can complicate the demonstration of in vitro enzymatic activity. Highly sensitive, yet technically accessible methods to assay these activities often rely on the use of radioactively labeled nucleotide substrates. This protocol employs a purified *Caulobacter crescentus* diguanylate cyclase (DGC), PleD*, to synthesize [^{32}P]-c-di-GMP from [α^{32}P]-GTP. The activity of DGCs is often regulated by posttranslational modifications such as phosphorylation, and can be allosterically inhibited by binding of the reaction product, c-di-GMP, to the I-site of the DGC. PleD* is a constitutively active allele of the *Caulobacter crescentus* protein PleD: it carries four point mutations (Asn/Thr 120, Ala/Thr 214, His/Pro 234, Tyr/Asn 357) [2] and exhibits specific activity approximately 100-fold greater than PleD [3]. A polyhistidine tag fused to the carboxy terminus of PleD* allows its purification by immobilized metal affinity chromatography. Incubation of

Karin Sauer (ed.), *c-di-GMP Signaling: Methods and Protocols*, Methods in Molecular Biology, vol. 1657,
DOI 10.1007/978-1-4939-7240-1_3, © Springer Science+Business Media LLC 2017

PleD* with [α^{32}P]-GTP in the presence of Mg^{2+} results in the near-quantitative conversion of GTP to [^{32}P]-c-di-GMP.

Although c-di-GMP can now be commercially purchased, this method allows for highly sensitive assays of phosphodiesterase activity. Thin-layer chromatography (TLC) is used for the detection of radiolabeled nucleotides, a methodology that may be more accessible to some labs than other techniques that are used for the qualitative and quantitative detection of these molecules, such as liquid chromatography-tandem mass spectrometry [4].

2 Materials

Prepare all solutions using deionized water (18 MΩ-cm at 25 °C) and analytical grade reagents. Prepare and store all reagents at room temperature (unless indicated otherwise).

2.1 Bacterial Culture

1. Lysogeny Broth (LB): 10 g tryptone, 5 g yeast extract, 10 g NaCl; resuspend in distilled or deionized water to a final volume of 1 L. Autoclave to sterilize.
2. Antibiotic stock solution: Dissolve ampicillin (1 g) in 10 mL of water to a final concentration of 0.1 g/L (1000×). Sterilize using a syringe filter and freeze (−20 °C) in sterile tubes as 1 mL aliquots. Store thawed aliquots at 4 °C and use within a week.
3. LB agar solid medium: Place LB broth constituents plus 15 g/L agar in a 2 L flask. Add 1 L of water, and a large magnetic stir bar. Autoclave to sterilize. Place hot media on a stirring plate, and gently stir to mix. When media has cooled to ca. 50–55 °C, add 1 mL of antibiotic stock using sterile technique, and allow it to mix. Pour ca. 40 mL of liquid agar per 10 cm Petri dish. Stack plates and allow them to solidify at room temperature.
4. IPTG (isopropyl β-D-1-thiogalactopyranoside, 0.5 M stock solution). Dissolve 1.452 g in 20 mL water, sterilize using a syringe filter, and freeze (−20 °C) in sterile tubes as small aliquots (0.5 mL).
5. Shaking incubator (set at 37 °C and 225 rpm).
6. Centrifuge capable of spinning 250–500 mL bottles at 7000 × *g* at 4–25 °C.

2.2 Protein Expression and Affinity Purification

1. Elution/Wash Buffer: 0.05 M Sodium phosphate pH 7.0, 0.3 M NaCl (store at 4 °C).
2. β-mercaptoethanol (BME): Commercially sold BME is a 14.4 M solution.

3. Phenylmethylsulfonyl fluoride (PMSF Mol wt 174): Prepare 0.2 M stock solution by dissolving 384 mg in 10 mL of isopropanol. Store at room temperature (RT) protected from light; stable for 9 months under these conditions. PMSF is highly unstable in aqueous solutions and must be added immediately prior to use.
4. Immobilized metal affinity resin, e.g., nickel-nitrilotriacetic acid (Ni^{2+}-NTA; available from Qiagen) or cobalt-carboxylmethylaspartate (Co^{2+}-CMA; Talon, available from Clontech).
5. Imidazole (Mol wt 68.1): Dissolve 0.5 g in 50 mL E/W buffer to make 0.15 M imidazole elution buffer. Store at 4 °C.
6. Buffer A: 0.05 M Tris–HCl, pH 8.0, 0.25 M NaCl, 0.005 M BME (store at 4 °C).
7. Dialysis membrane: Spectrum Medical Industries. Choose dialysis tubing of the appropriate dimension and molecular weight cutoff (MWCO) to prevent your protein of interest from equilibrating across the membrane.
8. French pressure cell press and French pressure cell. The French pressure cell should be chilled overnight in a cold room prior to use.
9. Centrifuge equipped with swinging-bucket rotor (4 °C, 700 × *g*).
10. Heating block set to 95 °C.
11. Disposable plastic columns (available from multiple vendors such as Thermo Scientific Pierce or Spectrum Chromatography).

2.3 Synthesis of [^{32}P]-c-di-GMP

1. [α^{32}P]-GTP (3000 Ci/mmol, 10 mCi/mL). Available from PerkinElmer.
2. $MgCl_2$: 0.1 M stock solution.

3 Methods

3.1 Growing a Bacterial Culture and Inducing Expression of the DGC, PleD*

A proper sterile technique should be used throughout.

1. Streak *Escherichia coli* BL21 transformed with pRP89 (pET11; *pleD**, C-terminal His_6 tag) [3] to LB plates supplemented with ampicillin (100 μg/mL).

   Incubate overnight at 37 °C (*see* **Note 1**).
2. Inoculate a 5 mL culture of LB plus ampicillin (100 μg/mL) with a single colony of *E. coli*/pRP89 and grow at 37 °C with shaking (250 rpm) overnight (14–18 h) to stationary phase.
3. Use the entire overnight culture to inoculate 500 mL of LB broth supplemented with ampicillin (100 μg/mL) in a 2 L

flask, and grow bacteria at 37 °C with aeration (225 rpm) to an OD_{600} of 0.8–1.0 (*see* **Note 2**).

4. Add IPTG (0.5 mM final concentration) to induce PleD*-His6 expression, and incubate culture for an additional 45 min at 37 °C with aeration.
5. Transfer culture to centrifuge bottles and spin for 20 min at 7000 × *g* (4–25 °C) to collect pellets. Pour off the supernatant and place bottles on ice (*see* **Note 3**).

3.2 Purification of PleD*

All the steps should be carried out at 4 °C or on ice, unless otherwise noted. Prechill the French pressure cell in a cold room prior to use.

1. Resuspend pellets in a total of 10 mL of cold E/W buffer (*see* **Note 4**). Add BME (final concentration, 5 mM) and the serine protease inhibitor PMSF (final concentration, 0.2 mM) (*see* **Note 5**).
2. Lyse bacteria by passing the resuspended pellet through a French press (14,000 psi). We routinely collect the lysate and pass it through the press a second time to achieve more complete disruption. Remove 0.1 mL of lysate to a microfuge tube and store at −20 °C for subsequent analyses. (This sample serves as a control for polyhistidine production induction.)
3. Centrifuge the lysate for 20 min at 20,000 × *g*. Remove 0.1 mL of cleared lysate to a microfuge tube and store at −20 °C for subsequent analyses. (This sample serves as a control to estimate the fraction of soluble polyhistidine protein.) (*see* **Note 6**).
4. Add the cleared lysate (supernatant) to 2 mL of prepared Ni^{2+}-NTA or Co^{2+}-CMA immobilized metal affinity resin (*see* **Note 7**) and incubate overnight at 4 °C on a nutator or other devices that will allow mixing of the beads and lysate.
5. Centrifuge the lysate-resin at 700 × *g* for 5 min. Remove the supernatant carefully and store at 4 °C for subsequent analyses. (This sample serves as a control for efficiency of affinity purification.)
6. Wash the resin twice with 20 column volumes (40 mL) of E/W buffer, centrifuging for 5 min at 700 × *g* each time.
7. Resuspend resin in E/W buffer and pour into a disposable column. (The volume of buffer used should be about 80–90% of the volume of the column.)
8. Wash column with 10 column volumes (20 mL) of E/W buffer.

9. Elute PleD* from column with 10 column volumes (20 mL) E/W buffer containing 0.15 M imidazole. Collect eluate in 0.5 mL fractions and store on ice.
10. Run 10 μL of each fraction on a 10% SDS-PAGE gel and stain with Coomassie Blue to identify fractions containing PleD*. Five microliter of the lysate (Subheading 3.2, **step 2**), cleared lysate (Subheading 3.2, **step 3**) and post-affinity chromatography lysate (Subheading 3.2, **step 5**) should be included on the gel (*see* **Note 8**).
11. Pool fractions containing eluted PleD*.
12. Dialyze pooled PleD* fractions against 1 L of Buffer A in the cold room. We routinely change the dialysis buffer after 4 h, and let the protein dialyze overnight in the cold room (*see* **Note 9**).

3.3 Synthesis of [^{32}P]-c-di-GMP

All personnel must be trained in and follow appropriate safety precautions when working with ^{32}P. The high-energy beta emissions from ^{32}P can present a substantial skin and eye dose hazard. Appropriate shielding, personal protective equipment and monitoring must be used when working with samples containing ^{32}P. All materials must be disposed of in accordance with your institution's radiation safety protocols.

1. Prepare the reaction mixture in a microfuge tube: combine 22 μL of dialyzed PleD* with 3 μL of 0.1 M $MgCl_2$ and 5 μL of [α^{32}P]-GTP (3000 Ci/mmol, 10 mCi/mL) (*see* **Note 10**).
2. Incubate 15 min at room temperature.
3. Stop the reaction by incubating at 95 °C for 5 min.
4. Centrifuge at room temperature for 2 min at 14,000 × *g*.
5. Store inactivated reaction mix at −20 °C.

Thin-layer chromatography (described in Chapter 22) can be used to confirm activity of PleD* and to monitor production of [^{32}P]-c-di-GMP.

4 Notes

1. A bacterial stock should be prepared by using a single colony (picked from a plate with appropriate antibiotic added to maintain plasmid) to inoculate 3 mL of LB plus antibiotic in a 14 mL culture tube. The culture should be grown overnight at 37 °C with aeration (shake at 250 rpm). Mix 700 μL of the dense culture with 300 μL of autoclaved 50% (v/v) glycerol in a sterile 1.5–2.0 mL tube and store at −80 °C. Scrape the frozen stock with a sterile loop or pipet to remove a small volume of frozen media, then touch to the surface of a culture plate to

melt. This drop can then be streaked out of for single colonies using the standard sterile technique.

2. For convenience, LB media can be prepared and sterilized in the flask that will be used for culturing bacteria. Store the sterilized media at RT until needed, and add appropriate antibiotics at the time of inoculation.
3. Bacterial pellets can be frozen at −80 °C at this time, if necessary. Thaw pellets on ice before proceeding with protein purification.
4. The easiest way to resuspend pellets is to vortex pellet vigorously prior to buffer addition until pellet is slightly loosened. Measure 10 mL of buffer using a 10 mL pipet, and gradually add buffer to pellet, pipetting up and down to resuspend bacterial cells. Carry out the procedure on ice. The pellet must be thoroughly resuspended for effective lysis.
5. Alternate protease inhibitors can be used, but EDTA must be avoided as it will interfere with subsequent immobilized metal-affinity chromatography steps.
6. A French press provides a highly efficient method for lysing bacterial cells without overheating samples (which can lead to loss of protein activity). If a French press is not available, bacterial cells can be disrupted by sonication, or lysed by incubation with cell wall disrupting agents (lysozyme, EDTA) and non-ionic detergent. A ready-to-use formulation such as B-PER™ Complete Bacterial Protein Extraction Reagent (Thermo Scientific) can be employed for this purpose.
7. Qiagen and Clontech are commercial sources for resins that can be used to purify polyhistidine-tagged proteins. Manufacturer's instructions should be used to prepare resin. A thorough discussion of immobilized metal-affinity chromatography can be found in [5].
8. Evaluation of these intermediate samples will help determine if polyhistidine protein production is highly induced (lysate); if it is largely soluble or insoluble (cleared lysate); and if it is effectively removed from lysate by affinity chromatography (post-chromatography lysate).
9. We routinely obtain yields of 0.5–1 mg/mL of PleD* using this protocol.
10. This reaction is set up with the enzyme, PleD* (200–400 pM), in significant molar excess relative to the substrate (17 pM), which ensures quantitative conversion of GTP to cyclic-di-GMP.

References

1. Romling U, Galperin MY, Gomelsky M (2013) Cyclic di-GMP: the first 25 years of a universal bacterial second messenger. Microbiol Mol Biol Rev 77(1):1–52. doi:10.1128/mmbr.00043-12
2. Aldridge P, Paul R, Goymer P, Rainey PB, Jenal U (2003) Role of the GGDEF regulator PleD in polar development of *Caulobacter crescentus.* Mol Microbiol 47:1695–1708, DOI: 10.1046/j.1365-2958.2003.03401.x
3. Paul R, Weiser S, Amiot NC, Chan C, Schirmer T, Giese B, Jenal U (2004) Cell cycle-dependent dynamic localization of a bacterial response regulator with a novel di-guanylate cyclase output domain. Genes Dev 18:715–727. doi:10.1101/gad.289504
4. Jain R, Behrens AJ, Kaever V, Kazmiercak BI (2012) Type IV pilus assembly in *Pseudomonas aeruginosa* over a broad range of cyclic-di-GMP concentrations. J Bacteriol 194:4285–4294. doi:10.1128/JB.00803-12
5. Bornhorst J, Falke J (2000) Purification of proteins using polyhistidine affinity tags. Methods Enzymol 326:245–254

Part II

Detection and Quantitation of c-di-GMP

Chapter 4

High-Performance Liquid Chromatography (HPLC)-Based Detection and Quantitation of Cellular c-di-GMP

Olga E. Petrova and Karin Sauer

Abstract

The modulation of c-di-GMP levels plays a vital role in the regulation of various processes in a wide array of bacterial species. Thus, investigation of c-di-GMP regulation requires reliable methods for the assessment of c-di-GMP levels and turnover. Reversed-phase high-performance liquid chromatography (RP-HPLC) analysis has become a commonly used approach to accomplish these goals. The following describes the extraction and HPLC-based detection and quantification of c-di-GMP from *Pseudomonas aeruginosa* samples, a procedure that is amenable to modifications for the analysis of c-di-GMP in other bacterial species.

Key words bis-(3′,5′)-cyclic dimeric guanosine monophosphate, c-di-GMP, Cyclic dinucleotides, Reversed phase, Ion exchange, High-performance liquid chromatography, Biofilm

1 Introduction

Key cellular behaviors in a wide variety of bacteria are regulated via the turnover of bis-(3′,5′)-cyclic dimeric guanosine monophosphate (c-di-GMP) [1, 2]. Modulation of c-di-GMP levels involves di-guanylate cyclases (DGCs) catalyzing the condensation of two guanosine-5′-triphosphate (GTP) molecules to form c-di-GMP, and phosphodiesterases (PDEs) driving the hydrolysis of c-di-GMP to guanosine monophosphate (GMP) via the linear intermediate 5′-phosphoguanylyl-(3′→5′)-guanosine (pGpG) [3–6]. The delicate balance between c-di-GMP synthesis and breakdown contributes to subtle changes that drive major alterations in bacterial cell physiology. Thus, reliable methods for measuring c-di-GMP levels and turnover are essential to investigate mechanisms of c-di-GMP regulation.

High-performance liquid chromatography (HPLC) was among the methods used in the seminal body of work that originally identified c-di-GMP as a secondary messenger molecule [3, 4, 7]. And HPLC is still commonly used today for the separation,

Karin Sauer (ed.), *c-di-GMP Signaling: Methods and Protocols*, Methods in Molecular Biology, vol. 1657, DOI 10.1007/978-1-4939-7240-1_4, © Springer Science+Business Media LLC 2017

detection, and quantitation of c-di-GMP [8–11]. It is fairly rapid and does not require extensive sample preparation. The method enables the quantitation of c-di-GMP in the picomolar, or even the femtomolar range, depending on the detection approach used [12–14]. Moreover, HPLC has the potential to separate c-di-GMP from other compounds likely present in cellular extracts and/or those that can interfere with proper quantitation of c-di-GMP (Fig. 1). While enabling this separation, HPLC can also detect compounds that are associated with c-di-GMP turnover, including GTP, GMP, and pGpG. Thus, in addition to simple quantitation of total cellular c-di-GMP levels, HPLC analysis also allows for time-course assays of c-di-GMP synthesis or breakdown, and has been successfully used to investigate DGC and PDE enzyme kinetics (Fig. 1) [16–18].

HPLC is a form of column chromatography that pumps at high pressure a sample (analyte) dissolved in a solvent (mobile phase) through a column with an immobilized chromatographic packing material (stationary phase). The properties of the sample and the solvent, as well as the nature of the stationary phase, determine the retention time of the analytes, or how fast they pass through the column. As the sample passes through the column, analytes having the strongest interactions with the stationary phase exit the column the slowest, meaning that they exhibit the longest retention times. In contrast, samples demonstrating little interaction with the column material elute quickly and are thus characterized by short retention times. Separation of compounds in a sample can be accomplished via an isocratic elution, where the composition of the mobile phase remains constant, or via a gradient elution, where the mobile phase composition is changed over the course of the separation toward conditions favoring analyte dissociation from the stationary phase. Upon exiting the column, the mobile phase passes through a detection module, such as a fluorimeter or a UV-absorbance detector. Selection of the appropriate detector and monitoring wavelengths is essential for optimizing the sensitivity of HPLC detection. The detector generates a signal correlating to the quantity of analyte emerging from the column, which is then transferred to and recorded by an HPLC control computer program, with the data available for subsequent analyses.

A few variations of HPLC protocols for the detection and quantitation of c-di-GMP have been developed. While studies requiring the separation of cyclic nucleotides and di-nucleotides have preferentially opted for reversed-phase HPLC [12, 19–21], which utilizes a nonpolar stationary phase and a moderately polar mobile phase, the detection approach has been more varied. Often, cyclic nucleotides and di-nucleotides are directly monitored by a UV detector module in the 250–275 nm wavelength range [21], with c-di-GMP exhibiting an absorption maximum at 253 nm. More recently, quantitation of c-di-GMP and other cyclic

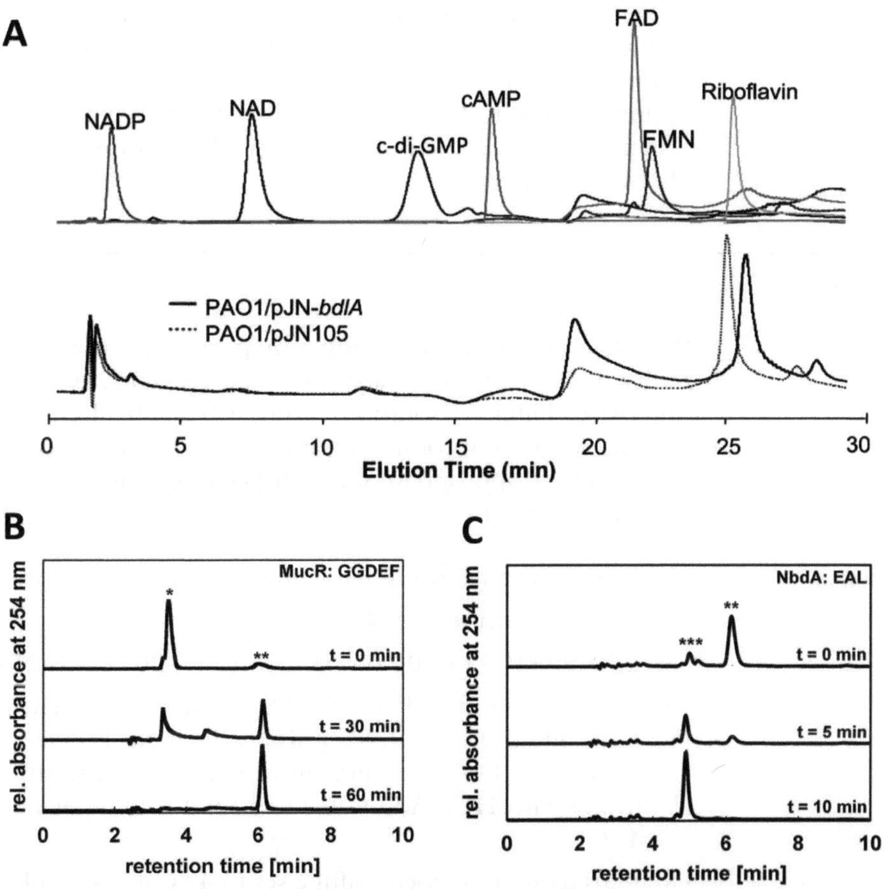

Fig. 1 Separation of c-di-GMP and other small molecules by HPLC. (**a**) Experiment demonstrating that the *Pseudomonas aeruginosa* chemotaxis protein BdlA does not bind FAD, FMN, NAD, NADP, cAMP, c-di-GMP, or riboflavin. The top panel demonstrates the HPLC elution profiles of FAD, FMN, NAD, NADP, cAMP, c-di-GMP, or riboflavin. The bottom panel demonstrates chromatograms of supernatants of purified BdlA (from *P. aeruginosa* PA01/pJN-*bdlA*) or from vector control extracts (PA01/pJN105). Reproduced from [15] with permission from American Society for Microbiology (ASM). (**b**) HPLC chromatograms demonstrating c-di-GMP synthesis by *P. aeruginosa* MucR. Peaks representing GTP (*) and c-di-GMP (**) are assigned. Reproduced from [16] with permission from ASM. (**c**) HPLC chromatograms showing c-di-GMP hydrolysis by the *P. aeruginosa* PDE NbdA. Peaks representing c-di-GMP (**) and pGpG (***) are assigned. Reproduced from [16] with permission from ASM. For all the experiments, separation was performed on a C18 reversed-phase column, with analytes monitored at 254 nm

nucleotides has been accomplished by coupling HPLC to tandem mass spectrometry (LC-MS/MS) or matrix-assisted laser desorption/ionization–time-of-flight (MALDI-TOF) analysis for improved detection of the messenger molecules [8, 11, 12, 14].

The following presents a specific variation of the approaches described above: an optimized method for the extraction of c-di-GMP from planktonic and biofilm cells of the opportunistic human pathogen *Pseudomonas aeruginosa* and subsequent separation and quantitation of c-di-GMP using reversed-phase HPLC with a UV detector module. The procedure has been adapted and modified from [8, 9]. While the method has been optimized using *P. aeruginosa* cells and an Agilent 1100 HPLC system, the procedure is easily adaptable for other organisms and HPLC systems.

2 Materials

Prepare all the solutions using nanopure water (18 MΩ-cm ionic purity at 25 °C) and analytical grade reagents, unless indicated otherwise. Filter all solutions using a 0.2 μm filter. Prepare and store all reagents at room temperature, unless indicated otherwise. Follow all the appropriate waste disposal regulations when disposing waste materials.

2.1 Extraction of c-di-GMP

1. Unpolished 1.5 mL microcentrifuge tubes (*see* **Note 1**).
2. Bacterial culture (*see* **Note 2**).
3. Ethanol (190 or 200 proof). Store at −20 °C (*see* **Note 3**).
4. Phosphate-buffered saline (PBS): To approximately 800 mL of water add 8.0 g NaCl, 0.2 g KCl, 2.56 g $Na_2HPO_4{\cdot}7H_2O$, and 0.24 g KH_2PO_4. Mix until all salts are dissolved. Adjust the pH to 7.4 with HCl. Add water to a total volume of 1 L. Store at 4 °C.
5. Refrigerated microcentrifuge set to 4 °C and capable of attaining speeds up to 16,000 × *g*.
6. Heating block set to 100 °C.
7. Vacuum concentrator/Centrifugal evaporator (e.g., SpeedVac).

2.2 HPLC Quantitation of c-di-GMP

1. HPLC system equipped with an autosampler, degasser, and UV/Vis detector (e.g., Agilent 1100 HPLC), and connected to a computer with HPLC system control software for HPLC peak analysis (e.g., ChemStation for LC, Agilent Technologies).
2. HPLC Solvent A: Add 0.385 g ammonium acetate (MS grade) to 500 mL of water for a final concentration of 10 mM. Do not adjust pH. Store at 4 °C until use (*see* **Note 4**).
3. HPLC Solvent B: Add 0.385 g ammonium acetate (MS grade) to 500 mL of methanol. Dissolve ammonium acetate salt in methanol. Do not adjust pH. Store at 4 °C until use (*see* **Note 4**).
4. Bis-(3′-5′)-cyclic diguanylic monophosphate (c-di-GMP) (Bio-log): Dilute with water to a final concentration of 200 pmol/μL. Store at −80 °C.
5. Syringes (1 mL).
6. HPLC Syringe filters (Hydrophobic PTFE, 0.45 μm).
7. Reverse-phase C_{18} column (2.1 × 40 mm, 5 μm).

2.3 Protein Quantitation

1. Refrigerated microcentrifuge set to 4 °C and capable of attaining speeds up to 16,000 × g.
2. Protein determination reagents (e.g., Modified Lowry Assay).
3. 0.5 M Ethylenediaminetetraacetic acid (EDTA) disodium salt solution (pH 8.0): Add 186.1 g of disodium EDTA·$2H_2O$ to 800 mL of water. Stir vigorously on a magnetic stirrer. Adjust the pH to 8.0 with NaOH (~20 g of NaOH pellets) (*see* **Note 5**).
4. TE buffer: To approximately 800 mL water add 1.21 g Tris and 2 mL of the 0.5 M EDTA (pH 8.0) solution. Mix to allow the Tris to dissolve. Adjust the pH to 8.0 with HCl. Store at 4 °C.
5. Spectrophotometer capable of measuring absorbance at 600 nm.
6. Sonicator capable of pulsed mode operation and 5 W output.

3 Methods

3.1 Extraction of c-di-GMP

1. Grow bacterial cells to desired growth stage under required experimental conditions. Proceed directly with the extraction, with no waiting periods or incubation of cells on ice, as this may drastically alter the c-di-GMP levels (*see* **Note 2**).
2. Determine the optical density of the bacterial culture at 600 nm (OD_{600}) (*see* **Note 6**).
3. Obtain a bacterial culture volume equivalent to 1 mL of $OD_{600} = 1.8$ (e.g., if the $OD_{600} = 0.9$, spin down 2 mL of culture) (*see* **Notes 2** and **7**).
4. Centrifuge (16,000 × *g*, 2 min, 4 °C) the respective culture volume. Discard the supernatant.
5. Wash the cell pellet with 1 mL ice-cold PBS (16,000 × *g*, 2 min, 4 °C). Discard the supernatant.
6. Repeat Subheading 3.1, **step 5**.
7. Resuspend the cell pellet in 100 μL ice-cold PBS and incubate at 100 °C for 5 min (*see* **Note 8**).
8. Add ice-cold ethanol (stored at −20 °C until use) to a final concentration of 65% (186 μL of 200 proof ethanol or 217 μL of 190 proof ethanol) and vortex for 15 s (*see* **Note 8**).
9. Centrifuge sample (16,000 × *g*, 2 min, 4 °C).
10. Remove and retain the supernatant containing extracted c-di-GMP in a new microfuge tube. Store the tube with the supernatant on ice until the next step. Retain the cell pellet.
11. Using the cell pellet, repeat twice the extraction procedure in Subheading 3.1, **steps 7–10**. Pool the supernatants obtained

from these repeat extractions into the microfuge tube from Subheading 3.1, **step 10** (i.e., supernatants from the repeated three extractions should be combined in one tube). Retain the cell pellet after the final extraction step, and store it at −20 °C until Subheading 3.2.3, **step 1**.

12. Dry the combined supernatants (total volume of ~900 μL) using a vacuum concentrator/centrifugal evaporator. The drying time can vary from 3 to 6 h (*see* **Note 9**).
13. Following evaporation, a white pellet should be visible. Store this sample, containing the extracted c-di-GMP, at −80 °C until Subheading 3.2.2, **step 1**.

3.2 Quantification of c-di-GMP

This procedure has been optimized for the detection of c-di-GMP using an Agilent 1100 HPLC equipped with an autosampler, degasser, pressure regulator, prefilter, and UV/Vis detector set to 253 nm. Separation was carried out using a reverse-phase C_{18} Targa column (2.1 × 40 mm; 5 μm) and a flow rate of 0.2 mL/min. Solvents containing methanol and ammonium acetate (*see* Subheading 2.2) were used. The following gradient was used to elute c-di-GMP: 0–9 min, 1% B (=1% Solvent B and 99% Solvent A); 9–14 min, 15% B; 14–19 min, 25% B; 19–26 min, 90% B; 26–40 min, 1% B. This gradient resulted in the elution of c-di-GMP at approximately 14–15 min (*see* **Note 10**).

Use at least 20 column volumes of Solvent A to equilibrate a newly installed HPLC column, and at least 10 column volumes of Solvent A to re-equilibrate the column between runs.

3.2.1 Generation of a c-di-GMP Standard Curve

1. Using the 200 pmol/μL c-di-GMP stock solution, prepare the following standards in nanopure water, preferably by serial dilution: 1, 2, 5, 10, and 20 pmol/μL.
2. Inject and run on the HPLC 20 μL of the 1 pmol/μL c-di-GMP standard.
3. Repeat Subheading 3.2.1, **step 2** for the 2, 5, 10, and 20 pmol/μL c-di-GMP standards.
4. Inject and run 20 μL of nanopure water as a negative control (0 pmol/μL c-di-GMP).
5. Use the HPLC analysis software (e.g., ChemStation) to obtain the peak area for each of the standard concentrations (*see* **Note 11**).
6. Prepare a standard curve by plotting the c-di-GMP amount in pmol (e.g., 200 pmol for the 20 μL of the 10 pmol/μL standard) vs. the peak areas (*see* Fig. 2 for an example of a standard curve).
7. Calculate the linear trendline and the R^2 coefficient of the plotted points (*see* Fig. 2 for an example) (*see* **Note 12**).

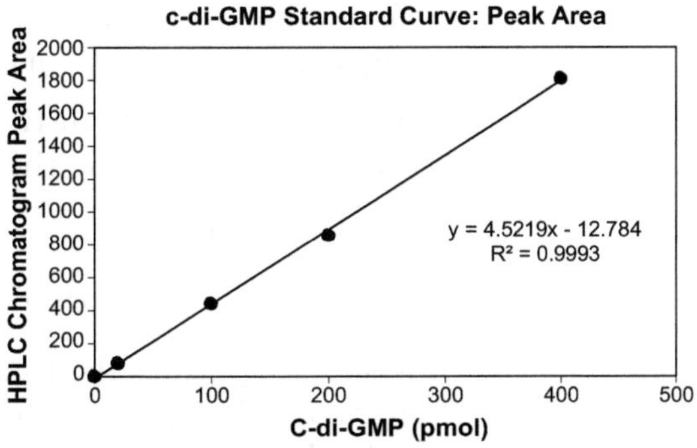

Fig. 2 Example of a c-di-GMP HPLC standard curve. The standard curve was generated by plotting the peak areas obtained following the separation of c-di-GMP standards versus the total c-di-GMP amounts (pmol) in the respective standard samples. 20 μL aliquots of the following c-di-GMP standards (0, 1, 5, 10, and 20 pmo/μL) were separated on a reverse-phase C_{18} Targa column. Peak areas were obtained using the ChemStation for LC software

3.2.2 HPLC Analysis of Extracted c-di-GMP Samples

1. Resuspend the dried extract from Subheading 3.1, **step 13** in 200 μL nanopure water.
2. Vortex for 1 min.
3. Centrifuge the solution at max speed ($\geq$16,000 × *g*) to remove insoluble material.
4. Carefully draw up the liquid sample into a 1 mL syringe, making sure to leave the pelleted debris in the tube.
5. Attach a 2 μm syringe filter to the 1 mL syringe with the extract.
6. Filter the sample supernatant into a new microfuge tube (*see* **Note 13**).
7. Analyze 20 μL per sample using the HPLC program used for the standards above.
8. Using the HPLC analysis software, determine the area of the c-di-GMP peak detected in the sample, ensuring that the peak is sharp and well separated from other detected compounds (*see* **Note 14**).
9. Repeat Subheading 3.2.2, **steps 1–8** for all c-di-GMP samples to be analyzed.
10. Determine the c-di-GMP amounts using the standard curve established with the commercially available c-di-GMP in Subheading 3.2.1.

3.2.3 Normalization of c-di-GMP Calculations

1. Resuspend the cell pellet from Subheading 3.1, **step 11** in 500 μL TE buffer.
2. Sonicate the suspension for a total of 2 min using 10 s bursts at 5 W followed by 15 s off. Sonication should be carried out on ice (*see* **Note 15**).

3. Determine the protein concentrations using a protein determination assay (e.g., Modified Lowry Assay).
4. Normalize the c-di-GMP levels to total cellular protein levels (i.e., pmol/mg) using the following calculations:

   Total c-di-GMP

$$T_{\text{c-di-GMP}} = 10 \times F_{\text{c-di-GMP}}$$

   where $T_{\text{c-di-GMP}}$ is the total c-di-GMP in the extracted sample in pmol, $F_{\text{c-di-GMP}}$ is the fraction of c-di-GMP (20 μL of 200 μL) quantified by the HPLC analysis and calculated in Subheading 3.2.2, **step 10**.

   Total protein

$$T_{\text{protein}} = 0.5\ \text{mL} \times C_{\text{protein}}$$

   where T_{protein} is the total cellular protein in the sample subjected to extraction, and C_{protein} is the protein concentration determined in Subheading 3.2.3, **step 3**.

   Normalized c-di-GMP

$$N_{\text{c-di-GMP}} = T_{\text{c-di-GMP}} / T_{\text{protein}}$$

   where $N_{\text{c-di-GMP}}$ is the cellular c-di-GMP level in pmol normalized per mg of cellular protein.

4 Notes

1. Polished, especially highly polished, microcentrifuge tubes should be avoided during c-di-GMP extraction, to prevent static build up and the migration of powder during the sample drying procedure in a vacuum concentrator/centrifugal evaporator in Subheading 3.1, **step 12**. Such static build up and dry sample powder migration can lead to reduced recovery and quantitation of c-di-GMP.
2. c-di-GMP can be extracted from cells grown as a planktonic culture or as biofilms. It is important to note, however, that extraction from cells in the early exponential phase may yield c-di-GMP amounts below the limit of detection for HPLC-UV/Vis analysis. The procedure described here has been successfully utilized for the analysis of c-di-GMP levels from *P. aeruginosa* cultures where the following growth variables have been altered: mid- to late-log and stationary phase planktonic cells; 3- to 6-day-old biofilm cells grown under flowing conditions; Lennox broth or Vogel-Bonner minimal medium (VBMM) [22]; wild-type and mutant strains; strains harboring plasmids and subjected to antibiotics for plasmid maintenance.

However, the culture conditions and cell numbers/optical density measurements used for c-di-GMP extraction may need to be optimized when other bacterial species and/or growth conditions are tested.

3. Both 190 and 200 proof ethanol can be utilized for the c-di-GMP extraction protocol, and the appropriate volumes for the respective concentrations are provided in the protocol. However, due to its hygroscopic nature, it is preferable to store the 200 proof ethanol in an airtight glass container.
4. The HPLC Solvents A and B should be remade at least every 5 days due to the volatile nature of ammonium acetate. Use of old solvents with reduced concentrations of ammonium acetate may negatively affect the reproducibility of the separations and c-di-GMP retention times.
5. While preparing the EDTA solution, it is important to note that the EDTA disodium salt will not go into solution until the pH is ~8.0. When adjusting the pH, solid NaOH pellets should be used, rather than a solution of NaOH, due to the high amount of NaOH required to reach pH 8.0.
6. If working with samples with significant aggregation and cell-cell adhesion (e.g., biofilms), including a step to homogenize (10 s on high) the cultures to disrupt cell aggregates prior to OD_{600} determination may be beneficial to ensure accurate OD measurements and subsequent c-di-GMP quantitation.
7. This biomass has been optimized for the analysis of c-di-GMP in *P. aeruginosa* strains PAO1 and PA14 planktonic and biofilm samples. Analysis of c-di-GMP levels in other strains or species may require the initial biomass harvested for extraction to be adjusted.
8. Following the incubation at 100 °C and consequent cell lysis, the remaining cellular debris may become viscous and may aggregate. Make sure that the pellet is vortexed vigorously upon addition of ethanol.
9. Depending on the vacuum concentrator/centrifugal evaporator used, the drying time for the extracted samples can vary significantly. The drying time also depends on whether the instrument has separate settings for evaporation of aqueous or alcohol solvents. In our experience, the average evaporation time is 4 h, with no less than 3 h required. Do not apply heat to accelerate the evaporation process.
10. Analysis of c-di-GMP levels using a different reverse-phase column and/or HPLC system may require optimization of HPLC separation gradients.
11. Ensure that the selected HPLC separation procedure results in a distinct sharp peak for c-di-GMP that demonstrates reproducible retention times for all of the standard concentrations

tested. Also, confirm that the peak retention time is >5 min, to ensure that c-di-GMP binds to the column and is not in the initial unbound flow-through.

12. The generated c-di-GMP standard curve should only be used with $R^2 > 0.98$. Coefficients of correlation smaller than 0.98 indicate issues with c-di-GMP standard dilution preparation or with HPLC c-di-GMP separation and quantitation.
13. Small sample volume loss may occur during the filtration of the resuspended c-di-GMP samples, but will not interfere with downstream application, as only a limited sample volume (20 μL out of 200 μL) is subjected to HPLC analysis.
14. This procedure has been optimized for *P. aeruginosa* PAO1 and PA14. Analysis of c-di-GMP levels in other strains or species may require the adjustment of sample volumes to account for significant differences in levels of c-di-GMP detected in the samples. Additionally, presence of other compounds eluting at similar times as c-di-GMP may hinder analysis, if distinct c-di-GMP peaks are no longer detectable. In this case, the HPLC conditions, including the gradient and potentially solvents, will have to be altered to accomplish sharp separation of the c-di-GMP peaks. If separation conditions have to be adjusted, the pure c-di-GMP standards will have to be re-analyzed using the new conditions.
15. To prevent overheating of the protein samples during sonication, the microfuge tubes containing the samples can be suspended in an ice water bath using floating foam tube racks for the duration of the sonication procedure.

References

1. Römling U, Simm R (2009) Prevailing concepts of c-di-GMP signaling. Contrib Microbiol 16:161–181
2. Sondermann H, Shikuma NJ, Yildiz FH (2012) You've come a long way: C-di-GMP signaling. Curr Opin Microbiol 15:140–146
3. Ross P, Aloni Y, Weinhouse C et al (1985) An unusual guanyl oligonucleotide regulates cellulose synthesis in Acetobacter xylinum. FEBS Lett 186:191–196
4. Ross P, Weinhouse H, Aloni Y et al (1987) Regulation of cellulose synthesis in Acetobacter xylinum by cyclic diguanylic acid. Nature 325:279–281
5. Schirmer T (2016) C-di-GMP synthesis: structural aspects of evolution, catalysis and regulation. J Mol Biol 428:3683–3701
6. Hengge R (2009) Principles of c-di-GMP signalling in bacteria. Nat Rev Microbiol 7:263–273
7. Ross P, Mayer R, Weinhouse H et al (1990) The cyclic diguanylic acid regulatory system of cellulose synthesis in Acetobacter xylinum. Chemical synthesis and biological activity of cyclic nucleotide dimer, trimer, and phosphothioate derivatives. J Biol Chem 265:18933–18943
8. Thormann KM, Duttler S, Saville RM et al (2006) Control of formation and cellular detachment from Shewanella oneidensis MR-1 biofilms by cyclic di-GMP. J Bacteriol 188:2681–2691
9. Ueda A, Wood TK (2009) Connecting quorum sensing, c-di-GMP, pel polysaccharide, and biofilm formation in Pseudomonas aeruginosa through tyrosine phosphatase TpbA (PA3885). PLoS Pathog 5:e1000483
10. Liu X, Beyhan S, Lim B et al (2010) Identification and characterization of a phosphodiesterase that inversely regulates motility and biofilm

formation in Vibrio cholerae. J Bacteriol 192:4541–4552
11. Hickman JW, Harwood CS (2008) Identification of FleQ from *Pseudomonas aeruginosa* as a c-di-GMP-responsive transcription factor. Mol Microbiol 69:376–389
12. Burhenne H, Kaever V (2013) Quantification of cyclic dinucleotides by reversed-phase LC-MS/MS. Methods Mol Biol 1016:27–37
13. Spangler C, Böhm A, Jenal U et al (2010) A liquid chromatography-coupled tandem mass spectrometry method for quantitation of cyclic di-guanosine monophosphate. J Microbiol Methods 81:226–231
14. Simm R, Morr M, Remminghorst U et al (2009) Quantitative determination of cyclic diguanosine monophosphate concentrations in nucleotide extracts of bacteria by matrix-assisted laser desorption/ionization-time-of-flight mass spectrometry. Anal Biochem 386:53–58
15. Petrova OE, Sauer K (2012) PAS domain residues and prosthetic group involved in BdlA-dependent dispersion response by *Pseudomonas aeruginosa* biofilms. J Bacteriol 194:5817–5828
16. Li Y, Heine S, Entian M et al (2013) NO-induced biofilm dispersion in Pseudomonas aeruginosa is mediated by an MHYT domain-coupled phosphodiesterase. J Bacteriol 195:3531–3542
17. Ryjenkov DA, Tarutina M, Moskvin OV et al (2005) Cyclic Diguanylate is a ubiquitous signaling molecule in bacteria: insights into biochemistry of the GGDEF protein domain. J Bacteriol 187:1792–1798
18. Schmidt AJ, Ryjenkov DA, Gomelsky M (2005) The ubiquitous protein domain EAL is a cyclic Diguanylate-specific Phosphodiesterase: enzymatically active and inactive EAL domains. J Bacteriol 187:4774–4781
19. Wojcik W, Olianas M, Parenti M et al (1981) A simple fluorometric method for cAMP: application to studies of brain adenylate cyclase activity. J Cyclic Nucleotide Res 7:27–35
20. Van Lookeren Campagne MM, Van Haastert PJM (1983) A sensitive cyclic nucleotide phosphodiesterase assay for transient enzyme kinetics. Anal Biochem 135:146–150
21. Martinez-Valdez H, Kothari RM, Hershey HV et al (1982) Rapid and reliable method for the analysis of nucleotide pools by reversed-phase high-performance liquid chromatography. J Chromatogr A 247:307–314
22. Vogel H, Bonner D (1956) Acetylornithinase of *Escherichia coli*: partial purification and some properties. J Biol Chem 218:97–106

Chapter 5

Identification and Quantification of Cyclic Di-Guanosine Monophosphate and Its Linear Metabolites by Reversed-Phase LC-MS/MS

Heike Bähre and Volkhard Kaever

Abstract

Cyclic dinucleotides such as bis-(3′,5′)-cyclic dimeric guanosine monophosphate (3′,3′-c-di-GMP) represent an important class of second messengers in bacteria and are involved in numerous (patho)physiological settings. Here, we describe a sensitive and specific quantification method for 3′,3′-c-di-GMP by HPLC-coupled tandem mass spectrometry (LC-MS/MS). Additionally, linear 3′,3′-c-di-GMP metabolites, i.e., 5′-phosphoguanylyl-3′,5′-guanosine (pGpG) and 5′-guanosine monophosphate (5′-GMP), as well as cyclic guanosine monophosphate (3′,5′-cGMP) and 3′,3′ c-di-GMP analogues (2′,3′-c-di-GMP and 2′,2′-c-di-GMP) can be simultaneously determined by this method.

Key words Cyclic di-GMP, HPLC, Tandem mass spectrometry

1 Introduction

Numerous low-molecular weight bacterial signaling molecules, either present in the cytosol or being secreted, have been described within the last years. Among them, linear and cyclic nucleotides play major roles as second messengers in the regulation of important bacterial functions, i.e., guanosine 3′-diphosphate, 5′-triphosphate (pppGpp), guanosine-3′,5′-bispyrophosphate (ppGpp), 3′,5′-cyclic adenosine monophosphate (3′,5′-cAMP) and 3′,5′-cyclic guanosine monophosphate (3′,5′-cGMP), bis-(3′,5′)-cyclic dimeric guanosine monophosphate (3′,3′-c-di-GMP), and bis-(3′,5′)-cyclic dimeric adenosine monophosphate (3′,3′-c-di-AMP) [1–3]. Especially 3′,3′-c-di-GMP has attracted interest as it has been identified in many bacterial species, regulating biofilm formation and virulence [4–7]. Due to their low concentrations and rapid metabolism, identification and quantification of these molecules is a highly challenging analytical task. We have developed sensitive and specific HPLC-coupled tandem mass spectrometry (LC-MS/MS) methods

Karin Sauer (ed.), *c-di-GMP Signaling: Methods and Protocols*, Methods in Molecular Biology, vol. 1657,
DOI 10.1007/978-1-4939-7240-1_5, © Springer Science+Business Media LLC 2017

with the main focus on unequivocal quantification of linear and cyclic nucleotides, cyclic dinucleotides, and their respective degradation products [8]. In this case, LC-MS/MS with inclusion of isotope-labeled internal standards and analysis of specific quantifier and several qualifier mass transitions represents the analytical method of choice. However, special care has to be taken regarding the initial sample preparation steps, the robustness of the HPLC methods, and a reliable MS/MS procedure. We and others have previously described specific LC-MS/MS methods for the detection of 3′,3′-c-di-GMP [8–10, 15]. In this protocol, we present our improved LC-MS/MS method for the simultaneous identification and quantification of 3′,3′-c-di-GMP and its linear metabolites pGpG and 5′-GMP, as well as the cyclic nucleotide monophosphate 3′,5′-cGMP. The established procedure can easily be upgraded for the analysis of the respective adenosine containing cyclic and linear nucleotides.

2 Materials

2.1 Chemicals

Use HPLC grade (water, methanol) or ultra gradient HPLC grade (acetonitrile) solvents for sample preparation and LC-MS/MS analysis (*see* **Note 1**).

1. Nucleotides.
   (a) 3′,3′-c-di-GMP stock solution (100 μM in water) stored at −20 °C.
   (b) pGpG stock solution (1 mM in water) stored at −20 °C.
   (c) 5-GMP stock solution (1 mM in water) stored at −20 °C.
   (d) 3′,5′-cGMP stock solution (10 mM in water) stored at −20 °C.
   (e) 2′,2′-c-di-GMP stock solution (100 μM in water) stored at −20 °C.
   (f) 2′,3′-c-di-GMP stock solution (100 μM in water) stored at −20 °C.
   (g) $^{13}C_{20}{}^{15}N_{10}$–3′,3′-c-di-GMP stock solution (50 μM in water) stored at −20 °C.
2. Analytes. The nucleotides 3′,3′-c-di-GMP, pGpG, 5′-GMP, 3′,5′-cGMP are referred here as analytes.
3. Calibrators (3′,3′-c-di-GMP, pGpG, 5′-GMP, 3′,5′-cGMP). Prepare solutions containing all four analytes at the following concentrations: 6.6/16.4/41/102/256/640/1600/4000/10,000/25,000 nM (corresponds to 0.066/0.164/0.410/1.02/2.56/6.4/16/40/100/250 pmol per sample). Store at 10 μL aliquots at −20 °C.

4. Quality control samples. Prepare quality control samples containing all four analytes at low, medium, and high concentrations (15/1000/20,000 nM). Prepare samples independently of the calibrators. Store 10 μL aliquots at −20 °C.
5. Internal Standard. $^{13}C_{20}{}^{15}N_{10}$–3′,3′-c-di-GMP (400 ng/mL) is utilized as an isotope-labeled internal standard (*see* **Note 2**).
6. Test mix. The test mix containing all analytes at a concentration of 500 nM should be prepared in water and be stored in glass vials at −20 °C.

2.2 Sample Preparation

1. 2.0 mL Safe-Seal micro vials.
2. 1.5 mL Safe-Lock micro vials.
3. 15 mL polypropylene tubes (*see* **Note 3**).
4. Extraction solution. Mix acetonitrile/methanol/water at 2/2/1 (v/v/v).

2.3 Liquid Chromatography

1. HPLC instrumentation. Configuration consists of a degasser, micro pumps, an autosampler fitted with cooling option, equipped with an 100 μL sample loop, and an HPLC-column oven (*see* **Note 4**). Instrument settings/parameters are detailed in Table 1.
2. Software. Analyst, version 1.5.2 (Sciex) software to control of the HPLC.
3. HPLC vials for autosampler: 2 mL injection vials and 200 μL micro glass inserts as well as screw caps.
4. HPLC Column. EC 50/3 Nucleodur C18 Pyramid 3 μ, 50 × 3 mm (Macherey-Nagel).
5. Security Guard. C18, 4 × 2 mm.
6. Column Saver. 0.5 μm.
7. HPLC solvent A: 10 mM NH_4OAc/0.1% HAc, v/v. Dissolve 1.54 g of ammonium acetate (NH_4OAc) in 2 L HPLC grade water and mix it with 2 mL acetic acid (*see* **Note 5**). Store at 4 °C.
8. HPLC solvent B: 100% MeOH (HPLC grade).

2.4 Mass Spectrometry

1. MS/MS instrumentation. The API 4000 triple quadrupole mass spectrometer (Sciex) is equipped with an electrospray ion source (ESI) (*see* **Note 6**). Instrument settings/parameters are detailed in Table 1.
2. Software. Analyst, version 1.5.2 (Sciex) software to control of the MS/MS systems as well as data sampling.
3. Nitrogen gas 5.0 (supplied from liquid nitrogen) as curtain and collision gas (*see* **Note 7**).

Table 1
HPLC and MS/MS instrumentation and parameters for the quantification of 3′,3′-c-di-GMP, pGpG, 5′-GMP, and 3′,5′-cGMP

Instrumentation	
HPLC system	Autosampler, Degasser, Pumps, Oven, Controller
HPLC column	EC 50/3 Nucleodur C18 Pyramid 3 μ, 50 × 3 mm (Macherey-Nagel)
Security guard	C18, 4 × 2 mm
Column saver	0.5 μm
Mass spectrometer	API 4000 (Sciex)

HPLC Parameters	
Sample solvent	H_2O (HPLC grade)
Injection volume	50 μL
Flow rate	0.6 mL/min
HPLC solvent A	10 mM NH_4OAc/0.1 % HAc, v/v
HPLC solvent B	MeOH (HPLC grade)
Injection needle flushing	MeOH/H_2O, 50/50, v/v
Temperature (column oven)	30 °C
Temperature (autosampler)	4 °C
Maximal column backpressure	2,200 psi
Analytes	3′,3′-c-di-GMP, pGpG, 5′-GMP, 3′,5′-cGMP
Calibrators	6.6/16.4/41/102/256/640/1,600/4,000/10,000/25,000 nM (working solution), corresponds to 0.066/0.164/0.410/1.02/2.56/6.4/16/40/100/250 pmol per sample
Internal standards	$^{13}C_{20}{}^{15}N_{10}$-3′,3′-c-di-GMP (400 ng/mL working solution)
Analysis time/sample	13 min
Volume eluent A/sample	6.1 mL
Volume eluent B/sample	0.5 mL

(continued)

Table 1 (continued)

MS/MS Parameters							
Ionisation mode	Positive						
Ion source parameters	Curtain gas:	30	Temperature:	650 °C			
	Collision gas:	6	Ion source gas 1:	60 psi			
	Ion spray voltage:	3,000 V	Ion source gas 2:	45 psi			
Analyte parameters	Analyte	m/z Precursor	m/z Fragment	DP [V]	CE [V]	CXP [V]	
	3′,3′-c-di-GMP $[M+H]^+$	690.9	152.1	81	47	12	
		690.9	539.9	81	29	18	
		690.9	248.0	81	35	8	
	3′,3′-c-di-GMP $[M+2H]^{2+}$	345.7	152.1	56	27	14	
		345.7	135.0	56	61	10	
		345.7	110.0	56	65	10	
	pGpG	708.9	152.1	66	49	14	
		708.9	558.0	66	52	14	
		708.9	97.1	66	52	14	
	5′-GMP	364.0	152.1	51	21	10	
		364.0	135.1	51	65	14	
		364.0	304.2	51	11	10	
	3′,5′-cGMP	345.7	152.1	56	27	14	
		345.7	135.1	56	61	10	
		345.7	110.0	56	65	10	
	$^{13}C_{20}{}^{15}N_{10}$-3′,3′-c-di-GMP $[M+H]^+$	721.1	162.1	71	59	16	
		721.1	559.9	71	31	18	
	$^{13}C_{20}{}^{15}N_{10}$-3′,3′-c-di-GMP $[M+2H]^{2+}$	360.8	162.1	66	29	12	

Dwell time: 40 ms, Entrance potential (EP): 10 V, *m/z* mass to charge Ratio, *DP* Declustering Potential, *CE* Collison Energy, *CXP* Collison Cell Exit Potential

3 Methods

3.1 Preparation of the Internal Standard $^{13}C_{20}{}^{15}N_{10}$–3′,3′-c-di-GMP, Calibrators and Quality Control Samples for Analysis

1. Prepare the internal standard $^{13}C_{20}{}^{15}N_{10}$–3′,3′-c-di-GMP at a concentration of 400 ng/mL.
2. Mix the 10 μL aliquot of each calibrator (Subheading 2.1, **item 3**) with 40 μL of HPLC grade water and centrifuge for 10 min at 4 °C at 20,800 × *g*.
3. Mix 40 μL of each calibrator sample with an equal volume of the internal standard (400 ng/mL) directly in the micro insert of the injection vial. Avoid air bubbles (*see* **Note 8**).
4. In addition, prepare quality control samples (containing all four analytes) independently of the calibrators at low, medium, and high concentrations.

3.2 Extraction and Preparation of Nucleotides from Bacterial Liquid Culture

The described protocol works well with Gram-negative bacteria. Modified extraction methods including a preincubation step with lysozyme or performing bacterial lysis with glass beads by a special instrumentation, e.g., FastPrep® system (MP Biomedicals) have already been published for Gram-positive species [11, 12]. Tissue species can also be homogenized and lysed similarly using the protocol described by Corrigan et al. [12].

1. Incubate the liquid bacterial culture until the desired optical density is reached (*see* **Note 9**).

   Perform all the procedures described in **steps 2–7** on ice or at 4 °C.
2. Take 1–5 mL of bacterial suspension and transfer it into a 15 mL polypropylene tube (*see* **Note 10**).
3. Centrifuge for 20 min at 4 °C at 2500 × *g*. Discard supernatant fluid (*see* **Note 11**).
4. Resuspend bacterial pellet with 2 × 500 μL culture medium and transfer suspension into 1.5 mL vials.
5. Centrifuge for 20 min at 4 °C at 2500 × *g*. Discard supernatant fluid (*see* **Note 11**).
6. Resuspend bacterial pellet with 300 μL extraction solution (*see* **Note 12**).
7. Incubate suspension on ice for 15 min (*see* **Note 13**).
8. Deactivate residual phosphodiesterase and phosphatase activities by heating the extracted suspension for 10 min at 95 °C, and then cool again on ice for at least 15 min (*see* **Note 14**).
9. Centrifuge for 10 min at 4 °C at 20,800 × *g* (*see* **Note 15**). Transfer supernatant fluid into 2.0 mL micro vial and keep aside on ice.
10. Repeat extraction (Subheading 3.2, **steps 6–9**) twice with 200 μL extraction solution but omit **step 8** (heating at 95 °C).
11. Combine supernatant fluids of the three extraction steps (about 700 μL, collected in Subheading 3.2, **step 9**).
12. Centrifuge for 10 min at 4 °C at 20,800 × *g* and then transfer supernatant fluid into a new 2.0 mL vial. The combined extracts can be stored at −20 °C (*see* **Note 16**) or directly be evaporated to dryness at 40 °C by a gentle nitrogen stream. Alternatively, an evaporation system can be used (*see* **Note 17**).
13. Store protein pellet at −20 °C for protein quantification. The protein content of the respective bacterial culture (here, protein pellet) should be determined using, e.g., a BCA protein assay (*see* **Note 18**) to normalize 3′,3′-c-di-GMP and further nucleotide amounts.

14. As an alternative for the tedious first centrifugation steps (Subheading 3.2), rapid filtering of the bacterial suspension followed by direct extraction of the metabolites from the filter may be applied [13].
15. In case of expected low analyte concentrations it may be beneficial to include an extended off-line sample preparation step by anion exchange or affinity chromatography to reduce matrix effects.
16. Reconstitute dried sample extracts (Subheading 3.2, **step 12**) with 200 μL of HPLC grade water by intensive vortexing for at least 10 s.
17. Centrifuge for 10 min at 4 °C at 20,800 × *g*.
18. Mix 40 μL of the supernatant of each sample with an equal volume of the internal standard directly in the micro glass insert of the injection vial. Avoid air bubbles (*see* **Note 8**).

3.3 HPLC Separation of Nucleotides

1. A test mix (Subheading 2.1, **item 6**) should be analyzed each time before a new set of calibrators and samples is injected. By this, the specific retention times and the actual sensitivity of the mass spectrometer (*see* Subheading 3.4) can be controlled.
2. Inject 50 μL of each sample (calibrators, quality controls, and biological samples) automatically into the HPLC system. HPLC parameters/settings are given in Table 1.
3. Analyte separation is performed on the HPLC column according to the applied HPLC gradient method listed in Table 2 (*see* **Notes 19–21**). Respective elution times are given in Table 3.
4. To prepare a calibration curve, inject 50 μL of each calibrator and analyze as described above.

Table 2
HPLC gradient method

Total time (min)	Flow rate (μL/min)	HPLC solvent A (%)	HPLC solvent B (%)
0.0	600	100	0
4.0	600	100	0
7.3	600	90	10
8.3	600	90	10
11.0	600	70	30
11.1	600	100	0
13.0	600	100	0

Table 3
Average retention times of nucleotides

Nucleotide	Average retention time
3′,3′-c-di-GMP	6.1 min
pGpG	5.6 min
5′-GMP	1.0 min
3′,5′-cGMP	4.7 min
$^{13}C_{20}{}^{15}N_{10}$–3′,3′-c-di-GMP	6.1 min

3.4 Analysis of Nucleotides by Tandem Mass Spectrometry

1. The applied tandem mass spectrometer is operated in the positive ionization mode. The ion source parameters (electrospray ionization), mass-to-charge ratios (m/z) for precursor ions and specific fragments (selected reaction monitoring, SRM), and mass spectrometer specific settings are listed in Table 1 (*see* **Note 22**). The most intensive mass transitions are used as quantifiers, whereas additional fragments serve as qualifiers (*see* Figs. 1–2).
2. Data interpretation of the MS/MS signals is carried out by calculating the ratios of the peak areas of the calibrators and samples in relation to the respective peak areas of the internal standard. The peak areas of the respective mass fragments from single ($[M + H]^+$) and double charged ($[M + 2H]^{2+}$) precursor ions (Figs. 1–2) should be summarized.
3. A calibration curve for 3′,3′-c-di-GMP (ranging from 0.066 to 250 pmol/sample) is shown in Fig. 3 (*see* **Note 23**). Comparable calibration curves can also be obtained for pGpG (retention time 5.6 min), 5′-GMP (retention time 1.0 min), and 3′,5′-cGMP (retention time 4.7 min) (data not shown).
4. Beware of misinterpretation of false-positive 3′,3′-c-di-GMP peaks by recording specific quantifier and at least two specific qualifier mass transitions (*see* **Note 24**).

4 Notes

1. Plastic labware should not be reused. All laboratory glassware should be intensely rinsed with deionized water after cleaning with dish washing liquid.
2. A method for enzymatic production of 3′,3′-c-di-GMP has been described elsewhere [14]. Usage of recombinant dinucleotide cyclases lacking product inhibition is recommended.

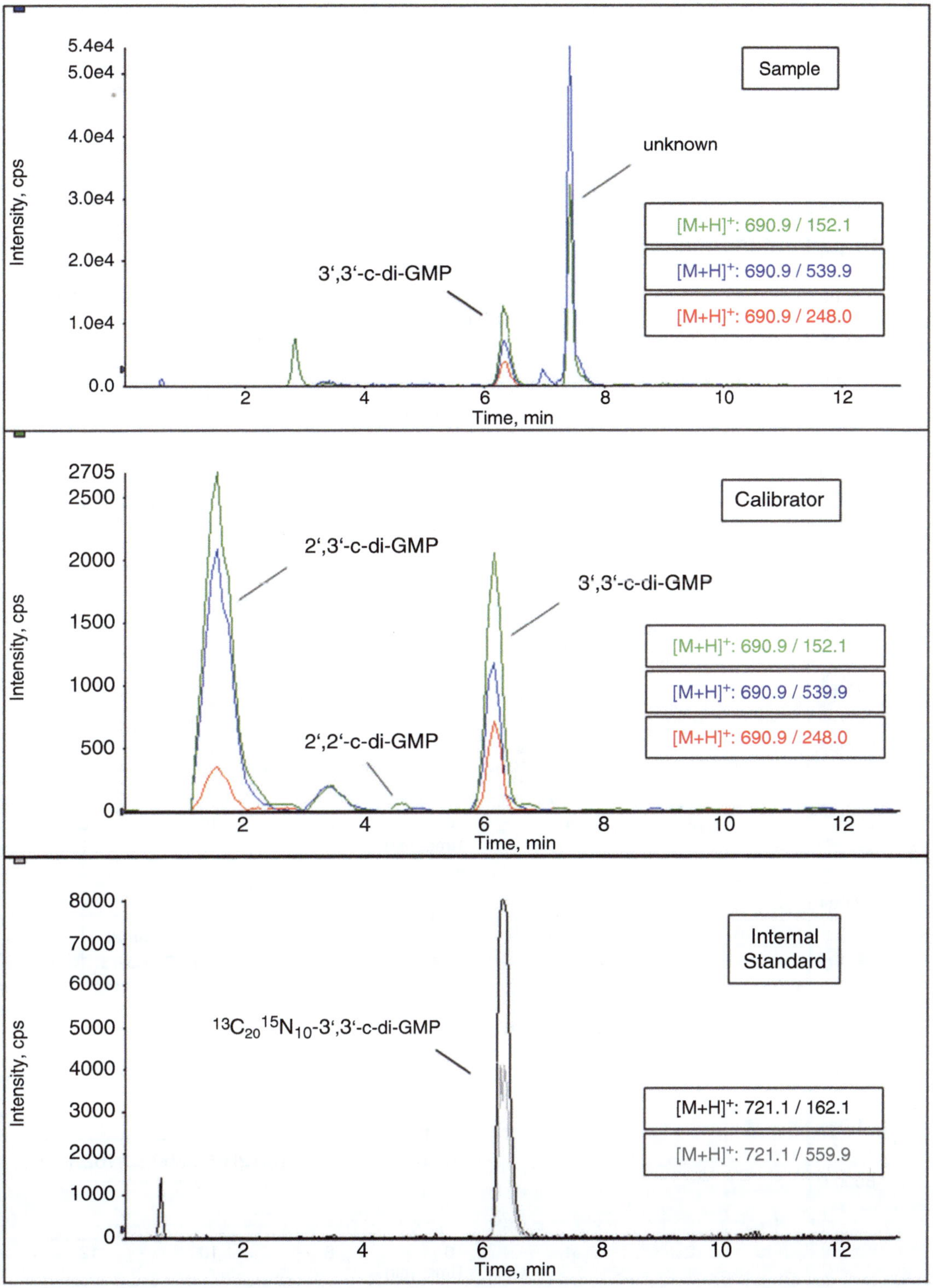

Fig. 1 Representative chromatogram of an extract from *Pseudomonas aeruginosa* (*upper panel*), c-di-GMP standards, 0.5 μM each (*middle panel*), i.e., 2′,3′-c-di-GMP (1.6 min), 2′,2′-c-di-GMP (4.6 min), and 3′,3′-c-di-GMP (6.1 min), and the internal standard $^{13}C_{20}{}^{15}N_{10}$–3′,3′-c-di-GMP (6.1 min) (*lower panel*) with their respective retention times specified in parentheses. Only mass transitions of single charged precursor ions $[M+H]^+$ are displayed

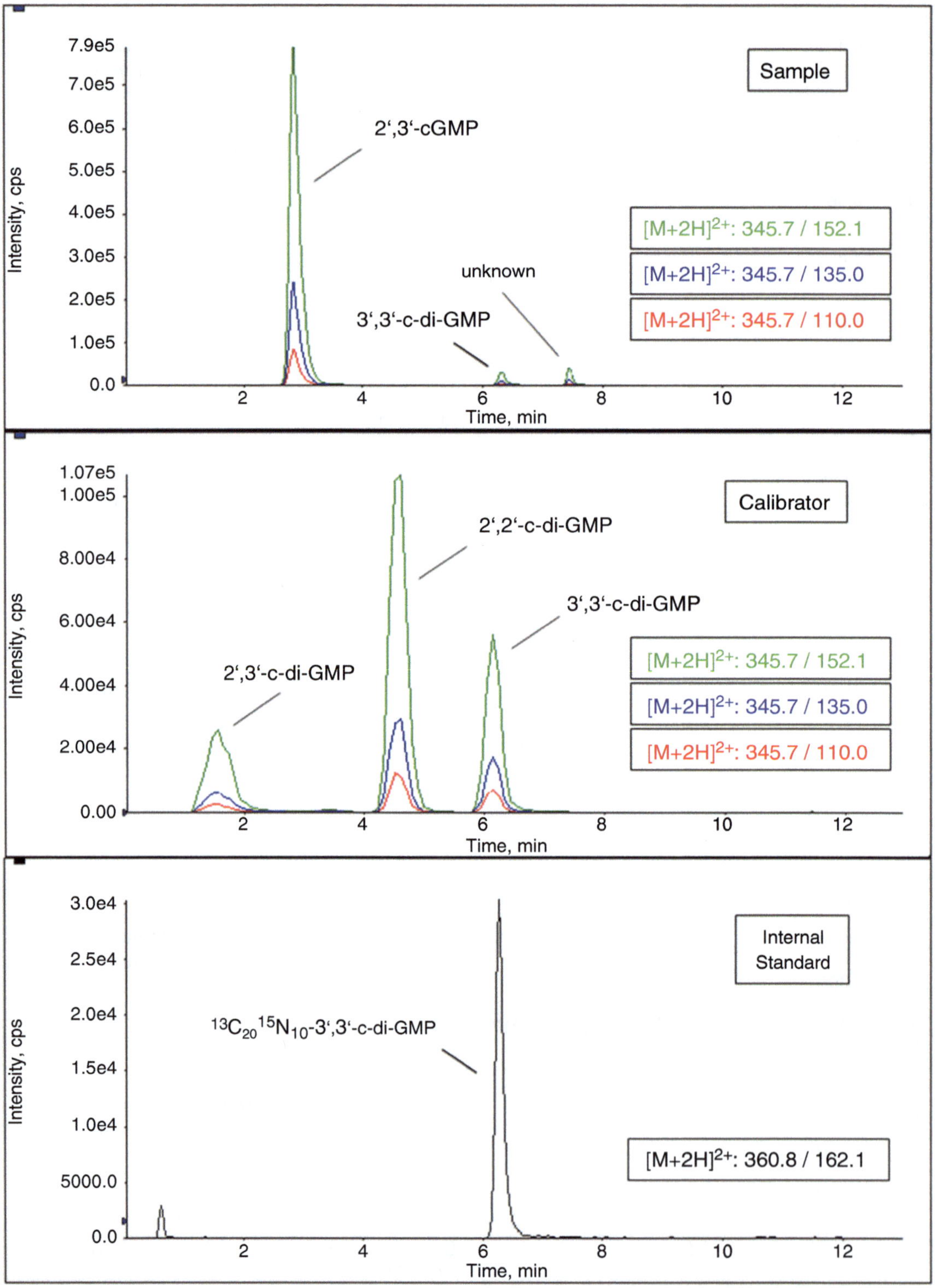

Fig. 2 Representative chromatogram of an extract from *Pseudomonas aeruginosa* (*upper panel*), c-di-GMP standards, 0.5 μM each (*middle panel*), i.e., 2′,3′-c-di-GMP (1.6 min), 2′,2′-c-di-GMP (4.6 min), and 3′,3′-c-di-GMP (6.1 min), and the internal standard $^{13}C_{20}{}^{15}N_{10}$–3′,3′-c-di-GMP (6.1 min) (*lower panel*) with their respective retention times specified in parentheses. Only mass transitions of double charged precursor ions $[M+2H]^{2+}$ are displayed

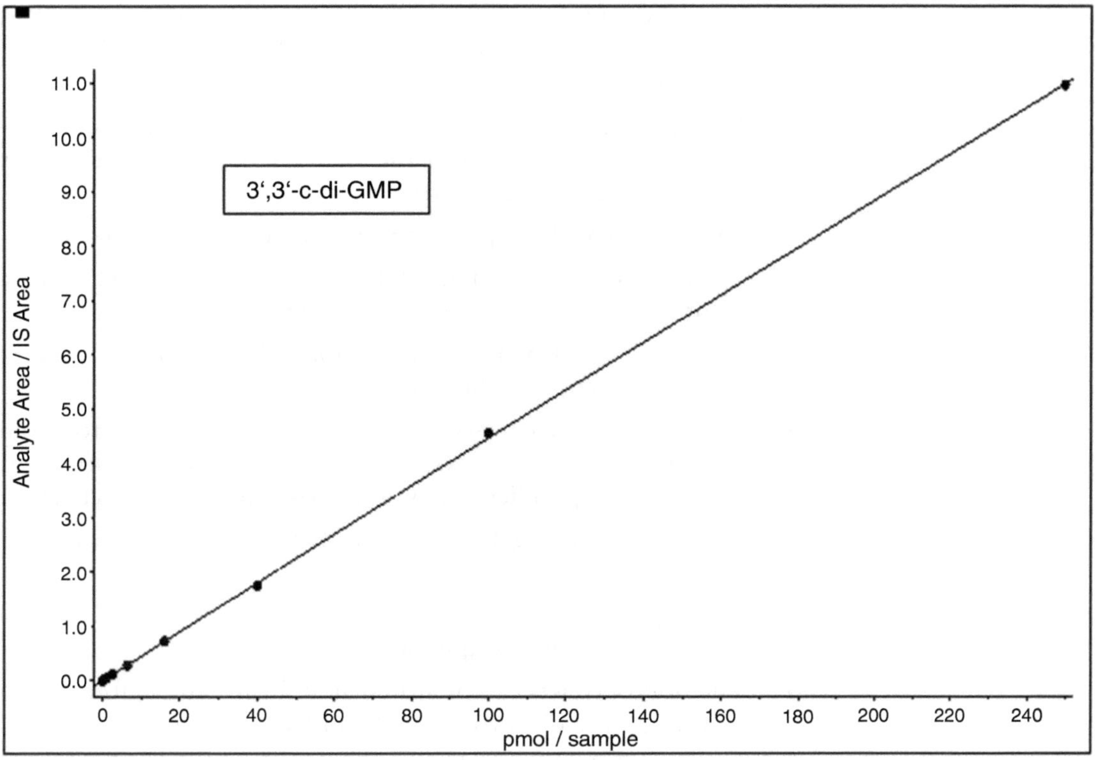

Fig. 3 3′,3′-c-di-GMP calibration curve prepared with standards ranging from 0.066 to 250 pmol/sample. For all calibrators and the internal standard, the peak areas of the respective mass fragments from single and double charged precursor ions were summarized

3. All vials should be checked for extractable residues or pre-washed with the extraction solution.
4. A versatile HPLC system that consists of at least two high pressure solvent pumps and a binary mixer for exact gradient mixture is needed to separate 3′,3′-c-di-GMP from its analogues 2′,3′-c-di-GMP and 2′,2′-c-di-GMP and from interfering metabolites. The described classical HPLC configuration is sufficient for this purpose but a further improvement of analyte separation may be achieved by UPLC methods.
5. HPLC solvent preparation should be performed under a fume hood. No pH adjustment is necessary.
6. A sensitive tandem mass spectrometer (triple quadrupol) system is recommended. However, quadrupole-time-of-flight or orbitrap systems may also be used.
7. A nitrogen generator may be applied as an alternative to liquid nitrogen.
8. Correct injection of the sample could be hampered in case of air bubbles in the micro insert of the injection vial. Therefore, the vials should be vortexed or slightly flipped with a finger.

9. Very often, highest 3′,3′-c-di-GMP concentrations are reached at the late exponential growth phase. Biological triplicates are recommended for analysis.
10. Increased culture volumes often lead to declined chromatographic separation of the analytes and unwanted matrix effects in mass spectrometry.
11. Take care that the supernatant fluid is completely removed.
12. The amount of extraction solution should be adapted if culture volumes >5 mL are chosen as starting sample.
13. Following resuspension, a prolonged incubation ensures proper extraction of the nucleotides.
14. Put an additional heavy cover on the lids in order to avoid spilling. Wear eye protective goggles as safety precaution. At this step phosphodiesterases and phosphatases are inactivated.
15. The high centrifugation force leads to sedimentation of precipitated proteins.
16. Overnight storage of the extraction solution at −20 °C leads to improved protein precipitation.
17. The dried extracts can be stored at ambient temperatures or at 4 °C as 3′,3′-c-di-GMP is stable under these conditions.
18. Resuspend the bacterial pellet in 800 μL of 0.1 N NaOH and heat it at 95 °C for 15 min for subsequent determination of the protein content. If the bacterial pellet is not completely resuspended after 15 min, the incubation time should be prolonged. In some cases, a centrifugation step will be necessary, in order to separate insoluble material. Perform a BCA or comparable protein assay.
19. MS-compatible HPLC solvents and additives have to be applied. Prior degassing of the solvents is only indicated if no degassing unit is integrated in the HPLC system. Several blank samples and an appropriate test mix should be analyzed to ensure trouble-free working of all instruments before the biological samples are measured.
20. A representative chromatogram of an extract from *Pseudomonas aeruginosa* is shown in Figs. 1 and 2 (upper panels). Note that by LC-MS/MS analysis additional peaks (annotated as "2′,3′-cGMP" and "unknown," respectively) are obvious at different retention times compared to 3′,3′-c-di-GMP (*see* **Note 24**).
21. This HPLC method can also be applied for the purification of 3′,3′-c-di-GMP after chemical or enzymatic synthesis. In this case, a simple UV detector (254 nm) would be sufficient.
22. The specific instrument settings will vary between mass spectrometry systems of different vendors and have to be adapted by the respective operator.

23. The established method should be validated in terms of precision and accuracy. The lower limit of detection (LOD) is defined at a signal-to-noise (S/N) ratio of >3. The lower limit of quantification (LLOQ) is specified by an S/N ratio of >10.
24. In Figs. 1 and 2 (upper panels) the chosen mass transition signals for 3′,3′-c-di-GMP from an extracted *Pseudomonas aeruginosa* sample are presented. A peak displaying all three specific mass transitions for 3′,3′-c-di-GMP is obvious at the retention time of authentic 3′,3′-c-di-GMP (middle panels) and the internal standard $^{13}C_{20}{}^{15}N_{10}$–3′,3′-c-di-GMP (lower panels) at 6.1 min. As can be seen in Figs. 1 and 2, an additional peak appears at a retention time of about 7.6 min at the quantifier mass transitions (m/z 690.9/152.1 and m/z 690.9/539.9, Fig. 1, and m/z 345.7/152.1 and m/z 345.7/135.0, Fig. 2). However, the third mass transitions (m/z 690.9/248.0, Fig. 1, and m/z 345.7/110.0), which are specific for 3′,3′-c-di-GMP, are not detectable for this additional peak from the biological sample. The occurrence of such a peak in bacterial extracts has already been described previously [8, 9, 15]. It has been stated that it represents a linear guanosine dinucleotide with a 2′,3′- phosphate linkage and was, therefore, termed 2′,3′-c-GpGp [15]. However, the true composition of this metabolite has not been proven, and nothing is known about a biological role, so far. We also tested authentic standards of 3′,3′-c-di-GMP analogues, i.e., 2′,2′-c-di-GMP and 2′,3′-c-di-GMP. As seen in Figs. 1 and 2 (middle panels), these analytes eluted clearly separated from 3′,3′-c-di-GMP and 2′,3- c-GpGp at retention times of 4.6 and 1.6 min, respectively. In Fig. 2 (upper panel) a further peak at a retention time of 2.8 min is obvious. All three mass transitions of the double charged 3′,3′,-c-di-GMP ion can be recorded. In comparison to an authentic standard, it turned out to represent 2′,3′-cGMP, which is obviously produced in high amounts in *Pseudomonas aeruginosa*. It has to be noted that 3′,5′-cGMP could not be detected in this sample.

These findings emphasize the importance of the application of appropriate internal standards in LC-MS/MS analyses, the establishment of efficient HPLC separation methods, and the inclusion of quantifier and at least two qualifier mass transitions in the MS/MS recordings.

Acknowledgments

We gratefully acknowledge the skillful technical assistance of Annette Garbe.

References

1. Pesavento C, Hengge R (2009) Bacterial nucleotide-based second messengers. Curr Opin Microbiol 12:170–176
2. Gomelsky M (2011) cAMP, c-di-GMP, c-di-AMP and now cGMP: bacteria use them all! Mol Microbiol 79:562–565
3. Kalia D, Merey G, Nakayama S, Zheng Y, Zhou J, Luo Y, Guo M, Roembke BT, Sintim HO (2013) Nucleotide, c-di-GMP, c-di-AMP, cGMP, cAMP, (p)ppGpp signalling in bacteria and implications in pathogenesis. Chem Soc Rev 42:305–341
4. Hengge R (2009) Principles of c-di-GMP signalling in bacteria. Nat Rev Microbiol 7:263–273
5. Römling U, Galperin MY, Gomelsky M (2013) Cyclic di-GMP: the first 25 years of a universal bacterial second messenger. Microbiol Mol Biol Rev 77:1–52
6. Whitely CG, Lee DJ (2015) Bacterial diguanylate cyclases: structure, function and mechanisms in exopolysaccharide biofilm development. Biotechnol Adv 33:124–141
7. Caly DL, Bellini D, Walsh MA, Dow JM, Ryan RP (2015) Targeting cyclic di-GMP signalling: a strategy to control biofilm formation? Curr Pharm Des 21:12–24
8. Burhenne H, Kaever V (2013) Quantification of cyclic dinucleotides by reversed-phase LC-MS/MS. Methods Mol Biol 1016:27–37
9. Spangler C, Böhm A, Jenal U, Seifert R, Kaever V (2010) A liquid chromatography-coupled tandem mass spectrometry method for quantitation of cyclic di-guanosine monophosphate. J Microbiol Methods 81:226–231
10. Irie Y, Parsek MR (2014) LC/MS/MS-based quantitative assay for the secondary messenger molecule, c-di-GMP. Methods Mol Biol 1149:271–279
11. Gundlach J, Mehne FM, Herzberg C, Kampf J, Valerius O, Kaever V, Stülke J (2015) An essential poison: synthesis and degradation of cyclic di-AMP in *Bacillus subtilis.* J Microbiol 197:3265–3274
12. Corrigan RM, Abbott JC, Burhenne H, Kaever V, Gründling A (2011) C-di-AMP is a new second messenger in *Staphylococcus aureus* with a role in controlling cell size and envelope stress. PLoS Pathog 7:e1002217
13. Liebeke M, Dörries K, Mwyer H, Lalk M (2012) Metabolome analysis of gram-positive bacteria such as *Staphylococcus aureus* by GC-MS and LC-MS. Methods Mol Biol 815:377–398
14. Spangler C, Kaever V, Seifert R (2011) Interaction of the diguanylate cyclase YdeH of *Escherichia coli* with 2′,(3′)-substituted purine and pyrimidine nucleotides. J Pharmacol Exp Ther 336:234–241
15. Gao X, Mukherjee S, Matthews PM, Hammad LA, Kearns DB, Dann CE III (2013) Functional characterization of core components of the *Bacillus subtilis* cyclic-di-GMP signalling pathway. J Bacteriol 195:4782–4792

Chapter 6

Detection of Cyclic Dinucleotides by STING

Xiao-Xia Du and Xiao-Dong Su

Abstract

STING (stimulator of interferon genes) is an essential signaling adaptor protein mediating cytosolic DNA-induced innate immunity for both microbial invasion and self-DNA leakage. STING is also a direct receptor for cytosolic cyclic dinucleotides (CDNs), including the microbial secondary messengers c-di-GMP (3′,3′-cyclic di-GMP), 3′,3′cGAMP (3′,3′-cyclic GMP-AMP), and mammalian endogenous 2′,3′cGAMP (2′,3′-cyclic GMP-AMP) synthesized by cGAS (cyclic GMP-AMP synthase). Upon CDN binding, STING undergoes a conformational change to enable signal transduction by phosphorylation and finally to active IRF3 (Interferon regulatory factor 3) for type I interferon production. Here, we describe some experimental procedures such as Isothermal Titration Calorimetry and luciferase reporter assays to study the CDNs binding and activity by STING proteins.

Key words STING, CDNs, c-di-GMP, cGAMP, ITC

1 Introduction

STING is an important adaptor protein in animal innate immune system playing key roles in cytosolic double strand DNA (dsDNA)-mediated type I interferon production [1–4]. The cyclic dinucleotides (CDNs) including c-di-GMP, c-di-AMP, and certain cGAMPs are used by bacteria and archaea as secondary messengers in many cellular processes, such as bacterial biofilm formation, mobility, and virulence [5–7]. Studies have revealed that cyclic dinucleotides are capable of stimulating innate immune responses against pathogen infection [8]. Previous studies including ours have shown that STING can directly bind c-di-GMP and then activate the downstream expression of IFNs (type I interferon) and cytokines [9].

The cyclic GMP-AMP synthase (cGAS) is a newly defined master sensor of nonspecific dsDNA in the mammalian cytoplasms regardless of the DNA's origin [10]. Binding of tiny amounts of dsDNA to cGAS will cause a conformational change, resulting in the activation of cGAS to catalyze cytosolic GTP and ATP to synthesize 2′,3′-cyclic GMP-AMP (2′,3′cGAMP) [11, 12]. Similar

Karin Sauer (ed.), *c-di-GMP Signaling: Methods and Protocols*, Methods in Molecular Biology, vol. 1657,
DOI 10.1007/978-1-4939-7240-1_6, © Springer Science+Business Media LLC 2017

to c-di-GMP, the endogenous 2′,3′cGAMP could bind even more potently to STING to activate the type I interferon production pathway [11, 13].

Upon binding of CDNs, STING will be activated by recruiting and activating IRF3 via STING's C-terminal domain (CTD), after a series of phosphorylation events mediated by TBK1, then IRF3 will dimerize and enter cell nucleus to activate transcription of relevant genes, resulting in type I interferon production [14, 15].

STING protein is an ER localized four-helices-transmembrane protein composed of N-terminal trans-membrane domain and C-terminal cytosolic domain (CTD) [1, 4]. STING CTD ($STING^{CTD}$) by itself can form dimer and is involved in binding with c-di-GMP and other CDNs [4, 9].

Our group together with some other groups have previously solved the crystal structure of $STING^{CTD}$ in apo form and CDNs-bound forms [13, 16–22]. We have expressed and purified the recombinant $STING^{CTD}$ in *E. coli* and studied its biochemical properties in detail. In particular, we have measured the dissociation constants and the thermodynamic parameters for $STING^{CTD}$'s binding with c-di-GMP and other relevant CDNs using ITC (Isothermal Titration Calorimetry), and have cocrystallized $STING^{CTD}$ with c-di-GMP [16] and with other CDNs to determine the crystal structures of the complexes. We have also tested the full-length STING function with relevant CDNs by cell-based IFNβ luciferase reporter assays [4]. Here, we describe Isothermal Titration Calorimetry and luciferase reporter assays to study the CDNs binding and activity by STING proteins.

2 Materials

Prepare all solutions using ultra-pure water and analytical grade reagents. Filter all liquids through a 0.22 μm filter before they are used (except LB medium). Prepare and store all reagents at room temperature (unless indicated otherwise).

1. *Escherichia coli* BL21 (DE3) competent cells.
2. Plasmids:
   (a) $STING^{CTD}$ (residues 140–379) subcloned into pET28a vector (*see* **Note 1**).
   (b) Full-length STING subcloned into pcDNA3.1 vector (*see* **Note 2** and **3**).
   (c) pGL3-mIFNβ-promotor-Luc construct (*see* **Note 2**).
   (d) pGL3-acitin-promotor-Luc construct (*see* **Note 2**).
3. HEK293T cell line.

4. Kanamycin: 50 mg/ml. Weigh 2.5 g Kanamycin and transfer to a glass beaker, add about 40 ml water. Dissolve and transfer to a graduated cylinder, make up to 50 ml with water. Mix and filter with a 0.22 μm filter. Store at −20 °C.
5. Isopropyl β-d-1-thiogalactopyranoside (IPTG): 1 M. Weigh 11.9 g IPTG and prepare a 50 ml solution as in the previous step. Store at −20 °C.
6. Luria Broth (LB) medium: 10 g/l NaCl, 10 g/l Tryptone, 5 g/l Yeast extract. Weigh 10 g NaCl, 10 g tryptone and 5 g yeast extract, transfer to a beaker, add 1 l water and mix, transfer to a 3 l conical flask and seal with a breathable film. Autoclave and cool to room temperature.
7. LB medium containing 50 μg/ml kanamycin.
8. LB agar plates: Prepare LB medium as previous, but add 15 g agar before autoclaving. After autoclaving, cool to about 55 °C, add 1 ml kanamycin, pour into petridishes. Wait to harden and store at 4 °C.
9. Buffer A: 20 mM Tris–HCl, pH 8.0, 500 mM NaCl. Weigh 2.42 g Tris–HCl and 29.22 g NaCl and transfer to a 1 l glass beaker. Add about 950 ml water and mix. Adjust pH with HCl. Make up to 1 l with water.
10. Buffer B: 20 mM Tris–HCl, pH 8.0, 500 mM NaCl, 500 mM immidazole. Weigh 2.42 g Tris–HCl, 29.22 g NaCl, 34.04 g immidazole and prepare a 1 l solution as in the previous step.
11. Buffer C: 20 mM Tris–HCl, pH 8.0, 200 mM NaCl. Weigh 2.42 g Tris–HCl and 11.69 g NaCl, prepare a 1 l solution as in the previous step.
12. c-di-GMP: Prepare a 50 mM solution in buffer C. Store at −20 °C.
13. cGAMPs (2′3′cGAMP, 3′3′cGAMP, 2′2′cGAMP): Prepare a 50 mM solution of each cGAMP in buffer C. Store at −20 °C.
14. Thrombin: 1 unit/μl solution in PBS buffer (*see* **Note 4**), store at −80 °C.
15. Dulbecco's modified eagle medium (DMEM).
16. Perfringolysin O (PFO).
17. Sonicator with 5 mm probe for breaking *E. coli* cells.
18. FPLC: ÄKTA pure system for protein purification.
19. HiTrap HP column: Ni Sepharose, 5 ml.
20. Centrifugal filter: cut off 10 kDa, 15 ml.
21. Superdex 200 column: 10 × 300 mm, 30 ml, composite of cross-linked agarose and dextran.
22. MicroCal ITC_{200} machine: Malvern, UK.

23. Syringe: 500 μl.
24. 48-Well double sample plates for Sitting Drop Crystallization Plate (XtalQuest Inc., China).
25. Crystal clear sealing tape (XtalQuest Inc., China).
26. Crystal screen kits (*see* **Note 5**).
27. 24-well tissue culture plates.
28. Dual-Luciferase® Reporter (DLR™) Assay System, Promega.
29. SDS-PAGE supply.

3 Methods

3.1 Preparation of STING CTD Protein

1. Transform the plasmid of pET28a-STING^{CTD} into *Escherichia coli* BL21 (DE3) cells. Plate on kanamycin selection plates and incubate overnight at 37 °C.
2. Resuspend a single colony in 20 ml LB medium with kanamycin (50 μg/ml), incubating for 16–18 h with shaking (220 rpm) at 37 °C.
3. Inoculate into 1 l LB medium containing 50 μg/ml kanamycin and incubate at 37 °C with shaking (220 rpm) until an OD_{600nm} of 0.6–0.8 has been reached.
4. Induce the bacterial culture by adding 0.5 ml of 1 M IPTG.
5. Incubate the induced bacterial culture for an additional 20 h at 18 °C with shaking (220 rpm) to express the STING protein.
6. Collect cells by centrifugation at 6500 rpm (Beckman Coulter Avanti J-25, JLA9.1000 rotor).
7. Resuspend the cell pellet in 30 ml buffer A.
8. Dissolute cells using ultrasonic lysis. The power is 300 W.
9. Remove cell debris by centrifugation at 22,000 rpm (Beckman Coulter Avanti J-25, JA25.50 rotor) for 1 h.
10. Filter supernatants through a 0.2 μm filter.
11. Connect a HiTrap HP column next to a peristaltic pump, equilibrate with 25 ml buffer A at a flow rate of about 3 ml/min.
12. Load the filtered supernatant onto the HiTrap HP column at a flow rate of about 1 ml/min.
13. Place the column into ÄKTA Pure system.
14. Wash impurities with 20–40 ml buffer A at a flow rate of 3–5 ml/min until UV absorption at 280 nm reaching baseline.
15. Then, add 10% buffer B to wash until UV absorption at 280 nm reaching baseline.
16. Elute the target proteins with a linear gradient buffer B from 10% to 100% within 50 ml (Fig. 1a). Collect target protein according to UV 280 nm absorption.

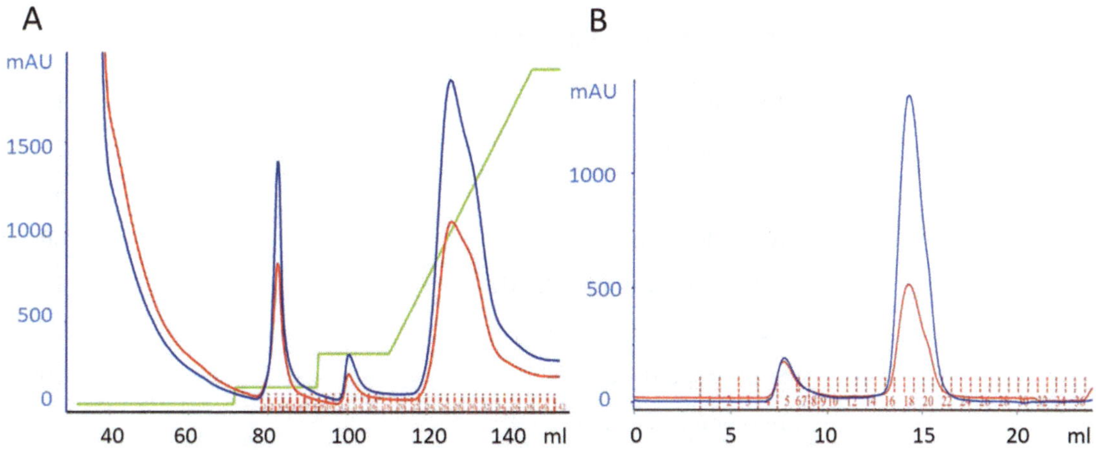

Fig. 1 Elution curve of STING^{CTD} in HiTrap HP column (**a**) and Superdex 200 column (**b**)

17. Check the peak fractions by SDS-PAGE (*see* **Note 6**).
18. Concentrate target protein to about 10 mg/ml using a centrifugal filter.
19. Add thrombin to target protein in proportion of 10 units/mg, digest at 4 °C overnight to remove the His-tag.
20. Place a Superdex 200 column into ÄKTA Pure system. Equilibrate with 30 ml buffer C at a flow rate of 0.5 ml/min.
21. Load target protein onto the Superdex 200 column to separate the oligomer and the dimer. The protein amount loaded should be less than 10 mg and volume less than 500 μl.
22. Elute with buffer C at a flow rate of 0.5 ml/min. The oligomer form of target protein should be eluted at about 7 ml, and the dimer form should be eluted at about 14.5 ml (Fig. 1b).
23. Check the peak fractions by SDS-PAGE (*see* **Note 6**) (Fig. 1).
24. Combine fractions containing the dimer form of target protein, and concentrate the pooled fractions to 10 mg/ml using a centrifugal filter.
25. Freeze protein in liquid nitrogen and store at −80 °C.

3.2 Measurement of the Dissociation Constants and Thermodynamic Parameters for STING^{CTD}'s Binding with CDNs Using Isothermal Titration Calorimetry (ITC)

1. Dilute CDNs to 1 mM and STING^{CTD} protein to 0.1 mM in buffer C.
2. Centrifuge at 16,000 × *g* and 25 °C for 10 min to remove precipitate and bubbles.
3. Switch on the MicroCal ITC_{200} machine and start MicroCal ITC_{200} software, set up experiment parameters as Table 1.
4. Wash sample cell and syringe using the wash module of the machine.

Table 1
Experiment parameters for ITC

Total injections	19
Cell temperature (°e)	25
Reference power	5
Initial delay (s)	60
Syringe concentration (mM)	1
Cell concentration (mM)	0.1
Stirring speed (RPM)	1000
Injection parameters	
Volume (μl)	2
Duration (s)	4
Spacing (s)	150
Filter period (s)	5

5. Inject about 300 μl buffer C to rinse the sample cell with a syringe and then remove buffer C, repeat for 5–10 times.
6. Inject about 280 μl protein solution to fill the sample cell.
7. Fill a syringe with 40 μl CDN, insert the pipette into the cell port, start titration.
8. Analyze data with software Origin v7.0 (MicroCal) to calculate Kd, ΔH and ΔG, *see* Fig. 2 and Table 2.

3.3 Cystallization of STING^{CTD}-c-di-GMP Complex

The STING^{CTD}-c-di-GMP complex crystals were obtained by the sitting-drop vapor diffusion method. Carry out all the procedures at 18 °C.

1. Mix 8 mg/ml protein with 3 mM c-d-GMP, incubate at 18 °C for 18 h.
2. Pipet 100 μl of crystallization solution into reservoir of XtalQuest 48 wells crystallization plate.
3. Pipet 1 μl of protein-c-di-GMP mixture into the sample cell of crystallization plate.
4. Pipet 1 μl of solution from the reservoir into the sample cell.
5. Repeat **steps 2–4** for each row of wells of XtalQuest 48-well crystallization plate, seal the rows with crystal clear sealing tape.
6. Repeat above **steps 2–5** to complete the whole 48-well crystallization plate. Place the crystallization plates at 18 °C.
7. Observe the growth process of crystal under a microscope everyday. Under the conditions tested, crystals appeared in

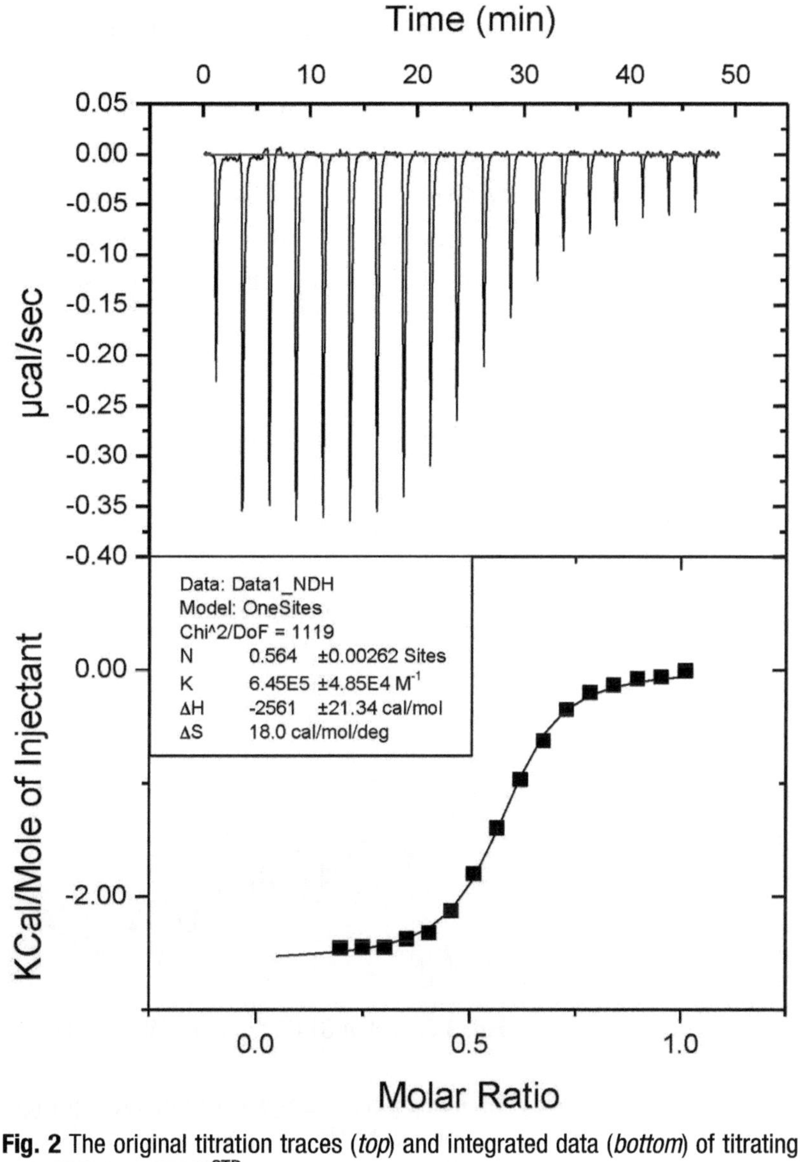

Fig. 2 The original titration traces (*top*) and integrated data (*bottom*) of titrating c-di-GMP into STING^{CTD}

the drop contained reservoir solution composed of 0.025 M $MgSO_4$, 0.05 M Tris–HCl pH 8.5, 1.8 M $AmSO_4$ (Fig. 3).

3.4 Luciferase Reporter Assay of STING Activation

1. Seed HEK293T cells (1×10^5) in 24-well tissue culture plates, and incubate for 12–20 h until cell density reaches about 70% confluent.
2. Transfect the cells with 50 ng of pcDNA3.1-STING, 50 ng of pGL3-mIFNβ-promotor-Luc construct, and 50 ng pGL3-acitin-promotor-Luc construct as an internal control using standard calcium phosphate precipitation method. Incubate for 12 h.

Table 2
Thermodynamic parameters for binding of CDNs to STING^{CTD}

Ligands	Kd (nM)	ΔH (cal/mol)	ΔS (cal/mol/deg)
2′3′cGAMP	5.4	−6182	17.1
3′3cGAMP	178.6	−1330	26.4
2′2′cGAMP	31.7	−2799	24.9
c-di-GMP	1550	−2561	18

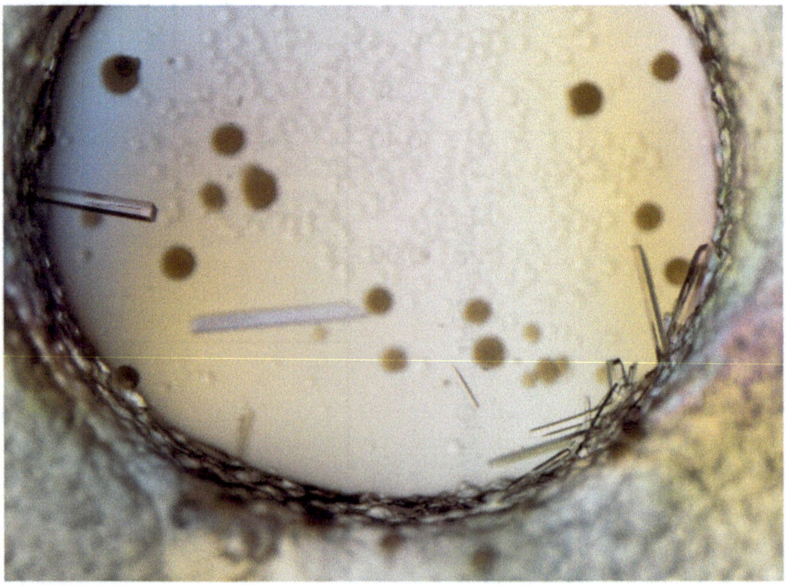

Fig. 3 Crystal of STING^{CTD}-c-di-GMP complex

3. Add perfringolysin O (PFO, 1.5 μg/μl) along with c-di-GMP or cGAMPs (5 μM) and incubate for permeabilize cell for 30 min to permeabilize the cells and thus, deliver c-di-GMP or cGAMPs.
4. Then replace with fresh medium, incubate for 12 h.
5. Harvest cells, lyse in reporter lysis buffer (*see* **Note** 7).
6. Measure the luciferase activity in the total cell lysate with the Dual-Luciferase Reporter Assay System (Promega) (*see* **Note 8**) (Fig. 4).
7. Analyze data with software GraphPad Prism 5 (Fig. 4).

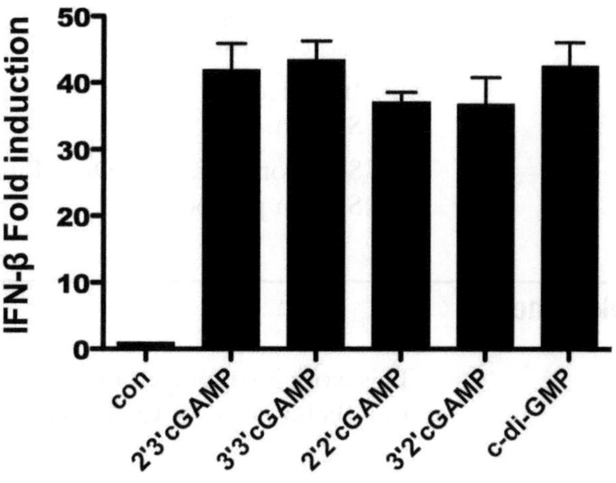

Fig. 4 Luciferase assay of STING stimulated by c-di-GMP and cGAMPs

4 Notes

1. Amino acid sequence of expressed STING^{CTD}: MGSSHHH HHHSSGLVPRGSHMASMTGGQQMGRGSMAPAEISAVC EKGNFNVAHGLAWSYYIGYLRLILPELQARIRTYNQHYN NLLRGAVSQRLYILLPLDCGVPDNLSMADPNIRFLDKLP QQTADRAGIKDRVYSNSIYELLENGQRAGTCVLEYATPL QTLFAMSQYSQAGFSREDRLEQAKLFCQTLEDILADAPE SQNNCRLIAYQEPADDSSFSLSQEVLRHLRQEEKEEVTV GSLKTSAVPSTSTMSQEPELLISGMEKPLPLRTDFS.
2. Get from Dr. Zhengfan Jiang's lab, Peking University.
3. Amino acid sequence of expressed full-length STING:

   MPHSSLHPSIPCPRGHGAQKAALVLLSACLVTLWGLGE PPEHTLRYLVLHLASLQLGLLLNGVCSLAEELRHIHSRY RGSYWRTVRACLGCPLRRGALLLLSIYFYYSLPNAVGPPF TWMLALLGLSQALNILLGLKGLAPAEISAVCEKGNFNVA HGLAWSYYIGYLRLILPELQARIRTYNQHYNNLLRGAVS QRLYILLPLDCGVPDNLSMADPNIRFLDKLPQQTADRA GIKDRVYSNSIYELLENGQRAGTCVLEYATPLQTLFAMS QYSQAGFSREDRLEQAKLFCQTLEDILADAPESQNNCR LIAYQEPDDSSFSLSQEVLRHLRQEEKEEVTVGSLKTSA VPSTSTMSQEPELLISGMEKPLPLRTDFSSRYPYDVPDYA
4. PBS: Phosphate-buffered saline. 137 mM NaCl, 2.7 mM KCl, 10 mM Na_2HPO_4, 1.8 mM KH_2PO_4, pH 7.4.
5. The crystal screen kits used in our lab include Crystal Screen1,2, Index and Salt RX from Hampton Research and Wizard Classic crystallization screen series from Regaku Reagents, Inc.

6. The loading quantity of sample should be decided according to the UV absorption. Usually, we load 1–5 μl of each fraction.
7. Included in the Dual-Luciferase® Reporter (DLR™) Assay System.
8. See more details in Dual-Luciferase® Reporter (DLR™) Assay System protocol.

Acknowledgment

This work is supported by a grant from Chinese Ministry of Science and Technology (MOST 2014CB910102 to XDS).

References

1. Ishikawa H, Barber GN (2008) STING is an endoplasmic reticulum adaptor that facilitates innate immune signalling. Nature 455 (7213):674–678. doi:10.1038/nature07317
2. Jin L, Waterman PM, Jonscher KR, Short CM, Reisdorph NA, Cambier JC (2008) MPYS, a novel membrane tetraspanner, is associated with major histocompatibility complex class II and mediates transduction of apoptotic signals. Mol Cell Biol 28(16):5014–5026. doi:10.1128/MCB.00640-08
3. Zhong B, Yang Y, Li S, Wang YY, Li Y, Diao F, Lei C, He X, Zhang L, Tien P, Shu HB (2008) The adaptor protein MITA links virus-sensing receptors to IRF3 transcription factor activation. Immunity 29(4):538–550. doi:10.1016/j.immuni.2008.09.003
4. Sun W, Li Y, Chen L, Chen H, You F, Zhou X, Zhou Y, Zhai Z, Chen D, Jiang Z (2009) ERIS, an endoplasmic reticulum IFN stimulator, activates innate immune signaling through dimerization. Proc Natl Acad Sci 106(21):8653–8658. doi:10.1073/pnas.0900850106
5. Witte G, Hartung S, Buttner K, Hopfner KP (2008) Structural biochemistry of a bacterial checkpoint protein reveals diadenylate cyclase activity regulated by DNA recombination intermediates. Mol Cell 30(2):167–178. doi:10.1016/j.molcel.2008.02.020
6. Davies BW, Bogard RW, Young TS, Mekalanos JJ (2012) Coordinated regulation of accessory genetic elements produces cyclic di-nucleotides for V. cholerae virulence. Cell 149 (2):358–370. doi:10.1016/j.cell.2012.01.053
7. Witte CE, Whiteley AT, Burke TP, Sauer JD, Portnoy DA, Woodward JJ (2013) Cyclic di-AMP is critical for *Listeria monocytogenes* growth, cell wall homeostasis, and establishment of infection. MBio 4(3): e00282-00213. doi:10.1128/mBio.00282-13
8. Woodward JJ, Iavarone AT, Portnoy DA (2010) c-di-AMP secreted by intracellular *Listeria monocytogenes* activates a host type I interferon response. Science 328 (5986):1703–1705. doi:10.1126/science.1189801
9. Burdette DL, Monroe KM, Sotelo-Troha K, Iwig JS, Eckert B, Hyodo M, Hayakawa Y, Vance RE (2011) STING is a direct innate immune sensor of cyclic di-GMP. Nature 478(7370):515–518. doi:10.1038/nature10429
10. Sun L, Wu J, Du F, Chen X, Che ZJ (2013) Cyclic GMP-AMP synthase is a cytosolic DNA sensor that activates the type I interferon pathway. Science 339(6121):786–791. doi:10.1126/science.1232458
11. Gao P, Ascano M, Wu Y, Barchet W, Gaffney BL, Zillinger T, Serganov AA, Liu Y, Jones RA, Hartmann G, Tuschl T, Patel DJ (2013) Cyclic [G(2′,5′)pA(3′,5′)p] is the metazoan second messenger produced by DNA-activated cyclic GMP-AMP synthase. Cell 153(5):1094–1107. doi:10.1016/j.cell.2013.04.046
12. Wu J, Sun L, Chen X, Du F, Shi H, Chen C, Chen ZJ (2013) Cyclic GMP-AMP is an endogenous second messenger in innate immune signaling by cytosolic DNA. Science 339(6121):826–830. doi:10.1126/science.1229963
13. Zhang X, Shi H, Wu J, Zhang X, Sun L, Chen C, Chen ZJ (2013) Cyclic GMP-AMP containing mixed phosphodiester linkages is an endogenous high-affinity ligand for STING. Mol Cell 51(2):226–235. doi:10.1016/j.molcel.2013.05.022

14. Tanaka Y, Chen ZJ (2012) STING specifies IRF3 phosphorylation by TBK1 in the cytosolic DNA signaling pathway. Sci Signal 5 (214):ra20–ra20. doi:10.1126/scisignal.2002521

15. Liu S, Cai X, Wu J, Cong Q, Chen X, Li T, Du F, Ren J, YT W, Grishin NV, Chen ZJ (2015) Phosphorylation of innate immune adaptor proteins MAVS, STING, and TRIF induces IRF3 activation. Science 347(6227):aaa2630. doi:10.1126/science.aaa2630

16. Huang YH, Liu XY, XX D, Jiang ZF, XD S (2012) The structural basis for the sensing and binding of cyclic di-GMP by STING. Nat Struct Mol Biol 19(7):728–730. doi:10.1038/nsmb.2333

17. Ouyang S, Song X, Wang Y, Ru H, Shaw N, Jiang Y, Niu F, Zhu Y, Qiu W, Parvatiyar K, Li Y, Zhang R, Cheng G, Liu ZJ (2012) Structural analysis of the STING adaptor protein reveals a hydrophobic dimer interface and mode of cyclic di-GMP binding. Immunity 36 (6):1073–1086. doi:10.1016/j.immuni.2012.03.019

18. Shang G, Zhu D, Li N, Zhang J, Zhu C, Lu D, Liu C, Yu Q, Zhao Y, Xu S, Gu L (2012) Crystal structures of STING protein reveal basis for recognition of cyclic di-GMP. Nat Struct Mol Biol 19(7):725–727. doi:10.1038/nsmb.2332

19. Shu C, Yi G, Watts T, Kao CC, Li P (2012) Structure of STING bound to cyclic di-GMP reveals the mechanism of cyclic dinucleotide recognition by the immune system. Nat Struct Mol Biol 19(7):722–724. doi:10.1038/nsmb.2331

20. Yin Q, Tian Y, Kabaleeswaran V, Jiang X, Tu D, Eck MJ, Chen ZJ, Wu H (2012) Cyclic di-GMP sensing via the innate immune signaling protein STING. Mol Cell 46(6):735–745. doi:10.1016/j.molcel.2012.05.029

21. Gao P, Ascano M, Zillinger T, Wang W, Dai P, Serganov AA, Gaffney BL, Shuman S, Jones RA, Deng L, Hartmann G, Barchet W, Tuschl T, Patel DJ (2013) Structure-function analysis of STING activation by c[G(2′,5′)pA(3′,5′)p] and targeting by antiviral DMXAA. Cell 154 (4):748–762. doi:10.1016/j.cell.2013.07.023

22. Zhang H, Han MJ, Tao J, Ye ZY, Du XX, Deng MJ, Zhang XY, Li LF, Jiang ZF, Su XD (2015) Rat and human STINGs profile similarly towards anticancer/antiviral compounds. Sci Rep 5:18035. doi:10.1038/srep18035

Chapter 7

Spectrophotometric and Mass Spectroscopic Methods for the Quantification and Kinetic Evaluation of In Vitro c-di-GMP Synthesis

Geoffrey B. Severin and Christopher M. Waters

Abstract

The expression and activity of diguanylate cyclase (DGC) and phosphodiesterase (PDE) enzymes are responsible for modulating and maintaining the intracellular concentration of the bacterial second messenger cyclic diguanosine-monophosphate (c-di-GMP). Here, we describe an in vitro method for the spectrophotometric detection and quantification of DGC catalyzed c-di-GMP synthesis through adaptation of the EnzChek® Pyrophosphate Assay Kit. We also outline a method for the quantification of c-di-GMP produced in this in vitro reaction using Ultra-Performance Liquid Chromatography tandem Mass Spectrometry (UPLC-MS/MS). These methods can be leveraged for a number of experimental applications including the evaluation of enzyme activity for the in vitro synthesis of c-di-GMP, examination of how molecular signals impact these activities, identifying the catalytic properties of hybrid DGC-PDE proteins, and the development of DGC inhibitors.

Key words c-di-GMP, Diguanylate cyclase, EnzChek, Pyrophosphate, Second messenger, WpsR, UPLC-MS/MS

1 Introduction

Diguanylate cyclases (DGCs) and Phosphodiesterases (PDEs) are the enzymes responsible for the anabolism and catabolism of the bacterial cyclic-dinucleotide second messenger c-di-GMP, respectively [1]. The DGC catalyzed synthesis of a single c-di-GMP molecule from two guanine triphosphates (GTP) involves the formation of two phosphodiester bonds resulting in the liberation of two molecules of inorganic pyrophosphate (PPi) [1]. Depending on the PDE species, the degradation of c-di-GMP involves the hydrolysis of one or both of these phosphodiester bonds resulting in the production of the linear dinucleotide 5′-phosphoguanylyl-(3′,5′)-guanosine (pGpG) or two molecules of guanosine monophosphate (GMP) [1]. Here, we present an in vitro method for the detection and quantification of DGC activity by adapting the

Karin Sauer (ed.), *c-di-GMP Signaling: Methods and Protocols*, Methods in Molecular Biology, vol. 1657,
DOI 10.1007/978-1-4939-7240-1_7, © Springer Science+Business Media LLC 2017

EnzChek® Pyrophosphate Assay Kit to monitor the release of PPi during c-di-GMP synthesis.

The EnzChek® Pyrophosphate Assay Kit, based on the coupling of reactions first described by Webb in 1992 for the detection of PPi [2], couples the activity of an inorganic pyrophosphatase and purine nucleoside phosphorylase (PNP). The series of reactions begins with the hydrolysis of PPi into two molecules of inorganic phosphate (Pi) by the enzyme inorganic pyrophosphatase. Free Pi in solution is then utilized by PNP to convert the PNP substrate 2-amino-6-mercapto-7-methyl-purine ribonucleoside (MESG) into one molecule of ribose 1-phosphate and one molecule of 2-amino-6-mercapto-7-methyl-purine. This reaction results in a shift in the solution absorbance from 330 nm to 360 nm as MESG is converted into 2-amino-6-mercapto-7-methyl-purine. This reaction can be monitored by continuously measuring the increase in absorbance at 360 nm.

Our adaptations to the manufacturer's protocol include the addition of purified DGC enzyme, GTP, and supplemental $MgCl_2$. We have also reduced the reaction volumes from 1 mL to 100 μL allowing for high-throughput monitoring of reaction progression in 96-well microtiter plates. By developing a standard curve based on known concentrations of PPi, one can use this assay to evaluate the activity and kinetics of DGC enzymes. As an example, we demonstrate the in vitro activity of the *Pseudomonas aeruginosa* DGC WspR (R272A) using the modified EnzChek® Pyrophosphate Assay Kit. While this example illustrates the end-point quantification of c-di-GMP synthesized over the course of an hour in the presence of 50 μM GTP and 10 mM $MgCl_2$, it could be easily modified to determine the kinetics of the c-di-GMP synthesis reaction by altering the concentrations of the DGC and GTP substrates.

We also briefly describe a complementary approach to measuring the activity of a DGC enzyme by the direct quantification of c-di-GMP produced in the in vitro reaction, described above, using UPLC-MS/MS [3]. DGC enzymatic reactions are monitored using the EnzChek® assay and direct detection of c-di-GMP by UPLC-MS/MS function in parallel to provide rapid quantification of enzyme kinetics while ensuring that the enzymes are generating the predicted products.

2 Materials

All solutions and buffers should be prepared using ultra-pure water and HPLC grade reagents. When generating compound standards using serial dilutions, vortex and briefly centrifuge each solution prior to performing the subsequent dilution. When pipetting to mix, utilize a volume that is at least 10% of the total solution volume (i.e., 10 μL pipette volume to mix a 100 μL total solution volume).

Ensure that any glassware and non-disposable materials used have been thoroughly rinsed with deionized water to minimize introduction of contaminating inorganic phosphate (Pi).

2.1 Plate-Based DGC Enzyme Assay

1. EnzChek® Pyrophosphate Assay Kit (Thermo-Fischer Scientific). The following components are provided in the EnzChek® Pyrophosphate Assay Kit and should be prepared as follows:
   (a) 1 mM MESG substrate: Add 20 mL of water to the bottle of 2-amino-6-mercapto-7-methyl-purine ribonucleoside (MESG) and vortex vigorously to mix at room temperature (*see* **Note 1**). Once dissolved, immediately aliquot the MESG in 250 μL and 500 μL portions and store at −20 °C (*see* **Note 2**).
   (b) PPi standards: Thaw the 50 mM pyrophosphate (PPi) solution on ice. Add 20 μL of 50 mM PPi to 980 μL water to make a 1 mM PPi stock. Vortex to mix. Prepare the PPi standards (100 μM–600 μM) as described below and store all PPi solutions at −20 °C (*see* **Note 3**).
      - 100 μM PPi standard: Combined 25 μL 1 mM PPi with 225 μL water.
      - 200 μM PPi standard: Combined 50 μL 1 mM PPi with 200 μL water.
      - 300 μM PPi standard: Combined 75 μL 1 mM PPi with 175 μL water.
      - 400 μM PPi standard: Combined 100 μL 1 mM PPi with 150 μL water.
      - 500 μM PPi standard: Combined 125 μL 1 mM PPi with 125 μL water.
      - 600 μM PPi standard: Combined 150 μL 1 mM PPi with 100 μL water.
   (c) 20× Reaction Buffer: Thaw the 20× Reaction Buffer on ice, divide into 500 μL aliquots, and store at −20 °C. This buffer is composed of 1.0 M Tris–HCl, 20 mM MgCl2, pH 7.5, containing 2 mM sodium azide.
   (d) 1× Reaction Buffer: Add 50 μL of 20× Reaction buffer to 950 μL water. Vortex to mix. Divide into 200 μL portions and store at −20 °C.
   (e) 100 U/mL PNP: The EnzChek® Pyrophosphate assay contains two 50 U vials of lyophilized PNP. Add 500 μL of water to one vial, gently pipette to mix. Store the 100 U/mL PNP solution at 4 °C and the remaining 50 U vial at −20 °C (*see* **Note 4**).
   (f) 30 U/mL Inorganic Pyrophosphatase: Add 200 μL of water to the 6 U vial of lyophilized inorganic pyrophosphatase. Mix with gentle pipetting and store at 4 °C (*see* **Note 5**).

2. 500 μM GTP Stock: Prepare a 250 mM GTP solution by weighing 65.4 mg of GTP and quantitatively transferring to a 15 mL conical tube. Add water to a final volume of 5 mL and vortex to mix. Make a 100 mM GTP solution by combining 400 μL of 250 mM GTP with 600 μL water. Prepare 500 μL of 1 mM GTP by combining 5 μL of 100 mM GTP with 495 μL of water. Finally, prepare a 500 μM stock solution by mixing 250 μL of 1 mM GTP and 250 μL of water. Divide 500 μM GTP into 100 μL portions and store all GTP solutions at −80 °C.
3. 100 mM $MgCl_2$: Prepare a 1 M $MgCl_2$ stock solution by weighing 0.476 g $MgCl_2$ and quantitatively transferring to a 15 mL conical tube. Add water to a final volume of 5 mL and vortex to mix. Make a working 100 mM $MgCl_2$ solution by diluting 100 μL 1 M $MgCl_2$ in 900 μL water, and vortex to mix. Divide the 100 mM $MgCl_2$ into 250 μL portions and store all $MgCl_2$ solutions at −20 °C.
4. Purified DGC protein: To be purified per each individual laboratory and is specific to each enzyme (*see* **Notes 6** and 7).
5. DGC Buffer: When purifying your DGC, reserve 5 mL of used dialysis buffer in 1 mL aliquots and store in the same conditions as the purified DGC protein (*see* **Note 8**).
6. Generic 96-well microtiter plate.
7. 0.2 mL PCR strip tubes and 1.5 mL microcentrifuge tubes.
8. Twelve-channel multichannel pipette capable of dispensing 10 μL volumes.
9. Microtiter plate reader capable of measuring absorbance at 360 nm.

2.2 c-di-GMP Quantification Using UPLC-MS/MS

1. c-di-GMP Buffer A: 15 mM Acetic Acid and 10 mM Tributylamine (TBA) in 97:3 Water:Methanol. In a chemical hood, measure 18 mL of HPLC grade Methanol in a glass graduated cylinder and add 1.43 mL of Tributylamine (TBA). Quantitatively transfer the TBA-MeOH solution into a 1 L graduated cylinder and fill to 600 mL with water. Pour the solution into a 1 L bottle, add 520 μL glacial acetic acid, and swirl to mix. Store at room temperature.
2. c-di-GMP Buffer B: 1 L HPLC grade Methanol in a 1 L bottle.
3. UPLC Column: ACQUITY UPLC BEH C18 Column, 130 Å, 1.7 μm, 2.1 mm × 50 mm (Waters Corp.).
4. c-di-GMP Standards. Add 1 mL of water to 1 micromol of c-di-GMP (Axxora) to make a 1 mM stock solution (*see* **Note 9**). Prepare the c-di-GMP standards (1.9 nM–250 nM) as described below. Prepare a 100 μM c-di-GMP stock by adding 100 μL of 1 mM c-di-GMP to 900 μL of water. Make a 1 μM

dilution of c-di-GMP by adding 10 μL of 100 μM c-di-GMP to 990 μL of water. To make a 250 nM stock for the highest concentration of c-di-GMP standard in the standard curve, add 250 μL of 1 μM c-di-GMP to 750 μL of water. Begin two-fold serial dilutions by adding 500 μL of 250 nM c-di-GMP to 500 μL of water. Perform six further twofold serial dilutions until a final concentration of 1.9 nM c-di-GMP has been achieved. The entire standard range will include 250 nM, 125 nM, 62.5 nM, 31.25 nM, 15.6 nM, 7.8 nM, 3.9 nM, and 1.9 nM c-di-GMP (*see* **Note 10**).

3 Methods

The following protocol outlines the preparation of a DGC activity Master Mix, reaction initiation and monitoring, and the use of UPLC-MS/MS to validate the spectrophotometric determination of c-di-GMP synthesis. The goal of the described experiment is to compare the end-point concentration of c-di-GMP as determined using the modified EnzChek® Pyrophosphate Assay Kit with the quantification of c-di-GMP in solution using UPLC-MS/MS. The described Master Mix is designed to accommodate twelve 100 μL reactions in total. This includes enough reagents for seven standard PPi reactions used to generate a PPi standard curve, two control DGC reactions that include either DGC enzyme or GTP substrate, one experimental DGC reaction that includes both GTP substrate and DGC enzyme, and two "spare" reactions to facilitate accurate division and dispensing of the Master Mix.

Prepare the reaction Master Mix and PPi Standard Reaction Mix and perform reactions at room temperature.

3.1 DGC Activity Master Mix and PPi Standard Reaction Mix

Prepare the reaction Master Mix. A master mix accommodates twelve 100 μL reactions in total. A single 100 μL reaction is composed of:

- 43 μL water.
- 10 μL 100 mM $MgCl_2$.
- 5 μL 20× Reaction Buffer.
- 20 μL of MESG substrate solution.
- 1 μL of 3 U/mL inorganic pyrophosphatase solution.
- 1 μL of 100 U/mL purine nucleoside phosphorylase (PNP).
- 10 μL of either DGC enzyme or DGC Buffer.
- 10 μL of substrate (This could be water for the negative control, 100 μM–600 μM PPi for the standards, or 500 μM GTP for the DGC reaction).

1. Turn on the 96-well microtiter plate reader (*see* **Note 11**).
2. Thaw the following reagent stocks on ice, occasionally vortexing to both encourage thawing and thoroughly mix solutions:
   (a) PPi standards (10 μM–60 μM), 250 μL each.
   (b) 500 μM GTP, 100 μL.
   (c) 100 mM $MgCl_2$, 250 μL.
   (d) 20× Reaction Buffer, 500 μL.
   (e) 1× Reaction Buffer, 200 μL.
   (f) DGC Buffer, 250 μL.
3. *Nearly* thaw a frozen 250 μL aliquot of MESG in a preheated heat block at 37 °C. Complete thawing by vigorously vortexing the solution (*see* **Note 12**). Store at room temperature.
4. Place the following components on ice. Do not vortex any enzymes or solutions containing enzymes:
   (a) 100 U/mL PNP stock.
   (b) 30 U/mL inorganic pyrophosphatase stock.
   (c) DGC enzyme solution (*see* **Note 13**).
5. Prepare 20 μL of a 3 U/mL inorganic pyrophosphatase working stock by adding 2 μL of 30 U/mL inorganic pyrophosphatase stock to 18 μL of 1× Reaction Buffer in a 0.2 mL PCR tube. Thoroughly mix with gentle pipetting.
6. Assemble the Master Mix at room temperature in a 1.5 mL microcentrifuge tube by adding the following components followed by vortexing to mix:
   (a) 516 μL of water.
   (b) 120 μL of 100 mM $MgCl_2$ (*see* **Note 14**).
   (c) 60 μL of 20× Reaction Buffer.
   (d) 240 μL of MESG substrate solution.
7. Add 12 μL of the 3 U/mL working stock of inorganic pyrophosphatase prepared in Subheading 3.1, **step 5** and 12 μL of 100 U/mL PNP to the Master Mix. Gently pipette to mix (*see* **Note 15**).
8. Dispense 80 μL of the Master Mix into three consecutive wells in a 96-well microtiter plate (i.e., wells A1–A3).
9. Add 10 μL of the following components to the designated wells and mix thoroughly by pipetting (*see* **Note 16**):
   (a) A1–DGC enzyme ("DGC Reaction") (*see* **Note 17**).
   (b) A2–DGC enzyme ("DGC Control") (*see* **Note 18**).
   (c) A3–DGC buffer ("GTP Control") (*see* **Note 19**).
10. Make a PPi Standard Reaction Mix by mixing 640 μL of the remaining Master Mix in a new 1.5 mL microcentrifuge tube with 80 μL of DGC buffer (*see* **Note 20**). Gently pipette to mix.

11. Dispense 90 μL of the PPi Standard Reaction Mix into seven consecutive wells in the 96-well microtiter plate in the same row as those prepared in Subheading 3.1, **step 9** (i.e., wells A4-A10) (*see* **Note 21**).
12. Label ten 0.2 mL PCR tubes according to the wells loaded in subheading 3.1, **step 9** and Subheading 3.1, **step 11** (i.e., A1–A10). Line these tubes up in a PCR tube rack in the same orientation as the reaction wells in the microtiter plate. This will facilitate the rapid addition of substrates when initiating the reactions in Subheading 3.2, **step 2**.
13. Add 15 μL of the following reaction substrates to the designated 0.2 mL PCR tube:
    (a) Tube A1–500 μM GTP ("DGC Reaction").
    (b) Tube A2–Water ("DGC Control").
    (c) Tube A3–500 μM GTP ("GTP Control").
    (d) Tube A4–Water ("No Substrate Control").
    (e) Tube A5–100 μM PPi (10 μM PPi Standard).
    (f) Tube A6–200 μM PPi (20 μM PPi Standard).
    (g) Tube A7–300 μM PPi (30 μM PPi Standard).
    (h) Tube A8–400 μM PPi (40 μM PPi Standard).
    (i) Tube A9–500 μM PPi (50 μM PPi Standard).
    (j) Tube A10–600 μM PPi (60 μM PPi Standard).
14. Solution preparations are now complete. Discard remaining MESG, Master Mix, PPi Standard Reaction Mix, and 3 U/mL inorganic pyrophosphatase working stock. Return all other reagents to the appropriate long-term storage conditions.

3.2 Reaction Initiation and Spectrophotometric Monitoring

1. Assemble the 96-well microtiter plate and the substrate PCR tubes close to the microtiter plate reader. If possible, place the microtiter plate into the plate reader loading position. This will minimize the time it takes to load the substrates and make the first absorbance measurement in Subheading 3.2, **step 3**.
2. Using a twelve-channel multichannel pipette, draw 10 μL of the substrate solutions from the 0.2 mL PCR tubes loaded in Subheading 3.1, **step 13,** and dispense the substrates into the corresponding wells. Mix thoroughly by pipetting (*see* **Notes 22** and **23**).
3. Immediately begin reading the absorbance at 360 nm every 30 s for approximately 1 h (*see* **Notes 24** and **25**).
4. After 1 h, collect samples for UPLC/MSMS analysis by performing 1:100 dilutions of the A1 "DGC Reaction" and A2 "DGC Control" reactions by adding 10 μL of the reaction solutions to 990 μL of water in 1.5 mL microcentrifuge tubes. Vortex to mix.

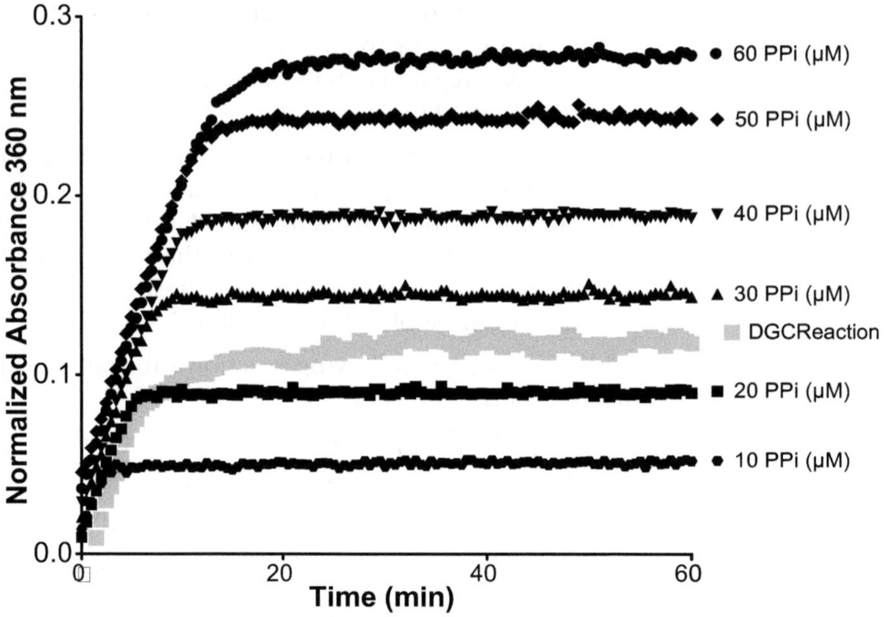

Fig. 1 Monitoring PPi Liberation During c-di-GMP Synthesis. The data represents the normalized absorbance at 360 nm following the addition of substrates in the "PPi Standard Reactions" (10 μM–60 μM PPi) and the "DGC Reaction" (280 μM *P. aeruginosa* WspR (R272A) with 50 μM GTP) recorded every 30 s for a total of 60 min

5. Heat the 1:100 dilutions at 90 °C for 5 min followed by centrifugation at 18,000 × *g* for 5 min to pellet denatured protein (*see* **Note 26**). Transfer the supernatant to a new 1.5 mL microcentrifuge tube. Discard the original tubes.
6. Add 100 μL of the 1:100 dilutions to 100 μL of water in new microcentrifuge tubes to make 200 μL of 1:200 dilutions from the original reactions (*see* **Note 27**). Store the 1:100 and 1:200 dilutions at 4 °C for up to 2 days or −80 °C for up to a month.

3.3 Spectrophotometric Analysis

1. Generate a reaction plot of ABS 360 nm vs. time (Fig. 1):
   (a) Normalize the PPi standard reaction measurements, wells A5 – A10, by subtracting the ABS 360 nm of the "No Substrate Control", well A4, for each time point.
   (b) Calculate the PPi and Pi contribution from spontaneous GTP hydrolysis at each time point by subtracting the ABS 360 nm of the"GTP Control", well A3, by the "No Substrate Control", well A4.
   (c) Calculate the ABS 360 nm contribution of PPi derived from c-di-GMP synthesis by subtracting both the ABS 360 nm of the "DGC Control", well A2, and phosphate contribution from spontaneous GTP hydrolysis (calculated previously in Subheading 3.3, **step 1b**) at each time point from the "DGC Reaction", well A1.

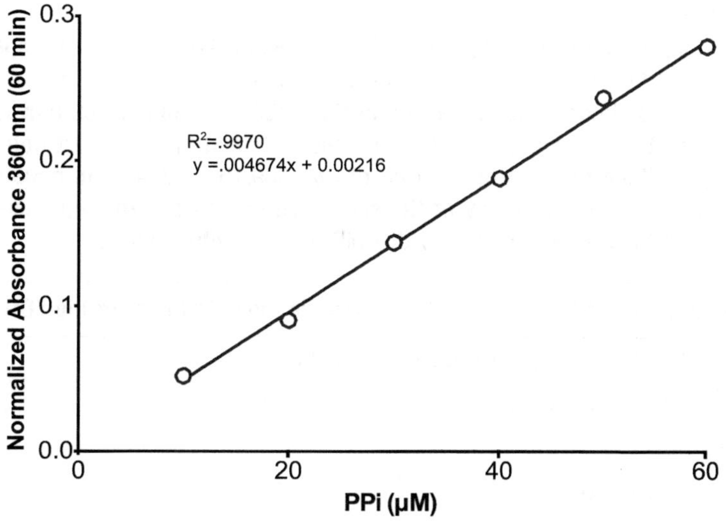

Fig. 2 PPi Standard Curve (60 min). The linear regression is calculated from the normalized absorbance at 360 nm of the "PPi Standard Reactions" after 60 min of incubation with substrate (10 µM–60 µM PPi)

2. Generate a PPi Standard Curve and calculate c-di-GMP synthesized (Fig. 2):
   (a) Plot the normalized PPi standard reaction ABS 360 nm measurements after 60 min (calculated previously in Subheading 3.2, **step 4**) vs. PPi standard concentration.
   (b) Generate a least squares regression line based on the plot.
   (c) Use the linear equation to calculate the concentration of PPi evolved from the normalized "DGC Reaction", well A1, after 60 min.
   (d) To determine the concentration of c-di-GMP synthesized after 60 min, divide the calculated concentration of PPi in the "DGC Reaction", well A1, by two. (Table 1) (*see* **Note 28**).

3.4 c-di-GMP Quantification Using UPLC-MS/MS

1. Thaw the c-di-GMP Standards (1.9 nM–250 nM) on ice. Vortex to mix and briefly centrifuge to collect solution. Keep c-di-GMP standards and the 1:200 dilutions of the "DGC Reaction" and "DGC Control" samples from Subheading 3.2, **step 6** at 4 °C.
2. Measure the concentration of c-di-GMP present in the 1:200 diluted "DGC Reaction" and "DGC Control" samples using the UPLC-MS/MS parameters described below [3] (*see* **Note 29**):

Table 1
c-di-GMP Quantification for DGC Reaction using the EnzChek® Pyrophosphate Assay Kit and UPLC-MS/MS. The spectrophotometric method for quantifying c-di-GMP is calculated from the normalized absorbance at 360 nm in the "DGC Reaction" at 60 min following the addition of 50 μM GTP substrate. The UPLC–MS/MS method is based on the direct detection of c-di-GMP in solution using multiple reaction-monitoring of *m/z* 689.16 → 344.31 in negative-ion mode. This example demonstrates the accuracy of the spectrophotometric method presented in the protocol to measure and monitor the synthesis of c-di-GMP in an in vitro reaction

Comparison of In Vitro c-di-GMP Quantification Methods for the "DGC Reaction"	
EnzChek® Spectrophotometric Quantification	
Normalized ABS 360 nm (60 min)	0.118
PPi (μM)	24.8
c-di-GMP (μM)	**12.4**
UPLC-MS/MS Quantification	
1:200 Dilution c-di-GMP (nM)	59.2
c-di-GMP (μM)	**11.8**

(a) MS Parameters:

- Electrospray ionization in negative-ion mode with multiple-reaction monitoring, *m/z* 689.16 → 344.31.
- Capillary voltage, 3.5 kV; cone voltage, 50 V; collision energy, 34 V.
- Source temperature, 110 °C; desolvation temperature, 350 °C.
- Cone gas flow (nitrogen), 50 L/h; desolvation gas flow (nitrogen), 800 L/h; collision gas flow (nitrogen), 0.15 mL/min.

(b) UPLC column and parameters:

- ACQUITY UPLC BEH C18 Column, 130 Å, 1.7 μm, 2.1 mm × 50 mm (Waters Corp.).
- Perform reverse-phase chromatography using c-di-GMP Buffer A ("A") and c-di-GMP Buffer B ("B") flowing at a rate of 0.3 mL/min in this 10 min step gradient:

3. t = 0 min; A-99%:B-1%, t = 2.5 min; A-80%:B-20%, t = 7.0 min; A-35%:B-65%, t = 7.5 min; A-5%:B-95%, t = 9.01 min; A-99%: B-1%, t = 10 min (end of gradient).
4. Following c-di-GMP detection (*see* **Note 30**), use your instrument's software to calculate a standard curve based on the peak areas of the c-di-GMP standards (*see* **Note 31**).
5. Use the calculated standard curve to determine the concentration of c-di-GMP present in the 1:200 dilution samples. Normalize the c-di-GMP concentration of the "DGC Reaction" by

subtracting the c-di-GMP present in the "DGC Control" sample (*see* **Note 32**) (Table 1).

4 Notes

1. This is not a rapid process, vortex continuously until flakes of undissolved MESG are no longer visible in the solution.
2. MESG is temperature sensitive and should be frozen immediately to minimize degradation. MESG solutions are stable for at least 1 year at −20 °C.
3. These concentrations will be diluted 1:10 when added into the reaction mixtures in Subheading 3.2, **step 2**. The EnzChek® Pyrophosphate Assay Kit protocol states that a linear standard curve can be generated from 0 μM –70 μM PPi, but in our experience the standard curve is most linear from 0 μM to 60 μM PPi.
4. Once reconstituted, PNP is stable for 1 month at 4 °C. To extend the life of the enzyme, we have found that reconstituted PNP can be stored at −20 °C in 20 μL aliquots. However, there is diminished activity following storage at −20°, therefore, kinetic experiments should be performed using freshly reconstituted PNP that has never been frozen.
5. Reconstituted inorganic pyrophosphatase is stable for 1 week at 4 °C. To extend the life of the enzyme, 10 μL aliquots can be stored at −20 °C. As with PNP, freezing diminishes the enzyme's activity and kinetic experiments should be performed using freshly reconstituted inorganic pyrophosphatase that has never been frozen.
6. The DGC utilized here to illustrate the assay is 2.8 mM *P. aeruginosa* WspR (R272A) dialyzed in 30 mM Tris buffer [pH 7.6] and 100 mM NaCl and stored in 50% glycerol.
7. Avoid purifying the DGC in buffers containing phosphates as they interfere with the ability to use this method for assessing DGC activity.
8. DGC Buffer is used in some reactions to control for contaminating Pi and contributions to ABS 360 nm that the buffer components may have. It is also used to control for any effects the buffer components may have on the activity of the EnzChek® assay enzymes.
9. Prior to the addition of water, briefly centrifuge the lyophilized c-di-GMP to collect the compound at the bottle of the vial.
10. Depending on instrument sensitivity, the range of this standard curve can be adjusted.

11. This will allow the instrument to warm up and the emission spectrum to stabilize.
12. MESG is temperature sensitive and its half-life decreases dramatically at warm temperatures. To minimize degradation of MESG, do not expose it to heat for more time than is necessary to nearly melt the solution; about 2 min should be sufficient. Any MESG that remains frozen after 2 min will quickly melt when vortexed. Do not refreeze any leftover MESG solution.
13. The data shown in this example reaction utilizes 2.8 mM *P. aeruginosa* WspR (R272A) dialyzed in 30 mM Tris buffer [pH 7.6] and 100 mM NaCl and stored in 50% glycerol.
14. While the standard EnzChek® reaction buffer contains 1 mM $MgCl_2$ we have found that supplementing with up to 10 mM $MgCl_2$ can greatly enhance DGC activity without dramatically affecting the activity of the EnzChek® Pyrophosphate Assay Kit enzymes.
15. Do not vortex the Master Mix after enzymes have been added and take care not to introduce bubbles while pipetting.
16. Avoid introducing bubbles. If bubbles are formed, briefly centrifuge the 96-well microtiter plate until removed.
17. This well ultimately receives both DGC enzyme and GTP substrate and is used to both spectrophotometrically and mass spectroscopically determine the concentration of c-di-GMP synthesized by the DGC.
18. This well ultimately receives DGC enzyme and water and serves to normalize for the contribution of buffer and protein absorbance at 360 nm. It is also used to normalize the c-di-GMP concentration determined by UPLC-MS/MS in the DGC Reaction as some contaminating c-di-GMP may be present in the DGC enzyme stock.
19. This well ultimately receives DGC Buffer and GTP substrate and serves to control for contaminating PPi and Pi resulting from spontaneous hydrolysis of GTP.
20. This PPi Standard Reaction Mix contains the equivalent of 10 μL DGC buffer per reaction to control for any contaminating PPi and Pi that may exist in the protein buffer. It also controls for any effects that the buffer components may have on the activities of the PNP and inorganic pyrophosphatase as well as any contribution the buffer may have to the solution absorbance at 360 nm. There will be about 90 μL of PPi Standard Reaction Mix and 80 μL of Master Mix remaining in the microcentrifuge tubes after all the reaction wells have been loaded.
21. Keeping all the reaction solutions, including those loaded in Subheading 3.1, **step 9**, in a single row facilitates the addition

of substrates using a multichannel pipette, allowing for better reaction synchronization.

22. Check that equivalent volumes of all solutions have been pulled into the pipette tips. Pipetting errors in this step can have dramatic effects on experimental results.
23. Be sure to mix the substrate solutions well in the reaction mixtures. Do this quickly and gently without creating bubbles.
24. Utilize a shaking option if available to agitate the reaction solutions and improve homogenization of reaction components.
25. The frequency and duration of the measurements are entirely dependent on the type of experiment being performed. For this experiment, we measured the concentration of c-di-GMP at the reaction end-point. In our experience, 30 s provides enough resolution to monitor reaction progression over time. 60 min of total measurement time is sufficient for both the PPi standard curve reactions and DGC reactions to reach their end-points. However, this is highly dependent on DGC activity and concentration. When experimentally determining the kinetics of a DGC reaction more frequent measurements for less total time may be desirable, along with varying the concentrations of GTP substrate and DGC enzyme.
26. We have found that briefly heating the sample does not result in decomposition of c-di-GMP.
27. This dilution is based on the assumption that the DGC reaction converted all available GTP into c-di-GMP yielding a final concentration of 250 nM in a 1:200 dilution, as 1 unit of c-di-GMP requires two GTP equivalents. Further dilutions may be necessary as 250 nM c-di-GMP is at the high edge of the c-di-GMP standard curve utilized in the UPLC-MS/MS process. This determination can be made following the spectrophotometric analysis of c-di-GMP synthesized in the reaction following Subheading 3.3, **step 2d**.
28. If the spectrophotometrically determined concentration of c-di-GMP is above 100 nM, a further 1:10 dilution of the 1:200 "DGC Reaction" and "DGC Control" reaction samples collected for UPLC-MS/MS should be performed.
29. The instruments we use for UPLC-MS/MS are a Quattro Premier XE mass spectrometer (Waters) coupled with an Acquity Ultra Performance LC system (Waters).
30. We typically find that c-di-GMP elutes from our column at approximately 5.50 min but this can vary by 0.1 min from day to day and depending on the age of the column.
31. We use Waters MassLynx™ Software.

32. This step removes the contribution of any contaminating c-di-GMP that may have been associated with the DGC protein following purification from the total c-di-GMP detected in the "DGC Reaction" sample.

References

1. Römling U, Galperin MY, Gomelsky M (2013) Cyclic di-GMP: the first 25 years of a universal bacterial second messenger. Microbiol Mol Biol Rev: MMBR 77:1–52
2. Webb MR (1992) A continuous spectrophotometric assay for inorganic phosphate and for measuring phosphate release kinetics in biological systems. Proc Natl Acad Sci U S A 89:4884–4887
3. Massie JP, Reynolds EL, Koestler BJ et al (2012) Quantification of high-specificity cyclic diguanylate signaling. Proc Natl Acad Sci 109:12746–12751

Part III

Visualizing c-di-GMP Levels Using Biosensors

Chapter 8

Gauging and Visualizing c-di-GMP Levels in *Pseudomonas aeruginosa* Using Fluorescence-Based Biosensors

Morten Rybtke, Song Lin Chua, Joey Kuok Hoong Yam, Michael Givskov, Liang Yang, and Tim Tolker-Nielsen

Abstract

Recent research has shown that the molecule c-di-GMP is an important second messenger regulating various functions in bacteria. In particular, the implication of c-di-GMP as a positive regulator of adhesion and biofilm formation has gained momentum as a highly relevant research topic, as detailed knowledge about the underlying regulatory mechanisms may enable the development of measures to control biofilms in both industrial and medical settings. Accordingly, it is in many cases of interest to measure the c-di-GMP level in bacteria under specific conditions or in specific mutant strains. We have developed a collection of fluorescence-based c-di-GMP biosensors capable of gauging the c-di-GMP level in *Pseudomonas aeruginosa* and closely related bacteria. Here, we describe protocols for the use of these biosensors in gauging and visualizing cellular c-di-GMP levels of *P. aeruginosa* both in in vitro setups such as continuous-culture flow-cell biofilms, and in in vivo settings such as a murine corneal infection model.

Key words c-di-GMP, Cyclic di-GMP, Biosensor, Fluorescence, Pseudomonas, Biofilm, Ocular infection

1 Introduction

Biofilms pose numerous problems both industrially and clinically due to their biofouling properties and their recalcitrance toward the immune system and antibiotics [1]. Over the past decade it has become clear that the bacterial second messenger c-di-GMP is a key signaling molecule in many bacteria, governing the transition from the planktonic motile life-form to the sessile life in biofilms [2]. Knowledge of c-di-GMP signaling pathways, from the environmental inputs to effector outputs, will ultimately allow us to develop measures to control biofilms in clinical and industrial settings.

The level of c-di-GMP in bacteria under specific physiologic conditions or in specific mutant strains is a key parameter for the study of c-di-GMP-mediated regulation. A golden standard for

Karin Sauer (ed.), *c-di-GMP Signaling: Methods and Protocols*, Methods in Molecular Biology, vol. 1657, DOI 10.1007/978-1-4939-7240-1_8, © Springer Science+Business Media LLC 2017

quantitating cellular c-di-GMP levels is analysis of bacterial extracts by LC/MS/MS [3]. Although precise and highly sensitive, there are drawbacks to the LC/MS/MS method, including end-point-only sampling as well as laborious and complex sample analysis. An alternative to the LC/MS/MS analysis is semiquantitative estimation of c-di-GMP levels using fluorescence-based biosensors that respond to fluctuations in the messenger level. These biosensors have the advantage of allowing nondestructive real-time measurements or visualizations of bacterial c-di-GMP levels in both in vitro and in vivo settings.

We have developed a range of fluorescence-based c-di-GMP biosensors for use in *P. aeruginosa* (Table 1) [4]. This bacterium is causing various persistent infections, and serves as a model organism for biofilm formation and c-di-GMP signaling. The biosensors are all based on a transcriptional fusion of the c-di-GMP-responsive *cdrA* promoter to *gfp*. The plasmid-based biosensors come in two different versions differentiated by a subscript C or S denoting their origin of creation (Copenhagen and Seattle, respectively). Furthermore, the C-set includes mini-Tn7-based integrative biosensors. Versions of the biosensors utilizing either stable Gfp(Mut3) or unstable Gfp(ASV) have been constructed. Transcription from the *cdrA* promoter is regulated by c-di-GMP through binding of the molecule to the transcriptional regulator FleQ. Utility of the biosensors may therefore be limited to *P. aeruginosa* and related bacterial species harboring close FleQ homologs.

Here, we describe protocols for the use of the biosensors in gauging and visualizing the cellular c-di-GMP levels in *P. aeruginosa* bacteria both in vitro using liquid batch cultures and continuous-culture flow-cell biofilms, and in vivo using a newly developed murine corneal infection model.

Table 1
Fluorescence-based c-di-GMP biosensors available (*see* Note 2)

Vector	Relevant genotype and characteristics	Reference
pCdrA-gfp^C	pUCP22Not-PcdrA-RBSII-gfp(Mut3)-T_0-T_1, Gm^r Amp^r	[4]
pCdrA-gfp^S	pUCP22Not-PcdrA-RBS-CDS-RNaseIII-gfp(Mut3)-T_0-T_1, Gm^r Amp^r	[4]
pCdrA-gfp(ASV)^C	pUCP22Not-PcdrA-RBSII-gfp(ASV)-T_0-T_1, Gm^r Amp^r	[4]
pCdrA-gfp(ASV)^S	pUCP22Not-PcdrA-RBS-CDS-RNaseIII-gfp(ASV)-T_0-T_1, Gm^r Amp^r	[4]
pTn7CdrA-gfp^C	miniTn7-PcdrA-RBSII-gfp(Mut3)-T_0-T_1 delivery vector, Gm^r (Amp^r)	[4]
pTn7CdrA-gfp (ASV)^C	miniTn7-PcdrA-RBSII-gfp(ASV)-T_0-T_1 delivery vector, Gm^r (Amp^r)	[4]

2 Materials

2.1 Gauging and Visualizing c-di-GMP Levels in P. aeruginosa *in In Vitro Systems*

1. Standard LB broth and LB agar for inoculum preparation.
2. A10 solution: 20 g/L $(NH_4)_2SO_4$, 60 g/L $Na_2HPO_4 \cdot 2H_2O$, 30 g/L KH_2PO_4, and 30 g/L NaCl in water. Upon preparation, pH of the solution should be 6.4. If the pH is not 6.4 the solution must be discarded. Do not adjust pH. Autoclave at 121 °C for 15 min.
3. $MgCl_2$ stock solution (1 M): 203.31 g/L $MgCl_2 \cdot 6H_2O$ in water.
4. $CaCl_2$ stock solution (1 M): 147.01 g/L $CaCl_2 \cdot 2H_2O$ in water.
5. Trace metal stock solution: 200 mg/L $CaSO_4 \cdot 2H_2O$, 200 mg/L $FeSO_4 \cdot 7H_2O$, 20 mg/L $MnSO_4 \cdot H_2O$, 20 mg/L $CuSO_4 \cdot 5H_2O$, 20 mg/L $ZnSO_4 \cdot 7H_2O$, 10 mg/L $CoSO_4 \cdot 7H_2O$, 12 mg/L $NaMoO_4 \cdot H_2O$, 5 mg/L H_3BO_3 in water.
6. BTrace solution: 1 mL $MgCl_2$ stock solution, 0.1 mL $CaCl_2$ stock solution and 0.1 μL trace metal stock solution per 900 mL of water. Autoclave at 121 °C for 15 min.
7. ABTrace medium: 100 mL A10 solution and 900 mL BTrace solution (*see* **Note 1**).
8. $FeCl_3$ stock solution (10 mM): 2.7 mg/mL $FeCl_3 \cdot 6H_2O$ in water. Filter sterilize through a 0.22 μm pore filter.
9. Glucose stock solution (10% w/v): 100 g/L glucose in water. Autoclave at 121 °C for 15 min.
10. Casamino acid stock solution (20% w/v): 200 g/L casamino acid in water. Autoclave at 121 °C for 15 min.
11. ABTraceFe growth medium supplemented with glucose and casamino acids for liquid batch culturing: 1 L ABTrace medium, 100 μL $FeCl_3$ stock solution, 20 mL glucose stock solution, and 25 mL casamino acid stock solution.
12. ABTrace growth medium supplemented with glucose for continuous culturing of biofilms in flow-cells: 1 L ABTrace medium supplemented with 100 μL trace metal stock solution, and 2.7 mL glucose stock solution.
13. NaCl solution (0.9% w/v): 9 g/L NaCl in water. Autoclave at 121 °C for 15 min.
14. A *cdrA-gfp* biosensor for gauging or visualizing c-di-GMP. *See* Table 1 and **Note 2** for a list and description, respectively, of the available biosensors.
15. A *P. aeruginosa* strain harboring the biosensor (*see* **Notes 3** and **4**).

16. 50 mL Erlenmeyer flasks.
17. Black 96-well microplates suitable for fluorescence measurements.
18. A plate reader (e.g., Victor X4, Perkin Elmer) equipped with a filter set for measuring green fluorescence (Ex 485 nm/Em 535 nm) and a filter for measuring light absorbance at 590 nm.
19. SYTO®62: 5 μM Red Fluorescent Nucleic Acid Stain (Molecular Probes, USA) dissolved in 0.9% (w/v) NaCl (working solution).
20. A flow-cell setup identical to the one described by Crusz et al. [5].
21. A confocal laser scanning microscope (e.g., Zeiss LSM710) with lasers equipped for visualizing Gfp (biosensor output) and Syto®62 (total biofilm biomass).
22. IMARIS image processing software (Bitplane AG, Switzerland).

2.2 Visualization of c-di-GMP Levels in P. aeruginosa *in a Murine Ocular Infection Model*

1. Standard LB broth and LB agar for inoculum preparation and CFU counts.
2. NaCl solution (0.9% w/v) for serial dilutions: 9 g/L NaCl in water. Autoclave at 121 °C for 15 min.
3. *P. aeruginosa* PAO1.
4. *P. aeruginosa* PAO1/pCdrA-gfp^{C}.
5. *P. aeruginosa* PAO1 Tn7::P$_{lac}$-gfp (constitutive *gfp* expression).
6. Female C57BL/6 mice (Invivos, Singapore) 7–8 weeks old (*see* **Note 5**). Maintain on water and standard mouse chow ad libitum.
7. Miniblade (BD-Beaver, Cat. No. 376400).
8. Tweezers.
9. Surgical scissors.
10. Pellet pestle (autoclavable).
11. Glass beads (5 mm diameter).
12. SYTO®62 Red Fluorescent Nucleic Acid Stain (Molecular Probes, USA). 5 μM in 0.9% (w/v) NaCl (working solution).
13. μ-Slide 8 Well glass bottom chamber slide.
14. A confocal laser scanning microscope (e.g., Zeiss LSM780) with lasers equipped for visualizing Gfp (biosensor output) and Syto®62 (total biofilm biomass).
15. A Zeiss Stemi-2000C dissecting microscope.
16. IMARIS image processing software (Bitplane AG, Switzerland).

3 Methods

3.1 Gauging c-di-GMP Levels in P. aeruginosa *in In Vitro Systems During Liquid Batch Culturing*

1. Prepare a starter culture of the *P. aeruginosa* PAO1 ΔwspF ΔpelA ΔpslBCD/pCdrA-gfp^{C} biosensor strain (15 mL ABTraceFe medium supplemented with glucose and casamino acids in a 50 mL Erlenmeyer flask). Use a couple of freshly grown colonies from an LB plate as inoculum and incubate the culture at 37 °C under well-shaken conditions (e.g., orbital shaking with a diameter of 30 mm and a speed of 200 rpm) for 16–18 h (*see* **Note 6**).
2. Prepare the experimental culture of the biosensor strain (15 mL ABTraceFe medium supplemented with glucose and casamino acids in a 50 mL Erlenmeyer flask). Measure OD_{600} of the outgrown starter culture (should be 3–4) and dilute it into the experimental culture to a final OD_{600} of 0.01.
3. Incubate the culture at 37 °C under well-shaken conditions (e.g., orbital shaking with a diameter of 30 mm and a speed of 200 rpm) for 14 h.
4. Sample 100 μL of the culture every hour and transfer the sample to an empty well of a black 96-well microplate suitable for fluorescence measurements in a plate reader.
5. Read the growth level (Abs590nm) and the biosensor output (Gfp fluorescence) of the most recent sample in the plate reader after each sampling.
6. Once the experiment is concluded, plot growth curves (Abs590 over time), absolute fluorescence curves (Gfp fluorescence over time) and relative fluorescence curves (Abs590-normalized Gfp fluorescence over time) of the culture and evaluate the relative c-di-GMP level of the culture (*see* **Note 7**). A typical result is shown in Fig. 1. In addition to the scatter plot data representation displaying relative monitor activity and growth as a function of time, data from a single representative time-point may be extracted and displayed in a column chart. This is particularly useful when comparing multiple different strains, growth conditions, treatments etc.

3.2 Visualizing c-di-GMP Levels in P. aeruginosa *in In Vitro Continuous-Culture Flow-Cell Biofilms*

1. Set up the flow-cell system according to the guidelines described by Crusz et al. [5].
2. Prepare a starter culture of *P. aeruginosa* PAO1 containing the Tn7-integrative biosensor encoding unstable Gfp(ASV) (Table 1) using 5 mL of LB broth in a standard test tube. Use a couple of freshly grown colonies from an LB plate as inoculum and incubate the culture at 37 °C with 200 rpm shaking for 16–18 h.

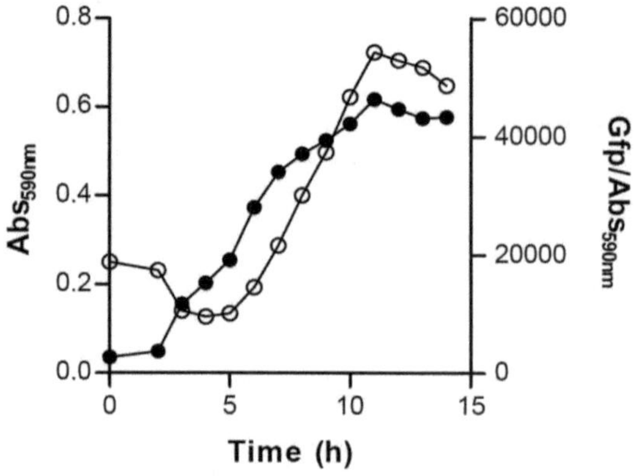

Fig. 1 Example of the biosensor output from *P. aeruginosa* during planktonic culturing. Growth (Abs590nm, closed circles) and relative fluorescence (Gfp/Abs590nm, closed circles) of a planktonic *P. aeruginosa* Δ*wspF* Δ*pelA* Δ*pslBCD*/pCdrA-gfp^C culture are plotted as a function of incubation time. The graph shows that the biosensor output is low during the early and mid-exponential phases and increases in the late exponential phase. The decrease in biosensor output during the early exponential growth phase is due to residual stable Gfp being present in cells from the outgrown culture used as inoculum. The high *PcdrA-gfp* gene dose from the plasmid-based reporter allows monitoring of c-di-GMP fluctuations in a growing culture of the Δ*wspF* Δ*pelA* Δ*pslBCD*/pCdrA-gfpC strain

3. Dilute the starter culture to an OD_{600} of 0.001 using 0.9% saline and inoculate each flow-cell chamber with 300 μL of the dilution according to Crusz et al. [5].
4. Incubate the system at 37 °C for the biofilm to develop. For *P. aeruginosa PAO1*, the flow-cell biofilm development during a standard experiment running for 96 h normally progresses through stages of substratum attachment, microcolony formation and maturation of the microcolonies (e.g., mushroom-type microcolony formation). A defined dispersal stage is not reached due to the continuous feed of nutrients. Experiments may last beyond the typical 96 h.
5. Remove the system from incubation at the desired time points to investigate the biosensor output and spatial location in the flow-cell-grown biofilms using confocal laser scanning microscopy.
6. Stain the total biofilm biomass immediately before conducting the microscopy by very gentle injection of 300 μL Syto®62 solution into the chamber under investigation in a manner similar to the initial inoculation of the flow-cell. Take great care not to disrupt the biofilm. Leave the chamber protected from light for 10 min before image acquisition for the stain to have effect. Once the biofilm has been stained with Syto®62 it should be excluded from the remainder of the experiment due to the nucleic acid binding properties of the molecule (*see* **Note 8**).

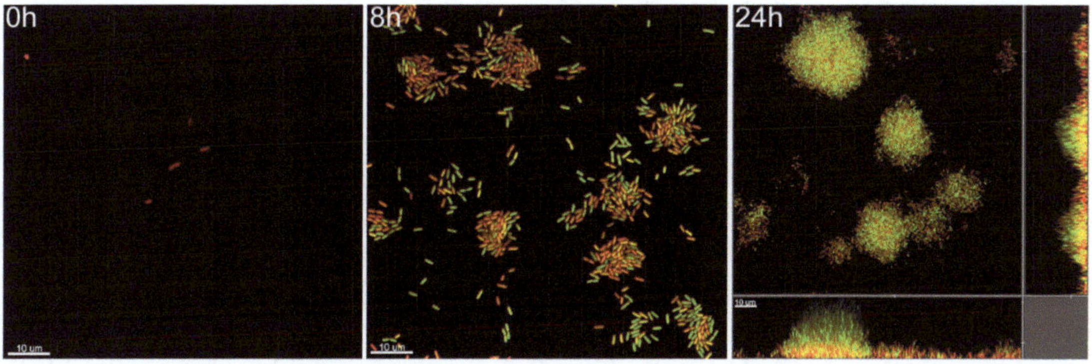

Fig. 2 Example of biosensor visualization during *P. aeruginosa* flow-cell biofilm development. Flow-cell grown biofilms of *P. aeruginosa* PAO1 Tn7::cdrA-gfp(ASV)^{C} were stained red with SYTO®62 and visualized by confocal laser scanning microscopy as described in Subheading 3.2. The biofilms shown are from 0, 8, and 24 h post attachment to the substratum. *Green* coloring depicts the biosensor output while *red* coloring depicts the total biofilm biomass. The 0 h and 8 h images display a top-down view while the 24 h image displays a central top-down view with flanking vertical cross sections. Scale bar equals 10 μm. The pictures show that biosensor output is absent in planktonic cells when first attaching to the substratum (0 h) indicating low cellular c-di-GMP levels. When the biofilm develops the monitor output from the bacteria increases indicating that the cellular levels of c-di-GMP increase (8 h and 24 h). Notice also the heterogeneity of the biosensor output in the biofilm cells. The single-gene dose of the chromosomally integrated *PcdrA-gfp* reporter is sufficient to render biofilm-grown PAO1 wild-type bacteria fluorescent, but does not give rise to fluorescence from PAO1 wild-type bacteria grown in planktonic culture, indicating that biofilm-grown *P. aeruginosa* bacteria have much higher c-di-GMP content than planktonically grown bacteria

7. Acquire suitable image stacks visualizing the biosensor output (Gfp fluorescence) and the total biofilm biomass (Syto®62 fluorescence) using the Zeiss LSM710 confocal laser scanning microscope. Take care to avoid pixel saturation (loss of fluorescence level information) for the detection of biosensor-derived Gfp fluorescence in order to be able to qualitatively analyze the spatio-temporal fluctuations of c-di-GMP levels within the cells of the biofilm. Use the Imaris image processing software (or similar) for the qualitative analysis of the different image stacks and to prepare publication-ready images. An example is shown in Fig. 2.

3.3 Gauging c-di-GMP Levels in P. aeruginosa Bacteria in a Murine Ocular Infection Model

3.3.1 Inoculum Preparation

1. Streak *P. aeruginosa* PAO1 or *P. aeruginosa* PAO1/pCdrA-gfp^{C} or P. *aeruginosa* PAO1 Tn7::P_{lac}-gfp on LB agar and grow at 37 °C for 16 h.
2. Pick a colony and grow in 2 mL LB broth at 37 °C and 200 rpm for 16 h. The CFU/mL of the overnight culture is estimated to be between 1×10^{9} and 1×10^{10}.
3. Transfer 1 mL of culture into a 1.5 mL microcentrifuge tube and centrifuge at 12,000 × *g* for 3 min.
4. Remove the supernatant and resuspend the cell pellet in 1 mL 0.9% NaCl.

5. Repeat the wash step twice.
6. Measure the optical density at 600 nm of the bacterial suspension and adjust it to 0.58 (5×10^8 CFU/mL) in 1 mL 0.9% NaCl.

3.3.2 Establishing the Experimental Keratitis

1. Examine the eyes of the mice under the slit lamp for potential corneal defects and abnormalities. If any defects or abnormalities are observed the mouse is excluded from the experiment.
2. Anaesthetize the mice with 120 μL of premixed xylene (10 mg/kg) and ketamine (40 mg/kg).
3. Using the dissecting microscope, scratch each corneal epithelia with a sterile miniblade creating four superficial scratches of 1–2 mm in length.
4. Apply 10 μL of *P. aeruginosa* suspension (from Subheading 3.3.1) topically to each scratched cornea.
5. Place the mice in labeled cages and allow the corneal infection to develop. Examine the corneas on a daily basis using the dissecting microscope.

3.3.3 Quantification of Colony Forming Units in Biofilms

1. After 2, 5, and 7 days of infection (*see* **Note 9**), anaesthetize the mice with 120 μL of premixed xylene (10 mg/kg) and ketamine (40 mg/kg). Kill the mice by cervical dislocation. Use the tweezers and surgical scissors to excise and extract the corneas.
2. Place the cornea in a 1.5 mL microcentrifuge tube and crush it using the pellet pestle.
3. Resuspend the cornea in 500 μL 0.9% NaCl and add in 10 glass beads. Disrupt the biofilm cells in the crushed cornea by vortexing at maximum speed for 5 min.
4. Perform serial dilutions on the bacterial cells by adding 100 μL of cell suspension to 900 μL 0.9% NaCl sequentially from dilution factor 10^0–10^{-7} (*see* **Note 10**). Spread 100 μL of the 10^{-5}, 10^{-6}, and 10^{-7} dilutions on LB agar plates. Incubate the plates at 37 °C for 16 h.
5. Count the number of colonies formed on each LB agar plate. Choose plates with a minimum of 30 and a maximum of 300 colonies for calculating the CFU/mL.

3.3.4 Imaging Biofilms on Cornea

1. Use *P. aeruginosa* PAO1/pCdrA-gfp^C and *P. aeruginosa* PAO1 Tn7::P_{lac}-gfp to establish the experimental keratitis as described above. *P. aeruginosa* Tn7::P_{lac}-gfp acts as a positive control for Gfp fluorescence.
2. After 2, 5, and 7 days of infection, anaesthetize the mice with 120 μL of premixed xylene (10 mg/kg) and ketamine

(40 mg/kg). Kill the mice by cervical dislocation. Use the tweezers and surgical scissors to excise and extract the corneas.

3. Place the excised cornea in 200 μL SYTO®62 solution on the μ-Slide 8-well glass bottom chamber slide.
4. Incubate the samples in the dark for 10 min at room temperature, to stain both bacterial and murine cells with SYTO®62.
5. Capture image stacks of biofilm on the cornea using the Zeiss LSM780 confocal laser scanning microscope equipped with a 63× oil objective. Visualize the biosensor output as Gfp fluorescence and visualize the total biofilm and murine cell biomass as Syto®62 fluorescence. Take care to avoid pixel saturation (loss of fluorescence level information) for the detection of biosensor-derived Gfp fluorescence in order to be able to qualitatively analyze the spatio-temporal fluctuations of c-di-GMP levels within the cells of the biofilm.
6. Process the images with the IMARIS software. Using the Section 2D display, present the Z-stack images as a central horizontal cross-section (X and Y axes) with flanking vertical cross-sections below (X and Z axes) and to the right (Y and Z axes) of the central section. Create images with both single (Gfp or SYTO®62) and merged channels (both Gfp and SYTO®62). An example is shown in Fig. 3.

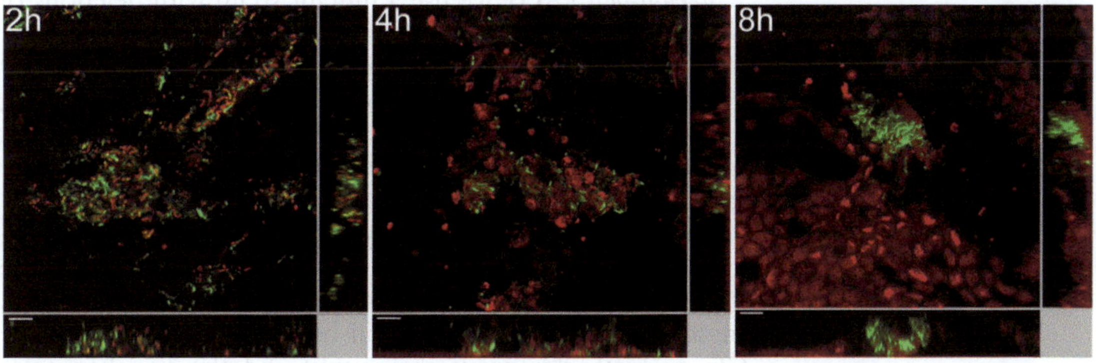

Fig. 3 An example of the biosensor output in corneal biofilms of *P. aeruginosa* obtained using a murine ocular infection*see also Pathogenicity Assay* model. The *P. aeruginosa* PAO1/pCdrA-gfp^{C} biofilms shown are from corneas removed 2, 4, and 8 h post infection. *Green* coloring depicts the biosensor output while *red* coloring depicts the total biofilm biomass in combination with host immune cells. The images were acquired using confocal laser scanning microscopy as described in Subheading 3.3.4 and displayed as central top-down views with flanking vertical cross sections. Scale bar equals 10 μm. At the early infection stage (2 h) with little or no biofilm formation, there is a high degree of heterogeneity of the monitor output indicating fluctuations in the c-di-GMP level across the population. At the late infection stage (8 h) a biofilm has developed displaying a high biosensor output with less heterogeneity and hence high cellular levels of c-di-GMP throughout the biofilm population

4 Notes

1. The A and BTrace parts of the ABTrace medium are prepared and sterilized separately to avoid the formation of undesirable precipitates that will form if the two solutions are mixed and autoclaved together.
2. The biosensors available are either plasmid-based or integrative. The plasmid-based (pUCP22Not) biosensors have the highest basal output and signal-to-noise ratio making them useful when working with plate readers or other equipment where the strength of the excitation source (such as a UV lamp with a mounted band pass filter) is modest. The single-copy integrative Tn7-based biosensors have a much lower basal output compared to the plasmid-based biosensors but are especially suitable for use in flow-cell biofilm experiments where the strong excitation and sensitive detection of fluorescence emission offered by confocal laser scanning microscopy is used as the visualization method. In addition, the biosensors come in two variants employing either stable Gfp or unstable Gfp(ASV). Using the biosensors with stable Gfp leads to the accumulation of the fluorescent signal which improves the output and the signal-to-noise ratio. However, employing unstable Gfp has the advantage of recording fluctuations in cellular c-di-GMP levels over time.
3. For gauging c-di-GMP levels in planktonic cultures of *P. aeruginosa* PAO1 we use a mutant strain with deletions of *wspF*, *pelA*, and *pslBCD* harboring the pCdrA-gfp^{C} biosensor. The *wspF* mutation results in a high basal level of c-di-GMP due to increased activation of the c-di-GMP synthesizing diguanylate cyclase WspR leading to a strong basal fluorescent output from the biosensor which increases the signal-to-noise ratio. The *pelA* and *pslBCD* mutations render the strain exopolysaccharide deficient thereby avoiding formation of aggregates during culturing, which would otherwise occur in the *wspF* deletion background and obscure growth measurements. In addition, we use the plasmid-based biosensor with stable Gfp to further improve the output and the signal-to-noise ratio. For visualization of c-di-GMP levels in flow-cell biofilms using confocal laser scanning microscopy, there are no restrictions on the *P. aeruginosa* strain type or biosensor type due to the strength of the laser-based excitation source and the eminent sensitivity of fluorescence emission detection offered by the microscope setup. We prefer, however, to use the integrative Tn7CdrA-gfp(ASV)^{C} biosensor with unstable Gfp to be able to monitor the fluctuations in reporter activity over time within the biofilm without the need for proper selection during the prolonged incubation of the experiment.

4. The plasmid-based biosensors are favorably introduced into *P. aeruginosa* by electroporation using the sucrose-based protocol described by Choi et al. [6]. To reduce the incidents of detrimental sample arching during electroporation, we use a slightly modified version of the protocol by using only 2 mL of outgrown liquid culture which is finally resuspended in 500 μL sucrose. Introduce the integrative Tn7-based biosensors into *P. aeruginosa* by four-parental mating conjugation as described by Klausen et al. [7].
5. Research involving animals such as mice requires permission from the proper authorities prior to being conducted. We encourage researchers to obtain such permissions and to perform the experiments according to any ethical guidelines that exist within the field of research.
6. The biosensors are sensitive to the growth conditions and require sufficient aeration for an optimal fluorescent output to be achieved.
7. Calculation of the relative fluorescence allows comparison of biosensor output between cultures differing in their growth characteristics. Still, care should be taken only to compare cultures that are in the same growth phase.
8. An alternative to invasive nucleic acid staining of the total biofilm biomass using Syto®62 is the use of a constitutively expressed fluorescent marker gene compatible with Gfp. An example of such a marker is the red-fluorescent protein mCherry. Using such a noninvasive genetic marker would allow for repeated visualization of the same biofilm at different time-points, thereby following the temporal fluctuations of cellular c-di-GMP levels down to the single-cell level in real time.
9. The corneas were observed on Days 2, 5, and 7 for the formation and development of biofilms. As previously described [8], bacteria adhered onto the corneal surface on Day 1, while bacteria were embedded in extracellular polymeric substances (EPS) on Day 2. By Days 5 and 7, biofilms of *P. aeruginosa* cells densely packed in EPS were observed.
10. As a viscous slurry of bacteria and cornea will be used for serial diluting, extreme caution must be taken to ensure that the correct volume will be transferred from one dilution to another with a pipette, so as to reduce errors in liquid transfer.

Acknowledgments

This work was supported by grants from and the Danish Council for Independent Research (DFF–1323-00177) to TTN, and from the Danish Strategic Research Council and the Lundbeck Foundation (R198-2015-486) to MG.

References

1. Costerton JW, Stewart PS, Greenberg EP (1999) Bacterial biofilms: a common cause of persistent infections. Science 284 (5418):1318–1322
2. Romling U, Galperin MY, Gomelsky M (2013) Cyclic di-GMP: the first 25 years of a universal bacterial second messenger. Microbiol Mol Biol Rev 77(1):1–52. doi:10.1128/MMBR.00043-12
3. Irie Y, Parsek MR (2014) LC/MS/MS-based quantitative assay for the secondary messenger molecule, c-di-GMP. Methods Mol Biol 1149:271–279. doi:10.1007/978-1-4939-0473-0_22
4. Rybtke MT, Borlee BR, Murakami K, Irie Y, Hentzer M, Nielsen TE, Givskov M, Parsek MR, Tolker-Nielsen T (2012) Fluorescence-based reporter for gauging cyclic di-GMP levels in Pseudomonas aeruginosa. Appl Environ Microbiol 78(15):5060–5069. doi:10.1128/AEM.00414-12
5. Crusz SA, Popat R, Rybtke MT, Camara M, Givskov M, Tolker-Nielsen T, Diggle SP, Williams P (2012) Bursting the bubble on bacterial biofilms: a flow cell methodology. Biofouling 28(8):835–842. doi:10.1080/08927014.2012.716044
6. Choi KH, Kumar A, Schweizer HP (2006) A 10-min method for preparation of highly electrocompetent Pseudomonas aeruginosa cells: application for DNA fragment transfer between chromosomes and plasmid transformation. J Microbiol Methods 64(3):391–397. doi:10.1016/j.mimet.2005.06.001. S0167-7012(05)00158-2 [pii]
7. Klausen M, Heydorn A, Ragas P, Lambertsen L, Aaes-Jorgensen A, Molin S, Tolker-Nielsen T (2003) Biofilm formation by Pseudomonas aeruginosa wild type, flagella and type IV pili mutants. Mol Microbiol 48(6):1511–1524
8. Saraswathi P, Beuerman RW (2015) Corneal biofilms: from planktonic to Microcolony formation in an experimental keratitis infection with Pseudomonas aeruginosa. Ocul Surf 13 (4): 331–345. doi: 10.1016/j.jtos.2015.07.001 [pii]

Chapter 9

Cyclic di-GMP-Responsive Transcriptional Reporter Bioassays in *Pseudomonas aeruginosa*

Bradley R. Borlee, Grace I. Borlee, Kevin H. Martin, and Yasuhiko Irie

Abstract

3′,5′-cyclic diguanosine monophosphate (cyclic di-GMP) is a bacterial secondary messenger molecule that regulates many important cellular activities and behaviors, such as motility and biofilm formation. While mass spectrometry protocols for quantitative analyses of intracellular cyclic di-GMP concentrations have been developed, they are time intensive, expensive, low-throughput, and incapable of directly monitoring dynamic changes in vivo. In this protocol, we provide a *Pseudomonas aeruginosa*-specific detailed methodology to assay the intracellular levels of cyclic di-GMP using biological reporters.

Key words Cyclic di-GMP, c-di-GMP, Luminescence assay, Fluorescence assay, *Pseudomonas aeruginosa*

1 Introduction

Prokaryotic gene expression is regulated at multiple levels: transcription, posttranscription, translation, and posttranslation. Bacteria possess intricate sensory and response mechanisms to adapt and respond to changes in their environment, and these processes affect their gene expression and gene product functions accordingly. In recent years, intensive research has revealed that secondary intracellular messenger molecules such as 3′,5′-cyclic diguanosine monophosphate (cyclic di-GMP) mediate several of these regulatory mechanisms, necessitating development of assay methods to monitor the dynamics of intracellular cyclic di-GMP levels.

Pseudomonas aeruginosa is a versatile opportunistic pathogen capable of establishing itself in various environments including soil, water, and eukaryotic hosts. Much of this versatility is attributed to its enhanced capacity to metabolize various nutrient sources and produce various exoproducts and virulence factors, all of which are ultimately reliant on multifaceted regulatory networks that alter the cellular processes in response to environmental perturbations [1]. Cyclic di-GMP is one of the regulatory factors that controls some

Karin Sauer (ed.), *c-di-GMP Signaling: Methods and Protocols*, Methods in Molecular Biology, vol. 1657,
DOI 10.1007/978-1-4939-7240-1_9, © Springer Science+Business Media LLC 2017

of the processes that are needed for adapting and surviving in diverse environments. Cyclic di-GMP is produced by diguanylate cyclase (DGC) enzymes containing GGDEF domains, and reciprocally degraded by phosphodiesterases (PDE) with EAL or HD-GYP domains [2]. *P. aeruginosa* possesses at least 17 GGDEF, 5 EAL, 3 HD-GYP, and 16 GGDEF + EAL domain containing proteins [3], and it can rapidly alter the intracellular concentrations of cyclic di-GMP by regulating the expression and/or activities of these enzymes in response to various environmental cues [4]. The specific identities of these cues or signals are not very well-characterized.

Several methods for the quantitative analyses of intracellular cyclic di-GMP using mass spectrometry have been described for *P. aeruginosa* [5, 6]. While many of these protocols are specifically designed to accurately measure the concentrations of intracellular cyclic di-GMP, mass spectrometry-based assays present a number of challenges for biological researchers. In particular, they are not suitable for real-time observations to approximate cyclic di-GMP levels. Mass spectrometry-based methods also limit the accessibility of cyclic di-GMP research to many scientists and high-throughput projects due to the time, expertise, and cost required to operate a mass spectrometer. To overcome these obstacles, the methodology outlined in these protocols will introduce detailed methods for determining the relative intracellular concentrations*Reporter* of cyclic di-GMP using transcriptional fusion bioassays. In *P. aeruginosa*, the FleQ transcription factor differentially regulates gene expression based on the concentrations of cyclic di-GMP (high vs. low) by directly binding to cyclic di-GMP. Downstream genes such as the biofilm-associated adhesin encoded by *cdrA* [7] are directly regulated by FleQ [8, 9]. Transcriptional fusions of the *cdrA* promoter with luciferase and GFP reporter genes allow for rapid and efficient monitoring of cyclic di-GMP-regulated activities through the quantification of luminescence or fluorescence output signals [7, 9–12]. These tools are designed to facilitate microscopic and/or high-throughput research. Since FleQ is a *Pseudomonas*-specific regulator, bioassays that are based on transcriptional fusions of the *cdrA* promoter are currently limited to studies in *P. aeruginosa*. Furthermore, these constructs only work under conditions where *cdrA* is expressed and our current knowledge of *cdrA* regulation is somewhat limited. If there are other cyclic di-GMP-independent factors that regulate *cdrA*, or if there are conditions where changes to cellular cyclic di-GMP levels somehow do not affect the FleQ-dependent activation of *cdrA* transcription, then users must be aware that there may be limitations to the effectivity of these tools and therefore how the experimental results are interpreted.

2 Materials

2.1 Growth Medium and Strains

1. Lysogeny broth (LB) medium: Lennox formulation consisting of 10 g/L tryptone (Fisher Scientific, Pittsburgh, PA), 5 g/L yeast extract (Becton, Dickinson and Company, Sparks, MD), and 5 g/L NaCl (Fisher Scientific, Pittsburgh, PA). This culture medium is recommended for luminescence assays:
2. LB broth cultures supplemented with 100 μg/mL kanamycin.
3. LB agar plates supplemented with 100 μg/mL kanamycin.
4. For fluorescence assays:
5. LB broth cultures supplemented with 100 μg/mL gentamicin.
6. LB agar plates supplemented with 100 μg/mL gentamicin.
7. Vogel-Bonner Minimal Medium (VBMM) 10× concentration: 2 g/L $MgSO_4 \cdot 7H_2O$ (0.976 g/L anhydrous), 20 g/L citric acid (30.62 g/L for sodium citrate), 35 g/L $NaNH_4HPO_4 \cdot 4H_2O$, and 100 g/L K_2HPO_4. Dissolve in the following order: $MgSO_4 \cdot 7H_2O$, citric acid, $NaNH_4HPO_4 \cdot 4H_2O$, and K_2HPO_4, adjust pH to 7.0, bring to 1 L. Filter sterilize the 10× stock solution through 0.22 μm vacuum PVDF filtration units. The medium is diluted with sterile deionized or double deionized water to 1× prior to use. This culture medium is recommended for fluorescence assays (*see* **Note 1**):
8. VBMM broth cultures supplemented with 100 μg/mL gentamicin.
9. VBMM agar plates supplemented with 100 μg/mL gentamicin.
10. Alternatively, bacterial cultures can also be grown in LB medium as described in Subheading 2.1, **item 1** for fluorescence assays (*see* **Note 1**).
11. Strains and plasmids: *See* Table 1.

2.2 Equipment and Materials

1. For fluorescence assays, multimode plate reader capable of measuring fluorescence (excitation = 485 nm, emission = 535 nm), and absorbance (optical density = 600 nm) (*see* **Note 2**).
2. Black 96-well or 384-well clear-bottom microtiter plates for simultaneous measurement of fluorescence and absorbance.
3. For luminescence assays, multimode plate reader with built-in luminescence detector and capability to read absorbance (*see* **Note 2**).
4. White 96-well or 384-well plates with clear-bottom wells for simultaneous measurement of luminescence and absorbance.

Table 1
Reporter plasmids and host strain description

Plasmids	Genotype	Reference/ Source
pMH489	pUCP22Not-RNase III-*gfp*(ASV)-T_0-T_1, Amp^r Gm^r	[9]
pCdrA*gfp*(ASV)	pUCP22Not-P_{cdrA}-RBS-CDS-RNaseIII-*gfp*(ASV)-T_0-T_1, Amp^r Gm^r	[9]
pMS402	Reporter plasmid carrying promoterless *luxCDABE* gene; *ori* of pRO1614, Km^R	[13]
pCdrA*lux*	P_{cdrA}-*luxCDABE*, Km^r (cloned into pMS402)	Described here
Host strain	**Strain description**	**Reference/ Source**
PAO1	Laboratory wild-type strain	[14]
PAO1 CTX:: P_{BAD}PA1120	Arabinose-inducible *PA1120* (*tpbB*)[a] integrated into the *attB* site on the chromosome	[9]
PAO1 Δ*fleQ*	In-frame deletion mutant of *fleQ*	[9]
PAO1 Δ*pelA* Δ*pslBCD*	Double mutant: polar mutants of *pelA* and *pslBCD*	[7]
PAO1 Δ*wspF* Δ*pelA* Δ*pslBCD*	Triple mutant: in-frame deletion of *wspF* gene with *pelA* and *pslBCD* polar mutations	[7]
PAO1 Δ*pelA* Δ*pslBCD* CTX:: P_{BAD}PA1120	Arabinose-inducible *PA1120* (*tpbB*)[a] integrated into the *attB* site of the PAO1 Δ*pelA* Δ*pslBCD* chromosome	[9]

[a]L-arabinose can be supplemented as needed in studies where the conditional expression of the DGC, PA1120, is evaluated in the P_{BAD}PA1120 strain background.

5. Glass test tubes (16 × 150 mm or 16 × 100 mm) or 15 mL conical tubes.
6. 1.7 or 2.0 mL microcentrifuge tubes.
7. Centrifuge with plate holder adaptors for microtiter plates capable of at least 4000 × *g* set to room temperature.
8. Benchtop microcentrifuge capable of at least 16,000 × *g* set to room temperature.
9. Static incubator set to 37 °C.
10. Orbital shaking incubator set to 250 rpm and 37 °C.

2.3 Microscopic Visualization

1. A fluorescence microscope capable of excitation and detection of emission established for GFP. Use of a confocal microscope is recommended to reduce background fluorescence and detect expression spatially in three dimensions. In the examples provided in this protocol, excitation is achieved using 488 nm

laser and emission is detected using variable bandpass filter set to detect wavelengths between 511 nm and 611 nm.

2. 8-well chamber μ-slides with glass bottoms for maintaining, growing, and visualizing living bacteria. Well dimensions (W × L × H) are 9.4 mm × 10.7 mm × 6.8 mm. Total volume per well, 300 μL. Alternatively, a variety of devices are commercially available or can be constructed for specific assays that allow reporter visualization with microscopy during the evaluation of cyclic di-GMP-regulated behaviors, which include motility and biofilm*Sodium Dodecyl Sulfate-Polyacrylamide Gel Electrophoresis* formation. The experimental device should be chosen based on the specific application and could include wet mount slides, chamber slides, channel slides, chemotaxis slides, flow cells, 35 mm dishes, and microtiter dishes. The device should incorporate optical grade plastic or glass with a known and appropriate thickness for microscopic visualization.
3. Software used for the acquisition, visualization, and analysis of the acquired images. We recommend using software that can render sequential image slices into three-dimensional composite images and measure voxel intensity for quantification purposes. A few examples include open source software packages such as Image J, Fiji, and Comstat, or commercially available packages that are similar to Volocity (Perkin Elmer).

3 Methods

To rapidly evaluate c-di-GMP levels, this method makes use of *cdrA* transcriptional activation by c-di-GMP. Specific examples for applications are listed in Table 2 that involve the identification of conditions, treatments, or cues/signals that alter the levels of reporter genes in response to increasing or decreasing intracellular levels of c-di-GMP. The relevant control strains for the various applications are also highlighted in Table 2. The *cdrA* promoter, as previously described [9], can be PCR amplified and cloned into the BamHI site of pMS402 [13] to create the plasmid, pCdrA*lux*, containing the P_{cdrA}–*luxCDABE* transcriptional fusion. The following primers can be used for PCR amplification: forward 5′-**GGATCC**CC-GAGGTCGAGGGAGGCAT-3′ and reverse 5′-**GGATCC**TGGC-TATCCGGACGGACC-3′. Bold nucleotides indicate BamHI site.

3.1 Luminescence-Based Plate Assays

1. Cultures of luminescence-based reporter strains are initiated from glycerol stocks maintained at −80 °C by streaking for the isolation on LB agar plates supplemented with 100 μg/mL kanamycin and incubated at 37 °C overnight (16–18 h).
2. LB broth cultures supplemented with 100 μg/mL kanamycin are inoculated from isolated individual colonies.

Table 2
Reporter strain description for luminescence assays

Reporter activity	Reporter strain[a]	Relevant characteristics and use	Cyclic di-GMP concentration of parental strain
Low cyclic di-GMP and low levels of reporter activity	PAO1 pCdrA*lux*	Wild-type *P. aeruginosa*. Allows screening for treatments that induce DGC activity.	~11 μM [6]
Constitutive control: High levels of reporter activity	PAO1 Δ*fleQ* pCdrA*lux*	Deletion of *fleQ*, results in a reporter strain that no longer responds to cyclic di-GMP levels. Allows monitoring the effects of treatments on bacterial growth and phenotypes unrelated to cyclic di-GMP metabolism.	~5 μM [8]
High cyclic di-GMP and high levels of reporter activity	PAO1 Δ*wspF* pCdrA*lux*	Constitutive expression of WspR _DGC that allows for monitoring inhibition under conditions of high cellular levels of cyclic di-GMP.	~34 μM [6]
Tunable cyclic di-GMP levels and analogous levels of reporter activity	PAO1 P_{BAD}PA1120 pCdrA*lux*	Arabinose inducible expression of the TpbB DGC (PA1120) under the control of the P_{BAD} promoter allows for inducible control of cellular levels of cyclic di-GMP by titrating varying amounts of arabinose inducer (Fig. 1). Allows for the simultaneous identification of treatments that induce or inhibit DGC activity during conditions of half-maximal induction of DGC activity. Additionally allows conditional control of reporter activity to use as a control.	not previously tested

[a]Analogous vector control strains that harbor pMS402, which does not contain the *cdrA* promoter that mediates cyclic di-GMP-dependent FleQ transcriptional regulation of *luxCDABE* reporter activity, should be used for determining activity. The Δ*pelA* Δ*pslBCD* mutant backgrounds can also be used to prevent cellular aggregation that may alter the accuracy and normalization of some analyses in addition to alleviating potential alteration of the bacterial physiology and gene expression as a direct result of aggregation.

3. Broth cultures are grown overnight (16–18 h) with aeration (shaking at 250 rpm) at 37 °C.
4. Overnight cultures are diluted 1:250 (2 mL added to achieve 500 mL total) into LB broth supplemented with 100 μg/mL kanamycin and grown with aeration (shaking at 250 rpm) at 37 °C.
5. Cultures are incubated until OD_{600} reaches 0.1 (approximately 3–4 h).

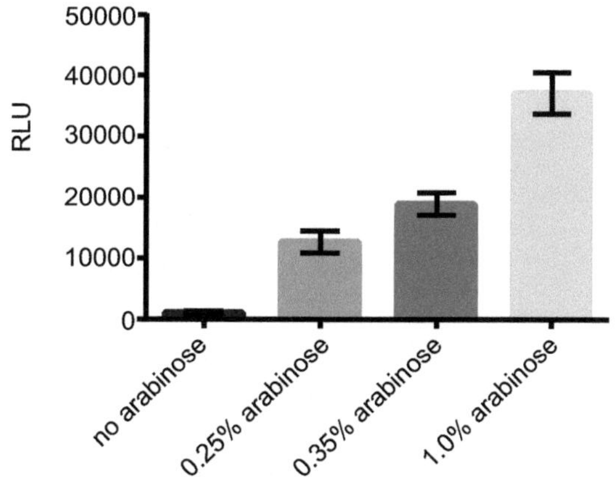

Fig. 1 Modulation of cyclic di-GMP levels by conditional expression of diguanylate cyclase as controlled by dose-dependent addition of L-arabinose to *P. aeruginosa* CTX::P_{BAD}PA1120 pCdrA/*lux*. Relative luminescence units (RLU) produced by the pCdrA/*lux* reporter increased in a dose-dependent manner with increasing concentrations of arabinose as would be expected for increased DGC activity and increased cellular levels of cyclic di-GMP. RLUs were calculated by dividing luminescence by absorbance (OD_{600})

6. Transfer 100 μL or 30 μL of the resulting bacterial suspension into 96- or 384-well white clear-bottom plates, respectively.
7. Gently tap spin (engage spin for 2 s) microtiter plates using a benchtop centrifuge to ensure that the bacteria are consistently dispensed into the bottom of each well.
8. Plates are statically incubated at 37 °C for 1–3 h (*see* **Note 3**), then luminescence and absorbance (OD_{600}) are quantified on a multimode plate reader (*see* **Note 4**).
9. Luminescence values are expressed as relative luminescence units (RLU) (*see* **Note 5**).
10. RLUs are calculated by dividing the luminescence value by the absorbance value (OD_{600}) measured in each individual well to account for pipetting and growth differences (*see* **Note 6**).

3.2 Fluorescence Plate-Based Assays

1. Cultures of bacterial reporter strains harboring GFP fusion reporters are initiated from glycerol stocks maintained at −80 °C by streaking for isolation on VBMM agar plates supplemented with 100 μg/mL gentamicin and incubated at 37 °C overnight (16–18 h) (*see* **Note 7**).
2. VBMM broth cultures supplemented with 100 μg/mL gentamicin are inoculated from isolated individual colonies.
3. After growth overnight (16–18 h) at 250 rpm, the strains are diluted (1:100) into fresh VBMM supplemented with 100 μg/mL gentamicin and grown to mid-logarithmic growth phase (OD_{600} = 0.5).

4. Aliquot 30 μL or 100 μL of the bacterial suspension into a 384 or 96-well black clear-bottom microtiter plate, respectively.
5. Read fluorescence using an excitation wavelength of 485 nm and an emission at 535 nm using a multimode plate reader (*see* **Note 6**).
6. Since the wells have clear bottoms, the same plate can then be used to read the absorbance (e.g., 600 nm) using the plate reader.
7. Fluorescence values are divided by the absorbance values measured and presented as relative fluorescence units (RFU), which are arbitrary fluorescence intensity units corrected for cell density.

3.3 Microscopic Visualization

1. Cultures of bacterial reporter strains harboring GFP fusion reporters are initiated from glycerol stocks maintained at −80 °C by streaking for isolation on LB agar plates supplemented with 100 μg/mL gentamicin and incubated at 37 °C overnight (16–18 h).
2. LB broth cultures supplemented with 100 μg/mL gentamicin are inoculated from isolated individual colonies.
3. After growth overnight (16–18 h) at 250 rpm, the strains are diluted 1:200 into VBMM supplemented with 100 μg/mL gentamicin and grown at 37 °C at 250 rpm.
4. Once the bacteria have reached mid-logarithmic growth phase ($OD_{600} = 0.3–0.5$), add 200 μL of the bacterial cell suspension to 8-well chamber μ-slides.
5. Allow bacterial cells to adhere for 1 h at 28 °C under static conditions.
6. Acquire fluorescence images (Fig. 2) using a microscope that allows for detection of fluorescence. It is essential to capture all images with equivalent exposures and settings during image acquisition.
7. Fluorescence images are further processed with image analysis software (*see* **Note 8**).

3.4 Other Fluorescence Applications

Fluorescent reporters can also be used in flow cytometers [10]. However, there are significant equipment-to-equipment performance variations and optimum conditions must be manually tested. In our experience, the biggest challenges generally lie in whether the analyses can distinguish bacteria from noise and other small particles such as dust. Determining the capability of the equipment, and setting up appropriate gating on forward scatter/side scatter plots using constitutively fluorescent strains and non-fluorescent strains are critical pilot experiments that we strongly recommend users to perform.

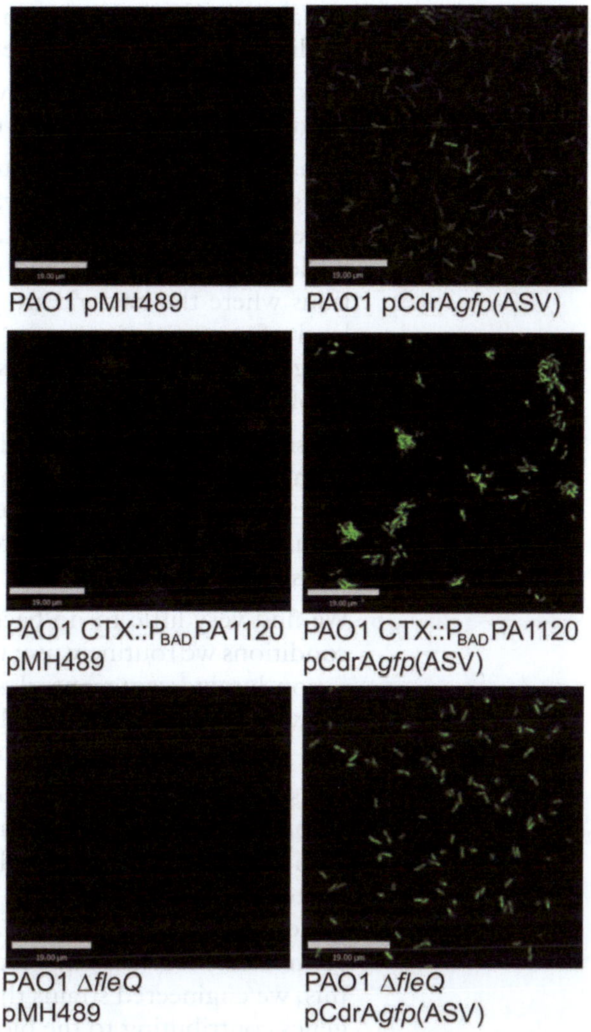

Fig. 2 Fluorescence visualization of cyclic di-GMP induction during bacterial attachment to a surface. GFP expression from the pCdrA*gfp* reporter is induced temporally in bacterial cells of wild-type PAO1 during attachment. GFP expression is conditionally (1% L-arabinose) or constitutively expressed in the P_{BAD}PA1120 and Δ*fleQ* strain backgrounds, respectively. Images were acquired with an Olympus IX81 FV1000 confocal imaging system with equivalent exposures and acquisition settings and processed with Volocity image analysis software by setting the GFP channel to 1× brightness. No other manipulations of the images were performed

4 Notes

1. Appropriate growth medium should be chosen with regard to the experimental design. For fluorescence assays, reducing background fluorescence is a key component to maximize the

signals. Under the conditions we tested, complex undefined media such as LB have higher autofluorescent properties.

2. Absorbance settings vary between equipment. Common wavelengths include 595 nm and 600 nm.
3. Dynamic range of the reporter assays is affected by the conditions used for bacterial cultivation. Factors that can contribute to reporter gene expression are temperature, aeration as controlled by shaking rpm, and growth media. In general, conditions where the bacteria are growing rapidly have the highest level of reporter expression. Both luciferase and GFP require oxygen, so all experiments using these reporters are limited to aerobic conditions.
4. Sensitivity on some plate readers can be adjusted to increase the dynamic range. Plate dimensions and reads should be optimized based on the manufacturer's recommendations to ensure accuracy and eliminate signal bleed over from adjacent wells on the microtiter plate.
5. We find very little to no background luminescence under the conditions we routinely use; therefore, it is not necessary to use a non-luminescent control strain to subtract out the noise. However, it is always advisable to include control treatments that harbor an empty vector control such as pMS402.
6. Cells with high intracellular levels of cyclic di-GMP often form visible aggregates due to their overexpression of biofilm factors, such as extracellular polysaccharides and adhesion proteins [7]. Bacterial aggregation alters the accuracy of the absorbance values that are used to normalize the fluorescence or luminescence measurements reported as arbitrary units. To circumvent this, we engineered strains that do not aggregate by disrupting genes contributing to the biosynthesis and corresponding production of the PEL and PSL polysaccharides (Δ*pelA* and Δ*pslBCD* respectively).
7. It is possible to use other culture media for monitoring reporter activity during cultivation. However, the effects of bacterial growth medium and growth conditions should be preliminarily evaluated for autofluorescence of the bacterial strains before proceeding. For example, if the bacterial cultures are grown in LB broth, the following additional steps should be added to Subheading 3.2 (Fluorescence plate-based assays) after **step 2**:
   (a) The bacterial cells should be pelleted by centrifugation using a benchtop centrifuge at maximum speed for 10 min at room temperature and the supernatant carefully removed.

(b) Bacterial pellets are resuspended in ½ the original volume in PBS. For example, if 1 mL of the bacterial culture was pelleted, then resuspend in 500 μL PBS.

(c) Aliquot 100 μL of the bacterial suspension into a 96-well black clear-bottom microtiter plate.

(d) Read fluorescence using a multimode plate reader with an excitation wavelength of 485 nm and an emission of 535 nm. If the bacterial cells in the negative control strain (pMH489) are still fluorescent, then repeat these steps again. This step is necessary to remove autofluorescent metabolites (e.g., pyoverdine) which share overlapping fluorescence excitation and emission spectra with GFP. These steps can be potentially avoided by the use of media and growth conditions that limit the production of fluorescent metabolites; however, this should be validated using strains that do not produce GFP under the conditions of the desired assay application. Once the bacterial cells have been washed free of autofluorescent metabolites, Subheading 3.2 (Fluorescence plate-based assays) can resume at **step 3**.

8. Bacterial cells can be counterstained or simultaneously imaged with phase contrast or DIC to merge multiple channels to differentiate between cells that are expressing reporter activity and those that are not. Reporter activity can also be quantified for statistical analysis. However, the analysis of images for quantification requires accuracy and precision during image capture to appropriately quantify and compare images. We recommend reviewing published methodologies before proceeding with image acquisition [15]. Software packages such as Image J, Fiji, Comstat, or Volocity (Perkin Elmer) can be used for quantification.

Acknowledgments

This work was funded in part by research support from the Boettcher Foundation's Webb-Waring Biomedical Research Program to BRB.

References

1. Stover CK, Pham XQ, Erwin AL, Mizoguchi SD, Warrener P, Hickey MJ, Brinkman FS, Hufnagle WO, Kowalik DJ, Lagrou M, Garber RL, Goltry L, Tolentino E, Westbrock-Wadman S, Yuan Y, Brody LL, Coulter SN, Folger KR, Kas A, Larbig K, Lim R, Smith K, Spencer D, Wong GK, Wu Z, Paulsen IT, Reizer J, Saier MH, Hancock RE, Lory S, Olson MV (2000) Complete genome sequence of *Pseudomonas aeruginosa* PAO1, an opportunistic pathogen. Nature 406(6799):959–964. doi:10.1038/35023079
2. Römling U, Galperin MY, Gomelsky M (2013) Cyclic di-GMP: the first 25 years of a universal

bacterial second messenger. Microbiol Mol Biol Rev 77(1):1–52. doi:10.1128/MMBR.00043-12

3. Kulasakara H, Lee V, Brencic A, Liberati N, Urbach J, Miyata S, Lee DG, Neely AN, Hyodo M, Hayakawa Y, Ausubel FM, Lory S (2006) Analysis of *Pseudomonas aeruginosa* diguanylate cyclases and phosphodiesterases reveals a role for bis-(3′-5′)-cyclic-GMP in virulence. Proc Natl Acad Sci U S A 103 (8):2839–2844
4. Coggan KA, Wolfgang MC (2012) Global regulatory pathways and cross-talk control *Pseudomonas aeruginosa* environmental lifestyle and virulence phenotype. Curr Issues Mol Biol 14 (2):47–70
5. Hickman JW, Tifrea DF, Harwood CS (2005) A chemosensory system that regulates biofilm formation through modulation of cyclic diguanylate levels. Proc Natl Acad Sci U S A 102 (40):14422–14427
6. Irie Y, Parsek MR (2014) LC/MS/MS-based quantitative assay for the secondary messenger molecule, c-di-GMP. Methods Mol Biol 1149:271–279. doi:10.1007/978-1-4939-0473-0_22
7. Borlee BR, Goldman AD, Murakami K, Samudrala R, Wozniak DJ, Parsek MR (2010) *Pseudomonas aeruginosa* uses a cyclic-di-GMP-regulated adhesin to reinforce the biofilm extracellular matrix. Mol Microbiol 75 (4):827–842
8. Hickman JW, Harwood CS (2008) Identification of FleQ from *Pseudomonas aeruginosa* as a c-di-GMP-responsive transcription factor. Mol Microbiol 69(2):376–389
9. Rybtke MT, Borlee BR, Murakami K, Irie Y, Hentzer M, Nielsen TE, Givskov M, Parsek MR, Tolker-Nielsen T (2012) Fluorescence-based reporter for gauging cyclic di-GMP levels in *Pseudomonas aeruginosa*. Appl Environ Microbiol 78(15):5060–5069. doi:10.1128/AEM.00414-12
10. Irie Y, Borlee BR, O'Connor JR, Hill PJ, Harwood CS, Wozniak DJ, Parsek MR (2012) Self-produced exopolysaccharide is a signal that stimulates biofilm formation in *Pseudomonas aeruginosa*. Proc Natl Acad Sci U S A 109 (50):20632–20636. doi:10.1073/pnas.1217993109
11. Pawar SV, Messina M, Rinaldo S, Cutruzzola F, Kaever V, Rampioni G, Leoni L (2016) Novel genetic tools to tackle c-di-GMP-dependent signalling in *Pseudomonas aeruginosa*. J Appl Microbiol 120(1):205–217. doi:10.1111/jam.12984
12. Rugjee KN, An SQ, Ryan RP (2016) Establishment of a high-throughput setup for screening small molecules that modulate c-di-GMP signaling in *Pseudomonas aeruginosa*. J Vis Exp (112). doi:10.3791/54115
13. Duan K, Dammel C, Stein J, Rabin H, Surette MG (2003) Modulation of *Pseudomonas aeruginosa* gene expression by host microflora through interspecies communication. Mol Microbiol 50(5):1477–1491
14. Holloway BW, Krishnapillai V, Morgan AF (1979) Chromosomal genetics of *Pseudomonas*. Microbiol Rev 43(1):73–102
15. Waters JC (2009) Accuracy and precision in quantitative fluorescence microscopy. J Cell Biol 185(7):1135–1148. doi:10.1083/jcb.200903097

Chapter 10

Live Flow Cytometry Analysis of c-di-GMP Levels in Single Cell Populations

Jongchan Yeo, Xin C. Wang, and Ming C. Hammond

Abstract

Second-generation RNA-based fluorescent biosensors have been developed that enable flow cytometry experiments to monitor the population dynamics of c-di-GMP signaling in live bacteria. These experiments are high-throughput, provide information at the single-cell level, and can be performed on cells grown in complex media and/or under anaerobic conditions. Here, we describe flow cytometry methods for three applications: (1) high-throughput screening for diguanylate cyclase activity, (2) analyzing c-di-GMP levels under anaerobic conditions, and (3) monitoring cell population dynamics of c-di-GMP levels upon environmental changes. These methods showcase RNA-based fluorescent biosensors as versatile tools for studying c-di-GMP signaling in bacteria.

Key words Cyclic dinucleotide, Cyclic di-GMP, Flow cytometry, RNA-based fluorescent biosensor, Spinach aptamer, Anaerobic growth, C-di-GMP signaling

1 Introduction

Cyclic di-GMP (c-di-GMP) is a near-universal bacterial signal that controls the transition from a free-living state to a surface-attached biofilm, a lifestyle decision that requires coordinated changes in motility, exopolysaccharide production, quorum sensing, and other behaviors [1]. To date, many analytical methods have been developed to measure cellular c-di-GMP levels. Direct in vitro methods include thin-layer chromatography (TLC) [2], high-performance liquid chromatography (HPLC) followed by mass spectrometry (MS) [3], detection of c-di-GMP G-quadruplexes [4], and detection based on binding to effector protein domains or riboswitch aptamers [5, 6]. Indirect methods include phenotypic assays for motility [7] and biofilm formation [8]. More recently, in vivo methods for screening and live cell imaging of c-di-GMP have been developed based on genetic reporters whose expression is controlled by c-di-GMP binding effectors [9], or engineered fluorescent biosensors whose fluorescence or Forster resonance energy

Karin Sauer (ed.), *c-di-GMP Signaling: Methods and Protocols*, Methods in Molecular Biology, vol. 1657,
DOI 10.1007/978-1-4939-7240-1_10, © Springer Science+Business Media LLC 2017

transfer (FRET) efficiency is changed upon c-di-GMP binding [10–12].

Our lab has focused on the design and development of RNA-based fluorescent biosensors for different bacterial signals and cofactors, including c-di-GMP [13, 14]. These biosensors are genetically encodable RNA sequences incorporating a riboswitch aptamer as the sensing domain that changes conformation upon binding the target ligand, e.g., c-di-GMP. This ligand-dependent conformational change is engineered to stabilize the second domain of the biosensor, the dye-binding Spinach aptamer, which binds and enhances the fluorescence of the dye molecule DFHBI (3,5-difluoro-4-hydroxybenzylidene imidazolinone) [15]. Thus, the biosensor system is comprised of a genetically encodable RNA sequence that is a riboswitch-Spinach fusion and an exogenously added dye compound [16]. DFHBI and related dyes are cell permeable and have sufficiently low background fluorescence that they can be kept in the media during imaging experiments.

Our first-generation c-di-GMP biosensor was the Vc2 GEMM-I riboswitch aptamer fused to the Spinach aptamer in a tRNA scaffold [12, 13]. Recently, we developed second-generation c-di-GMP biosensors with a broader range of binding affinities, including sensitivity down to a few c-di-GMP molecules per cell, higher fold fluorescence turn-on and overall brightness, and faster turn-on kinetics [17]. These biosensors, called Ct, Dp, and Pl-B [17], are comprised of different GEMM-I riboswitch aptamers fused to the Spinach2 aptamer, a second-generation dye-binding domain that works with DFHBI and DFHBI-1T (4-(3,5-difluoro-4-hydroxybenzylidene)-2-methyl-1-(2,2,2-trifluoroethyl) imidazolone) [18], in a tRNA scaffold (Table 1). With the improvements in biosensor properties stated above, these RNA-based fluorescent biosensors are optimized for use in live cell flow cytometry experiments to measure c-di-GMP in single cell populations. In this chapter, we describe flow cytometry methods for three applications: (1) high-throughput screening for diguanylate cyclase activity, (2) analyzing c-di-GMP levels under anaerobic conditions, and (3) monitoring cell population dynamics of c-di-GMP upon environmental changes. We encourage the reader to consider these methods as a starting foundation for other biosensor applications using flow cytometry, including screening for phosphodiesterase activity and analyzing c-di-GMP levels or dynamics in other conditions or in other organisms.

2 Materials

2.1 Equipment

1. Micropipettor.
2. Vortex mixer.
3. Microcentrifuge.

Table 1
Sequences of c-di-GMP biosensors (*ITALICS* = T7 promoter; UNDERLINED = tRNA scaffold; CAPS = Spinach2 sequence; BOLD = c-di-GMP riboswitch aptamer sequence (Dp, Ct or Pl-B); *BOLD ITALICS* = T7 terminator)

Biosensor	Sequence
Dp	*TAATACGACTCACTATAGGG*GCCCGGATAGCTCAGTCGGTAGAGCAGCGGCCG GATGTAACTGAATGAAATGGTGAAGGACGGGTCCA**CUUCUCGACAAAGGCA AACCCUCCGCGAGGGGGGGACGCAAAGCCCACGGAACUCCGCUGCUC CGCUCUUCUCUCAGGGCAGCACGGAAGUUGGCCGGGCCACCGAAAGA AG**TTGTTGAGTAGAGTGTGAGCTCCGTAACTAGTTACATCCGGCCGCGGTC CAGGGTTCAAGTCCCTGTTCGGGCGCCA***TAGCATAACCCCTTGGGGCCTCT AAACGGGTCTTGAGGGGTTTTTTG***
Ct	*TAATACGACTCACTATAGGG*GCCCGGATAGCTCAGTCGGTAGAGCAGCGGCCG GATGTAACTGAATGAAATGGTGAAGGACGGGTCCA**ATGAAACAGGGCAAAA TCACCGAAAGGTGATGACGCAAAGCCATGGGTCTACTGTTTTAAAACAA TGTTTTAAAGCTATGATCGCCAGGCTGCCAT**TTGTTGAGTAGAGTGTGAG CTCCGTAACTAGTTACATCCGGCCGCGGTCCAGGGTTCAAGTCCCTGTTCG GGCGCCA***TAGCATAACCCCTTGGGGCCTCTAAACGGGTCTTGAGGGGTTTT TTG***
Pl-B	*TAATACGACTCACTATAGGG*GCCCGGATAGCTCAGTCGGTAGAGCAGCGGCCG GATGTAACTGAATGAAATGGTGAAGGACGGGTCCA**CTTCGATAACGGCAAA CTTGTCGAAAGATAAGGACGCAAAGCCACAGGGCCTTCTTGATGAACCG TCAATGGCAGCCTGGCTACCGAAG**TTGTTGAGTAGAGTGTGAGCTCCGTA ACTAGTTACATCCGGCCGCGGTCCAGGGTTCAAGTCCCTGTTCGGGCGCCA ***TAGCATAACCCCTTGGGGCCTCTAAACGGGTCTTGAGGGGTTTTTTG***
Spinach2	*TAATACGACTCACTATAGGG*GCCCGGATAGCTCAGTCGGTAGAGCAGCGGCCG GATGTAACTGAATGAAATGGTGAAGGACGGGTCCATTGTTGAGTAGAGTGTG AGCTCCGTAACTAGTTACATCCGGCCGCGGTCCAGGGTTCAAGTCCCTGTT CGGGCGCCA***TAGCATAACCCCTTGGGGCCTCTAAACGGGTCTTGAGGGGT TTTTTG***

4. PCR thermocycler.
5. PCR thermocycler equipped with 96-well platform.
6. Incubator shaker set to 37 °C (Maximum shaking speed of 325 rpm or higher).
7. Benchtop orbital shaker.
8. Timer.
9. Flow cytometer equipped with 488 nm laser and autosampler.
10. Hand-operated crimper for aluminum seals.

2.2 Supplies

1. Micropipettor tips.
2. 1.5 mL microcentrifuge tubes.

3. Polyethersulfone membrane sterile syringe filter units (25 mm diameter, 0.2 μm pore size).
4. 14 mL culture tubes.
5. Petri dishes.
6. 12-well suspension culture plate, sterile, with lid.
7. Plating glass beads, sterile.
8. Anaerobic culture tubes.
9. Chlorobutyl rubber stoppers.
10. Aluminum seals for culture tubes.
11. Argon gas cylinder.
12. Plastic tubing (inner diameter = 1/4 in.) and tubing connectors (straight and Y-shaped).
13. 1 mL syringes.
14. Syringe needles (18G, 1.5 in.).
15. Round-bottom, nontreated, sterile polypropylene 96-well plates.
16. 96-well PCR plates, clear.
17. 96-well deep well microplates (well volume ~ 2.2 mL).
18. Gas-permeable sealing membrane for microtiter plates, sterile.

2.3 Reagents

1. Sterile water: sterile filter-sterilized pyrogen-, nuclease-, protease-, and bacteria-free water, 18.2 MΩ.
2. 1× TAE buffer: 40 mM Tris–HCl, 20 mM acetic acid, 1 mM EDTA, pH 8.4.
3. Carbenicillin: 50 mg/mL stock concentration, filtered through a 0.2 μm nitrocellulose filter.
4. Kanamycin: 50 mg/mL stock concentration, filtered through a 0.2 μm nitrocellulose filter.
5. Luria Broth (1% tryptone, 0.5% yeast extract, 1% sodium chloride), autoclave-sterilized.
6. LB/Carb: Luria Broth containing 50 μg/mL carbenicillin, autoclave-sterilized.
7. LB/Carb/Kan: Luria Broth containing 50 μg/mL carbenicillin and 50 μg/mL kanamycin, autoclave-sterilized.
8. LB/Carb agar: Pour warm LB/Carb solution containing 1.5% agar into 100 mm petri dishes and cool the plates to solidify.
9. LB/Carb/Kan agar: Pour warm LB/Carb/Kan solution containing 1.5% agar into 100 mm petri dishes and cool the plates to solidify.

10. 12-well plate LB/Carb/Kan agar: Pour warm LB/Carb/Kan containing 1.5% agar solution into a 12-well suspension culture plate and let it cool to solidify.
11. IPTG: dissolve isopropyl β-D-1-thiogalactopyranoside in sterile water to the final concentration of 0.5 M filtered through a 0.2 μm nitrocellulose filter.
12. AI/Carb/Kan: ZYP-5052 autoinduction media (50 mM KH_2PO_4, 50 mM Na_2HPO_4, 25 mM $(NH_4)_2)$, SO_4, 1 mM $MgSO_4$, 1% tryptone, 0.5% yeast extract, 0.05% g glucose, 0.2% g α-lactose, 0.5% glycerol) supplemented with 50 μg/mL carbenicillin and 50 μg/mL kanamycin, autoclave-sterilized.
13. AI/Carb: ZYP-5052 autoinduction media supplemented with 50 μg/mL carbenicillin, autoclave-sterilized.
14. S.O.C. medium: 2% tryptone, 0.5% yeast extract, 10 mM NaCl, 2.5 mM KCl, 10 mM $MgCl_2$, 10 mM $MgSO_4$, and 20 mM glucose, autoclave-sterilized.
15. DFHBI-1T (4-(3,5-difluoro-4-hydroxybenzylidene)-2-methyl-1-(2,2,2-trifluoroethyl) imidazolone): stock concentration of 20 mM DFHBI-1T in 100% DMSO, stored in 50 μL aliquots at −20 °C.
16. 1 M $ZnCl_2$: Prepare a 1 M $ZnCl_2$ solution in sterile water, filtered through a 0.2 μm nitrocellulose filter (freshly prepared stock).
17. 1× PBS: 137 mM NaCl, 2.7 mM KCl, 10 mM Na_2HPO_4, 1.8 mM KH_2PO_4, pH 7.4 in sterile water, filtered through a 0.2 μm nitrocellulose filter.
18. *E. coli* BL21(DE3) Star chemically competent cells.
19. Biosensor constructs in pET31b plasmid (*see* Table 1 for construct sequences).
    (a) Ct biosensor (Addgene #79158, pET31b-T7-Spinach2-Ct, expresses Spinach2-Ct biosensor under a T7 promoter).
    (b) Dp biosensor (Addgene #79159, pET31b-T7-Spinach2-Dp, expresses Spinach2-Dp biosensor under a T7 promoter).
    (c) Pl-B biosensor (Addgene #79161, pET31b-T7-Spinach2-Pl-B, expresses Spinach2-Pl-B biosensor under a T7 promoter).
    (d) Spinach2 control (Addgene #79783, pET31b-T7-Spinach2, expresses Spinach2 biosensor under a T7 promoter).

20. Enzyme constructs in pCOLA plasmid.
    (a) Diguanylate cyclase WspR WT (Addgene #79162, pCOLA-T7-WspR, expresses WspR WT under a T7 promoter).
    (b) Inactive mutant WspR G249A (Addgene #79163, pCOLA-T7-WspR:G249A, expresses WspR G249A under a T7 promoter).
    (c) Diguanylate cyclase WspR D70E (Addgene #79164, pCOLA-T7-WspR:D70E, expresses constitutively active WspR D70E under a T7 promoter).
    (d) Phosphodiesterase YhjH (Addgene #79165, pCOLA-T7-YhjH, Expresses YhjH under a T7 promoter).
    (e) pCOLA empty vector.

3 Methods

3.1 Screening for Diguanylate Cyclase Activity

We have used RNA-based fluorescent biosensors to screen 29 candidate GGDEF enzyme genes for c-di-GMP and/or c-AMP-GMP synthase activity using flow cytometry [19]. Fig. 1 shows the results of the screen for diguanylate cyclase (c-di-GMP synthesis) activity. Enzymatic activity of these signaling enzymes can be assayed in vivo even in the absence of activating signal because overexpression often drives dimerization to the active state. In this section, we describe the procedure for performing the high-throughput screen in a 96-well format.

3.1.1 High-Throughput Generation of E. coli Strains

1. To prepare cells co-expressing biosensor and enzyme library members, aliquot 10 μL of BL21(DE3) Star *E. coli* chemically competent cells into a 96-well PCR plate on ice and add ~50 ng of each plasmid DNA (Dp biosensor in pET31b and enzyme genes of interest in pCOLA) to each well. Also, prepare a negative control (Dp biosensor in pET31b and pCOLA empty vector) and positive control (Dp biosensor in pET31b and diguanylate cyclase WspR WT in pCOLA). Incubate the plate of cells with DNA on ice for 30 min.
2. Turn on the PCR thermocycler equipped with 96-well platform. Create and run a program to keep the sample holder temperature at 42 °C.
3. Transfer the 96-well PCR plate from ice to the heated sample holder for 1 min to apply heat shock.
4. Place the 96-well PCR plate back on ice for 3 min.
5. In the meantime, prepare recovery plate by adding 250 μL of S.O.C. media to each well of a 96-well deep well microplate.
6. Gently add transformed cells to each well of the recovery plate containing S.O.C. media.

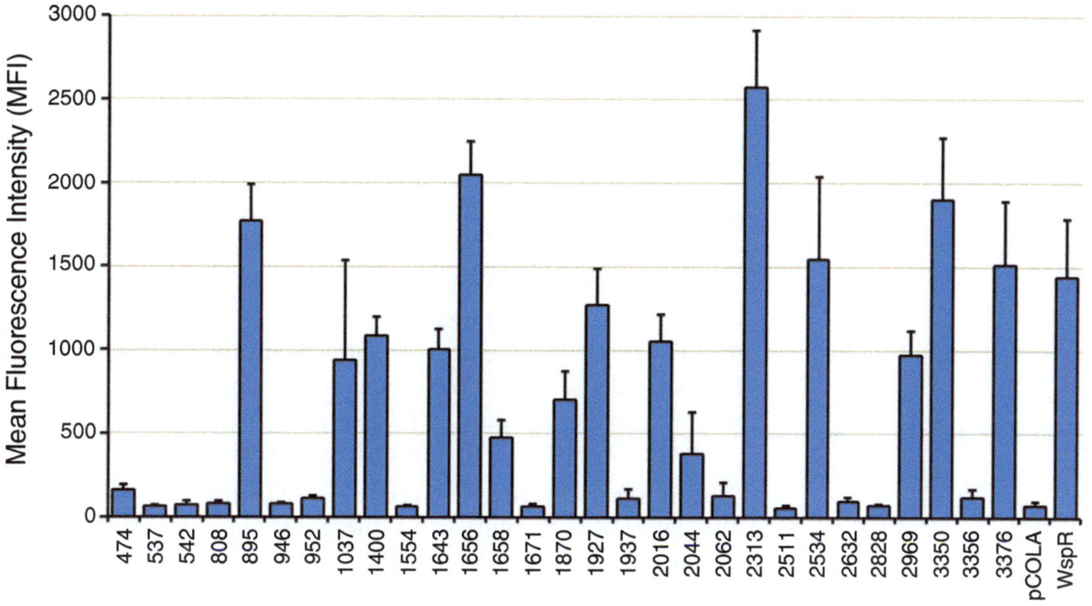

Fig. 1 In vivo fluorescent biosensor screen of GGDEF genes from *Geobacter sulfurreducens* for diguanylate cyclase activity. Mean fluorescence intensity (MFI) values were measured by flow cytometry analysis of *E.coli* BL21 (DE3) Star cells co-expressing the Dp biosensor and GGDEF enzymes from *G. sulfurreducens* strain PCA, empty pCOLA vector, or diguanylate cyclase WspR. *Error bars* represent the standard deviation from three independent biological replicates

7. Cover the plate with a gas-permeable sealing membrane and shake for 1 h at 37 °C and 225 rpm to recover.
8. In the meantime, prepare 12-well plate LB/Carb/Kan agar plates.
9. Transfer 30 μL of recovered cells onto LB/Carb/Kan agar in each well of 12-well plates. Spread cells by adding two or three plating glass beads in each well and rolling the beads around in a circular motion for about 30 s.
10. Cover plates with covers and incubate inverted at 37 °C for 18 h.

3.1.2 Preparation of Samples for Flow Cytometry

1. Pick single colonies from the 12-well plate and inoculate in 500 μL of LB/Carb/Kan in a high-wall 96-well plate, including positive and negative controls. Grow cells by shaking at 325 rpm 37 °C, 18 h until cells reached an $OD_{600} > 3$ (*see* **Note 1**).
2. Transfer 1 μL of cell cultures to 500 μL of AI/Carb/Kan in another high-wall 96-well plate and grow by shaking at 325 rpm for 16 h at 37 °C. At this step, cells will induce expression of biosensor and enzyme.

3. Prepare 1× PBS containing 50 μM DFHBI-1T and protect from light (*see* **Note 2**). Aliquot 70 μL of the solution to each well of a round-bottom 96-well polypropylene plate (the flow plate). Protect the plate from light by wrapping the plate with aluminum foil before use.

3.1.3 Setup of Flow Cytometry Parameters

1. Start up the flow cytometer equipped with 488 nm laser and associated software. Run performance test following manufacturer's protocol if necessary.
2. Transfer 1 μL of the negative control and positive control cell cultures from the induction plate in Subheading 3.1.2, **step 2** into wells of the flow plate containing PBS-DFHBI-1T prepared in Subheading 3.1.2, **step 3**. Incubate for 5 min to allow the DFHBI-1T dye to diffuse into cells.
3. Do test runs with negative control and positive control cells in PBS-DFHBI-1T solution in the flow plate to establish the forward scatter (FSC) and side scatter (SSC) regions and to optimize the voltage gain settings. In this experiment, negative controls are cells expressing Dp biosensor in pET31b but harbor the pCOLA empty vector while positive controls are cells expressing Dp biosensor in pET31b and wild-type WspR in pCOLA.
    (a) On the software, create a 96-well plate experiment (*see* **Note 3**).
    (b) On the workspace window, load three dot plots and one histogram. Set up the axes for these plots—FSC-Area/SSC-Area (both log axes), FSC-Height/FSC-Area (both log axes), and Time (linear axis)/FSC-Area (log axis) for the three dot plots, and count/GFP(530/30-A) for the histogram, respectively (Fig. 2).
    (c) Set up the flow rate, draw volumes, and stop settings. Recommended settings are 12.5 μL/min flow rate,

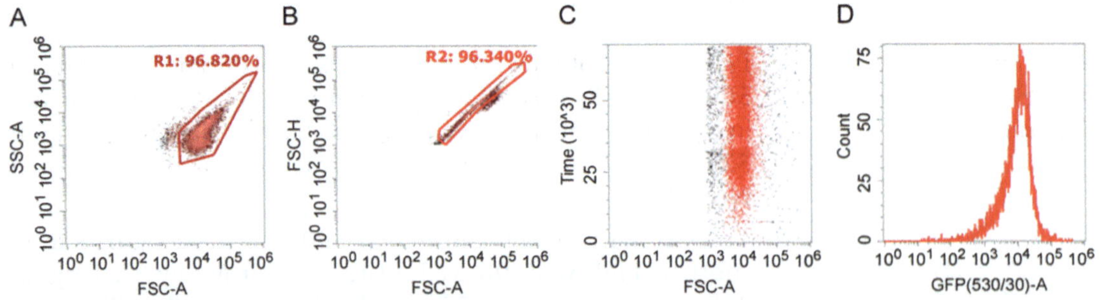

Fig. 2 Analysis for setting up flow cytometry gate parameters. Examples of (**a**) FSC-A/SSC-A (both log axes), (**b**) FSC-H/FSC-A (both log axes), (**c**) time (linear axis)/FSC-A (log axis), and (**d**) histogram for GFP(530/30)-A (log axis), along with two gates (labeled as R1 and R2) are shown

65 μL draw volumes (for sample volumes of 70 μL), and at least 10,000 readings before stop.

(d) Place the negative control sample in the sample holder.

(e) Load the negative control sample and run the flow cytometry analysis to obtain data for the four plots in **step b**.

(f) On the FSC/SSC dot plot, generate a gate to exclude reading any signals from debris (Fig. 2a).

(g) On the FSC-A/FSC-H dot plot, display readings only from the gate in **step f**. You should see a diagonal pattern. Generate a narrow diagonal gate to exclude readings off the diagonal. This gate excludes multiplet readings frequently generated from cell clumps (Fig. 2b).

(h) Check the Time/FSC-A dot plot to check that the flow of cells is stable. If any areas show a poor flow rate, generate a gate to exclude readings from that area (Fig. 2c).

(i) On the histogram, display readings only from the gates in **steps f–h** (Fig. 2d).

(j) Load the positive control sample and run the flow cytometry analysis. While running the sample, the four plots described in step f to i are displayed in real time by the software.

(k) Adjust the voltage gain settings and repeat analysis of positive and negative controls so that the readings from negative control and positive control appear around 10^2 and 10^4 of the fluorescence intensity, respectively.

(l) Save all settings as a template file to use for analysis of samples from the same experiment.

3.1.4 Analysis of Enzyme Library by Flow Cytometry

1. Transfer 1 μL of sample cell cultures from the induction plate in Subheading 3.1.2, **step 2** into wells of the flow plate containing PBS-DFHBI-1T prepared in Subheading 3.1.2, **step 3**. Incubate for 5 min to allow the DFHBI-1 T dye to diffuse into cells.
2. Load the flow plate onto the autosampler of the flow cytometer.
3. On the heat map, highlight the wells you are analyzing and create experiments. Typically, two technical replicates are performed by analyzing each well twice.
4. Run flow cytometry analysis of the samples using parameter settings from Subheading 3.1.3, **step 3**.

3.1.5 Determination of Mean Fluorescence Intensity (MFI)

1. Export all the results to create FCS files for analysis.
2. Open the FCS files with a flow cytometry analysis software (e.g., FlowJo).

3. Double-click on a sample to open FSC-Area/SSC-Area plot and repeat generating gates as described in Subheading 3.1.3. As the first step, change both axis settings to log, and generate a gate to exclude any signals from debris as in Subheading 3.1.3, **step 2f**.
4. Double-click on the new created subset file. Change axis settings to FSC-Height and FSC-Area (both log axes). Generate a thin diagonal gate to exclude multiplet readings as in Subheading 3.1.3, **step 2g**.
5. Double-click on the new created subset file. Change axis settings to histogram and GFP(530/30)-A (log axis) as in Subheading 3.1.3, **step 2b**.
6. Repeat Subheading 3.1.5, **steps 3–5** for all samples. Replicate all gates and settings to all samples by clicking on the first subset containing the FSC-Height and FSC-Area plot (usually named as "Lymphocytes" by default) and dragging it onto the other samples.
7. To analyze the mean fluorescence intensity (MFI) values, open histograms in the layout editor.
8. Normalize the y-axis of histograms to mode.
9. Perform statistical analysis for "Σ Mean: BL1-A" to display the MFI results on the layout window.
10. Calculate the standard deviation for MFI values of at least three independent biological replicates (*see* **Note 1**).

3.2 Analysis of c-di-GMP Levels Under Anaerobic Conditions

Some diguanylate cyclases and phosphodiesterases harbor oxygen-sensing domains, suggesting that oxygen regulates c-di-GMP levels [20–22]. In these experiments, we describe using an RNA-based fluorescent biosensor to analyze c-di-GMP levels in *E. coli* grown under anaerobic conditions and after oxygen recovery. When Ct biosensor is co-expressed with either inactive (G249A) or constitutively active (D70E) diguanylate cyclase WspR, clear differences in fluorescence intensities are observed that correspond to the expected low and high c-di-GMP levels, respectively (Fig. 3). These results demonstrate that the biosensor functions similarly with or without oxygen and that DFHBI and related dyes remain fluorescent under anaerobic conditions [17]. In contrast, fluorescent proteins in the GFP family that are commonly employed in protein biosensor construction require oxygen for chromophore maturation [23, 24], and are therefore not suitable for anaerobic analysis.

3.2.1 Generation of E. coli Strains

1. To prepare cells co-expressing biosensor and enzyme, add ~50 ng of each plasmid (e.g., Ct biosensor in pET31b and diguanylate cyclase WspR D70E in pCOLA) to BL21(DE3) Star *E. coli* chemically competent cells and transform the

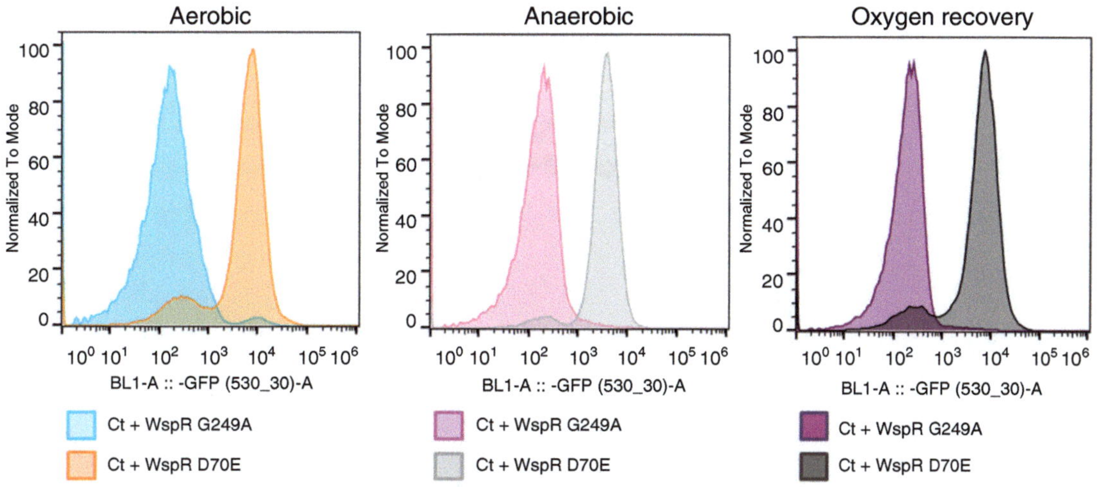

Fig. 3 Flow cytometry histograms of *E. coli* BL21(DE3) Star cells co-expressing Ct biosensor and indicated enzymes (inactive WspR G249A or constitutively active WspR D70E) under aerobic, anaerobic, and oxygen recovery conditions

plasmids following the manufacturer's protocol. Also, prepare positive control (diguanylate cyclase WspR G249A in pCOLA and Spinach2 in pET31b) and negative control (diguanylate cyclase WspR G249A in pCOLA and empty pET31b) cells for setting up flow cytometry parameters. The positive control is the constitutively fluorescent Spinach2 and the negative control is the non-fluorescent empty vector. The controls also co-express diguanylate cyclase WspR G249A, which is an inactive mutant, so that the controls have similar expression levels and growth media conditions as the experimental samples.

2. Plate cells on LB/Carb/Kan agar plates and incubate for 12–16 h at 37 °C. The plates can be stored at 4 °C for 2 weeks (*see* **Note 4**).

3.2.2 Preparation and Growth of Anaerobic Cultures

1. Autoclave anaerobic culture tubes (Balch tubes) and chlorobutyl rubber stoppers separately before use.
2. Add 3 mL AI/Carb/Kan to each anaerobic culture tube.
3. Inoculate the AI/Carb/Kan media with a single colony (*see* **Notes 1**, **5**, and **6**).
4. Close the culture tubes with chlorobutyl rubber stoppers. It is important to completely push the stoppers all the way down before proceeding to next step (*see* **Note 7**).
5. Cap the top of the culture tubes with aluminum seals and crimp using a hand-operated crimper (*see* **Note 8**). It is very important to ensure that all the tubes are completely sealed before proceeding to next step.

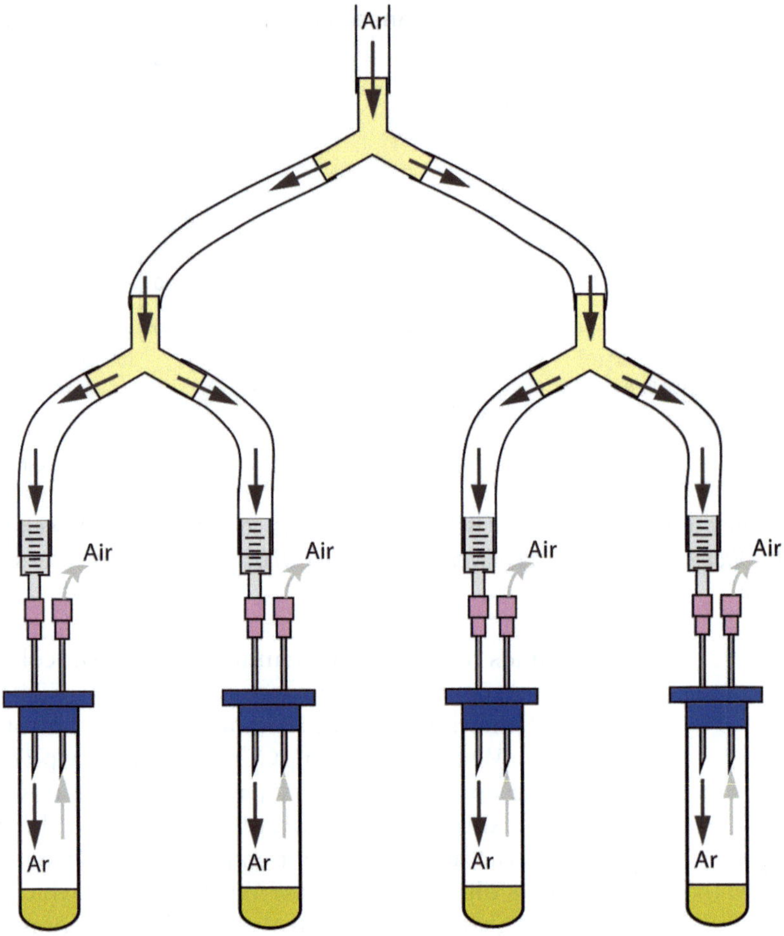

Fig. 4 Schematic for the setup of a multi-channel argon gas sparging system. The main gas flow is split into four or more paths using Y-shaped tubing connectors, and syringe heads are connected to each end. All connections should be sealed with parafilm. Two needles are stuck through rubber stoppers and each channel is connected to one of two needles to allow argon gas to flow inside and the air is expelled out

6. To prepare argon apparatus to sparge the headspace, connect one side of plastic tubing to the argon gas cylinder, and the other side of tubing to a Y-shaped tubing connector. Use additional tubing and Y-shaped connectors to set up multi-channel sparging system (*see* Fig. 4 for the illustrations of the setup).
7. To create a connection between the rubber tubing and needle, use a syringe head, which is obtained by cutting 1 mL syringes, at each end of channel. Seal all connections firmly with parafilm.
8. Stick two syringe needles into each culture tube through the rubber stopper so that the tips of syringe needles are in the head-space of the culture inside tubes. To make the later **step 10** easier, try to place two needles parallel to each other.

9. Connect one of two needles on each tube to the syringe heads from **step 8** and flow argon gas to sparge culture tubes with argon for >15 min. While sparging, the air is effused out through the other needle.
10. Keeping the argon gas flow on, quickly remove the two needles simultaneously to minimize the introduction of external air. Never leave argon flowing into the sealed tube without an outlet, as that could cause the pressurized tube to break.
11. Place all culture tubes in a tube rack and grow in an incubator shaker at 37 °C for 18–24 h until cells reach an $OD_{600} > 3$ (*see* **Note 9**).

3.2.3 Analysis of Anaerobic Cultures by Flow Cytometry

1. Prepare the 96-well flow plate containing PBS-DFHBI-1T as in Subheading 3.1.2, **step 3**.
2. Start up the flow cytometer.
3. Take out anaerobic culture tubes of positive and negative controls from the incubator shaker and place on ice.
4. While keeping the culture tubes on ice, break the aluminum seal of negative control culture (*see* **Note 10**).
5. Open the rubber stopper and immediately take out 1 μL of the negative control culture to dilute into the PBS-DFHBI-1T in the 96-well flow plate.
6. Load the flow plate onto the autosampler of the flow cytometer and analyze the sample on the flow plate.
7. Repeat Subheading 3.2.3, **steps 4–6** for the positive control culture.
8. Establish the FSC and SSC regions to optimize the voltage gain settings as described in Subheading 3.1.3.
9. Take out anaerobic culture tubes of sample cultures from the incubator shaker and place on ice.
10. On the heat map, highlight the well you are analyzing and create an experiment. Typically, two technical replicates are performed by analyzing each well twice.
11. Repeat Subheading 3.2.3, **steps 4–6** for each sample in turn.

3.2.4 Analysis of Oxygen Recovery Cultures by Flow Cytometry

1. For oxygen recovery, agitate the opened cultures from Subheading 3.2.3, **steps 3** and **9** by tapping the side of tubes and place them at 4 °C for 2 h. The temperature is kept low to ensure that the enzyme and biosensor expression levels in cells do not change throughout the oxygen recovery step. If possible, gently shake the tubes on an orbital shaker (<100 rpm).
2. On the heat map, highlight the well you are analyzing and create an experiment. Typically, two technical replicates are performed by analyzing each well twice.

3. Take out 1 μL of each oxygen-recovered culture and dilute into the PBS-DFHBI-1T in the 96-well flow plate.
4. Load the flow plate onto the autosampler of the flow cytometer and analyze the samples on the flow plate using the settings described in Subheading 3.2.3, **step 8** (*see* **Note 11**).
5. Once all samples have been analyzed, export all your results to create FCS files and analyze the MFI values for each sample using FlowJo software as described in Subheading 3.1.5.

3.3 Monitoring Cell Population Dynamics of c-di-GMP Upon Environmental Changes

Flow cytometry can be a powerful method to monitor the cell population dynamics of c-di-GMP upon environmental changes. One advantage is that the experiments are high-throughput and provide information at the single-cell level. In addition, experiments monitoring c-di-GMP dynamics can be performed on cells grown in complex media (e.g., LB or autoinduction media), which cannot be done using fluorescence microscopy due to the high autofluorescence of media components. It should be noted that flow cytometry does not track c-di-GMP levels of an individual cell over time, because cells are discarded after they are flowed through the analysis chamber. However, this can also be considered an advantage, because cells are not repeatedly subjected to light excitation and media changes. Instead, single-cell fluorescence of a cell population is analyzed by flow cytometry over time. When used in combination with our fluorescent biosensors, analysis over time enables monitoring the dynamics of c-di-GMP levels.

In this section, we describe a representative experiment in which the Pl-B biosensor was used to monitor cellular c-di-GMP increase upon Zn(II) depletion (Fig. 5) [25], which is dependent on the zinc-sensing diguanylate cyclase DgcZ in *E. coli* [26]. One critical aspect to performing the depletion experiment is the preparation of spent media, which is required to avoid changing media composition when cells are necessarily diluted for flow cytometry analysis. The preparation instructions for spent media are provided below (*see* Fig. 6 for schematic illustration). This method can be applied to monitor the effect of other input compounds added to the media.

3.3.1 Preparation of Initial Cultures

1. To prepare cells expressing biosensor, add ~50 ng of Pl-B in pET31b plasmid to BL21(DE3) Star *E. coli* chemically competent cells and transform the plasmids following the manufacturer's protocol. Also, prepare positive control (Spinach2 in pET31b vector) and negative control (empty pET31b vector) cells for setting up flow cytometry parameters.
2. Plate cells on LB/Carb agar plates and incubate for 12–16 h at 37 °C. Plates can be stored at 4 °C up to 2 weeks (*see* **Note 4**).

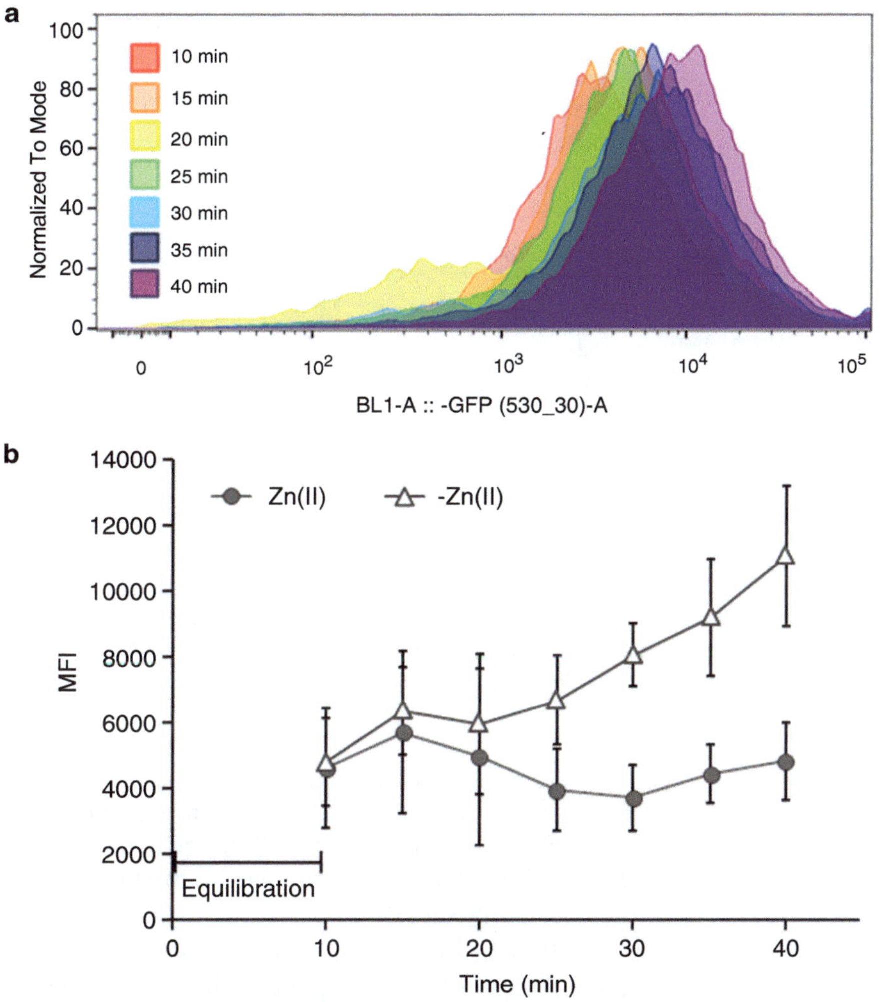

Fig. 5 Flow cytometry analysis of cell population dynamics of c-di-GMP. (**a**) Representative flow cytometry histograms for wild-type BL21(DE3) Star cells expressing PI-B biosensor upon Zn(II) depletion. (**b**) Dynamic changes in single-cell fluorescence of BL21(DE3) Star cells expressing c-di-GMP biosensor in the presence of Zn(II) (+ Zn(II), closed symbols) and upon Zn(II) depletion at time 0 (– Zn(II), open symbols). MFIs were obtained at indicated time points with flow cytometry. *Error bars* represent the standard deviations for the results from three independent biological replicates

3. Inoculate 3 mL AI/Carb with a single colony of the transformed cells (*see* **Note 1**). Shake cell cultures at 37 °C for 2 h.
4. Split each cell cultures into two tubes (1.5 mL each). Add 1.5 μL of freshly prepared 1 M $ZnCl_2$ stock solution to one of the tubes to a final concentration of 1 mM. Shake cell cultures for additional 18–24 h until cultures reach an $OD_{600} > 3$.

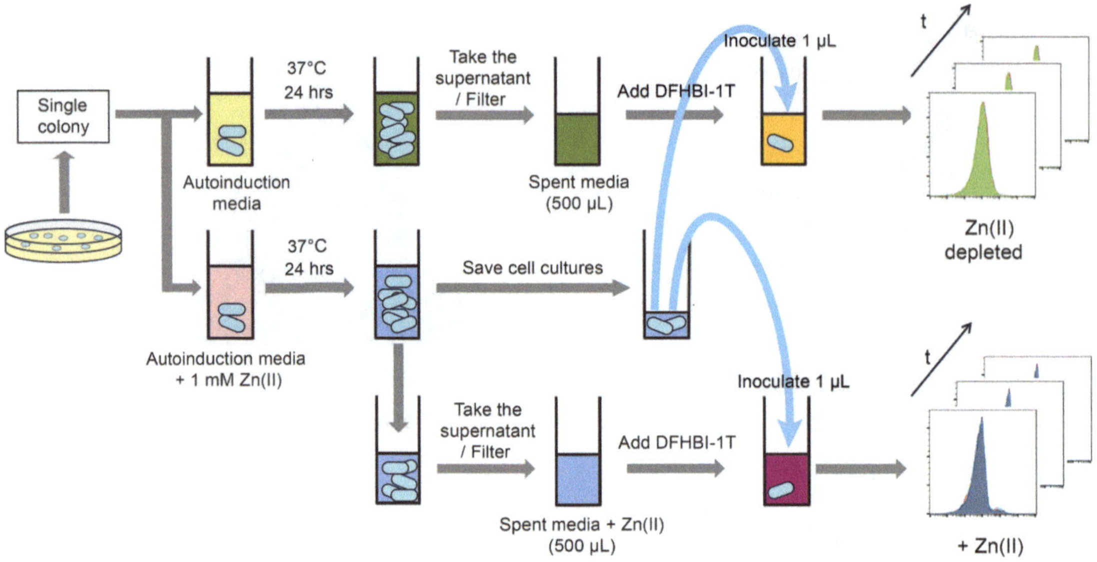

Fig. 6 Experimental schematic for the Zn(II) depletion experiment. After overnight growth in autoinduction media for biosensor expression, cells need to be diluted for flow cytometry experiments. To avoid changing media composition and centrifugation, cell cultures are simply diluted into spent media, which is prepared by sterile filtration of the culture supernatant from the same overnight cultures, with the addition of DFHBI-1T, the dye used by the biosensor

3.3.2 Preparation of Spent Media

1. Take 1.2 mL of each cell culture (from Subheading 3.3.1, **step 4**) and centrifuge at 2800 × *g* for 5 min at room temperature. Make sure to save remaining 0.3 mL cell culture containing Zn (II) at 37 °C. The cell culture will be further used in Subheading 3.3.3, **steps 3–4**.
2. Carefully take the supernatant and filter through a 0.22 μm sterile filter to obtain the spent media with and without Zn(II).
3. Supplement each spent media with DFHBI-1T to the final concentration of 25 μM. Keep the media away from light.

3.3.3 Flow Cytometry Experiments

1. Start up the flow cytometer and do test runs with PBS-DFHBI-1T solution with negative control and positive control cells to establish the forward scatter (FSC) and side scatter (SSC) regions and to optimize the voltage gain settings as described in Subheading 3.1.3. In this experiment, the negative control refers to cells expressing empty pET31b vector and the positive control to cells expressing Spinach2 in pET31b vector.
2. For Zn(II)-depletion experiment, add 1 μL of cells grown with Zn(II) from cell culture saved from Subheading 3.3.2, **step 1** to 500 μL spent media-DFHBI-1T solution prepared in Subheading 3.3.2, **step 3** without Zn(II) and start timer (*see* **Note 12**).
3. For no-depletion control experiment, add 1 μL of cells grown with Zn(II) saved from Subheading 3.3.2, **step 1** to 500 μL

spent media-DFHBI-1T solution with Zn(II), which was obtained by sterile filtration of the same culture in Subheading 3.3.2, **step 3** (*see* **Note 13**).

4. Shake the diluted cultures at 37 °C for 5–9 min to allow equilibration with DFHBI-1T.
5. Take 70 μL aliquots from each sample and analyze by flow cytometry. These samples will represent the first time-point (=10 min).
6. Take 70 μL aliquots every 5 min and repeat the analysis for up to 40 min.
7. Obtain MFI values at each time point using FlowJo software as described in Subheading 3.1.5.

4 Notes

1. To test independent biological replicates, pick different single colonies to inoculate. It is recommended to analyze at least three biological replicates.
2. The PBS buffer stock should be filter sterilized before use. It is highly recommended to prepare fresh PBS-DFHBI-1T solution each time. The approximate total volume of the solution needed is ~70 μL for each independent sample in the experiment. DFHBI can be used in place of DFHBI-1T.
3. The steps described here are based on the software specific to the Attune instrument. The software for different flow cytometry instruments may have different instructions for setup.
4. Do not use colonies from old plates. Old colonies from high overexpression strains such as BL21(DE3) Star *E. coli* tend to result in flow cytometry histogram with multiple humps and exhibit low fluorescence intensity in flow cytometry analysis. Similarly, we have seen variable results from frozen glycerol stocks. Preparing fresh transformants on agar plates also helps to obtain biological replicates as described in **Note 1**.
5. This procedure is suitable for facultative anaerobic bacteria, or may be adapted for obligate anaerobic bacteria using a glove box.
6. For facultative anaerobic bacteria, an alternative to argon sparging as described below to establish an anaerobic culture is to shake the sealed cultures at 37 °C for 16 h or longer so that residual oxygen gets rapidly used up by the bacteria. To assist with this, more media can be added to reduce the headspace in the tube.
7. Usually, it is difficult to cap the tubes properly just by pushing the stoppers down with fingers. Instead, first push down the

stopper gently and then rotate the stopper as if screwing closed the lid of a water bottle.

8. If the aluminum seal crimps loosely, remove the seal and repeat crimping using a new aluminum seal. If the problem persists, uncap/recap the rubber stopper and retry crimping.

9. Since *E. coli* cells tend to grow slower in anaerobic conditions, longer incubation time than 18 h may be needed to reach proper culture density for induction of expression. In trial experiments, control cultures can be prepared in Balch tubes and opened to check the incubation time required to reach the proper OD_{600}.

10. Because it is important to rapidly read the fluorescence before bacteria are exposed to oxygen, the seals must be broken individually (in other words, one tube at a time) before flow cytometry analysis. Do not break seals of two or more tubes at the same time. Alternatively, it may be possible to sample the culture through the stopper with a gas-tight syringe equipped with a very long needle.

11. You can also monitor dynamics during oxygen recovery by taking 1 μL of each sample at various time points and analyzing with the flow cytometer.

12. With the 1:500 dilution into autoinduction media without Zn (II) supplementation, cells change from a growth environment with 1 mM Zn(II) to less than or equal to 2 μM Zn(II). Media components may contribute additional trace amount of Zn (II).

13. To monitor c-di-GMP dynamics upon additionreporter) of an input compound instead of depletion, add the compound after dilution into spent media then start the timer. For example, this can be achieved by adding cells grown without Zn(II) saved from Subheading 3.3.2, **step 1** to 500 μL spent media-DFHBI-1T solution with Zn(II). This dilution procedure avoids needing to centrifuge the cells to be analyzed, and is necessary to obtain sample cell density in the proper range (~10^6 cells/mL) for flow cytometry.

Acknowledgment

The work on which this chapter is based was supported by NIH grant DP2 OD008677 (to M.C.H.). The authors thank Zachary Hallberg for assistance with figures.

References

1. Romling U, Galperin MY, Gomelsky M (2013) Cyclic di-GMP: the first 25 years of a universal bacterial second messenger. Microbiol Mol Biol Rev 77(1):1–52. doi:10.1128/MMBR.00043-12
2. Tischler AD, Camilli A (2004) Cyclic diguanylate (c-di-GMP) regulates vibrio cholerae biofilm formation. Mol Microbiol 53(3):857–869. doi:10.1111/j.1365-2958.2004.04155.x
3. Spangler C, Bohm A, Jenal U, Seifert R, Kaever V (2010) A liquid chromatography-coupled tandem mass spectrometry method for quantitation of cyclic di-guanosine monophosphate. J Microbiol Methods 81(3):226–231. doi:10.1016/j.mimet.2010.03.020
4. Stelitano V, Brandt A, Fernicola S, Franceschini S, Giardina G, Pica A, Rinaldo S, Sica F, Cutruzzola F (2013) Probing the activity of diguanylate cyclases and c-di-GMP phosphodiesterases in real-time by CD spectroscopy. Nucleic Acids Res 41(7):e79. doi:10.1093/nar/gkt028
5. Kellenberger CA, Sales-Lee J, Pan Y, Gassaway MM, Herr AE, Hammond MC (2015) A minimalist biosensor: quantitation of cyclic di-GMP using the conformational change of a riboswitch aptamer. RNA Biol 12(11):1189–1197. doi:10.1080/15476286.2015.1062970
6. Roelofs KG, Wang J, Sintim HO, Lee VT (2011) Differential radial capillary action of ligand assay for high-throughput detection of protein-metabolite interactions. Proc Natl Acad Sci U S A 108(37):15528–15533. doi:10.1073/pnas.1018949108
7. Wolfe AJ, Berg HC (1989) Migration of bacteria in semisolid agar. Proc Natl Acad Sci U S A 86(18):6973–6977
8. O'Toole GA, Pratt LA, Watnick PI, Newman DK, Weaver VB, Kolter R (1999) Genetic approaches to study of biofilms. Methods Enzymol 310:91–109
9. Koestler BJ, Waters CM (2014) Bile acids and bicarbonate inversely regulate intracellular cyclic di-GMP in vibrio cholerae. Infect Immun 82(7):3002–3014. doi:10.1128/IAI.01664-14
10. Ho CL, Chong KS, Oppong JA, Chuah ML, Tan SM, Liang ZX (2013) Visualizing the perturbation of cellular cyclic di-GMP levels in bacterial cells. J Am Chem Soc 135(2):566–569. doi:10.1021/ja310497x
11. Christen M, Kulasekara HD, Christen B, Kulasekara BR, Hoffman LR, Miller SI (2010) Asymmetrical distribution of the second messenger c-di-GMP upon bacterial cell division. Science 328(5983):1295–1297. doi:10.1126/science.1188658
12. Kellenberger CA, Wilson SC, Sales-Lee J, Hammond MC (2013) RNA-based fluorescent biosensors for live cell imaging of second messengers cyclic di-GMP and cyclic AMP-GMP. J Am Chem Soc 135(13):4906–4909. doi:10.1021/ja311960g
13. Kellenberger CA, Hallberg ZF, Hammond MC (2015) Live cell imaging using Riboswitch-Spinach tRNA fusions as metabolite-sensing fluorescent biosensors. Methods Mol Biol 1316:87–103. doi:10.1007/978-1-4939-2730-2_8
14. Kellenberger CA, Hammond MC (2015) In vitro analysis of riboswitch-Spinach aptamer fusions as metabolite-sensing fluorescent biosensors. Methods Enzymol 550:147–172. doi:10.1016/bs.mie.2014.10.045
15. Paige JS, Wu KY, Jaffrey SR (2011) RNA mimics of green fluorescent protein. Science 333(6042):642–646. doi:10.1126/science.1207339
16. Paige JS, Nguyen-Duc T, Song W, Jaffrey SR (2012) Fluorescence imaging of cellular metabolites with RNA. Science 335(6073):1194. doi:10.1126/science.1218298
17. Wang XC, Wilson SC, Hammond MC (2016) Next-generation RNA-based fluorescent biosensors enable anaerobic detection of cyclic di-GMP. Nucleic Acids Res 44(17):e139. doi:10.1093/nar/gkw580
18. Song W, Strack RL, Svensen N, Jaffrey SR (2014) Plug-and-play fluorophores extend the spectral properties of Spinach. J Am Chem Soc 136(4):1198–1201. doi:10.1021/ja410819x
19. Hallberg ZF, Wang XC, Wright TA, Nan B, Ad O, Yeo J, Hammond MC (2016) Hybrid promiscuous (Hypr) GGDEF enzymes produce cyclic AMP-GMP (3′, 3′-cGAMP). Proc Natl Acad Sci U S A 113(7):1790–1795. doi:10.1073/pnas.1515287113
20. Tuckerman JR, Gonzalez G, Sousa EH, Wan X, Saito JA, Alam M, Gilles-Gonzalez MA (2009) An oxygen-sensing diguanylate cyclase and phosphodiesterase couple for c-di-GMP control. Biochemistry 48(41):9764–9774. doi:10.1021/bi901409g
21. Lacey MM, Partridge JD, Green J (2010) *Escherichia coli* K-12 YfgF is an anaerobic cyclic di-GMP phosphodiesterase with roles in cell surface remodelling and the oxidative stress response. Microbiology 156(Pt 9):2873–2886. doi:10.1099/mic.0.037887-0
22. An S, Wu J, Zhang LH (2010) Modulation of Pseudomonas Aeruginosa biofilm dispersal by a

cyclic-di-GMP phosphodiesterase with a putative hypoxia-sensing domain. Appl Environ Microbiol 76(24):8160–8173. doi:10.1128/AEM.01233-10

23. Reid BG, Flynn GC (1997) Chromophore formation in green fluorescent protein. Biochemistry 36(22):6786–6791. doi:10.1021/bi970281w
24. Craggs TD (2009) Green fluorescent protein: structure, folding and chromophore maturation. Chem Soc Rev 38(10):2865–2875. doi:10.1039/b903641p
25. Yeo J, Dippel, AB, Wang, XC, Hammond, MC Manuscript in preparation
26. Zahringer F, Lacanna E, Jenal U, Schirmer T, Boehm A (2013) Structure and signaling mechanism of a zinc-sensory diguanylate cyclase. Structure 21(7):1149–1157. doi:10.1016/j.str.2013.04.026

Part IV

Indirect Detection of c-di-GMP Levels

Chapter 11

Experimental Detection and Visualization of the Extracellular Matrix in Macrocolony Biofilms

Diego O. Serra and Regine Hengge

Abstract

By adopting elaborate three-dimensional morphologies that vary according to their extracellular matrix composition, macrocolony biofilms offer a unique opportunity to interrogate about the roles of specific matrix components in shaping biofilm architecture. Here, we describe two methods optimized for *Escherichia coli* that profit from morphology and the high level of structural organization of macrocolonies to gain insight into the production and assembly of amyloid curli and cellulose—the two major biofilm matrix elements of *E. coli*—in biofilms. The first method, the macrocolony morphology assay, is based on the ability of curli and cellulose—either alone or in combination—to generate specific morphological and Congo Red-staining patterns in *E. coli* macrocolonies, which can then be used as a direct visual readout for the production of these matrix components. The second method involves thin sectioning of macrocolonies, which along with in situ staining of amyloid curli and cellulose and microscopic imaging allows gaining fine details of the spatial arrangement of both matrix elements inside macrocolonies. Beyond their current use with *E. coli* and related curli and cellulose-producing *Enterobacteriaceae*, both the methods offer the potential to be adapted to other bacterial species.

Key words Amyloid, Bacteria, Biofilm, C-di-GMP, Cellulose, Congo red, Curli, *Escherichia coli*, Exopolysaccharide, Thioflavine

1 Introduction

To form structurally complex communities known as biofilms, microorganisms rely on a self-produced extracellular matrix composed of secreted proteins, amyloid fibers, exopolysaccharides, and extracellular DNA collectively known as extracellular polymeric substances (EPS) [1]. Due to its direct involvement in promoting surface adhesion and cell cohesion and because of its contribution to the strong resistance of biofilms against antibiotics and multiple stresses [1], EPS have become a major focus of research in the biofilm field. In particular, several recent advances in understanding the role of EPS in shaping the biofilm architecture have come from studies done with macrocolonies [2–7], which are air-exposed biofilms growing for extended times on nutrient-providing agar.

Karin Sauer (ed.), *c-di-GMP Signaling: Methods and Protocols*, Methods in Molecular Biology, vol. 1657,
DOI 10.1007/978-1-4939-7240-1_11, © Springer Science+Business Media LLC 2017

As a remarkable feature, macrocolony biofilms can adopt intricate morphologies commonly known as "wrinkled," "rugose," or "rdar" (for red, dry, and rough) that include distinct morphological form elements such as "ridges," "wrinkles," and "rings" [3, 7–10]. It is now clear that these morphological phenotypes depend on specific EPS components [3, 11–13]. This relationship is the underlying principle of agar-based macrocolony morphology assays, which have been particularly exploited in the case of *E. coli* and *Salmonella* to qualitatively score the production of amyloid curli and the exopolysaccharide cellulose, the two major EPS elements in macrocolony biofilms of these bacteria [3, 4, 7, 14–17]. When produced alone or in combination within macrocolonies, curli and cellulose can generate visually distinctive morphological patterns (Fig. 1) [3, 7]. The macrocolony morphology assay additionally takes advantage of Congo Red (CR), a diazo dye added to the agar medium that binds to both curli and cellulose with no noticeable influence on bacterial growth or biofilm morphogenesis [12, 18–20]. Due to differences in its absorbance spectrum upon curli or cellulose binding, CR leads to specific color patterns on macrocolonies depending on whether these components are being produced together or alone or are absent from the macrocolonies (Fig. 1). Most commonly, Coomassie brilliant blue (CB) is also included in the assay [4, 20, 21], i.e., a dye that stains proteins and thus can enhance color discrimination between curli and cellulose.

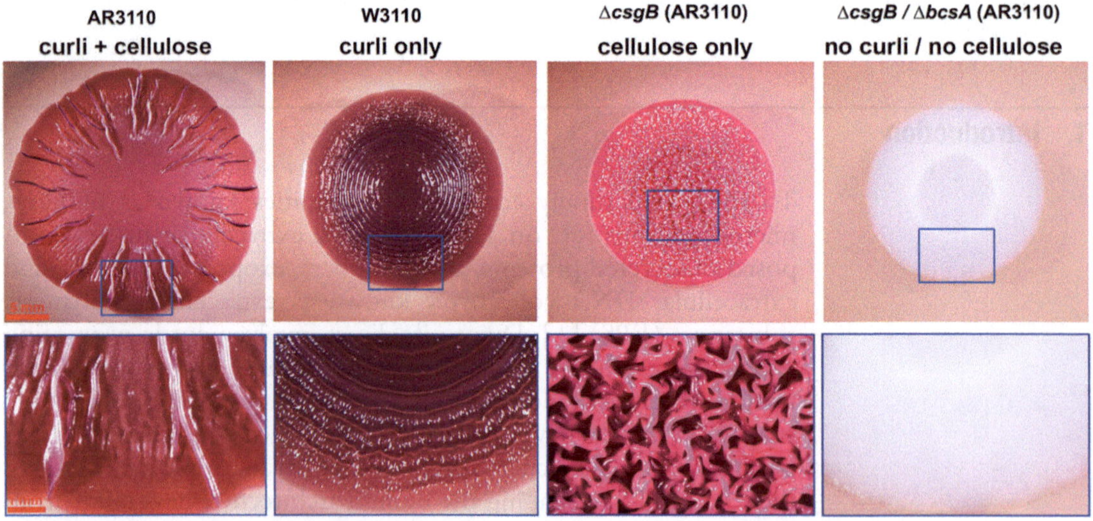

Fig. 1 Top view of 5-day-old macrocolonies of *E. coli* K-12 strains W3110, AR3110 and derivative mutants displaying distinctive morphological and CR/CB staining patterns based on the presence and/or absence of amyloid curli and cellulose. Macrocolonies were set on CR/CB-supplemented NaCl-free LB plates using the protocols described in Subheading 3.1. *Lower panels* show enlarged views of boxed areas in the macrocolonies in the *top row*

Since the macrocolony morphology assay can easily be scaled up by using large or multiple agar plates, it can also be used for large screens of mutant libraries or collections of clinical/environmental isolates. On the other hand, it should be advantageous beyond its current use with *E. coli*, *Salmonella* or other curli and cellulose-producing *Enterobacteriaceae*. Based on the ability of CR to bind amyloid fibers and to several (though not all) exopolysaccharides, the assay has been successfully adapted to the analysis of EPS components in macrocolonies of other bacteria, e.g., the amyloid TasA and the exopolysaccharide of *B. subtilis* [11] or the PEL and PSL polysaccharides of *Pseudomonas aeruginosa* strains [8, 13]. However, due to the ability of CR to bind to multiple EPS constituents care must be taken when assigning CR an indicator role for particular matrix elements. Thus, appropriate controls such as mutants lacking particular matrix components must be included in the assay.

Since macrocolony biofilms exhibit a remarkably high level of structural organization at the microscale, they are a highly valuable model system to study not only the production but also the three-dimensional architectural assembly of EPS within a biofilm [4, 22]. This was shown in *B. subtilis* [6] and more recently in *E. coli* where it was possible to observe in fine detail the specific spatial arrangement of curli and cellulose inside macrocolonies (Fig. 2e, f) [3, 4, 22]. Such close inspection into the macrocolony anatomy was possible by the combination of cryosectioning and microscopic imaging. While cryosectioning was applied quite early in the context of biofilm research [23], its use to studying macrocolonies started relatively recently only [3–6, 24]. By performing thin cuttings across the macrocolony, the main restriction that macrocolonies pose to most optical microscopy techniques, i.e., their massive compactness, can be overcome.

Analysis of EPS elements in macrocolony cross-sections can also be facilitated by the use of fluorescent dyes that allow in situ staining of matrix components. Fluorescent dyes that stain curli and cellulose with variable affinities and different fluorescent colors and that can be incorporated into the agar medium, are listed in Table 1. Thioflavin S (TS), for instance, shows strong green fluorescence upon binding to curli and cellulose, thereby allowing the display of fine details of the spatial arrangement of both elements in cross-sections of *E. coli* macrocolonies (Fig. 2e, f) [3, 4]. While in situ staining simplifies sample preparation, the fact that cryosections are mounted on glass slides offers the possibility of using them for a variety of approaches, including immunostaining or in situ hybridization. In addition, the cryosectioning/microscopy approach can directly be applied to visually examine spatially controlled gene expression in macrocolony biofilms by using strains expressing fluorescent reporter gene fusions (e.g., to Gfp, Yfp, or mCherry).

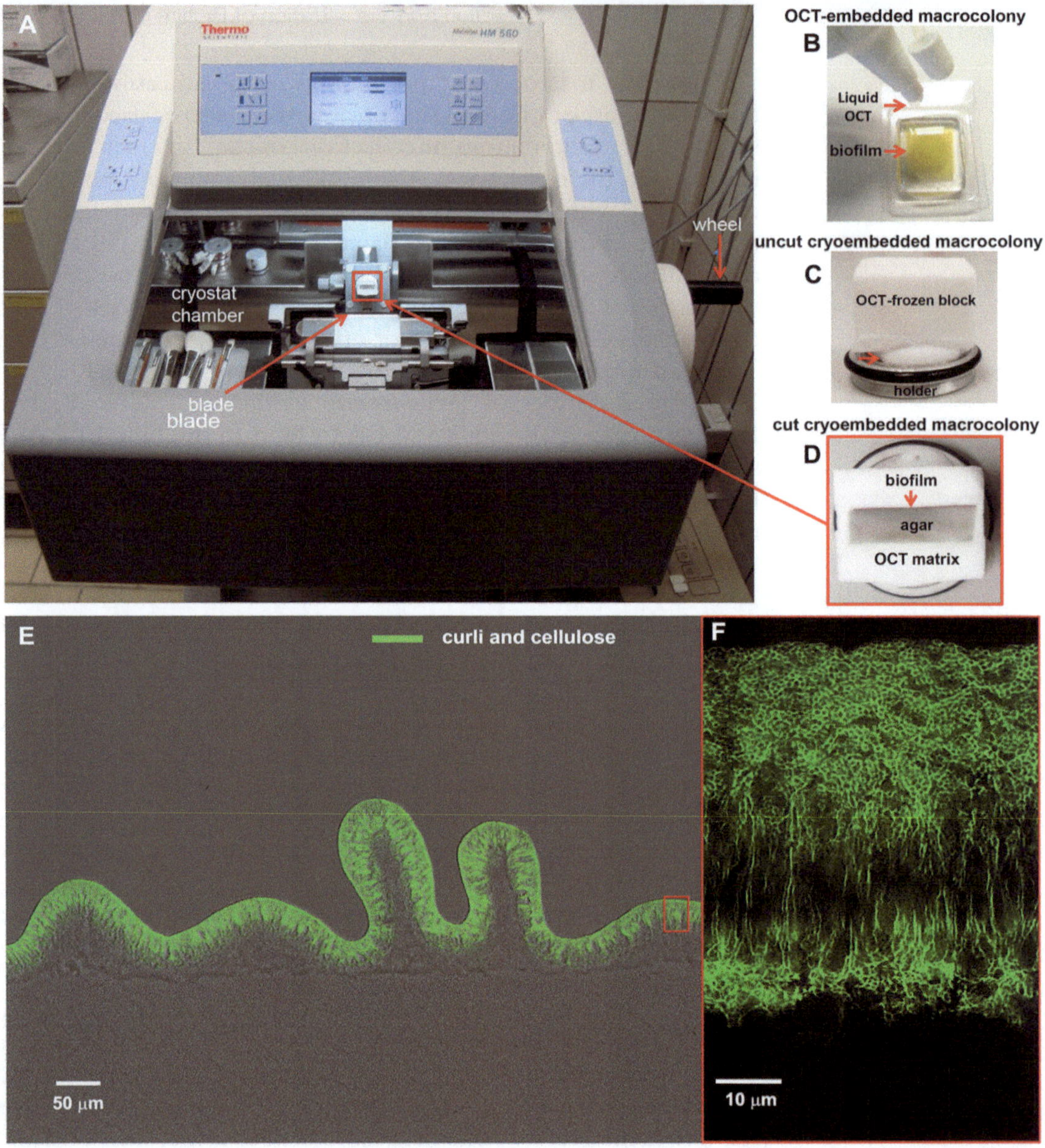

Fig. 2 (a) View of a HM560 cryostat (Thermo Scientific Microm) and its work stage. The image shows a cross-sectioned OCT block with a cryoembedded macrocolony placed in the sample stage of the cryostat. (**b**) View of a macrocolony placed in a Tissue-Tek mold embedded with liquid OTC. (**c**) View of an uncut OCT-frozen block (i.e., the cryoembedded macrocolony) attached to a specimen holder. (**d**) View of a cross-sectioned OCT block with a cryoembedded macrocolony. (**e**) Bright-field/fluorescence merged image of a 5-μm-thin vertical section of an AR3110 macrocolony grown on NaCl-free LB agar plates supplemented with TS. The section was prepared following the protocol described in Subheading 3.2 and visualized at low magnification by overlaying bright-field and fluorescence microscopic images. Green fluorescence in the merged image corresponds to TS-stained amyloid curli and cellulose, which are located exclusively in the top layer of the macrocolony, as previously reported [3, 4]. Due to its tissue-like properties, the top macrocolony layer can fold and buckle up giving rise to the emergence of wrinkles or ridges. (**f**) Enlarged view of the TS fluorescence pattern revealing fine details of the spatial arrangement of amyloid curli and cellulose in the top layer of a flat area of the macrocolony. The upper zone of the top layer depicts a "honeycomb"-like pattern associated with the presence of curli and cellulose, while the lower zone shows filamentous arrangements more characteristic of cellulose only [3]

Table 1
Fluorescent dyes assayed for in situ detection of amyloid curli and cellulose in *E. coli* macrocolonies or macrocolony cross-sections

Fluorescent dyes	Final Conc. (in agar)	Stained Matrix Component	Sample	Visualization mode/Readout	Reference
Thioflavin S (TS)	40 μg/mL	Curli and cellulose	Macrocolony cross-sections	Fluorescence microscopy/green fluorescence upon curli and cellulose binding (TS excitation 430 nm, emission 550 nm)	[3, 4]
Calcofluor White (CW)	100–250 μg/mL	Cellulose (preferentially) and curli (to a minor extent)	Agar-based macrocolonies	UV transillumination/light blue fluorescence when cellulose is produced; background fluorescence when curli only is produced.	[3, 12, 16]
			Macrocolony cross-sections	Fluorescence microscopy/bright blue fluorescence upon cellulose binding; attenuated blue fluorescence upon curli binding (CW excitation 370 nm, emission 440 nm)	
Pontamine fast scarlet 4b (P4b)	40 μg/mL	Cellulose (preferentially) and curli	Macrocolony cross-sections	Fluorescence microscopy/bright red fluorescence upon cellulose binding; red fluorescence upon curli binding (P4b excitation 560 nm, emission >600 nm)	Serra and Hengge (unpublished data)
Congo red (CR)	40 μg/mL	Curli and cellulose	Macrocolony cross-sections	Fluorescence microscopy/red fluorescence upon curli and cellulose binding (CR excitation 497 nm, emission >600 nm)	Serra and Hengge (unpublished data)

In this chapter, we describe the macrocolony morphology assay, which despite being conceptually relatively simple frequently leads to variable results that affect intra- and inter-laboratory reproducibility. In particular, we describe the procedure optimized for *E. coli* using K-12 W3110, which produces curli only, and AR3110, a W3110 derivative with restored capacity to produce cellulose. Like many other *E. coli* strains, W3110 and AR3110 produce curli and/or cellulose at temperatures below 30 °C, optimally on agar plates without salt [3, 4]. Finally, we describe the technical procedure to prepare macrocolony thin sections based on work performed with W3110 and AR3110 macrocolonies.

2 Materials

Prepare all solutions using bidestilled water and analytical grade reagents. Diligently follow all safety recommendations when handling toxic or potentially harmful reagents as indicated by manufacturers as well as all waste disposal regulations when disposing waste materials.

2.1 Materials Used in the Macrocolony Morphology Assay

1. Sodium chloride-free LB agar medium: weigh 10 g bacto trpytone, 5 g yeast extract, and 18 g bacto agar and transfer them to the glass media storage bottle. Make up to 1 L with water. Sterilize by autoclaving. Keep stirring in a heat block at 46–48 °C until pouring (*see* **Note 1**).
2. Congo Red (CR)/Coomassie Brilliant Blue G (CB) stock solution (CR/CB): weigh 200 mg CR and 100 mg CB and transfer them to a graduated cylinder. Add 100 mL 70% ethanol and dissolve by stirring. Sterilize by filtration using membrane filters with a pore size of 0.22 μm (*see* **Note 2**). Store protected from light at 4 °C.
3. Thioflavin S (TS) stock solution: weigh 200 mg TS and transfer it to a graduated cylinder. Add 100 mL 70% ethanol and dissolve by stirring. Sterilize by filtration as indicated in **item 2**. Store protected from light at 4 °C.
4. Calcofluor white (CW): weigh 500 mg CW (e.g., Fluorescent Brightener 28; Sigma) and transfer it to a graduated cylinder. Add 100 mL of water and dissolve by stirring. Sterilize by filtration as indicated in **item 2**. Store protected from light at 4 °C.
5. Pontamine fast scarlet 4b (P4b) stock solution: weigh 200 mg P4b and transfer to a graduated cylinder. Add 100 mL water and dissolve by stirring. Sterilize by filtration as indicated in **item 2**. Store protected from light at 4 °C.
6. Standard (94 × 16 mm) disposable petri dishes. Alternatively, small (35 × 10 mm) or large (135 × 20 mm) petri dishes can be used (*see* **Note 3**).

7. Visualization and image recording: the use of a stereomicroscope coupled to a digital camera is recommended. Alternatively, a digital camera with macro function can be used.

2.2 Materials Used for Thin-Sectioning Macrocolony Biofilms

2.2.1 Cryoembedding of Macrocolonies

1. Fixative: 10% Neutral-Buffered Formalin. Alternatively, 4% paraformaldehyde or 2% glutaraldehyde can be used.
2. Phosphate-Buffered Saline (PBS): weigh 8 g NaCl, 0.2 g KCl, 1.44 g Na_2HPO_4, and 0.24 g KH_2PO_4 and transfer to a graduated cylinder. Add water to a volume of 800 mL and dissolve by stirring. Adjust pH to 7.4. Make up to 1 L with water. Sterilize by autoclaving. Store at 4 °C.
3. Scalpel.
4. Tweezers.
5. Tissue-Tek cryomold (Sakura, 25 mm × 20 mm × 5 mm).
6. Silicon ice cube cryomold (25 mm × 25 mm × 20 mm).
7. Thermo-Flask liquid nitrogen.
8. Ethanol 70%.
9. Dry ice. To prepare the dry ice/ethanol bath, add pieces of dry ice to the Thermo flask. Then, carefully add the 70% ethanol solution to the flask until it covers the dry ice. Wear gloves during this procedure.
10. Liquid nitrogen (alternative to dry ice/ethanol).
11. Cryoembedding agent: Tissue-Tek optimum cutting temperature (OCT) compound (Sakura).

2.2.2 Thin-Sectioning of Cryoembedded Macrocolonies

1. Disposable cryotome blades (e.g., Thermo Scientific Microm SEC35p).
2. Cryostat Specimen Holders.
3. Superfrost Microscope Slides (Thermo Scientific).
4. Tissue-Tek OCT compound.
5. Cryostat (e.g., HM 560, Thermo Scientific Microm).

3 Methods

3.1 Macrocolony Morphology Assay

The assay described here has been optimized for assessing curli and cellulose production by *E. coli* strains. However, it can also be used with other curli and cellulose-producing *Enterobacteriaceae* and it can be adapted to other bacterial species. Besides temperature, which—depending on the *E. coli* strain background—can greatly influence morphological patterns in macrocolonies [15, 17], other factors such as ionic strength, osmolarity, availability of nutrients in the medium and humidity in the atmosphere can also sensibly modify macrocolony architecture. Therefore, to achieve reproducibility of morphological and CR/CB-based staining patterns of macrocolonies, the multiple steps described below have to be performed as precise and exactly reproducible as possible (with particular attention to **Notes 4–10**).

3.1.1 Pouring NaCl-Free LB Agar Plates Supplemented with CR/CB (or Alternative Dyes)

Before starting, clean the work area on the laboratory bench with disinfectant to minimize possible contamination. Set up a Bunsen burner and work carefully within the sterile field area created by the updraft of the flame.

1. Supplement 1 L of NaCl-free LB agar medium with 20 mL of CR/CB stock solution (*see* **Note 4**). Mix well avoiding the formation of air bubbles.
2. Pour 25 mL of CR-containing NaCl-free LB agar medium into each standard petri dish (94 × 16 mm) (*see* **Note 5**).
3. Keep the plates on the bench until the agar is completely solidified. Store the plates at 20–25 °C. Agar plates are typically prepared 1 day before inoculation (*see* **Note 6**).

This procedure is equally applicable to the preparation of agar media supplemented with other curli/cellulose staining dyes as those listed in Table 1 and for whom recipes are also described in Subheading 2.1.

3.1.2 Setting Up Macrocolonies

When working with bacteria set up a Bunsen burner and work carefully within the sterile field area created by the updraft of the flame.

1. Pick single colonies of *E. coli* strains from fresh stock plates and inoculate LB liquid cultures. Grow the bacteria at 37 °C overnight (*see* **Note** 7).
2. Set up macrocolonies by inoculating 5 μL of overnight *E. coli* cultures on the agar of NaCl-free LB plates supplemented with CR/CB (*see* **Note 8**). Allow the inocula to dry for 15–20 min.
3. Seal the petri dishes by wrapping Parafilm around the dish. Incubate at 26–28 °C for up to 5 days or longer if necessary (*see* **Notes 9** and **10**).
4. Check visually every day the morphology of the macrocolony biofilms. For visualization and image recording it is recommended to use a stereomicroscope coupled to a digital camera. Alternatively, a digital camera with macro mode can be used. Figure 1 shows examples of morphological and CR/CB staining patterns in macrocolonies of *E. coli* K-12 strains W3110, AR3110 and derivatives after 5 days of incubation.

3.2 Thin-Sectioning of Macrocolony Biofilms

The following protocol has been used to prepare cryosections of *E. coli* W3110 and AR3110 macrocolony biofilms grown in NaCl-free LB agar plates supplemented with the fluorescent dyes listed in Table 1 or macrocolonies of these strains harboring GFP reporter fusions [3, 4].

3.2.1 Cryoembedding of Macrocolonies

Before starting, clean the work area on the laboratory bench. Work under a fume hood when handling fixatives. Wear proper gloves and goggles when handling dry ice/ethanol or liquid nitrogen.

1. Fix single macrocolony biofilms directly on the agar plate. To do so, add 10% Neutral-Buffered Formalin to the agar plate containing the macrocolony so that the fixative overlays the entire biofilm (*see* **Note 11**). Allow it to fix for 30 min.
2. Remove the fixative and wash the macrocolony biofilm four times with PBS buffer (*see* **Note 12**). Remove excess of PBS from the plate after the last wash.
3. Cut out from the plate an agar block containing the macrocolony biofilm and transfer it into a cryomold (4557 Tissue-Tek) with the macrocolony in position right-side up (*see* **Note 13**).
4. Embed the macrocolony by overlaying it along with the underlying agar with Tissue-Tek OCT compound (*see* Fig. 2b and **Note 14**).
5. Snap-freezing OCT-embedded macrocolony in dry ice/ethanol bath (or alternatively in liquid nitrogen): using tweezers, bring the cryomold holding the OCT-embedded macrocolony into contact with the fluid (ethanol or alternatively liquid nitrogen) (*see* **Note 15**).
6. Remove the OCT-frozen block from the mold by pressing out the mold so that the block detaches and releases. Store at −80 °C. At this step the macrocolony is cryoembedded and ready for cryosectioning, as described in Subheading 3.2.2. Alternatively, a second embedding/freezing step can be performed to enlarge the thickness of the block (*see* **Note 16** and **following steps 7–10**).
7. Place the OCT-frozen block into a silicon ice cube mold that is at least 25 mm long, 20 mm wide, and 20 mm deep (*see* **Note 17**).
8. Fill the silicon mold holding the OCT-frozen block with liquid Tissue-Tek OCT compound.
9. Snap-freeze the OCT-embedded block in dry ice/ethanol bath (or alternatively in liquid nitrogen). Using tweezers bring the silicon mold in OCT into contact with the fluid.
10. Remove the OCT-frozen block from the mold. Press out the silicon mold so that the block detaches and releases. Store at −80 °C (*see* **Note 18**).

3.2.2 Thin-Sectioning of Cryoembedded Macrocolonies

1. Set the cryostat chamber temperature at −20 °C and place a new disposable cryotome blade (e.g., Microm SEC35p) in the cryostat blade holder. Figure 2a shows a view of a HM560 cryostat with the chamber open.
2. Remove the OCT-frozen block (i.e., the cryoembedded macrocolony) from the −80 °C freezer, bring it inside the cryostat chamber and allow it to equilibrate to the cryostat chamber temperature for approximately 10 min.
3. Mount the OCT-frozen block on a cryostat specimen holder using OCT compound as a mounting medium (*see* arrow in Fig. 2c) and place the holder on the sample stage of the cryostat.
4. Cut 5-μm-thick sections across the macrocolony at −20 °C (*see* **Note 19**). Figure 2d shows the example of a cross-sectioned OCT block with a cryoembedded macrocolony.
5. Transfer the thin section to a room temperature Superfrost microscope slide by touching the slide to the section.
6. Allow the thin section to air-dry. Sections can be stored in a sealed slide box at −80 °C for later use.
7. Sections can be further processed for other applications (e.g., immunostaining) or prepared for imaging by bright-field/fluorescence microscopy (Fig. 2e, f). For imaging, add mounting medium (e.g., ProLong 9 Gold antifade reagent, Thermo Scientific) on the slide and place a coverslip over it. Avoid the formation of air bubbles.

4 Notes

1. The absence of NaCl in LB medium favors maximal expression of curli and cellulose. In many *E. coli* strains high salt content reduces curli and cellulose production.
2. Make sure CR and CB are dissolved before filtering; otherwise CR and CB particles can block 0.22 μm membrane filters.
3. Small (e.g., 35 × 10 mm) petri dishes allow growing single macrocolonies and help to save expensive/precious additives. Large (e.g., 135 × 20 mm) petri dishes are useful for large screenings allowing setting up to 25 macrocolonies.
4. It is recommended to freshly prepare the agar medium, to sterilize and keep it (until use) pre-warmed at 45–48 °C under stirring conditions in a heat block. Handle the CR/CB stock solution carefully avoiding spills.
5. Use a sterile pipette, a graduated cylinder or a 50-falcon tube to dispense *precisely the same volume* of medium to each petri dish (this is crucial for final water content/humidity of the agar,

which is an essential parameter determining macrocolony morphology). Use 25 mL for standard petri dishes (94 × 16 mm). Alternatively, dispense 3 or 80 mL of medium when using small (35 × 10 mm) or large (135 × 20 mm) petri dishes, respectively. Since the presence of condensate water on the lid of the petri dish can later influence macrocolony morphology, it is desirable to minimize water condensation by pouring the agar when it is relatively cool, but still liquid.

6. Storage for one 1 day at 20–25 °C contributes to reduce humidity (condensed water) inside the petri dish, which favors the formation of macrocolony morphology patterns such as the concentric-ring pattern characteristic of curli-only producing strains like W3110 or radially oriented folds (ridges) in macrocolonies of curli and cellulose-producing strains such as AR3110 (Fig. 1).
7. Overnight cultures are directly used as inocula of macrocolony biofilms. Many *E. coli* strains do not produce curli or cellulose at 37 °C. Since curli and cellulose promote cell aggregation, reduced clumping is expected in liquid cultures of these strains at 37 °C compared to those at 26–28 °C. Low aggregation will generate a more homogeneous bacterial inoculum on the agar plate (*see* Δ*csgB*/Δ*bcsA*, Fig. 1).
8. Before inoculating the agar plates, check that there is no substantial water condensation on the lid of the petri dish. Otherwise, open the petri dish and let the agar and the lid to dry in proximity to the Bunsen burner.
9. Macrocolonies of *E. coli* strains typically develop their distinctive morphological patterns within 2–5 days. Sealing plates with parafilm is used to reduce dehydration of the agar over this extended incubation time. While the relatively dry atmosphere is necessary to obtain certain macrocolony morphologies (*see* **Note 6**), prolonged dehydration—which leads to concentration of media constituents—can be detrimental to morphogenesis.
10. As pointed out, many *E. coli* strains (e.g., W3110 and AR3110) produce curli and/or cellulose at temperatures below 30 °C (Fig. 1). Thus, to observe curli and cellulose-dependent morphologies on macrocolonies for those strains it is required to set the temperature of the incubator/room at 26–28 °C.
11. Fixation is recommended to stabilize and preserve the macrocolony biofilm structure. Alternative fixative formulations such as 4% paraformaldehyde or 2% glutaraldehyde can also be used.
12. Although fixation confers stability to the macrocolony structure, washing can still disturb this structure, especially if cohesiveness of the biofilm matrix elements is not strong enough. Therefore, it is recommended to perform the washes carefully.

13. The 4557 Tissue-Tek cryomold is 25 mm long, 20 mm wide, and 5 mm deep, so the final agar OCT-frozen block obtained after **step 5** will have equivalent dimensions.
14. The OCT compound is viscous at room temperature and miscible with H_2O, but freezes into a solid support at −20 °C. When overlaying the macrocolony with OCT, make sure no air bubbles are formed (Fig. 2b).
15. Avoid direct contact of liquid OCT with the freezing fluid as it may generate air bubbles.
16. By using the 4557 Tissue-Tek cryomold, the formed OCT-frozen block is 25 mm long, 20 mm wide and 5–10 mm thick. Since this block is relatively thin, a second embedding/freezing step can be used to enlarge its thickness.
17. Place the block in vertical orientation (with the thin section facing up) into the center of the mold so that at both the sides of the OCT block thin section there is space to be filled with OCT. Do not wait too long at this step as OCT melts at room temperature.
18. At this step, a thicker OCT-frozen block (about 25 mm × 20 mm × 20 mm) containing the macrocolony is formed (*see* Fig. 2c).
19. If necessary, adjust thickness of the sections up to 10 μm or the temperature of the cutting chamber ±5 °C. Subtle variations in temperature can influence the stiffness of the OCT-frozen block and thereby the cryosectioning.

Acknowledgment

Financial support was provided by the European Research Council under the European Union's Seventh Framework Programme (ERC-AdG 249780 to RH), the Deutsche Forschungsgemeinschaft (He 1556/20-1 to RH) and the Alexander von Humboldt Foundation (postdoc fellowship to DOS).

References

1. Flemming HC, Wingender J (2010) The biofilm matrix. Nat Rev Microbiol 8(9):623–633. doi:10.1038/nrmicro2415
2. Serra DO, Klauck G, Hengge R (2015) Vertical stratification of matrix production is essential for physical integrity and architecture of macrocolony biofilms of Escherichia coli. Environ Microbiol 17:5073–5088. doi:10.1111/1462-2920.12991
3. Serra DO, Richter AM, Hengge R (2013) Cellulose as an architectural element in spatially structured Escherichia coli biofilms. J Bacteriol 195(24):5540–5554. doi:10.1128/JB.00946-13
4. Serra DO, Richter AM, Klauck G, Mika F, Hengge R (2013) Microanatomy at cellular resolution and spatial order of physiological differentiation in a bacterial biofilm. mBio 4 (2):e00103-00113. doi:10.1128/mBio.00103-13
5. Hobley L, Ostrowski A, Rao FV, Bromley KM, Porter M, Prescott AR, MacPhee CE, van

Aalten DM, Stanley-Wall NR (2013) BslA is a self-assembling bacterial hydrophobin that coats the Bacillus Subtilis biofilm. Proc Natl Acad Sci U S A 110(33):13600–13605. doi:10.1073/pnas.1306390110
6. Vlamakis H, Aguilar C, Losick R, Kolter R (2008) Control of cell fate by the formation of an architecturally complex bacterial community. Genes Dev 22(7):945–953. doi:10.1101/gad.1645008
7. Römling U (2005) Characterization of the rdar morphotype, a multicellular behaviour in Enterobacteriaceae. Cell Mol Life Sci 62 (11):1234–1246. doi:10.1007/s00018-005-4557-x
8. Mann EE, Wozniak DJ (2012) Pseudomonas biofilm matrix composition and niche biology. FEMS Microbiol Rev 36(4):893–916. doi:10.1111/j.1574-6976.2011.00322.x
9. Aguilar C, Vlamakis H, Losick R, Kolter R (2007) Thinking about Bacillus Subtilis as a multicellular organism. Curr Opin Microbiol 10(6):638–643. doi:10.1016/j.mib.2007.09.006
10. Lim B, Beyhan S, Meir J, Yildiz FH (2006) Cyclic-diGMP signal transduction systems in vibrio cholerae: modulation of rugosity and biofilm formation. Mol Microbiol 60 (2):331–348. doi:10.1111/j.1365-2958.2006.05106.x
11. Romero D, Aguilar C, Losick R, Kolter R (2010) Amyloid fibers provide structural integrity to Bacillus Subtilis biofilms. Proc Natl Acad Sci U S A 107(5):2230–2234. doi:10.1073/pnas.0910560107
12. Zogaj X, Nimtz M, Rohde M, Bokranz W, Römling U (2001) The multicellular morphotypes of Salmonella typhimurium and Escherichia coli produce cellulose as the second component of the extracellular matrix. Mol Microbiol 39(6):1452–1463
13. Friedman L, Kolter R (2004) Two genetic loci produce distinct carbohydrate-rich structural components of the Pseudomonas Aeruginosa biofilm matrix. J Bacteriol 186(14):4457–4465. doi:10.1128/JB.186.14.4457-4465.2004
14. White AP, Surette MG (2006) Comparative genetics of the rdar morphotype in Salmonella. J Bacteriol 188(24):8395–8406. doi:10.1128/JB.00798-06
15. Bokranz W, Wang X, Tschape H, Römling U (2005) Expression of cellulose and curli fimbriae by Escherichia coli isolated from the gastrointestinal tract. J Med Microbiol 54 (Pt 12):1171–1182. doi:10.1099/jmm.0.46064-0
16. Zogaj X, Bokranz W, Nimtz M, Römling U (2003) Production of cellulose and curli fimbriae by members of the family Enterobacteriaceae isolated from the human gastrointestinal tract. Infect Immun 71(7):4151–4158
17. Richter AM, Povolotsky TL, Wieler LH, Hengge R (2014) Cyclic-di-GMP signalling and biofilm-related properties of the Shiga toxin-producing 2011 German outbreak Escherichia coli O104:H4. EMBO Mol Med 6(12):1622–1637. doi:10.15252/emmm.201404309
18. Chapman MR, Robinson LS, Pinkner JS, Roth R, Heuser J, Hammar M, Normark S, Hultgren SJ (2002) Role of Escherichia coli curli operons in directing amyloid fiber formation. Science 295(5556):851–855. doi:10.1126/science.1067484
19. Reichhardt C, Jacobson AN, Maher MC, Uang J, McCrate OA, Eckart M, Cegelski L (2015) Congo red interactions with Curli-Producing E. coli and native Curli amyloid fibers. PLoS One 10(10):e0140388. doi:10.1371/journal.pone.0140388
20. Hammar M, Arnqvist A, Bian Z, Olsen A, Normark S (1995) Expression of two csg operons is required for production of fibronectin- and congo red-binding curli polymers in Escherichia coli K-12. Mol Microbiol 18(4):661–670
21. Hammar M, Bian Z, Normark S (1996) Nucleator-dependent intercellular assembly of adhesive curli organelles in Escherichia coli. Proc Natl Acad Sci U S A 93(13):6562–6566
22. Serra DO, Hengge R (2014) Stress responses go three dimensional - the spatial order of physiological differentiation in bacterial macrocolony biofilms. Environ Microbiol 16(6):1455–1471. doi:10.1111/1462-2920.12483
23. Yu FP, Callis GM, Stewart PSS, Griebe T, Mcfeters GA (1994) Cryosectioning of biofilms for microscopic examination. Biofouling 8:85–91
24. Werner E, Roe F, Bugnicourt A, Franklin MJ, Heydorn A, Molin S, Pitts B, Stewart PS (2004) Stratified growth in Pseudomonas Aeruginosa biofilms. Appl Environ Microbiol 70 (10):6188–6196. doi:10.1128/AEM.70.10.6188-6196.2004

Chapter 12

Congo Red Stain Identifies Matrix Overproduction and Is an Indirect Measurement for c-di-GMP in Many Species of Bacteria

Christopher J. Jones and Daniel J. Wozniak

Abstract

Congo red is a diazo textile dye that has been used to visualize the production of amyloid fibers for nearly a century. Microbiological applications were later developed, especially in identifying strains that produce amyloid appendages called curli and overexpressing polysaccharides in the biofilm matrix. The second messenger cyclic diguanylate (c-di-GMP) regulates the production of biofilm matrix polysaccharides, and therefore Congo red staining of samples can be utilized as an indirect measurement of elevated c-di-GMP production in bacteria. Congo red allows the identification of strains producing high c-di-GMP in an inexpensive, quantitative, and high-throughput manner.

Key words Congo red, Cyclic diguanylate, Biofilm, Aggregation, Biofilm matrix, Differential medium

1 Introduction

For hundreds of years, textile*Biofilm matrix* dyes were derived from natural products. Due to advances in chemistry in the mid to late 1800s, there was a rapid discovery of many synthetic aniline dyes. These dyes were a significant improvement over natural dyes, however staining fabrics with aniline dyes was costly and labor intensive [1].

The German chemist Paul Bottiger synthesized the diazo dye Congo red in 1883 while working for the Friedrich Bayer Corporation (Fig. 1). This was the first direct dye, which was able to stain fabrics without the use of a mordant to fix the dye to the textile. In spite of this utility, the Bayer Corporation was uninterested in the dye due to its sensitivity to acids and Bottiger obtained a patent independently. He later sold the patent to the Bayer competitor AGFA, which had great commercial success in marketing the dye. As a result of AGFA obtaining such a large market share, Bayer came upon hard financial times. Bayer developed its own lucrative direct red dye, which turned out to be a slight modification of

Karin Sauer (ed.), *c-di-GMP Signaling: Methods and Protocols*, Methods in Molecular Biology, vol. 1657,
DOI 10.1007/978-1-4939-7240-1_12,

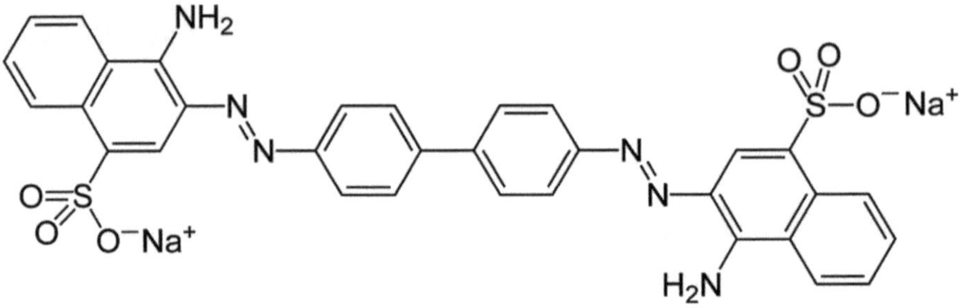

Fig. 1 Chemical structure of Congo red. Courtesy of https://commons.wikimedia.org/wiki/File:Congo-red-2D-skeletal.png#file

Table 1
Applications for Congo red dye

Application	Stained component	Reference
Textile dye	Cellulose of textile	[1]
Histopathology for amyloidosis	Amyloid proteins	[2–4]
Differential stain for *Leptospira*	*Leptospira spp.*	[5]
Curli detection in bacteria	Amyloid curli fibers	[6, 27, 28]
Polysaccharide production	Cellulose, polysaccharide components of matrix	[12–17]
Differential staining of virulent and avirulent strains		[7–11]
Indirect sensor of elevated c-di-GMP signaling	Biofilm matrix	[18–23]

Congo red. To avoid a lengthy patent lawsuit, Bayer and AGFA agreed to split the profits from their dye divisions [1].

After the success of Congo red in the textile industry, it was soon discovered that there were applications for diagnostic histology (Table 1). Congo red has been used for decades to monitor the production of amyloid fibers formed by collagen and other proteins [2–4]. In fact, Congo red staining of tissue samples is still diagnostic for the disease amyloidosis [1, 4]. The initial microbiological application for Congo red staining was to identify *Leptospira* in smears, which was a significant improvement over counting organisms by darkfield microscopy [5]. It was later discovered that Congo red staining was not unique to collagen amyloid fibers, but could stain amyloid fibers in general due to their conserved structures and properties. This property of Congo red was used to observe the production of amyloid curli fibers in *Escherichia coli* [6]. In addition to staining amyloid fibers, several groups recognized that Congo red can be utilized as a differential stain to determine virulence in a variety of species, such as *Shigella flexneri, Vibrio cholerae, Yersinia pestis, Escherichia coli, and Neisseria meningiditis* [7–11]. Congo red is used to stain several bacterial polysaccharides, including cellulose and other species-specific polysaccharides [12–17]. Several groups began utilizing Congo red as a reporter of elevated c-di-GMP signaling, which is a cyclic dinucleotide second messenger that often results in polysaccharide overexpression and enhanced Congo red binding [18–23]. This application has become popular, as it allows for rapid screening of libraries for mutations or compounds that alter c-di-GMP production. The examples presented here highlight the utility and flexibility of this dye in a wide variety of research applications.

2 Materials

2.1 Materials for Microbiological Applications of Congo Red

1. Vogel-Bonner Minimal Medium (VBMM): To make a 10× stock, add the following ingredients (in order) to 800 mL ddH_2O: 2.0 g $MgSO_4$ $7H_2O$, 20 g citric acid, 35 g $NaNH_4HPO_4$ $4H_2O$, and 100 g K_2HPO_4. Adjust pH to 7.0 and add ddH_2O to 1 L. Filter sterilize. Prior to use, dilute to 1× with sterile ddH_2O. For VBMM solid medium, make a 1.1× stock of agar by combining 10 g of agar with 900 mL of H_2O and autoclave. Add 100 mL of 10× VBMM stock and desired dyes prior to pouring plates (*see* **Note 1**).
2. Congo*Biofilm matrix* red: Make a stock solution of 1 mg/mL in H_2O. Filter sterilize and store at room temperature. Working concentration of Congo red in liquid and solid medium is 40 μg/mL.
3. Brilliant blue R: Make a stock solution of 1 mg/mL in H_2O. Filter sterilize and store at room temperature. The working concentration of Brilliant blue R in liquid and solid medium is 15 μg/mL.

2.2 Equipment Required

1. Spectrophotometer capable of determining optical density at 490 nm. Capability to read microplates is an advantage, but not required.
2. Shaking incubator or tube roller.
3. Camera.

3 Methods

3.1 Liquid Growth and Aggregation

Congo red can be utilized as a direct measurement of biofilm matrix production or as an indirect measurement of intracellular c-di-GMP concentrations. Growth of *P. aeruginosa* is one of the most clear examples of the use of Congo red to monitor biofilm matrix

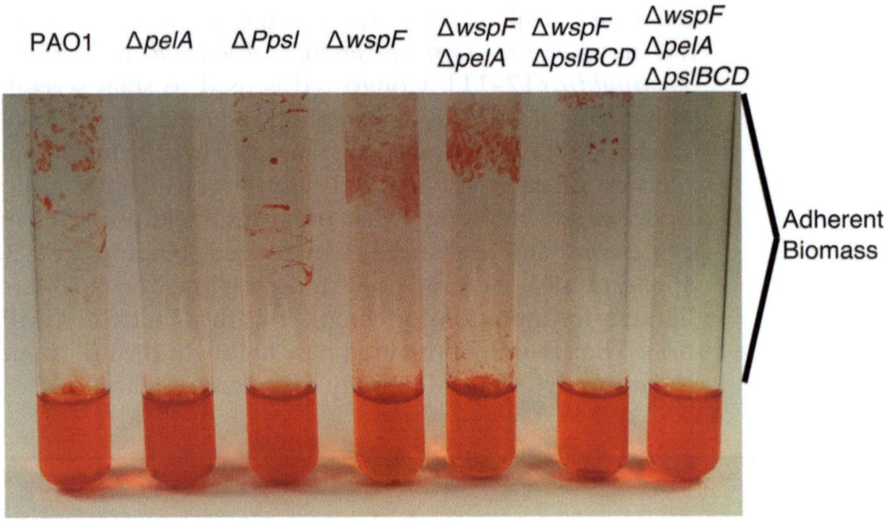

Fig. 2 Congo red stains adherent biomass in liquid culture. The indicated strains were grown in VBMM with 40 μg/mL of Congo red at 37 °C in a rolling drum incubator for 18 h. The adherent biomass on the culture tube is stained with Congo red from the medium

production. Here, we have included a canonical example by modulating the expression of the matrix polysaccharides Psl and Pel in *P. aeruginosa*. In this example, the wild-type strain PAO1 produces Psl and Pel (Fig. 2). The biomass adheres to the side of the tube and stains with both Congo red, which is often included in VBMM to visualize matrix overproducing strains ([14, 16, 18], *see* **Note 1**). In contrast, a strain unable to produce Psl or Pel adheres poorly to the culture tube and does not bind Congo red. This is evident from the dye remaining in the liquid culture [14]. The Δ*wspF* strain overproduces the signaling molecule c-di-GMP, and as a result overproduces Psl and Pel resulting in a small colony phenotype [18]. This hyper-aggregative strain adheres to the culture tube and binds the dye, resulting in a substantial depletion of the Congo red from the medium. The Δ*wspF* Δ*pelA* or Δ*wspF* Δ*pslBCD* strains, which are unable to produce Pel or Psl, respectively do not differ from the parental Δ*wspF* strain with regards to the amount of biomass adhered to the culture tube and depletion of Congo red from the medium. The Δ*wspF* Δ*pelA* Δ*pslBCD* triple mutant, which does not produce either polysaccharide, has less adherent biomass and clearing of the medium. These results demonstrate the utility of Congo red as an indicator of biofilm matrix production.

1. Add Congo red to 1× VBMM to a final concentration of 40 μg/mL.
2. Inoculate 4 mL of VBMM medium containing Congo red with bacteria and incubate overnight at 37 °C with aeration.
3. Growth in a rolling drum aerator spinning at ~100 RPM will generate the highest proportion of stained bacteria adhered to

the glass wall of the test tube, while growth in a shaking incubator at ~200 RPM will produce better aggregation "threads" in the culture (*see* **Note 2**).

3.2 Quantification of Congo Red Binding in Liquid

To quantify the dye binding, 200 μL of the liquid from the samples in Fig. 2. were placed into a 96-well*Biofilm matrix* flat-bottom plate and the OD_{490} was recorded, which monitors the amount of Congo red remaining in the medium (Fig. 3). Here, we observe that growth of PAO1 results in a reduction of the OD_{490} due to the removal of Congo red from the medium. The Δ*pelA* mutant does not form an adherent biofilm, therefore there is no loss of Congo red. The strain unable to produce Psl removes a similar amount of Congo red as PAO1, indicating that Psl is not responsible for the binding and removal of Congo red from the medium. The Δ*wspF* strain, as well as the single polysaccharide mutants in a Δ*wspF* background, removes significantly more dye than the WT PAO1 strain. The Δ*wspF* Δ*pelA* Δ*pslBCD* triple mutant removes less Congo red from the medium than PAO1, suggesting that there are additional roles for Psl in a high c-di-GMP background strain. Each of these results concurs with the visual observation in Fig. 2.

1. Pipette 200 μL of bacterial suspension (obtained from Subheading 3.1) into a flat-bottom 96-well plate. Three biological and three technical replicates are recommended.
2. Include uninoculated media prepared in Subheading 3.1, **step 1** as control.

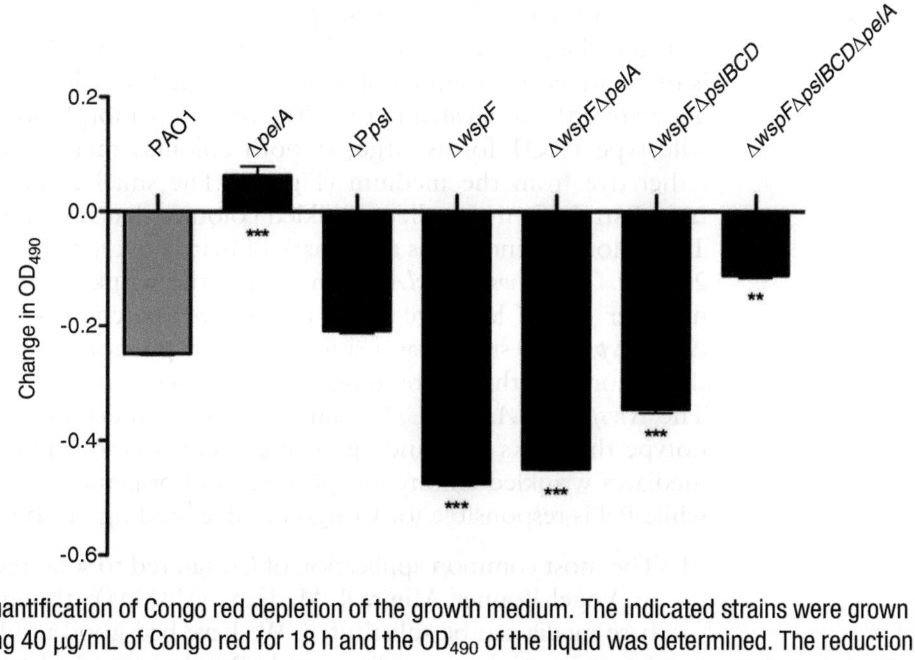

Fig. 3 Quantification of Congo red depletion of the growth medium. The indicated strains were grown in VBMM containing 40 μg/mL of Congo red for 18 h and the OD_{490} of the liquid was determined. The reduction of OD_{490} is plotted for each strain. Two biological replicates were performed in triplicate. Statistical significance was determined with a Oneway ANOVA followed by Dunnett's Multiple Comparison Test (**$p \leq 0.01$, ***$p \leq 0.001$)

3. Measure the absorbance at 490 nm.
4. Congo red binding by bacteria can be observed by monitoring the OD_{490} of the liquid culture. As Congo red is depleted from the medium by binding to the bacteria and matrix, the loss of free dye in the medium can be observed by a reduction of the OD_{490}.
5. Blank the spectrophotometer to the OD_{490} of uninoculated media. The resulting OD_{490} from the test samples will be reported as negative values, indicating the loss of free dye from the media.
6. Culture growth can be monitored independently by the OD_{600} [14], though this can lose accuracy in aggregative strains (*see* **Note 2**)
7. Alternatively, the absorbance can be determined individually using 1 mL cuvettes.
8. Blank the spectrophotometer to the OD_{490} of uninoculated media.
9. Determine the absorbance of the bacterial suspension. Three biological and three technical replicates are recommended.
10. The resulting OD_{490} from the test samples will be reported as negative values, indicating the loss of free dye from the media.

3.3 Solid Growth

Congo red can also be added to solid medium to visualize overproduction of bacterial matrix components and correlate this dye binding with colony morphology. Bacterial*Biofilm matrix* suspensions were spotted onto VBMM plates containing Congo red, brilliant blue, or both dyes and incubated at 37 °C. Brilliant blue is often added in conjunction with Congo red to solid medium to generate differential staining*biofilms)* of colony morphologies. The wild-type PAO1 forms large, smooth colonies that do not bind either dye from the medium (Fig. 4). The small colony variant Δ*wspF* strain forms smaller, wrinkled colonies that bind both dyes. This colony phenotype is a hallmark of matrix overproduction [18, 24–26]. The Δ*wspF*Δ*pelA* strain retains the wrinkled phenotype, however binds less dye than the Δ*wspF* parental strain. The Δ*wspF*Δ*pslBCD* strain has a smooth colony phenotype but retains the majority of the dye binding characteristics of the Δ*wspF* strain. The Δ*wspF*Δ*pelA*Δ*pslBCD* strain produces a smooth colony phenotype that lacks dye binding. These results demonstrate that Psl mediates wrinkled colony morphology and brilliant blue binding, while Pel is responsible for Congo red dye binding on solid media.

1. The most common application of Congo red to solid medium is in Vogel-Bonner Minimal Medium (VBMM), though other base media can be substituted. We have had good results using Jensen's minimal medium and *Pseudomonas* Isolation Agar (PIA).

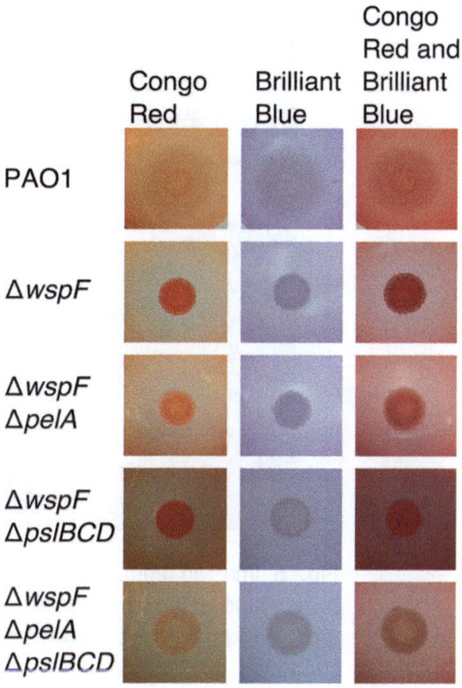

Fig. 4 Colony phenotypes and dye-binding characteristics on solid medium. 3 μL of culture was spotted onto a VBMM plate containing the indicated dyes. Colony morphology images were taken after 18 h of growth at 37 °C

2. Solid agar plates are prepared as described in Subheading 3 (*see* **Note 1**).
3. Cultures can be struck on the surface of the plate to observe the phenotype of single colonies. As an alternative, 2–5 μL of dilute culture can be spotted on the surface to allow observation of the colony biofilm phenotype after overnight growth. Colony phenotypes can be observed under varied growth conditions and duration.

3.4 Imaging Setup

One of the challenging aspects of this assay is obtaining high quality representative images of tubes and plates. We have optimized our imaging setup in order to accurately represent the data with minimal reflections and artifacts. Imaging against a pure*Biofilm matrix* white background enhances contrast and visualization of the Congo red dye.

1. For imaging colonies on plates place the plate on a white sheet of paper.
2. For imaging culture tubes, it is a little more complicated. The background is best if there is not a corner or crease between the vertical and horizontal surfaces. This is achieved by bending the paper in a smooth curve, so the transition from vertical to horizontal is gradual (Fig. 5).

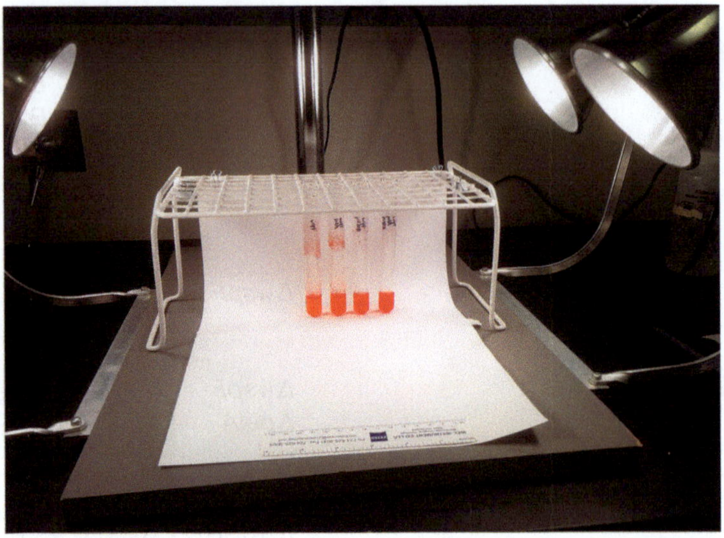

Fig. 5 Imaging apparatus example. This is an example of the lighting and background setup for optimal imaging of liquid cultures in tubes. A smooth, white background and multiple raking lights are the key to obtaining good images

3. Next, utilize an image acquisition system with four raking lights, which provides sufficient illumination while minimizing glare and reflection.
   (a) We have found that direct overhead lighting is a poor choice, as it leads to reflections that prevent the accurate representation of the strains.
   (b) It is not necessary to utilize a particular make or model of camera as a wide range of modern digital cameras capture sufficient detail and contrast to represent the data.
4. Images are often cropped and presented in a figure along with quantification of dye binding (Figs. 3 and 4). Additional optional analyses include colony size, dye adsorption, colony rugosity, and pattern formation.

4 Notes

1. This dye scheme will likely be appropriate with any chemically defined medium. We have found VBMM and Jensen's minimal medium to be well suited for use with *P. aeruginosa*.
2. Bacterial aggregate formation is frequently observed on these assays, with aggregates sticking to the walls of the culture tube. While the aggregate formation aids in sequestering Congo red out of solution, it also results in a decrease in the OD600. The decrease in OD600 is not indicative of a strain not growing, but rather adherent biomass.

References

1. Steensma DP (2001) "Congo" red: out of Africa? Arch Pathol Lab Med 125:250–252
2. Bennhold H (1922) Eine spezifische Amyloidfärbung mit Kongorot [Specific staining of amyloid with Congo red]. Münch Med Wochenschr 69:1537–1538
3. Elghetany MT, Saleem A, Barr K (1989) The congo red stain revisited. Ann Clin Lab Sci 19:190–195
4. Bély M, Makovitzky J (2006) Sensitivity and specificity of Congo red staining according to Romhányi. Comparison with Puchtler"s or Bennhold"s methods. Acta Histochem 108:175–180
5. HOYER BH (1956) A procedure for counting leptospirae based upon the Congo red negative stain. J Bacteriol 72:719–720
6. Olsén A, Jonsson A, Normark S (1989) Fibronectin binding mediated by a novel class of surface organelles on Escherichia coli. Nature 338:652–655
7. Surgalla MJ, Beesley ED (1969) Congo red-agar plating medium for detecting pigmentation in Pasteurella Pestis. Appl Microbiol 18:834–837
8. Surgalla MJ, Beesley ED (1971) Infectivity and virulence of nonpesticinogenic Pasteurella Pestis. Infect Immun 4:416–418
9. Payne SM, Finkelstein RA (1977) Detection and differentiation of iron-responsive avirulent mutants on Congo red agar. Infect Immun 18:94–98
10. Qadri F, Hossain SA, Ciznár I et al (1988) Congo red binding and salt aggregation as indicators of virulence in Shigella species. J Clin Microbiol 26:1343–1348
11. Sankaran K, Ramachandran V, Subrahmanyam YV et al (1989) Congo red-mediated regulation of levels of Shigella flexneri 2a membrane proteins. Infect Immun 57:2364–2371
12. Zogaj X, Nimtz M, Rohde M et al (2001) The multicellular morphotypes of Salmonella typhimurium and Escherichia coli produce cellulose as the second component of the extracellular matrix. Mol Microbiol 39:1452–1463
13. Friedman L, Kolter R (2004) Two genetic loci produce distinct carbohydrate-rich structural components of the Pseudomonas Aeruginosa biofilm matrix. J Bacteriol 186:4457–4465
14. Ma L, Jackson KD, Landry RM et al (2006) Analysis of Pseudomonas Aeruginosa conditional psl variants reveals roles for the psl polysaccharide in adhesion and maintaining biofilm structure post attachment. J Bacteriol 188:8213–8221
15. Ben Abdallah F, Chaieb K, Zmantar T et al (2009) Adherence assays and slime production of Vibrio alginolyticus and Vibrio Parahaemolyticus. Braz J Microbiol 40:394–398
16. Colvin KM, Irie Y, Tart CS et al (2012) The Pel and Psl polysaccharides provide Pseudomonas Aeruginosa structural redundancy within the biofilm matrix. Environ Microbiol 14:1913–1928
17. Kaiser TDL, Pereira EM, Dos Santos KRN et al (2013) Modification of the Congo red agar method to detect biofilm production by Staphylococcus Epidermidis. Diagn Microbiol Infect Dis 75:235–239
18. Hickman JW, Tifrea DF, Harwood CS (2005) A chemosensory system that regulates biofilm formation through modulation of cyclic diguanylate levels. Proc Natl Acad Sci U S A 102:14,422–14,427
19. Christen B, Christen M, Paul R et al (2006) Allosteric control of cyclic di-GMP signaling. J Biol Chem 281:32,015–32,024
20. Antoniani D, Bocci P, Maciag A et al (2010) Monitoring of diguanylate cyclase activity and of cyclic-di-GMP biosynthesis by whole-cell assays suitable for high-throughput screening of biofilm inhibitors. Appl Microbiol Biotechnol 85:1095–1104
21. Merritt JH, Ha D-G, Cowles KN et al (2010) Specific control of Pseudomonas Aeruginosa surface-associated behaviors by two c-di-GMP diguanylate cyclases. mBio 1: e00183–10–e00183–18
22. Jones CJ, Newsom D, Kelly B et al (2014) ChIP-Seq and RNA-Seq reveal an AmrZ-mediated mechanism for cyclic di-GMP synthesis and biofilm development by Pseudomonas Aeruginosa. PLoS Pathog 10:e1003984
23. Kuchma SL, Delalez NJ, Filkins LM et al (2015) Cyclic di-GMP-mediated repression of swarming motility by Pseudomonas Aeruginosa PA14 requires the MotAB stator. J Bacteriol 197:420–430
24. Kirisits MJ, Prost L, Starkey M et al (2005) Characterization of colony morphology variants isolated from Pseudomonas Aeruginosa biofilms. Appl Environ Microbiol 71:4809–4821
25. Starkey M, Hickman JH, Ma L et al (2009) Pseudomonas Aeruginosa rugose small-colony variants have adaptations that likely promote persistence in the cystic fibrosis lung. J Bacteriol 191:3492–3503
26. Haussler S (2003) Highly adherent small-colony variants of Pseudomonas Aeruginosa in cystic fibrosis lung infection. J Med Microbiol 52:295–301

27. Römling U, Bian Z, Hammar M et al (1998) Curli fibers are highly conserved between salmonella typhimurium and Escherichia coli with respect to operon structure and regulation. J Bacteriol 180:722–731
28. Bokranz W, Wang X, Tschäpe H et al (2005) Expression of cellulose and curli fimbriae by Escherichia coli isolated from the gastrointestinal tract. J Med Microbiol 54:1171–1182

Chapter 13

Type IV Pili-Dependent Motility as a Tool to Determine the Activity of c-di-GMP Modulating Enzymes in *Myxococcus xanthus*

Dorota Skotnicka and Lotte Søgaard-Andersen

Abstract

The nucleotide-based second messenger bis-(3′-5′)-cyclic dimeric GMP (c-di-GMP) regulates multiple processes in bacteria including cellular motility. The rod-shaped *Myxococcus xanthus* cells move in the direction of their long axis using two distinct motility systems: type IV pili (T4P)-dependent motility and gliding motility. Manipulation of the c-di-GMP level by expression of either an active, heterologous diguanylate cyclase or an active, heterologous phosphodiesterase causes defects in T4P-dependent motility without affecting gliding motility. As both an increased and a decreased level of c-di-GMP affect T4P-dependent motility, *M. xanthus* represents a good model system to assess enzyme activity of diguanylate cyclases and phosphodiesterases using T4P-dependent motility as a readout. Here, we describe the assay, which allows correlating diguanylate cyclase and phosphodiesterase activity with T4P-dependent motility in *M. xanthus*.

Key words C-di-GMP, *Myxococcus xanthus*, Motility, Type IV pili, Diguanylate cyclase, Phosphodiesterase

1 Introduction

c-di-GMP (cyclic di-3′,5′-guanosine monophosphate) is a ubiquitous nucleotide-based second messenger in bacteria and a regulator of a wide variety of processes most of which are associated with lifestyle changes in response to environmental changes [1]. Generally, elevated c-di-GMP levels are associated with inhibition of motility, increased adhesion and biofilm formation while low levels of c-di-GMP are associated with motile, free-living cells [1]. c-di-GMP is synthesized from two GTP molecules by diguanylate cyclases (DGCs) that contain a catalytic GGDEF domain [2, 3], and degraded to pGpG or GMP by phosphodiesterases (PDEs) that contain either a catalytic EAL domain or a HD-GYP domain [4, 5]. Changes in c-di-GMP levels are sensed by effectors, which in turn interact with downstream targets.

Karin Sauer (ed.), *c-di-GMP Signaling: Methods and Protocols*, Methods in Molecular Biology, vol. 1657,
DOI 10.1007/978-1-4939-7240-1_13, © Springer Science+Business Media LLC 2017

One of the processes in which c-di-GMP-dependent regulation has been extensively studied is cellular motility. A well-understood example involves the PilZ domain protein YcgR in *Escherichia coli* and *Salmonella* that upon c-di-GMP binding interacts with the flagella basal body to interfere with flagella rotation [6, 7]. c-di-GMP has also been reported to regulate gliding motility in *Bdellovibrio bacteriovorus* [8] and type IV pili (T4P) formation in *Pseudomonas aeruginosa* and *Xanthomonas axonopodis* pv citri [9].

Myxococcus xanthus is a model organism to study social behavior and multicellular development in bacteria [10]. Both these two processes depend on cellular motility. *M. xanthus* cells move using two motility systems. Gliding is the movement of single cells and is favored on hard surfaces. It depends on motility complexes that are distributed along the cell body and involves the deposition of slime [11]. The second motility system, which is equivalent to twitching motility in *Neisseria* and *Pseudomonas* species, is favored on soft surfaces and depends on T4P. T4P are highly dynamic structures that pull a cell forward by undergoing cycles of extension, adhesion, and retraction [12]. In *M. xanthus* T4P-dependent motility is characteristic for cells moving in groups. In *M. xanthus* 5–10 T4P are found exclusively at the leading cell pole [13]. The T4P machinery consists of 10 highly conserved proteins, which localize to the outer membrane, periplasm, inner membrane, and cytoplasm [14]. PilA is the major pilin that assembles into the filamentous pili. Mutation of any gene encoding a protein of the T4P machinery leads to loss or significant impairment of T4P-motility.

Manipulation of the c-di-GMP level in *M. xanthus* by overexpression of either an active, heterologous diguanylate cyclase or an active, heterologous phosphodiesterase causes defects in T4P-dependent motility and development without affecting gliding motility [15, 16]. On soft agar, which is favorable to T4P-dependent motility, wild-type (WT) forms long flares of cells spreading out from the colony while colonies of strains lacking T4P-dependent motility display a smooth edge (Fig. 1). To create the strains with increased or decreased c-di-GMP levels, we expressed in the WT DK1622 strain the DGC DgcA from *Caulobacter crescentus* or the PDE PA5295 from *P. aeruginosa* as well as their active site variants DgcA^{D164A} and PA5295^{E328A} [15]. The genes encoding these proteins were cloned under the control of *pilA* promoter in the plasmid pSW105, which integrates at the phage Mx8 *attB* attachment site [17]. The *pilA* promoter was chosen to ensure high and constitutive expression of the heterologous genes [18]. All four genes were fused to the StrepII-tag and, importantly, we were able to show with immunoblot analysis using a Streptactin-HRP conjugate that all protein variants accumulate in vivo [15]. c-di-GMP measurement in vivo confirmed that strains expressing DgcA^{WT} and PA5295^{WT} accumulate significantly more and less c-di-GMP than WT, respectively, while the c-di-GMP levels

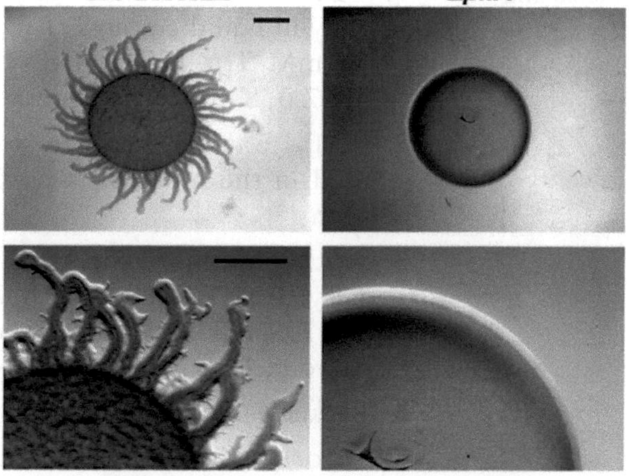

Fig. 1 Assay for T4P-dependent motility for WT (DK1622) and the negative control Δ*pilA* strain (DK10410). Motility was analyzed on motility plates with 0.5% Select agar. Pictures were taken with a stereomicroscope equipped with a camera using 8× and 25× magnification. *Scale bars*, 500 μm

in the strains containing the active site variants were unchanged in comparison to WT.

Strains expressing $DgcA^{WT}$ and $PA5295^{WT}$ show reduced T4P-dependent motility while the strains expressing the active-site variants show WT level of T4P-dependent motility. Thus, both an increase and a decrease in the level of c-di-GMP interfere with T4P-dependent motility. This effect can be quantified by measuring the increase in colony diameter and the length of the flares at the edge of the colony.

In this chapter, we describe the method for reliably assaying T4P-dependent motility in *M. xanthus*. This assay can be very useful for assaying the DGC and PDE activity of heterologous enzymes expressed in *M. xanthus* and in this way for indirectly determining changes in c-di-GMP levels. As a proper positive control, we suggest the WT strain DK1622. A proper negative control would be the Δ*pilA* strain DK10410, which is unable to assemble T4P due to mutation of the gene encoding the major pilin subunit PilA.

2 Materials

Prepare all solutions using ultrapure, distilled deionized water. Sterilize them (autoclave 20 min at 121 °C) prior to use and keep at room temperature (unless indicated otherwise).

2.1 Chemicals and Equipment

1. CTT liquid medium to grow *M. xanthus* cells for the motility assay in liquid cultures: 1 g/100 mL Bacto Casitone powder, 10 mM Tris–HCl pH 8, 1 mM K_2HPO_4/KH_2PO_4 buffer pH 7.6, 8 mM $MgSO_4$.
2. CTT 1.5% agar plates to streak out *M. xanthus* strains to be used in the motility assay: CTT liquid medium, 1.5% Difco agar.
3. Stock solutions of chemicals used to prepare plates for motility assay: 1 M Tris–HCl pH 8.0, 1 M K_2HPO_4/KH_2PO_4 buffer pH 7.6, 0.8 M $MgSO_4$. Additionally, 1.25 g Bacto Casitone powder, 1.25 g Select agar powder (*see* **Note 1**).
4. Beaker: 500 mL.
5. Autoclavable bottle: 500 mL.
6. Autoclavable flask: 50 mL.
7. Automatic pipettor: 1–50 mL.
8. Disposable, sterile serological pipettes: 10, 25, 50 mL.
9. Pipettes and sterile 1000, 200, and 20 μL tips.
10. Large square petri dishes: sterile, 120 × 120 × 17 mm.
11. 2 mL disposable, sterile microcentrifuge tubes.
12. Orbital shaker: 32 °C, 230 rpm.
13. Spectrophotometer, 550 nm.
14. 1 cm path length polystyrene cuvettes.
15. Bacteriological incubator, 32 °C.
16. Plastic box, nontransparent.
17. Benchtop centrifuge, 2075 × *g*, room temperature.
18. Water bath, 50 °C.
19. Stereomicroscope with 8× and 25× magnification and equipped with camera.

2.2 Strain Preparation

1. Streak the strains of interest (WT DK1622, Δ*pilA* DK10410, strains overexpressing DGC and/or PDE as well as their active site variants, *see* **Note 2**) 2–3 days before the planned experiment on CTT 1.5% Difco agar plates supplemented with the relevant antibiotic (50 μg/mL of kanamycin, in case of plasmid pSW105) and incubate at 32 °C in the darkness (*see* **Note 3**).
2. Transfer by pipette 4 mL of the CTT medium into sterile 25 mL flasks. Prepare as many flasks as strains to examine.
3. Prepare 2 mL sterile microcentrifuge tubes with 1 mL of CTT medium. Prepare as many tubes as strains to examine.
4. Harvest a single colony from the CTT 1.5% Difco agar plate using a sterile plastic pipette tip, place in the microcentrifuge tube with CTT medium (*see* **Note 4**).

5. Resuspend a colony in the CTT medium by pipetting until no cellular clumps are visible (*see* **Note 5**).
6. Transfer 1 mL of cell suspension into the 25 mL flask with 4 mL of CTT medium.
7. Incubate in the darkness at 32 °C, 230 rpm.
8. Grow cells to optical density (OD_{550}) of 0.5–0.9 (*see* **Note 6**).

2.3 Preparation of Plates for Motility Assay

1. Transfer 1.25 g Bacto Casitone powder to a beaker; add 200 mL double-distilled H_2O, 2.5 mL 1 M Tris–HCl pH 8.0, 0.25 mL 1 M K_2HPO_4/KH_2PO_4 buffer pH 7.6 and 2.5 mL 0.8 M $MgSO_4$. Then adjust to a final volume of 250 mL with double-distilled H_2O and dissolve Bacto Casitone powder using a magnetic stirrer.
2. Add 1.25 g Select agar to 500 mL bottle; transfer casitone solution to the bottle with agar and gently mix to disperse agar clumps.
3. Autoclave for 20 min at 121 °C and cool down in water bath (*see* **Note 7**).
4. Transfer by pipette 30 mL of the medium onto the plate; swirl gently to spread the agar evenly (*see* **Note 8**). Close the plate immediately.
5. Leave agar to set and keep plates at room temperature overnight (*see* **Note 9**).

3 Methods

1. Measure OD_{550} of the strains. All strains should have a similar OD_{550}, optimally 0.5–0.9.
2. Harvest 2 mL of the cells by centrifugation at 2075 × *g* for 10 min at room temperature (*see* **Note 10**).
3. Carefully remove the supernatant (*see* **Note 11**).
4. Resuspend the cells to 7×10^9 cells/mL ($OD_{550} = 7$) in fresh CTT medium (*see* **Notes 5** and **12**).
5. Spot 5 μL of this suspension in triplicate on above-mentioned plates for motility assay (*see* **Note 13**).
6. Immediately close the plates to avoid evaporation.
7. Incubate closed plates with agar side down at room temperature until the cell spots are dry. Mark the edges of the spots using a thin marker (*see* **Note 14**) to be able to measure initial diameter of colony.
8. Remove the lid from the plate. Take photos of colonies using a stereomicroscope. Use 8× magnification to take photo of the entire colony and use 25× magnification to take photo of the edge of the colony.

9. Incubate afterward at 32 °C with agar side up for 24 h (*see* **Note 15**).
10. Remove the lid from the plate. Take photos using a stereomicroscope with 8× (whole colony) and 25× (edge of colony) magnification.
11. Using the 8× magnification photos measure the colony diameter immediately after spotting and after 24 h of incubation, in triplicates (*see* **Note 16**); calculate the increase in colony diameter for each strain and analyze the differences between them. Use Student's T-test to test for significant differences between strains.
12. Using the 25× magnification photos measure the length of the flares at the edge of the colonies in triplicates. Use Student's T-test to test for significant differences between strains.

4 Notes

1. The type of agar can significantly affect the result of the experiment. In our lab, we routinely use Difco agar to grow *M. xanthus* strains on plates and Select agar for motility assays. Select agar is characterized by good clarity and controlled melting and gelation temperature. Alternatively, different agars should be tested with control strains.
2. It is important to check if the proteins of interest accumulate in *M. xanthus* by using immunoblot analysis. We usually tag the proteins with a Strep-tag for immunodetection using a Strep-tactin-HRP conjugate.
3. Illumination of *M. xanthus* cells with blue light leads to the induction of carotenoid synthesis and may influence the motility behavior. Therefore, plates and cultures should not be exposed to light for a prolonged period of time.
4. Inoculate from fresh plates (not older than 3 days) on which selection pressure has been maintained by including the relevant antibiotic. Always start the culture with a single, yellow colony.
5. Cells must be fully resuspended by pipetting and no clumps of cells should be visible.
6. Cultures should have similar OD_{550}. If this is not the case, dilute the cultures to the same OD_{550} and let them grow again. When diluting, make sure to use liquid CTT medium from the same batch in which the strains were originally inoculated. Keep the cells growing exponentially by diluting the cultures in fresh medium before they reach $OD_{550} = 1$. This will keep cells in steady state and prevent them from clumping, which typically happens in stationary phase.

7. Agar should cool down as much as possible prior to pouring plates to avoid evaporation, which could change final agar concentration in the plates but be careful to avoid premature solidifying. We usually wait until the bottle is still warm but not too hot to be easily held in hands (~ 50 °C). This is best achieved by allowing the agar to cool down in a water bath at 50 °C.
8. Make sure that the surface of the bench is straight by using a bubble level. It is very important that the plates have an even agar surface. Avoid aggressive mixing, foaming and air bubbles in the agar. Do not get rid of the bubbles by using Bunsen burner as this may locally affect agar concentration and harden the surface. The agar surface should be smooth, flat, and regular. Do not use plates showing irregularities or patches of solidified agar.
9. It is best to prepare the plates for the motility assay the day before the experiment. Store plates overnight at room temperature in a stack. Put an empty plate at the top and at the bottom of the stack to prevent increased water condensation.
10. Before harvesting the cells, make sure that all tubes, plates, and centrifuges are ready. Do not leave the cell suspensions waiting at the bench or in the centrifuge. Cells of some strains need longer to pellet than the others at 2075 × *g*, therefore we routinely use 10 min as a centrifugation time.
11. Some of the strains (for example Δ*pilA*) form a very loose pellet; therefore, it is important to gently remove the supernatant using a pipette without disturbing the pellet.
12. It is sufficient to calculate the resuspension volume; it is not necessary to re-measure the OD_{550} at this point. It is difficult to reliably measure the OD of highly concentrated cells. The correct volume of CTT medium in which to resuspend the cells can be calculated using the following equation:

    volume harvested/(wanted final OD_{550}/initial OD_{550}) = resuspension volume.
    Example for 2 mL of harvested culture with $OD_{550} = 0.6$ resuspended to $OD_{550} = 7$:

    2 mL/ (7/0.6) = 0.172 mL = 172 μL.
13. Prepare the plates in duplicates with controls spotted on every plate. To quantify the increase in colony diameter, spot the cell suspension in triplicates. Spot the WT in few replicates on different parts of the plate to be sure that the there is no difference in the agar surface across the plate. Pay attention to keeping a proper distance (~ 2 cm) between the spots as the colonies will spread (Fig. 2). The colony diameter of WT

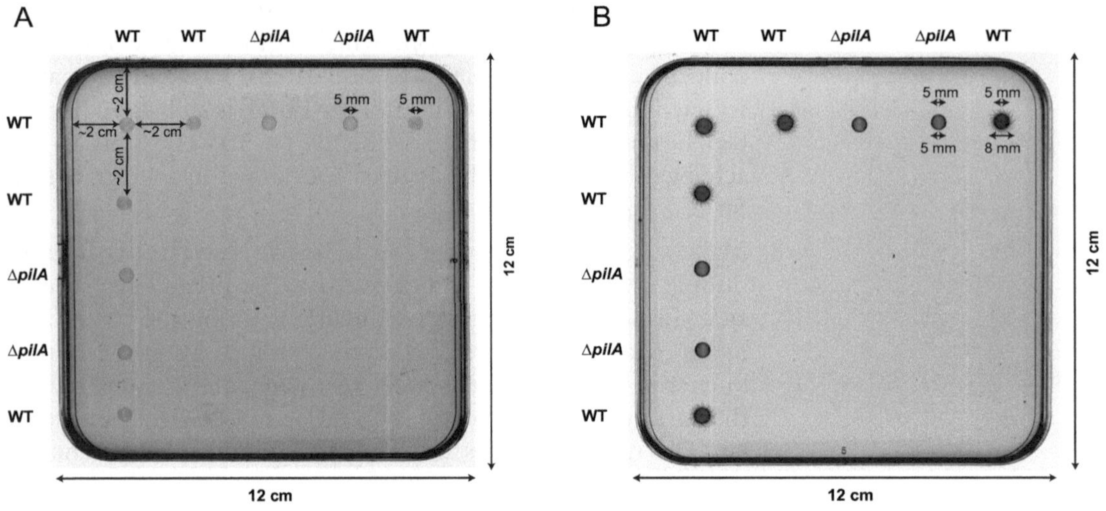

Fig. 2 Example of plate with motility assay of the control strains immediately after spotting the cultures (**a**) and after 24 h of incubation (**b**). Distances between spots and between spot and edge of the plate are indicated. Colony diameters of the motile strain (WT) and non-motile strain (Δ*pilA*) before and after 24 h of incubation are indicated

increases approximately 3–4 mm in 24 h. Therefore, do not spot more than five colonies in one row and not more than 25 on the whole plate. Make sure that you do not touch the agar surface with the top of the pipette during spotting. This will lead to visible agar breakage in the middle of the colony. Keep the pipet tip close to the agar surface and let the drop freely touch the agar. Avoid splashing.

14. It is important to mark the edge of the spot to be able to measure the initial diameter of colony. In this way, the increase in colony dimeter can be determined.
15. Incubating the plates with the agar side up prevents condensate from dripping onto the agar surface.
16. This time can be prolonged to 48 h if differences are not clear; nevertheless, growth of the colony and changes in agar concentration need to be taken into account.

Acknowledgment

This work was supported by the German Research Council within the framework of the Collaborative Research Center 987 "Microbial Diversity in Environmental Signal Response" and the Max Planck Society.

References

1. Römling U, Galperin MY, Gomelsky M (2013) Cyclic di-GMP: the first 25 years of a universal bacterial second messenger. Microbiol Mol Biol Rev 77(1):1–52. doi:10.1128/Mmbr.00043-12
2. Paul R, Weiser S, Amiot NC, Chan C, Schirmer T, Giese B, Jenal U (2004) Cell cycle-dependent dynamic localization of a bacterial response regulator with a novel di-guanylate cyclase output domain. Genes Dev 18 (6):715–727. doi:10.1101/gad.289504
3. Ryjenkov DA, Tarutina M, Moskvin OV, Gomelsky M (2005) Cyclic diguanylate is a ubiquitous signaling molecule in bacteria: insights into biochemistry of the GGDEF protein domain. J Bacteriol 187(5):1792–1798. doi:10.1128/jb.187.5.1792-1798.2005
4. Schmidt AJ, Ryjenkov DA, Gomelsky M (2005) The ubiquitous protein domain EAL is a cyclic diguanylate-specific phosphodiesterase: enzymatically active and inactive EAL domains. J Bacteriol 187(14):4774–4781. doi:10.1128/jb.187.14.4774-4781.2005
5. Ryan RP, Fouhy Y, Lucey JF, Crossman LC, Spiro S, He Y-W, Zhang L-H, Heeb S, Cámara M, Williams P, Dow JM (2006) Cell–cell signaling in *Xanthomonas campestris* involves an HD-GYP domain protein that functions in cyclic di-GMP turnover. Proc Natl Acad Sci U S A 103(17):6712–6717. doi:10.1073/pnas.0600345103
6. Paul K, Nieto V, Carlquist WC, Blair DF, Harshey RM (2010) The c-di-GMP binding protein YcgR controls flagellar motor direction and speed to affect chemotaxis by a "backstop brake" mechanism. Mol Cell 38(1):128–139. doi:10.1016/j.molcel.2010.03.001
7. Böhm A, Kaiser M, Li H, Spangler C, Kasper CA, Ackermann M, Kaever V, Sourjik V, Roth V, Jenal U (2010) Second messenger-mediated adjustment of bacterial swimming velocity. Cell 141:107–116. doi:10.1016/j.cell.2010.01.018
8. Hobley L, Fung RK, Lambert C, Harris MA, Dabhi JM, King SS, Basford SM, Uchida K, Till R, Ahmad R, Aizawa S, Gomelsky M, Sockett RE (2012) Discrete cyclic di-GMP-dependent control of bacterial predation versus axenic growth in *Bdellovibrio bacteriovorus*. PLoS Pathog 8(2):e1002493. doi:10.1371/journal.ppat.1002493
9. Guzzo CR, Dunger G, Salinas RK, Farah CS (2013) Structure of the PilZ–FimXEAL–c-di-GMP complex responsible for the regulation of bacterial type IV pilus biogenesis. J Mol Biol 425(12):2174–2197. doi:10.1016/j.jmb.2013.03.021
10. Konovalova A, Petters T, Søgaard-Andersen L (2010) Extracellular biology of *Myxococcus xanthus*. FEMS Microbiol Rev 34(2):89–106. doi:10.1111/j.1574-6976.2009.00194.x
11. Islam ST, Mignot T (2015) The mysterious nature of bacterial surface (gliding) motility: a focal adhesion-based mechanism in *Myxococcus xanthus*. Semin Cell Dev Biol 46:143–154. doi:10.1016/j.semcdb.2015.10.033
12. Pelicic V (2008) Type IV pili: *pluribus unum*? Mol Microbiol 68:827–837. doi:10.1111/j.1365-2958.2008.06197.x
13. Kaiser D (1979) Social gliding is correlated with the presence of pili in *Myxococcus xanthus*. Proc Natl Acad Sci U S A 76:5952–5956. doi:10.1073/pnas.76.11.5952
14. Chang YW, Rettberg LA, Treuner-Lange A, Iwasa J, Søgaard-Andersen L, Jensen GJ (2016) Architecture of the type IVa pilus machine. Science 351(6278):aad2001. doi:10.1126/science.aad2001
15. Skotnicka D, Petters T, Heering J, Hoppert M, Kaever V, Søgaard-Andersen L (2015) C-di-GMP regulates type IV pili-dependent-motility in *Myxococcus xanthus*. J Bacteriol 198 (1):77–90. doi:10.1128/JB.00281-15
16. Skotnicka D, Smaldone GT, Petters T, Trampari E, Liang J, Kaever V, Malone JG, Singer M, Søgaard-Andersen L (2016) A minimal threshold of c-di-GMP is essential for fruiting body formation and sporulation in *Myxococcus xanthus*. PLoS Genet 12(5):e1006080. doi:10.1371/journal.pgen.1006080
17. Jakovljevic V, Leonardy S, Hoppert M, Søgaard-Andersen L (2008) PilB and PilT are ATPases acting antagonistically in type IV pilus function in *Myxococcus xanthus*. J Bacteriol 190 (7):2411–2421. doi:10.1128/JB.01793-07
18. Wu SS, Kaiser D (1997) Regulation of expression of the *pilA* gene in *Myxococcus xanthus*. J Bacteriol 179(24):7748–7758. doi:10.1128/JB.01793-07

Part V

Modulation of c-di-GMP Levels and Bacterial Responses

Chapter 14

Using Light-Activated Enzymes for Modulating Intracellular c-di-GMP Levels in Bacteria

Min-Hyung Ryu, Anastasia Fomicheva, Lindsey O'Neal, Gladys Alexandre, and Mark Gomelsky

Abstract

Signaling pathways involving second messenger c-di-GMP regulate various aspects of bacterial physiology and behavior. We describe the use of a red light-activated diguanylate cyclase (c-di-GMP synthase) and a blue light-activated c-di-GMP phosphodiesterase (hydrolase) for manipulating intracellular c-di-GMP levels in bacterial cells. We illustrate the application of these enzymes in regulating several c-di-GMP-dependent phenotypes, i.e., motility and biofilm phenotypes in *E. coli* and chemotactic behavior in the alphaproteobacterium *Azospirillum brasilense*. We expect these light-activated enzymes to be also useful in regulating c-di-GMP-dependent processes occurring at the fast timescale, for spatial control of bacterial populations, as well as for analyzing c-di-GMP-dependent phenomena at the single-cell level.

Key words Cyclic dimeric GMP, Optogenetics, Photoactivation, Motility, Biofilm, Chemotaxis, Diguanylate cyclase, Phosphodiesterase

1 Introduction

Cyclic dimeric GMP (c-di-GMP) regulates various aspects of bacterial physiology and behavior including motility, surface attachment, formation and dispersion of biofilms, cell cycle, differentiation, production of secondary metabolites, and virulence [1]. The ability to manipulate intracellular c-di-GMP levels in real time may provide important insights into these processes. Inducible expression of the genes encoding enzymes involved in c-di-GMP synthesis and hydrolysis, i.e., diguanylate cyclases (DGCs) or c-di-GMP phosphodiesterases (PDEs), may not be adequate for this purpose because of the delays associated with gene expression. Further, removing gene expression inducer to reverse the process is difficult without disturbing the system. In contrast, the DGCs and PDEs whose activities are turned on by light of specific wavelengths can be controlled at high temporal resolution. These enzymes are also well suited for interrogating biological processes at high spatial

Karin Sauer (ed.), *c-di-GMP Signaling: Methods and Protocols*, Methods in Molecular Biology, vol. 1657, DOI 10.1007/978-1-4939-7240-1_14, © Springer Science+Business Media LLC 2017

resolution because light can be focused on a subpopulation of bacteria or even on individual cells. We describe the use of the recently engineered light-responsive DGC and a PDE whose activities are turned on by red and blue light, respectively [2–4]. Either one or both enzymes can be expressed in the desired bacterial cells or other types of cells for unidirectional or bidirectional modulation of c-di-GMP levels.

At the beginning of the chapter, we describe the use of the photoactivated enzymes for modulating c-di-GMP concentrations in *Escherichia coli* that result in changes in motility and biofilm phenotypes. These static assays, which are easy to perform, intend to train an inexperienced user in handling light-sensitive bacteria and observing straightforward light-dependent phenotypic changes. The more challenging applications of these enzymes involve manipulating c-di-GMP-dependent processes that occur at fast (seconds-to-minutes) timescales, which are illustrated by the effect of changing c-di-GMP levels on oxygen-dependent chemotaxis (aerotaxis) in the alphaproteobacterium *Azospirillum brasilense* [5]. The photo-activated enzymes involved in c-di-GMP turnover can also be adapted for regulating gene expression [3]; however, such modality is not discussed here.

The two light-activated enzymes described in this chapter are BphS and EB1. BphS [3] is a red/near-infrared light controlled DGC, which synthesizes c-di-GMP from the intracellular pool of GTP. Light enhances DGC activity of BphS by ~11-fold (at room temperature in vitro), compared to the protein activity in the dark. As a photoreceptor of the bacteriophytochrome class, BphS uses biliverdin IXα as a light absorbing chromophore [6]. Biliverdin IXα is a linear tetrapyrrole derivative of heme. Most bacteria, including *E. coli* and *A. brasilense*, synthesize heme; however, they lack a specific heme oxygenase for making biliverdin IXα. Therefore, biliverdin IXα has to be added to the media, or a heme oxygenase gene involved in its production needs to be introduced along with the *bphS* gene. In the experiments described below, we use the heme oxygenase gene *bphO* from *Rhodobacter sphaeroides*, the organism from which the photoreceptor module of BphS was derived [2]. BphS has spectral absorption maxima in the red-to-near-infrared and violet-to-blue spectral regions. The optimal wavelength for its activation is 712 nm [3]; however, commonly available sources of red light (630–660 nm) are perfectly suitable. Upon irradiation, a fraction of BphS undergoes a conformational change resulting in a lit form with an absorption maximum of 756 nm and increased DGC activity [3]. This form gradually decays to the less active, dark state.

The blue light-activated PDE, EB1 [4], hydrolyzes c-di-GMP to a linear dimeric GMP (pGpG), which is subsequently degraded to GMP by cellular nucleases [7, 8]. Upon irradiation, the PDE activity of the MBP-EB1 fusion protein, where MBP is maltose binding protein [9], is increased by >34-fold (in vitro at room temperature). Currently, this is the highest dynamic range for a photoactivated c-di-GMP PDE. As a photoreceptor of the BLUF domain family, EB1 captures light via a flavin chromophore (FAD or FMN) [10, 11]. Since all cells synthesize flavins, neither chromophore addition nor introduction of flavin synthesis genes is necessary. The absorption maxima of flavins lie in the UV-B-to-blue spectrum. Prolonged exposure of cells to such light is deleterious [12]. To avoid cell damage, induction of the PDE activity is accomplished by irradiation with short pulses of blue light. Importantly, at the wavelength of approximately 465 nm, the spectral overlap between BphS and EB1 is minimal, which allows for simultaneous use of these two enzymes for the dichromatic, bidirectional modulation of intracellular c-di-GMP levels [4].

2 Materials

2.1 Strains and Plasmids

1. *E. coli* MG1655[DE3] [3], a motile derivative of strain MG1655 that expresses a T7 Polymerase gene from an IPTG-inducible promoter (*see* **Note 1**).
2. *E. coli* MG1655 Δ*yhjH* (Km^r) [13], a MG1655 mutant with high intracellular c-di-GMP levels due to the lack of the major c-di-GMP PDE, YhjH/PdeH [14].
3. *E. coli* BL21 [DE3] (NEB), strain producing curli fimbriae in a c-di-GMP-dependent manner; contains a T7 Polymerase gene expressed from an IPTG-inducible promoter.
4. *A. brasilense* Sp7 (ATCC 29145).
5. *A. brasilense tlp1* (strain SG323), a mutant derived from strain Sp7, impaired in the chemotaxis receptor Tlp1 [15].
6. Plasmids expressing light-activated DGC and PDE (Fig. 1) (*see* **Note 2**).
   (a) pETBphSO [3]: pET23a(+) containing the *bphS-bphO* operon downstream of the T7 promoter; used for elevating c-di-GMP levels with red light.
   (b) pET23a(+) (EMD Millipore): vector control to be used in conjunction with pETBphSO.
   (c) pMal_EB1 [4]: pMal-c5x expressing an MBP-EB1 fusion; used for lowering c-di-GMP levels with blue light.
   (d) pMAL-c5× (NEB): vector control to be used in conjunction with pMal_EB1.
   (e) pMQbSHY3 [3]: pMQ56 [16] encoding BphS, BphO, and YhjH (RBS3) under the control of T7 promoter.

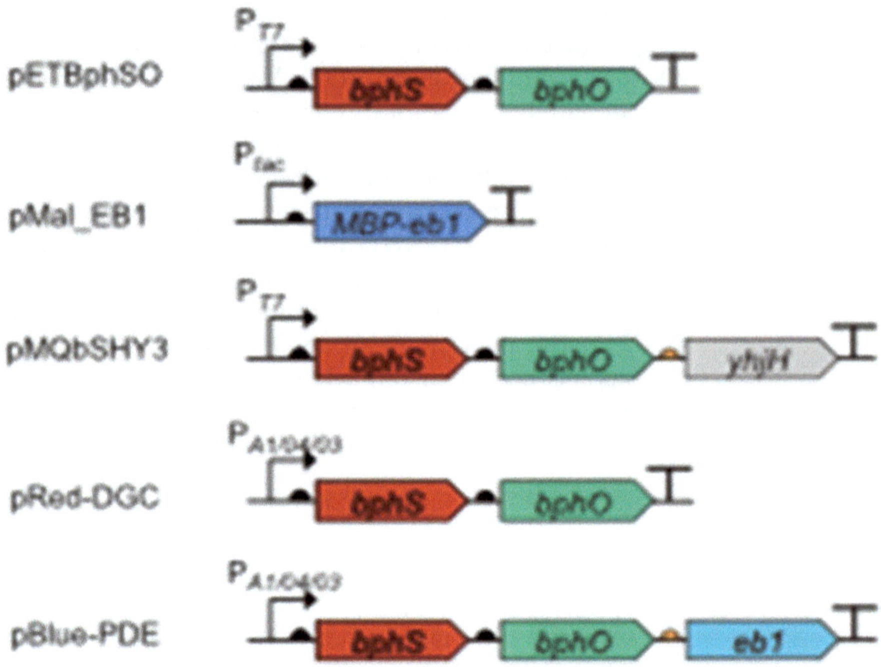

Fig. 1 Organization of the genes encoding the light-activated enzymes, DGC (BphS) and PDE (EB1 and MBP-EB1), in the plasmids described in this chapter

(f) pRed-DGC [5]: pIND4 containing the *bphS-bphO* operon downstream of the IPTG-inducible promoter.

(g) pBlue-PDE [5]: pIND4 containing the *bphS-bphO-eb1* operon downstream of the IPTG-inducible promoter.

(h) pIND4 [17]: broad-host range vector to be used in conjunction with pRed-DGC and/or pBlue-PDE.

2.2 Equipment and Supplies for Experiments in *E. coli*

1. Incubation room at 30 °C and illuminated with safe green light (*see* **Note 3**).
2. Shaking incubator setup to 200 rpm and 30 °C unless indicated otherwise.
3. Centrifuge(s) capable of handling 50 mL and 2 mL centrifugation tubes.
4. Two metal wire shelving units (racks).
5. Light panels (*see* **Note 4**)
   (a) All-red (660 nm) LED Grow light panels 225 (30.5 × 30.5 cm; LED Wholesalers).
   (b) All-blue (465 nm) LED Grow light panel 225 (30.5 × 30.5 cm; LED Wholesalers).
6. Light-impenetrable shield (e.g., cardboard, plastic, metal sheet).

7. Timer: CT-1 Digital Short Cycle Timer (Innovative Grower) (*see* **Note 5**).
8. Aluminum foil.
9. Sterile toothpicks.
10. 50 mL and 2 mL centrifugation tubes.

2.3 Equipment and Supplies for Experiments in A. brasilense

1. Phase contrast microscope (e.g., Nikon E200) equipped with a digital camera (e.g., Nikon Coolpix).
2. Hollow rectangular glass capillary tubes, 0.1 × 1 × 50 mm. These tubes are commercially available from Vitro Dynamics, Inc., Rockaway, NJ.
3. Gas equilibration chamber, 3.8 × 8 × 0.4 cm, with a cover slip top and glass slide bottom. The gas equilibration chamber was 3-D printed at The University of Tennessee following the specifications in ref. [18].
4. Compressed N_2 gas.
5. Remotely controlled Magic Light LED light bulb.
6. Optical filters: green (505–575 nm), red (610–730 nm), and blue light (450 nm) (Andover, Inc.). These filters are placed over the microscope white light source.

2.4 Media and Solutions for Motility and Biofilm Experiments in E. coli

Prepare all solutions using deionized water. Autoclave media and solutions at 121 °C for 20 min. Add antibiotics and chemical inducers to the autoclaved media when it is cooled to approximately 55 °C. Store all media and solutions at 4 °C, unless indicated otherwise.

1. LB medium: Add 5 g Yeast extract, 10 g Tryptone, 5 g NaCl into 1 L of deionized water. Autoclave.
2. LB agar: Add 15 g/L agar to LB medium. Autoclave. Add 100 μg/mL ampicillin.
3. SOC medium: Add 15 g of SOC powder into 1 L of deionized water. Autoclave.
4. Congo red dye solution: Dissolve 200 mg Congo red powder (Sigma-Aldrich) in 10 mL deionized water. Do not autoclave. Store at room temperature.
5. Congo red agar: To LB agar add 20 μg/mL Congo red dye solution, 100 μg/mL ampicillin, and 10 μM IPTG.
6. Ampicillin 1000× stock solution: Dissolve 1 g of ampicillin powder in 10 mL sterile deionized water making 100 mg/mL stock solution. Store in aliquots at −20 °C.
7. IPTG stock 1 M solution: Dissolve 2.38 g IPTG in 10 mL deionized water. Store in aliquots at −20 °C.

8. Semisolid agar for *E. coli* motility assays: Add 5 g NaCl, 10 g Tryptone and 2.5 g agar in 1000 mL of deionized water. Autoclave. Add 100 μg/mL ampicillin. For motility assays involving BphS, also add 10 μM IPTG (final concentration) (*see* **Notes 6** and 7).
9. 100 mM $CaCl_2$: Dissolve 1.11 g $CaCl_2$ in 100 mL deionized water.
10. 100 mM $CaCl_2$ + 15% glycerol solution: Dissolve 1.11 g $CaCl_2$ in 85 mL deionized water; add 15 mL glycerol. Autoclave.

2.5 Media and Solutions for c-di-GMP Extraction

1. C-di-GMP extraction buffer: 40% methanol (v/v), 40% acetonitrile (v/v) in 0.1 N formic acid. Do not sterilize (*see* **Note 8**).
2. Neutralization solution: 15% NH_4HCO_3 (w/v). Do not sterilize. Store at room temperature for no longer than 1 week.

2.6 Media and Solutions for Aerotaxis Experiments in A. brasilense

1. Minimal medium containing malate minus salts (MMAB-salts). Dissolve 3 g K_2HPO_4, 1 g NaH_2PO_4, 0.15 g KCl, 0.05 g Na_2MoO_4, 5 g malate, 1 g NH_4Cl in 995 mL deionized water. Adjust pH to 6.85–7.00 with NaOH. Autoclave.
2. MMAB medium: Prepare the following three salt solutions, autoclave them separately, and sterilely add the specified volumes to 1000 mL of the MMAB-salts medium prepared in Subheading 2.6, **step 1** to make MMAB medium. (1) 60 g $MgSO_4$ in 1000 mL water; add 5 mL; (2) 0.631 g $FeSO_4 \times 7\ H_2O$ and 0.592 g EDTA in 50 mL water; add 250 uL; (3) 20 g $CaCl_2$ in 1000 mL water; add 500 μL.
3. MMAB medium with 200 μg/mL ampicillin and 30 μg/mL kanamycin.
4. MMAB agar: Add 15 g agar to 1000 mL MMAB medium. Autoclave.
5. Tryptone-yeast medium: Dissolve 10 g Tryptone and 5 g Yeast extract in 1000 mL deionized water. Adjust pH to 7.00 with NaOH. Autoclave.
6. Tryptone-yeast agar. Add 15 g agar to 1000 mL Tryptone-yeast medium. Autoclave.
7. Chemotaxis buffer: Dissolve 1.7 g K_2HPO_4, 1.36 g KH_2PO_4 in 900 mL deionized water. Add EDTA to 0.1 mM (final concentration). Adjust pH to 6.85–7.00 with NaOH. Add water to 1000 mL. Autoclave.
8. Chemotaxis buffer with 1 mM malate: Add 0.134 g malate to 1000 mL of chemotaxis buffer prepared as described in Subheading 2.6, **step 7**. Autoclave.

3 Methods

3.1 Preparation of Chemically Competent E. coli *Cells and Transformation*

1. Grow an overnight culture of *E. coli* in LB medium at 37 °C with shaking. This method is applicable to all *E. coli* strains and plasmids described in this chapter (*see* **Note 9**).
2. Transfer 1 mL of the overnight culture into 100 mL fresh LB medium (use a 750 mL flask). Grow bacteria at 37 °C with shaking until cell optical density reaches OD_{600}, 0.6.
3. Precool the centrifuge capable of handling 50 mL conical tubes to 4 °C.
4. Transfer bacterial culture on ice and incubate for 15 min. In all subsequent steps keep cells on ice and perform all manipulations in a cold room, if available.
5. Transfer cells into the precooled 50 mL conical tubes.
6. Collect cells by centrifugation at 4500 × *g* for 10 min at 4 °C. Discard the supernatant as completely as possible.
7. Add 10 mL 100 mM $CaCl_2$ solution in each 50 mL tube, resuspend cell pellets.
8. Combine cells into a single tube and incubate it on ice for 15 min.
9. Spin cells down by centrifugation at 4500 × *g* for 10 min. Discard the supernatant.
10. Resuspend the pellet in 3 mL of cold 100 mM $CaCl_2$ + 15% glycerol solution.
11. Aliquot 100 μL into 1.5 mL tubes and proceed to **step 14** for immediate transformation. Alternatively, flash freeze tubes by dropping them into −80 °C ethanol or liquid nitrogen.
12. Store aliquots of chemically competent cells at −80 °C.
13. For transformation, defrost competent cells on ice for approximately 5 min.
14. Add 10–20 ng of plasmid DNA to competent cells, mix gently, and incubate on ice for 5 min.
15. Heat shock cells by placing the tube into a 42 °C water bath for 30 s, then place the tube back on ice for 60 s.
16. Add 900 μL SOC medium and incubate for 1 h at 37 °C.
17. Plate 20 μL on LB plates containing ampicillin. Incubate the plate at 37 °C until colonies appear.

3.2 Incubation Room Setup for Light Experiments Using E. coli

1. In a 30 °C incubation room, set up two metal wire shelving units (racks), preferably at a distance from each other. Place All-red Grow panel(s) on a shelf of one unit, and All-blue Grow panel(s) on a floor or a lower shelf of another unit (Fig. 2a).

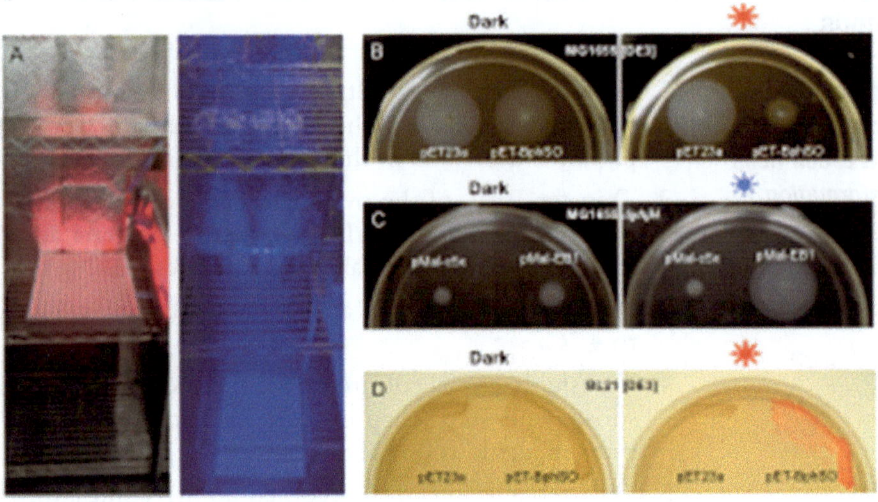

Fig. 2 Light regulation of c-di-GMP-dependent motility and biofilm phenotypes in *E. coli*. (**a**) Setup of the red and blue light panels. (**b**) Inhibition of swimming in the semisolid agar in strain MG1655[DE3] expressing the red light-activated DGC, BphS (plasmid pETBphSO). (**c**) Restoration of swimming in the semisolid agar in strain MG1655 Δ*yhjH* expressing the blue light-activated PDE, MBP-EB1 (plasmid pMal-EB1). (**d**) Induction of curli fimbria formation in BL21[DE3] expressing the red light-activated DGC, BphS, (plasmid pETBphSO)

2. Place a light-impenetrable shield between the shelving units to prevent cross-irradiation between blue and red light panels.
3. Program timers for the following irradiation regiments: 30 s light/120 s dark cycles for red light; and 10 s light/ 50 s dark for blue light.
4. Set up green light via flexible green LED strip for illuminating the incubation room (*see* **Note 10**).

3.3 Light-Regulated Motility Assays in E. coli

A red light-induced increase in c-di-GMP levels in a motile strain MG1655[DE3] inhibits its ability to swim in the semisolid (soft) agar [3]. Similarly, a blue light-induced decrease in intracellular c-di-GMP levels restores motility in the MG1655 *yhjH* mutant that has approximately tenfold elevated intracellular c-di-GMP levels, compared to MG1655, and therefore is impaired in swimming in the semisolid agar [4]. The experiments described below illustrate the effect of BphS and EB1 on modulating motility patterns in these strains.

1. Inoculate colonies of the freshly grown MG1655[DE3] transformants containing pETBphSO or pET23a (negative control) into the culture tubes containing 3 mL LB + ampicillin.
2. Inoculate colonies of the freshly grown MG1655 *yhjH* transformants containing pMal_EB1 or pMal-c5× (negative control) into the culture tubes containing 3 mL LB + ampicillin.

3. Grow cultures overnight (~16 h) at 30 °C in the dark (cover with aluminum foil) on a shaker (200 rpm) or culture wheel.
4. Drop 2 μL from each overnight bacterial culture onto two motility agar plates for subsequent incubation under red, blue, and no light [dark] conditions. Keep approximately 2 cm distance between the drops to avoid overlap between the neighboring swim zones (*see* **Note 11**).
5. For exposure to red light, place a plate (upside up) on a shelf of a red light wire shelving unit approximately 30 cm above the shelf with the All-red Grow panel.
6. For exposure to blue light, place a plate on a shelf of the blue light wire shelving unit, approximately 100 cm away from the All-blue Grow panel.
7. Wrap the control "dark" plate in aluminum foil to eliminate access to light and place it next to the plate exposed to light.
8. Turn the light panels on (via pre-programmed timers) and incubate plates at 30 °C for 8–10 h (*see* **Note 12**).
9. Observe diameters of swim zones (Fig. 2b, c). For quantitative analysis, measure swim zone diameters from at least three transformants and average the results. Repeat the experiment for obtaining statistically significant results.

3.4 Light-Regulated Curli Fimbriae Formation in E. coli *BL21[DE3]*

In *E. coli* strains producing curli fimbriae, such as BL21[DE3], red light-induced increase in c-di-GMP levels results in red-pigmented colonies grown on the agar medium in the presence of the Congo red dye [3]. All the steps below are carried out in the incubation room at 30 °C.

1. Using sterile toothpicks, streak colonies of the freshly grown BL21[DE3] transformants containing pETBphSO or pET23a onto two Congo red agar plates (for incubation under red light and dark conditions) (*see* **Note 13**).
2. For exposure to red light, place a plate (inverted, i.e., upside down) on a shelf of a red light wire shelving unit approximately 30 cm above the shelf with the All-red Grow panel.
3. Wrap the second plate ("dark" control) in aluminum foil to eliminate access to light and place it next to the plate exposed to light.
4. Turn the light panels on (via pre-programmed timers) and incubate plates at 30 °C for 3–4 days until red colony pigmentation develops (Fig. 2d).

3.5 Manipulating Intracellular c-di-GMP Levels in Liquid-Grown E. coli *Cultures Using a Red Light-Activated DGC*

In this section, we describe a system for red light-dependent control of intracellular c-di-GMP levels in liquid-grown *E. coli*. To overpower the background activity of BphS in the dark, the construct used in these experiments (pMQbSHY3) contains the *bphS-bphO* operon along with a gene for a constitutive c-di-GMP PDE, YhjH/PdeH (the *bphS-bphO-yhjH* operon) [3]. The role of

YhjH/PdeH is to keep the intracellular c-di-GMP levels at nearzero levels in the absence of red light (*see* **Note 14**). All the steps below are carried out in the incubation room at 30 °C.

1. Inoculate a colony of MG1655[DE3] (pMQbSHY3) expressing the *bphS-bphO-yhjH* operon in 3 mL liquid LB medium containing ampicillin.
2. Wrap the tube in aluminum foil and grow on a shaker (200 rpm) set at 30 °C overnight.
3. Place an All-red Grow light panel close to the shaker (~10 cm away) so that tubes can be exposed to red light (*see* **Note 15**). Do not turn on the light yet.
4. In a room illuminated with safe green light, measure OD_{600} of the overnight culture, and inoculate six tubes with 3 mL fresh LB medium containing ampicillin to OD_{600}, 0.2.
5. Wrap all tubes in foil.
6. Place wrapped tubes in a wire rack on a shaker (200 rpm) and grow for an additional 1 h at 30 °C.
7. Following the 1 h incubation, take two tubes off the shaker, place then on ice and proceed to nucleotide extraction. These two cultures will correspond to the initial (0 h irradiation) time point.
8. At the same time point, unwrap two tubes and turn the light panel on via a timer (preset for the following regimen: 30 s light, 120 s dark). Grow remaining cultures upon shaking.
9. At each desired time point during irradiation (e.g., 3 h and 6 h), collect two cultures—one wrapped (dark), another one exposed to light and proceed with nucleotide extraction (*see* **Note 16**).
10. Repeat the experiments several times for statistical analysis.

3.6 Nucleotide Extraction from Bacterial Biomass and c-di-GMP Measurements

1. Weigh empty 2 mL microtubes (one tube per bacterial sample) and record their weights.
2. Centrifuge 2 mL of cells prepared in Subheading 3.5 for 1 min at 13,000 × *g* at 4 °C. Discard the supernatant and weigh the tubes with wet biomass (*see* **Note 17**).
3. Preserve the remaining culture (approximately 1 mL) on ice for determining viable cell count, which will be needed for normalizing c-di-GMP concentrations.
4. Add 100 μL of extraction buffer per 50 mg of wet pellet from **step 1**. Vortex vigorously to fully resuspend cells.
5. Incubate slurries for 30 min at −20 °C.

6. Spin down the suspension at 4 °C, 13,000 g, for 3 min.
7. Place the supernatant into a new tube and neutralize it with 4 µL 15% NH_4HCO_3 per 100 µL of sample.
8. Extract remaining nucleotides from the insoluble material by repeating **steps 4–7**.
9. Combine two neutralized supernatants and measure the sample volume.
10. Send samples to a LC-MS/MS facility for c-di-GMP measurements (*see* **Note 18**).
11. Calculate approximate intracellular c-di-GMP levels using c-di-GMP values determined by mass-spectrometry, numbers of viable cells, and average cell volume (*see* **Note 19**).

3.7 Spatial Gradient Assay for Aerotaxis in A. brasilense

In a capillary containing liquid medium, *A. brasilense* forms a stable aerotactic band at a specific position corresponding to the preferred oxygen concentration [19] (Fig. 3a). The aerotactic band forms because *A. brasilense* senses both high and lower oxygen concentrations as repellents while the preferred, low oxygen concentration (~0.4% dissolved oxygen) as an attractant [20]. Intracellular c-di-GMP levels can modulate aerotaxis by affecting activity of at least one chemotaxis receptor, Tlp1 [21]. Expression of BphS, the red light activated DGC from a broad-host-range plasmid, pRed-DGC, or a blue light activated PDE from plasmid pBlue-PDE can be used to characterize the role of intracellular c-di-GMP levels in aerotactic behavior in *A. brasilense* [5]. These assays can be adapted to characterize other chemotaxis responses or other reversible behaviors occurring in chemical gradients.

1. Grow *A. brasilense* strains Sp7 and *tlp1* containing pRed-DGC, pBle-PDE, or pIND4 plasmids in MMAB medium with 200 µg/mL ampicillin and 30 µg/mL kanamycin in the dark, at 28 °C, with shaking at 180 rpm to OD_{600}, 0.6 (*see* **Notes 10, 20 and 21**).
2. Collect 1 mL of cells at OD_{600}, 0.6 by centrifugation at 13,000 × *g* for 3 min. at room temperature, decant the supernatant and suspend cells in 1 mL of chemotaxis buffer. After suspending, centrifuge again to wash and pellet cells. Repeat wash three times.
3. Suspend the washed cell pellet in 100 µL chemotaxis buffer with 1 mM malate by vortexing.
4. Take a small aliquot and observe cells under the light microscope, with 200× (or higher) magnification to ensure that they are motile. Over 80% motile cells are desirable.
5. Let washed cells equilibrate in the dark for 10 min, at room temperature, to avoid induction of enzymes by exposure to white light.

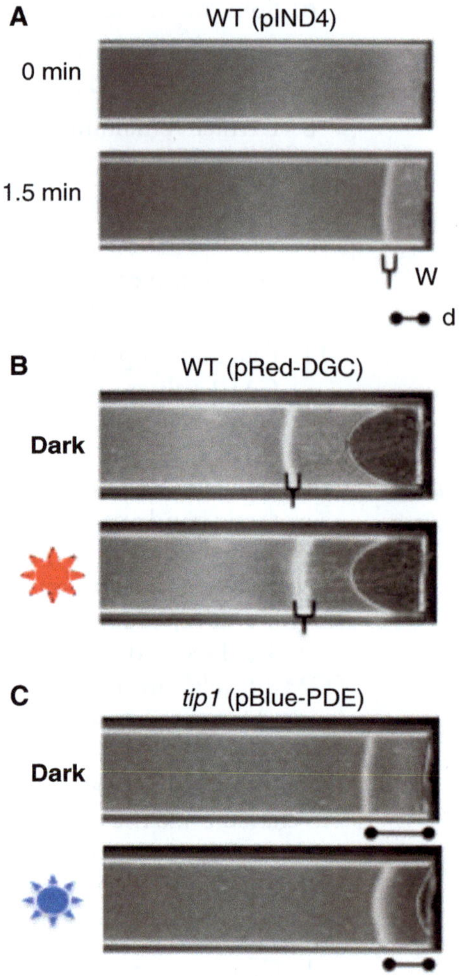

Fig. 3 C-di-GMP effects on aerotactic band formation in *A. brasilense*. (**a**) Aerotactic band formation after introduction of an oxygen gradient in the wild-type strain Sp7 containing an empty vector pIND4. Two parameters, distance from the band to the meniscus (*d*) and band width (*w*), are used to characterize oxygen sensitivity. (**b**) The red light-activated DGC, BphS, expressed from pRed-DGC increases band thickness. (**c**) The blue light-activated PDE, EB1, expressed from pBlue-PDE changes band position in regard to the meniscus in the *tlp1* mutant

6. Use forceps to place capillary tube into the suspension of washed motile cells. Allow capillary action to draw up the cell suspension into the capillary tube.
7. Wipe the outside of the capillary tube to remove excess liquid and place the filled capillary tube, horizontally, on a microscope slide, in the gas equilibration chamber [22].
8. Turn on N_2 gas and allow cells to equilibrate under an atmosphere of gaseous N_2 for 3 min (Fig. 3a).

9. Keep cells in the dark on the microscope stage. Then place the green filter over a microscope white light source.
10. Turn on the digital camera and start recording cells' behavior within the capillary tubes if downstream subsequent motion or image analysis is performed.

3.7.1 Modulating c-di-GMP Levels Before Air Gradient Establishment

We describe two experimental conditions. In Subheading 3.7.1, the effects of light-induced changes in intracellular c-di-GMP levels are assayed prior to exposure to the gradient and thus before aerotactic band formation. In Subheading 3.7.2, the effects of light-induced changes in intracellular c-di-GMP levels on an aerotactic band are assayed after the band has formed. These two assays discriminate between the response of cells newly exposed to an air gradient (Subheading 3.7.1) and the response of cells adapted to specific aeration conditions (Subheading 3.7.2).

1. Prepare capillary tubes as described in Subheading 3.7 and allow cells to equilibrate under an atmosphere of gaseous N_2 (Fig. 3a).
2. Expose cells to red or blue light, using appropriate optic filters, during last 10 s of N_2 equilibration. Avoid exposing the cells suspension to white light.
3. After an exposure of 10 s to red or blue light, return to green light using the appropriate optic lens filter. Avoid exposing the cells suspension to white light.
4. After 3 min, turn gas off and allow air to flow into the chamber and record the time to formation of the aerotaxis band and its position relative to the meniscus. For a given cell density, carbon source, and physiological stage, the aerotactic band forms at a similar position and within the same time frame (*see* **Note 22** and Fig. 3b).

3.7.2 Modulating c-di-GMP Levels After Establishment of the Air Gradient

1. Prepare capillary tubes as described in Subheading 3.7 and allow cells to equilibrate under an atmosphere of gaseous N_2 for 3 min (Fig. 3a).
2. After 3 min, turn off gas and allow air to flow into the chamber.
3. Start digital recording of cells' behavior within the capillary tube.
4. Allow the aerotaxis band to form and record time to formation and position of the aerotactic band relative to meniscus.
5. After the band has formed and is stable, expose cells to red of blue light once or twice for 10 s and continue recording under green light (Fig. 3c). Avoid exposing the cells suspension to white light, which could induce enzyme activity.

4 Notes

1. Intracellular c-di-GMP levels in *E. coli* strains MG1655 and MG1655 Δ*yhjH* differ by ~tenfold [23], which make them convenient for illustrating the effect of light-activated DGCs and PDEs. In the described assays, we use MG1655[DE3], as opposed to MG1655, to drive expression of the *bphS-bphO* operon from the T7 promoter. However, alternative promoters for *bphS* expression can and have been successfully used in *E. coli*, in other bacterial species [24] and in mammalian cells [25].
2. Various other light-activated enzymes involved in c-di-GMP synthesis and hydrolysis have been described [26–29]; however, in our opinion, they are not as convenient for optogenetic application as BphS and EB1. For further discussion of this issue, *see* ref. [4].
3. Place the shelving units, shaker, and timer in the incubation room. In the absence of such a room, the experiments in *E. coli* can be performed in a regular plate incubator. The disadvantage of an incubator is a risk of overheating. If performing an experiment in the incubator, place a light panel inside, turn it on in the desired irradiation regimen before the experiment, and test whether or not pulsed irradiation overheats the incubator beyond acceptable temperature fluctuation. Placing a small fan inside the incubator may improve air mixing and solve the overheating issue.
4. We use light panels because they are affordable and convenient for uniform irradiation of a large number of Petri dishes. However, various alternative sources of red (635–660 nm) or blue (465 nm) light are appropriate.
5. More affordable timer models can also be used. To avoid the inhibitory effect of light on bacterial growth, chose a timer allowing short (less than 1 min) irradiation periods.
6. For assays involving red light-induced motility inhibition, add 10 μM IPTG to the semisolid agar medium to induce *bphS-bphO* operon expression. If experiments are performed in different *E. coli* strains, optimizing IPTG concentration may be necessary. Adding IPTG for blue light-dependent induction of swimming involving pMal_EB1 is unnecessary because leaky MBP-EB1 expression from pMal_EB1 is sufficient.
7. For testing multiple strains, use large, 15 cm diameter, Petri dishes. Store semisolid agar plates at room temperature upright for no longer than 2 days prior to use.
8. It is advisable to prepare extraction buffer fresh for every extraction to avoid methanol evaporation. Alternatively, store this buffer in a gas-tight container.

9. This method can be replaced with any other method of preparing chemically competent or electroporation ready cells.
10. Avoid exposing bacteria in liquid cultures or on plates to white light. Perform handling of bacterial cultures in safe green light.
11. Do not invert semisolid (motility) agar plates. Keep them upside up (lid on the top) at all times.
12. For experiments performed at different temperatures, e.g., room temperature or 37 °C, incubation periods will differ, e.g., approximately 12–14 h and 4–6 h, respectively.
13. Note that curli fimbriae in strain BL21[DE3] as well as in most other *E. coli* strains are not expressed at temperatures higher than 30 °C. It is possible to perform these experiments at room temperature but not at 37 °C.
14. Several *bphS-bphO-yhjH* operons have been engineered where expression of *yhjH/pdeH* varies due to different ribosome-binding site strength [3]. It is expected that a proper construct can be found to keep intracellular c-di-GMP levels at the near-zero levels in a desired *E. coli* strain.
15. Secure the light panel at a safe distance from a shaker. Note that a red light panel can be replaced with a red light bulb; however, it would be important to ensure to even light fluency of the irradiated cultures.
16. In our hands intracellular c-di-GMP concentrations rose from ~4 nM in the dark at time point 0 h to ~84 nM after 3 h of irradiation (>20-fold increase) and ~210 nM after 6 h of irradiation (>50-fold increase) [3].
17. This method is based on [30]. It is applicable to both *E. coli* or *A. brasilense*. Alternative methods for calculating c-di-GMP concentrations that rely on c-di-GMP detection via specific chemicals or riboswitches have been described [31, 32]. Furthermore, c-di-GMP-specific ELISA kits are commercially available.
18. We and other laboratories have been successfully using the Michigan State University Mass Spectrometry and Metabolomics Core Facility. If using a different facility you may be asked to supply c-di-GMP standards for quantification purposes. C-di-GMP is available from various vendors.
19. Use the remaining *E. coli* culture to determine viable cell count, e.g., by plating serial dilutions onto LB agar containing ampicillin. Use the number of cells submitted for LC-MS/MS to calculate approximate intracellular c-di-GMP concentration. *Example*: Assume that c-di-GMP was extracted from 5×10^{10} cells (this corresponds to ~50 mg of wet biomass). Assume that c-di-GMP concentration in the 200 μL (2×10^{-4} L) buffer submitted for LC-MS/MS is 50 nM (50×10^{-9} M). Based on

these data, the total molar c-di-GMP amount extracted from bacterial sample is $(50 \times 10^{-9}$ M$) \times (2 \times 10^{-4}$ L$) = 1 \times 10^{-11}$ mole. Since 1 mole corresponds to 6.02×10^{23} molecules (Avogadro number), 1×10^{-11} mole corresponds to $(6.02 \times 10^{23}$ molecules/M$) \times (1 \times 10^{-11}$ moles$) = 6.02 \times 10^{12}$ c-di-GMP molecules. By dividing the number of molecules by the number of cells from which c-di-GMP was extracted, we obtain an approximate number of c-di-GMP molecules per cell: $(6.02 \times 10^{12}$ molecules$)/(5 \times 10^{10}$ cells$) = 120$ molecules/cell. Based on the assumption that 1 molecule in an average *E. coli* cell corresponds to ~1 nM concentration [33], the approximate intracellular concentration of c-di-GMP is 120×1 nM $= 120$ nM. An alternative method of normalizing c-di-GMP concentration may use total soluble protein, which can be measured based on the remaining 1 mL of *E. coli* culture from Subheading 3.6. Protein concentration can be determined using various commercial kits.

20. While pBlue-PDE contains both BphS and EB1, expression levels of BphS in *A. brasilense* are apparently low in the absence of induction; therefore, pBlue-PDE can be used as a source of the blue light-dependent PDE [5]. It is important to note that for bidirectional control of c-di-GMP levels, expression levels for BphS and EB1 need to be carefully optimized for any given strain. Further, MBP-EB1 is preferred over EB1 because of its superior dynamic range (fold-activation) by blue light.
21. Strains carrying the pRed-DGC grow slower than strains carrying pIND4 or pBlue-PDE. The cause for this effect on growth is unknown. Strains carrying pRed-DGC should thus be inoculated at least 1 day prior to strains carrying empty vectors or pBlue-PDE to ensure equivalent growth at the planned time for the experiment.
22. Aerotaxis bands typically form in less than 2 min after exposure to the air gradient in the wild-type strain. The bands are pretty stable and cells remain motile for at least 25 min under the conditions described here.

Acknowledgments

This work was supported in part by National Science Foundation grants MCB1052575 (to MG) and MCB13330344 (to GA). LON was supported by the National Institutes of Health grant R25GM086761.

References

1. Römling U, Galperin MY, Gomelsky M (2013) Cyclic di-GMP: the first 25 years of a universal bacterial secondary messenger. Microbiol Mol Biol Rev 77:1–52
2. Tarutina M, Ryjenkov DA, Gomelsky M (2006) An unorthodox bacteriophytochrome from *Rhodobacter sphaeroides* involved in turnover of the second messenger c-di-GMP. J Biol Chem 281:34751–34758
3. Ryu MH, Gomelsky M (2014) Synthetic second messenger module controlled by near-infrared window light. ACS Synth Biol 3:802–810
4. Ryu MH, Fomicheva A, Moskvin OM, Gomelsky M (2017) Optogenetic module for dichromatic control of c-di-GMP signaling. J Bacteriol pii: JB.00014-17. doi: 10.1128/JB.00014-17. [Epub ahead of print]
5. O'Neal L, Ryu MH, Gomelsky M, Alexandre G (2017) Optogenetic manipulation of c-di-GMP levels reveals the role of c-di-GMP in regulating aerotaxis receptor activity in Azospirillum brasilense. J Bacteriol pii: JB.00020-17. doi: 10.1128/JB.00020-17. [Epub ahead of print]
6. Rockwell NC, Su YS, Lagarias JC (2006) Phytochome structure and signaling mechanisms. Annu Rev Plant Biol 57:837–858
7. Orr MW, Donaldson GP, Severin GB, Wang J, Sintim HO, Waters CM, Lee VT (2015) Oligoribonuclease is the primary degradative enzyme for pGpG in *Pseudomonas aeruginosa* that is required for cyclic-di-GMP turnover. Proc Natl Acad Sci U S A 112:E5048–E5057
8. Cohen D, Mechold U, Nevenzal H, Yarmiyhu Y, Randall TE, Bay DC, Rich JD, Parsek MR, Kaever V, Harrison JJ, Banin E (2015) Oligoribonuclease is a central feature of cyclic diguanylate signaling in *Pseudomonas aeruginosa*. Proc Natl Acad Sci U S A 112:11359–11364
9. Terpe K (2003) Overview of tag protein fusions: from molecular and biochemical fundamentals to commercial systems. Appl Microbiol Biotechnol 60:523–533
10. Gomelsky M, Klug G (2002) BLUF: a novel FAD-binding domain involved in sensory transduction in microorganisms. Trends Biochem Sci 27:497–500
11. Conrad KS, Manahan CC, Crane BR (2014) Photochemistry of flavoprotein light sensors. Nat Chem Biol 10:801–809
12. Dai T (2017) The antimicrobial effect of blue light: what are behind? Virulence 4:1–4
13. Fang X, Gomelsky M (2010) A post-translational, c-di-GMP-dependent mechanism regulating bacterial flagellar motility. Mol Microbiol 76:1295–1305
14. Hengge R, Galperin MY, Ghigo JM, Gomelsky M, Green J, Hughes KT, Jenal U, Landini P (2015) Systematic nomenclature for GGDEF and EAL domain-containing c-di-GMP turnover proteins of *Escherichia coli*. J Bacteriol 198:7–11
15. Greer-Phillips S, Stephens B, Alexandre G (2004) An energy taxis transducer promotes root colonization by *Azospirillum brasilense*. J Bacteriol 186:6595–6604
16. Shanks RMQ, Kadouri DE, MacEachran DP, O'Toole GA (2009) New yeast recombineering tools for bacteria. Plasmid 62:88–97
17. Ind AC, Porter SL, Brown MT, Byles ED, de Beyer JA, Godfrey SA, Armitage JP (2009) Inducible-expression plasmid for *Rhodobacter sphaeorides* and *Paracoccus denitrificans*. Appl Environ Microbiol 75:6613–6615
18. Laszlo DJ, Taylor BL (1981) Aerotaxis in Salmonella typhimurium: role of electron transport. J Bacteriol 145:990–1001
19. Scharf BE, Hynes MF, Alexandre GM (2016) Chemotaxis signaling systems in model beneficial plant–bacteria associations. Plant Mol Biol 90:549–559
20. Zhulin IB, Bespalov VA, Johnson MS, Taylor BL (1996) Oxygen taxis and proton motive force in *Azospirillum brasilense*. J Bacteriol 178:5199–5204
21. Russell MH, Bible AN, Fang X, Gooding JR, Campagna SR, Gomelsky M, Alexandre G (2013) Integration of the second messenger c-di-GMP into the chemotactic signaling pathway. mBio 4:e00001–e00013
22. Taylor BL, Watts KJ, Johnson MS (2007) Oxygen and redox sensing by two-component systems that regulate behavioral responses: behavioral assays and structural studies of Aer using in vivo disulfide cross-linking. Methods Enzymol 422:190–232
23. Boehm A, Kaiser M, Li H, Spangler C, Kasper CA, Ackermann M, Kaever V, Sourjik V, Roth V, Jenal U (2010) Second messenger-mediated adjustment of bacterial swimming velocity. Cell 141:107–116
24. Hu Y, Wu Y, Mukherjee M, Cao B (2017) A near-infrared light responsive c-di-GMP module-based AND logic gate in *Shewanella oneidensis*. Chem Commun (Camb) 53:1646–1648
25. Folcher M, Oesterle S, Zwicky K, Thekkottil T, Heymoz J, Hohmann M, Christen M, Daoud El-Baba M, Buchmann P, Fussenegger M

(2014) Mind-controlled transgene expression by a wireless-powered optogenetic designer cell implant. Nat Commun 5:5392

26. Barends TRM, Hartmann E, Griese J, Beitlich T, Kirienko NV, Ryjenkov DA, Reinstein J, Shoeman RL, Gomelsky M, Schlichting I (2009) Structure and mechanism of a bacterial light-regulated cyclic nucleotide phosphodiesterase. Nature 459:1015–1018
27. Savakis P, De Causmaecker S, Angerer V, Ruppert U, Anders K, Essen LO, Wilde A (2012) Light-induced alteration of c-di-GMP level controls motility of *Synechocystis* sp. PCC 6803. Mol Microbiol 85:239–251
28. Enomoto G, Nomura R, Shimada T, Ni-Ni-Win, Narikawa R, Ikeuchi M (2014) Cyanobacteriochrome SesA is a diguanylate cyclase that induces cell aggregation in *Thermosynechococcus*. J Biol Chem 289:24801–24809
29. Enomoto G, Ni-Ni-Win, Narikawa R, Ikeuchi M (2015) Three cyanobacteriochromes work together to form a light color-sensitive input system for c-di-GMP signaling of cell aggregation. Proc Natl Acad Sci U S A 112:8082–8087
30. Massie JP, Reynolds EL, Koestler BJ, Cong JP, Agostoni M, Waters CM (2012) Quantification of high-specificity cyclic diguanylate signaling. Proc Natl Acad Sci U S A 109:12746–12751
31. Tsuji G, Sintim HO (2016) Cyclic dinucleotide detection with riboswitch-G-quadruplex hybrid. Mol Biosyst 12:773–777
32. Kellenberger CA, Sales-Lee J, Pan Y, Gassaway MM, Herr AE, Hammond MC (2015) A minimalist biosensor: quantitation of cyclic di-GMP using the conformational change of a riboswitch aptamer. RNA Biol 12:1189–1197
33. Moran U, Phillips R, Milo R (2010) SnapShot: key numbers in biology. Cell 141:1262–1262.e1

Chapter 15

Analysis of c-di-GMP Levels Synthesized by a Photoreceptor Protein in Response to Different Light Qualities Using an In Vitro Enzymatic Assay

Veronika Angerer, Lars-Oliver Essen, and Annegret Wilde

Abstract

Diguanylate cyclases are enzymes that use two GTP molecules to produce one molecule cyclic dimeric guanosine monophosphate (c-di-GMP). This cyclic dinucleotide is an ubiquitous prokaryotic second messenger that controls a variety of cell functions. Several proteins have been described which contain a photoreceptor domain fused to a diguanylate cyclase. The cyanobacterial light sensor Cph2 is responsible for the blue-light induced synthesis of c-di-GMP in *Synechocystis* sp. PCC 6803. Here, we provide a detailed protocol for an in vitro enzymatic assay with a purified photoreceptor protein using light as the crucial reaction parameter for c-di-GMP synthesis. The assay is accomplished under continuous illumination with light of different quality with inactivation of the enzyme by heat denaturation. Analytics are performed using HPLC-UV.

Key words c-di-GMP, Second messenger, Cyclic nucleotide, Cyclic diguanylate, Enzyme-activity assay, Light-dependent enzymatic activity, Photosensor, Photoswitch, In vitro enzymatic method

1 Introduction

The small nucleotide second messenger molecule cyclic dimeric guanosine monophosphate (c-di-GMP) controls a variety of cell functions like the production of exopolysaccharides [1], the change between planktonic and sessile lifestyle [2], motility [3] and virulence [4]. It is therefore important to know how the cellular c-di-GMP level is regulated. Detailed in vitro investigations often give mechanistic insights into the biochemical function of a certain enzyme. An enzyme is a biological macromolecule (or here, a protein) that catalyzes a chemical reaction. The second messenger c-di-GMP is catalytically produced by diguanylate cyclases (DGCs) [5, 6]. These proteins contain a GGDEF domain, which is responsible for the enzymatic production of c-di-GMP. Degradation is mediated by phosphodiesterases (PDEs) using either EAL or HD-GYP domains [5, 6].

Karin Sauer (ed.), *c-di-GMP Signaling: Methods and Protocols*, Methods in Molecular Biology, vol. 1657,
DOI 10.1007/978-1-4939-7240-1_15, © Springer Science+Business Media LLC 2017

Here, we focus on a method for in vitro enzymatic studies of c-di-GMP metabolizing enzymes. The standard procedure for in vitro enzymatic measurements is to incubate the purified enzyme with the substrate. Different reaction conditions influencing the rate of enzymatic activity like incubation time, temperature, substrate concentration, substrate type, metal ion dependence or additives like inhibitors or competitors are standard parameters to be tested. However, light as a reaction parameter influencing the rate of enzymatic activity is rarely subject of enzymatic investigations.

Photoreceptor proteins are able to sense light using chromophores and transduce this signal to downstream signaling components [7, 8]. Typical photosensory domains include the flavin binding light-oxygen-voltage (LOV) domain of phototropins [9], the blue light sensing using FAD (BLUF) domain [10], the domain module PAS-GAF-PHY (PerARNT Sim, cGMP phosphodiesterase Adenylyl cyclase FhlA, Phytochrome) of classical phytochromes [11] or the bilin-binding domains of cyanobacteriochromes (CBCR) [12]. Interestingly, Agostoni et al. [13] showed that in cyanobacteria, photoreceptor modules are the second most frequent domains fused to c-di-GMP-related domains. The photoreceptor Cph2 from *Synechocystis* sp. PCC 6803 is one example: the C-terminal part of this protein contains a CBCR domain (domain 5) detecting green and blue light, fused to a GGDEF domain (domain 6) (Fig. 1a). Savakis et al. showed that this Cph2 module consisting of domains 5 and 6 (in the following Cph2 (5, 6)) produces c-di-GMP in a light-dependent manner: upon blue light

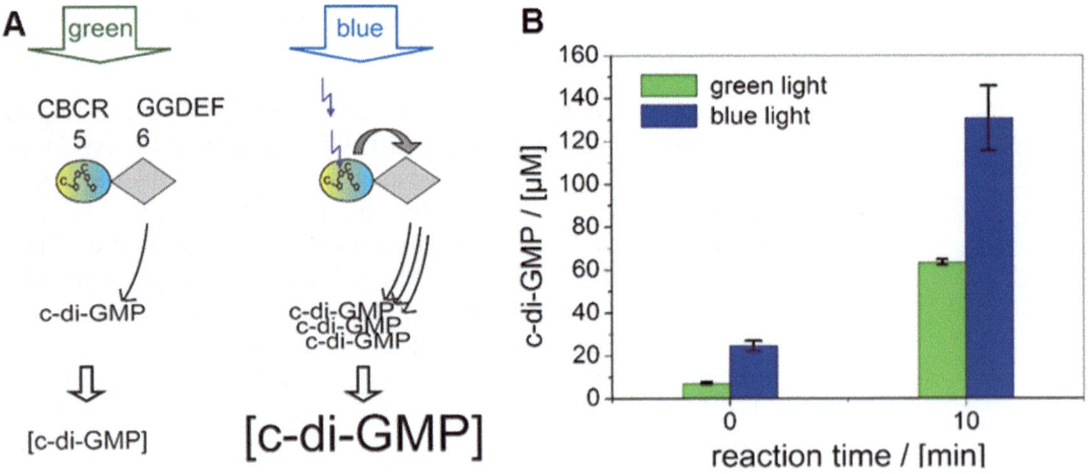

Fig. 1 (**a**) Scheme of enzymatic activity of domains 5 and 6 (CBCR and GGDEF) of Cph2 from *Synechocystis* sp. PCC 6803; when illuminating with *green light*, only a basal catalytic activity of the C-terminal GGDEF domain is detected; illumination with *blue light* stimulates the enzyme and c-di-GMP production is enhanced; (**b**) In vitro enzymatic activity of Cph2 (5, 6): under *blue light* illumination, a twofold higher c-di-GMP production is detected; data from [5]

illumination, the C-terminal GGDEF domain of Cph2 gets activated and shows a twofold higher enzymatic activity in comparison to green light illumination (Fig. 1b) [3].

This combination of photosensory and c-di-GMP domains in one protein is not unique to cyanobacteria. For instance, the protein YcgF from *E. coli* consists of a blue light sensing BLUF domain connected to an EAL domain [14]. Gomelsky et al. [15] have discussed more examples of combinations between photoreceptor and c-di-GMP domains. Here, we describe a method to measure light-dependent c-di-GMP production in an in vitro enzymatic assay using photoreceptor proteins expressed in *E. coli*.

2 Materials

This chapter assumes that a light-controlled enzyme is already purified to an acceptable extent. It also assumes that the photochemical properties of the protein have been determined beforehand (*see* **Note 1**). Also, the standard handling of an HPLC-UV/VIS machine should be known. If not, please refer to appropriate manuals, as this method paper is focusing mainly on the realization of the enzymatic reaction itself.

2.1 Sample Preparation

1. Spectrophotometer in a place that can be darkened completely.
2. Quartz cuvette.
3. Light source of specific wavelength (*see* **Note 2**).
4. Stand with post and clamp for fixing the light source above the cuvette.
5. Ruler or pocket rule to determine distance between light source and cuvette.
6. Protein solution with purified, light-responsive enzyme.
7. Assay buffer for enzymatic reaction, buffer exchange, and/or dilution of protein: 50 mM Tris/HCl, 50 mM NaCl, 10 mM $MgCl_2$, 0.5 mM EDTA, pH 7.5.

2.2 Enzymatic Assay

1. Room that can be darkened completely (*see* **Note 3**).
2. GTP (fresh aliquot) (100 mM) (*see* **Note 4**).
3. c-di-GMP (*BioLog*) dissolved in assay buffer (2 mM) (*see* **Note 5**).
4. Assay buffer.
5. Enzyme solution containing purified, light-sensing protein in assay buffer.
6. Two thermomixers: one at reaction temperature (e.g., 25 °C) and the other for heat inactivation at 98 °C.
7. Light source of specific wavelength (*see* also Subheading 2.1, **item 3** or **Note 2**).

8. Stand with post and clamp for fixing the light source above the reaction vial.
9. Ruler or pocket rule to determine distance between light source and reaction vial.
10. Safe lock reaction vials fitting into thermomixer.
11. Timer.

2.3 Analysis via HPLC-UV/VIS

1. Microcentrifuge (room temperature, min. 12,000 × *g*).
2. Ultrafree-MC GV centrifugal filter units, hydrophilic PVDF membrane with pore size 0.22 μm; for use in rotors for 1.5 mL tubes (*Merck Millipore*).
3. HPLC equipped with column heater, autosampler, and UV detector.
4. HPLC column: Nucleodur C18 Pyramid (*Macherey-Nagel*) with particle size 3 μm, column length 150 mm and inner diameter 2 mm; EC standard column.
5. Eluent A: 18 mM ammonium acetate, 1 mM acetic acid, pH 5.9; filter-sterilized.
6. Eluent B: methanol HPLC grade.
7. c-di-GMP (*BioLog*) dissolved in assay buffer for calibration plot of HPLC-system (2 mM) (*see* **Note 5**).
8. HPLC vials and caps.

3 Methods

Keep your light-sensing protein sample cooled on ice (4 °C) and protected from light (*see* **Note 6**) until use.

3.1 Sample Preparation and Logistics

The enzyme of interest has to be dissolved in the assay buffer. If the assay buffer is different from the purification buffer (*see* **Note 7**), a buffer exchange has to be done (*see* **Note 8**). Depending on the stability of the protein in the assay buffer, it might be necessary to perform the buffer exchange freshly every day or every time before the assay is performed. Before starting the enzymatic assay, it is crucial to check on the photochemical properties of the protein of interest in the assay buffer, especially if assay buffer and purification buffer differ from each other (*see* **Note 9**). The protein should display the photochemical behavior known from earlier studies, or studies should directly be presented in assay buffer. This also includes the determination of the illumination time needed for full photoconversion of the photoreceptor. For instance, the protein Cph2 (5, 6) does not significantly change its photostate after 15 min of illumination (Fig. 2). Thus, 15 min illumination is suitable for spectroscopic measurements as well as for enzymatic

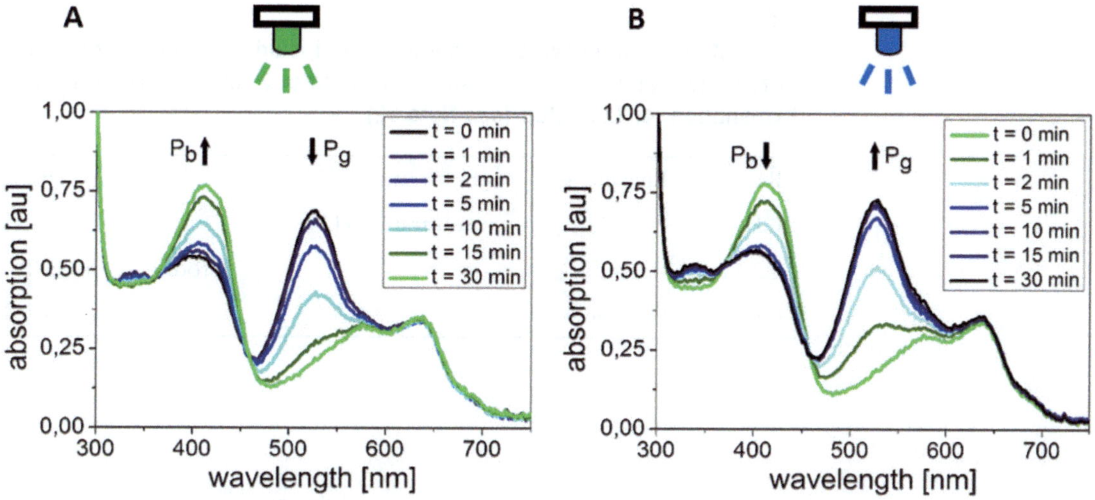

Fig. 2 Absorption spectra of Cph2 (5, 6) from *Synechocystis* sp. PCC 6803 after different times of illumination. (**a**) The protein was converted to the *green light* absorbing P_g photostate before illumination with *green light*; during illumination with *green light*, the absorption maximum of P_g at 525 nm decreases and the maximum of the *blue light* absorbing P_b photostate at 410 nm increases with illumination time; the protein solution contains a mixture of photoswitchable Cph2 (5, 6) ($\lambda_{max} = 525$ and 410 nm) and photochemically inactive forms of Cph2 (5, 6) (shoulders at 575 and 640 nm); (**b**) Photoswitch of the Cph2 (5, 6) module from P_b to P_g

activity tests of Cph2 (5, 6). The same parameters have to be used for the enzymatic assay (*see* **Note 10**).

1. To calculate the volume of a master mix, the minimal volume needed for the subsequent analysis should be determined. For the reaction setup described here a 100 µL sample is required for the final analysis (*see* **Note 11**). Next it has to be considered how many time points will be analyzed. Moreover, the number of different wavelengths to be tested is important, as one master mix containing the protein stock solution should be split into different aliquots for different light treatments to ensure that identical protein concentrations are tested under different light conditions. Lastly, the number of controls needs to be considered when determining the total volume of the master mix (*see* **Note 12**). An example of a test setup is presented in Table 1 and an example of calculation in **Note 13**. For a reaction setup including all controls, two different light qualities, five different time points per light quality and 100 µL aliquots taken per time point, a total volume of 2000 µL is needed (*see* **Note 13**).
2. Determine the concentration of the protein stock solution using protein assays (e.g., BCA assay). Prepare a dilution from the protein stock solution in assay buffer with a final concentration of 1.01 µM. Determine the required volume of the protein solution according to Subheading 3.1, **step 1** and **Note 13**.

Table 1
Test setup for an enzymatic activity assay including control reactions; enzymatic reaction (Rx); control 1 (C1), control 2 (C2), control 3 (C3). Explanation for the setup (*see* Note 12)

Rx	Enzyme +	Substrate (GTP)	
C1	–	Substrate (GTP)	
C2	–	–	Product (c-di-GMP)
C3	Enzyme +	–	

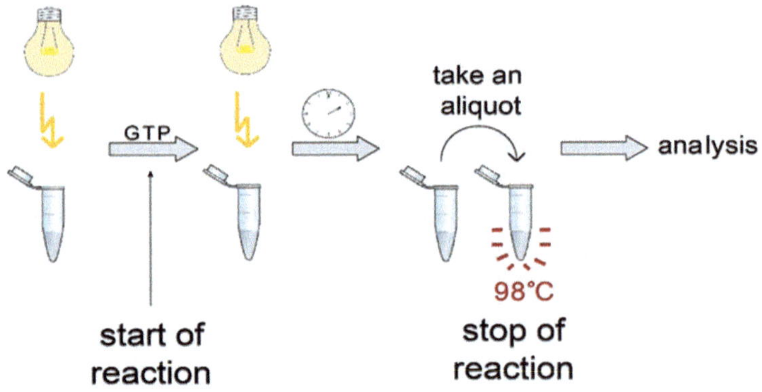

Fig. 3 Experimental sequence of a light-dependent enzymatic in vitro assay

3. The experiment must be carried out in a place that can be completely darkened. The setup for the experiment should be as follows: place the two thermomixers next to each other, with one set to the desired reaction temperature (e.g., 25 °C), the other to 98 °C for heat inactivation. The light source should be installed above the thermomixer in which the reaction will take place. Light should shine directly from the top into the vials, without any angle (or with same angle as for spectroscopic characterization). Distance between light source and thermomixer must be the same as for the spectroscopic analysis. Labeled safe lock vials (*see* **Note 14**) should be prepared and placed into the thermomixer at 98 °C (keep the lids open). Pipettes, tips, and a waste container shall be installed. The protein master mix of correct concentration should be stored on ice and protected from light. Likewise, store a fresh aliquot of GTP (100 mM) and a solution of c-di-GMP (2 mM) on ice. Additionally, prepare a timer to count the time once the enzymatic reaction has started. A schematic overview of the assay procedure is shown in Fig. 3.

Table 2
Example for final reaction parameters

Assay buffer	50 mM Tris/HCl 50 mM NaCl 10 mM $MgCl_2$ 0.5 mM EDTA pH 7.5
Enzyme concentration	1 μM
Substrate concentration (GTP)	1 mM
Reaction temperature	25 °C

3.2 Enzymatic Assay

The final reaction parameters and solution composition of the enzymatic reaction are summarized in Table 2. Prepare microfuge tubes for the enzymatic reaction as well as for the controls. To conduct the enzymatic assay, the different components are mixed as follows:

1. Add the enzyme solution to a final concentration of 1 μM according to the scheme of Table 1 to the reaction vial (Rx) and the vial for control 3 (C3) in a final volume calculated according to Subheading 3.1, **step 1**. No GTP as substrate is yet added in this step. For control reactions 1 and 2, add buffer to the respective vials considering the calculated final reaction volume.
2. Transfer the microfuge tube to the thermomixer at 25 °C. 10 min of incubation is suitable to bring the solution to the required reaction temperature of 25 °C.
3. Ensure that the area where the enzymatic assay will be carried out is completely dark. The only light should come from the light source with specific wavelength mounted above the thermomixer set to the reaction temperature (e.g., 25 °C) (*see* **Note 15**) and will be kept on during the whole process until the last aliquot has been taken (Subheading 3.2, **step 17**).
4. The enzyme solution is illuminated with light of one specific wavelength for the time needed to convert the enzyme into one specific photostate. To provide similar conditions as in the spectroscopic measurement (where a cuvette is used), ensure that the lid of the microcentrifuge tube is open during the time of illumination. The duration of the illumination is enzyme dependent and will need to be determined prior to this experiment (*see* Subheading 3.1).
5. All the actions described in the following steps should be carried out as fast as possible (according to the pipetting scheme of **Note 13**).

6. Add 20 μL water to the vial of control 3 containing 1980 μL protein solution. Mix by pipetting up and down 3–4 times.
7. Add 20 μL c-di-GMP (final concentration 20 μM) to the vial of control 2 containing 1980 μL assay buffer. Mix by pipetting up and down 3–4 times.
8. Add 20 μL GTP (final concentration 1 mM) to the vial of control 1 containing 1980 μL assay buffer. Mix by pipetting up and down 3–4 times.
9. Start the enzymatic reaction by adding GTP. The reaction solution should immediately be mixed by pipetting up and down 3–4 times.
10. Start the timer to count the reaction time.
11. Take an aliquot for the time point $t = 0$ min from the vial of the enzymatic reaction. Transfer the aliquot into the preheated microfuge tube located in the thermomixer set to 98 °C for heat inactivation.
12. Close the lid of the microfuge tube to avoid evaporation.
13. Then take as well an aliquot for the time point $t = 0$ min from the vial of control 1, control 2 and control 3 each. Transfer the aliquots into the respective preheated microfuge tube located in the thermomixer set to 98 °C.
14. Close the lids of the microfuge tube to avoid evaporation.
15. Incubate the aliquots at 98 °C for at least 3 min.
16. Following heat inactivation, store the microfuge tubes on ice until all time points have been taken. If samples are not processed right away they can be stored overnight or for several days at −20 °C.
17. For additional time points, repeat Subheading 3.2, **steps 11–16**.
18. For enzymatic test of a different light quality, repeat Subheading 3.2, **steps 1–17** using a different light source.

3.3 Analysis

1. Centrifuge the samples at 12,000 × g for 15 min at room temperature (RT) to precipitate denatured protein.
2. Filter the supernatant using Ultrafree-MC GV centrifugal filter units (0.22 μm). This step ensures the removal of precipitated protein from the denaturation step as well as residual particles left and prepares the sample for HPLC injection.
3. Transfer a suitable amount of sample into an HPLC vial. For most HPLC machines, 25 μL of sample are suitable to ensure the injection of 10 μL once into the machine. However, the sample volume might differ depending on the number of injections (*see* **Note 16**) and the dead volume of the used HPLC machine (*see* **Note 17**).

4. Perform HPLC-UV analysis, e.g., using the following parameters:
   (a) Column type: EC 150/2 Nucleodur C18 Pyramid 3 μm (*Macherey-Nagel*).
   (b) Column temperature during run: 40 °C.
   (c) Flow rate: 0.2 mL/min.
   (d) Detection at wavelength: 254 nm.
   (e) Injection volume: 10 μL.
   (f) Injections per sample: 1.
   (g) Elution takes place using the gradient presented in Table 3. Retention time is 10.5 min for c-di-GMP, 4 min for GTP and buffer components elute at 2.5 min (Fig. 4).

Table 3
Elution profile for HPLC analysis

Time (min)	Eluent A (%)	Eluent B (%)
0	100	0
10	80	20
15	5	95
18	5	95
20	100	0

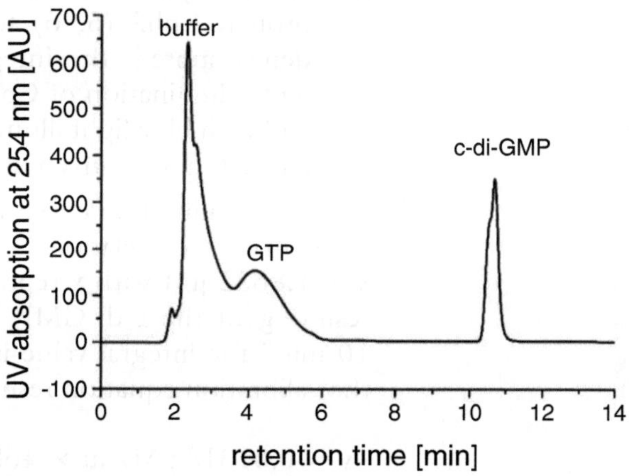

Fig. 4 Chromatogram of HPLC analysis of an enzymatic reaction mixture of a DGC protein

5. Using the HPLC analysis software, determine the area of the c-di-GMP peak detected in the chromatogram (at ~10.5 min) of the respective sample.
6. Repeat Subheading 3.3, **steps 1–5** for all enzymatic samples to be analyzed.

3.4 Generation of a c-di-GMP Standard Curve

1. Using the 2 mM c-di-GMP stock solution, prepare the following standards in assay buffer: 10, 50, 100, 150, and 200 μM (*see* **Note 18**).
2. Inject and run on the HPLC 10 μL of the 10 μM c-di-GMP standard using same parameters as described in Subheading 3.3, **step 4**.
   (a) Repeat to create a technical duplicate.
3. Repeat Subheading 3.4, **step 2** for the 50, 100, 150, and 200 μM c-di-GMP standards.
4. Inject and run 10 μL of assay buffer as a negative control (0 μM c-di-GMP).
5. Use the HPLC analysis software (e.g., ChemStation) to obtain the peak area for each of the standard concentrations.
6. Prepare a standard curve by plotting the peak areas versus the c-di-GMP amount in μM (*see* Fig. 5 for an example).
7. Calculate the linear trend line and the R^2 coefficient of the plotted points (*see* Fig. 5 for an example).

3.5 Calculations

1. Use the obtained equation from the calibration to calculate the amount of c-di-GMP present in the samples by inserting the integral value of UV absorption from measured samples into equation.
   (a) An exemplary calculation of c-di-GMP content of the protein Cph2 (5, 6) tested at green and blue light is demonstrated: the integrals of the samples with green light illumination of Cph2 (5, 6) after 0 min and 10 min and with blue light illumination of Cph2 (5, 6) after 0 min and 10 min are given in Table 4.
2. Using the calibration curve shown in Fig. 5, the fitting gives a mathematic correlation of $y = 0.01313\ \mu M/au \times x + 1.8632\ \mu M$ with x representing the integrals value and y resulting in the c-di-GMP content. For the sample "green 10 min" the integral value is 4685 au and inserting this into the calibration equation results in

$$y = 0.01313\ \mu M/au \times 4685\ au + 1.8632\ \mu M = 63.38\ \mu M$$

   (a) Thus, the protein Cph2 (5, 6) under green light illumination produced 63.38 μM c-di-GMP in 10 min.

a

Concentration of c-di-GMP [μM]	Integral 1 [au]	Integral 2 [au]	mean value integrals [au]	standard deviation [au]
200	15178	15153	15165.5	17.68
150	11446	11457	11451.5	7.78
100	6956	6962	6959	4.24
50	3813	3801	3807	8.49
10	761	763	762	1.41

b

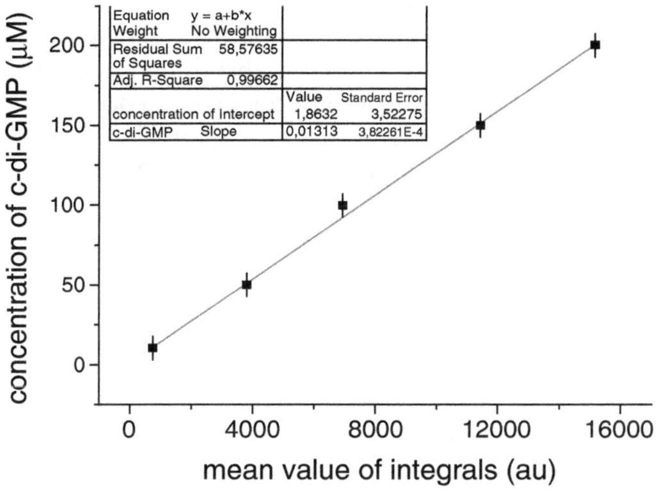

Fig. 5 (**a**) Example of integral values obtained for two independent measurements of the respective c-di-GMP standards; au = arbitrary unit; (**b**) Example of a graph showing mean value of integrals plotted versus c-di-GMP concentration; errors are standard deviation; a linear trend line was fitted giving a correlation of y = 0.01313 μM/au × x + 1.8632 μM and an R^2 of 0.99

Table 4
Integrals of c-di-GMP measurements of Cph2 (5, 6) under green or blue light illumination

Sample (min)	Integral 1 (au)	Integral 2 (au)	Mean value of integrals (au)	Concentration of c-di-GMP (μM)
Green 0	394	406	400	7.12
Green 10	4677	4692	4685	63.38
Blue 0	1726	1731	1729	24.57
Blue 10	9787	9811	9799	130.53

(b) In the same way, the c-di-GMP content can be calculated for the other time points. It should be noted that there might be a delay before the enzyme is completely deactivated.

(c) Thus, it is possible that the c-di-GMP concentration is not 0 at $t = 0$ min (*see* **Note 19**).

3. All further c-di-GMP contents of the different time points and reaction conditions of any desired DGC can be determined using this method.
   (a) For the calculation of standard enzymatic parameters like the maximum reaction rate v_{max}, the turnover number k_{cat}, the Michaelis constant K_M, dissociation constant K_D, 1/2 v_{max}, etc. based on the obtained c-di-GMP production rates, please refer to standard manuals of enzymatic kinetics [16, 17].
4. In order to assign the enzymatic activity under different light sources to a specific light-absorbing form of the photoreceptor, the ratio between the photostates, a possible photoequilibrium mixture or the presence of photochemically inactive contaminations should be considered (*see* **Note 1**).

4 Notes

1. Before starting the enzymatic assay with a photosensor domain protein, it is mandatory to know its photochemical behavior. Thus, spectroscopic and photochemical studies must be performed in advance. Interesting parameters are λ_{max} of photostates, time of illumination needed for photoconversion and stability of the photocycle: how often can the protein be switched from one photostate to the other? Is dark reversion happening? When basic photochemical parameters have been determined, investigation of the enzymatic activity can be started. Further information can be found in manuals for the spectroscopic characterization of photoreceptors [7, 8]. However, photosensors do not necessarily convert to 100% between different photochemical states but might consist of a ratio of, e.g., 80% one photostate plus some residual 20% of the other photostate. One example is the cyanobacterial phytochrome 1 (Cph1): due to the spectral overlap of the red-absorbing state P_r (λ_{max} ~ 660 nm) and the far red absorbing state P_{fr} (λ_{max} ~ 720 nm), the photoconversion achieved by saturating irradiation with red light results in a photoequilibrium consisting of maximally 70% P_{fr} and residual 30% P_r state [18]. The calculation of the photostate-dependent enzymatic activity should then consider the ratio of background activity obtained from photoequilibrium mixtures and/or photochemically inactive species (e.g., like has been shown for the kinase activity of Cph1 [19]). The information about ratios of photochemical states as well as the percentage of a possible photochemically inactive species can be achieved by deconvolution of spectra and subsequent calculations. Manuals concerning this topic can be found in text books [7, 8] and in specific literature on photoreceptor research [20–23].

2. Samples should be illuminated with light resembling λ_{max} of the respective absorption peaks of the photostates. Thus, for every photoreceptor, at least two different light sources are needed. For example, for Cph2 (5, 6) λ_{max} have been determined at 525 nm for the blue light absorbing state and at 410 nm for the green light absorbing state (Fig. 2) [3]. Consequently, for this protein a light source with λ_{max} at 525 nm and a second light source with λ_{max} at 410 nm should be used for detailed spectroscopic and enzymatic analysis. In contrast, for the N-terminal module of Cph2 λ_{max} is 644 nm for the red-light absorbing form and 695 nm for the far-red light absorbing form and therefore light sources for these two wavelengths are recommended for Cph2 (1, 2) [24]. Preliminary bioinformatics analysis of the primary protein sequence, its domain organization and literature search can often give a first hint to which group of photoreceptor belongs a specific protein. For first preliminary spectroscopic studies following light qualities have been established: BLUF –> blue; LOV –> blue; phytochrome –> red/far red; CBCR –> it has not been possible to predict the photocycle of a CBCR from its primary amino acid sequence so far except for two groups of DXCF CBCRs (for further information *see* ref. [25]). As light sources LEDs are recommended because of the defined light input. LEDs with specific emission characteristics (specific λ_{max} of emission and small half width) can easily be obtained commercially (e.g., from Roithner LaserTechnik). LEDs need to be installed on an electronic device or simply by replacing white light LEDs of a commercial flashlight.
3. If there is no dark room available in your laboratory, you can convert a temporally free laboratory into a dark room by darkening windows with blinds and covering or turning off monitors and machines with screens to eliminate any source of light.
4. GTP can dephosphorylate to GDP or GMP, e.g., by repeated thaw-freeze cycles. To avoid any competing effects of GDP or GMP, a fresh aliquot of GTP should be used. The easiest possibility to guarantee freshness is to aliquot GTP immediately after receiving from the manufacturer. Dissolving on ice (only needed for lyophilized GTP) and subsequently freezing aliquots is recommended. When needed, a new aliquot can be taken. Repeated thaw-freeze cycles should be avoided.
5. Dissolve the lyophilized c-di-GMP in assay buffer to obtain a 2 mM stock solution. Vortex the solution and let it incubate for 2 h at room temperature to make sure the c-di-GMP is properly dissolved. Vortex again, spin down residual drops from the lid, and use this stock solution for the calibration curve. It is important that c-di-GMP is in the same buffer as the samples, as different buffers have effect on retention time in HPLC

analysis, thus correlation of calibration curve and samples would not be possible.

6. While it is rarely mentioned in the literature, photosensor proteins (and proteins in general) produced in a heterologous host like *E. coli* are unstable over time, upon prolonged exposure to light or when repeatedly switched between the photostates [7, 26, 27]. Photoinactive species can evolve as a result of instability. Therefore, it is recommended to keep the light-sensing protein under conditions that protect the protein. For this purpose, proteins are usually kept in a cold, dark place while the experiment is carried out, by storing the enzyme on ice in an ice bucket with a lid. To protect the enzyme during handling and incubation steps, the reaction vial containing the protein solution can be wrapped with aluminum foil. Alternatively, black non-transparent reaction vials can also be used. However, black reaction vials impede a quick look for detecting the solutions color (due to the chromophore, solutions of photoreceptor proteins often are colored; some photoreceptors change color depending on whether they are present in the one or the other photostate) or for the occurrence of precipitates of the protein.
7. It has been shown that DGCs require the binding of two Mg^{2+} or Mn^{2+} cations for catalytic activity [6]. Therefore, buffers used to test enzymatic activity of GGDEF domain proteins must contain Mg^{2+} or Mn^{2+} cations. However, it should be checked whether manganese can be avoided during the purification procedure because of its relatively high price and heavy metal waste production. For examples of successfully used assay buffers see specific publications [28–30].
8. Buffer exchange can be performed in different ways, e.g., using PD-10 columns (*GE Healthcare*) or centrifugal filter devices such as centricons (*Merck Millipore*).
9. It is possible that the protein of interest exhibits altered photochemical properties following buffer exchange. To ensure that the protein's photochemical properties are the same prior to and post buffer exchange, record spectra and compare. Also, consider that the purified and/or rebuffered protein solution might be stored for a certain time (e.g., overnight). Before the enzymatic activity is tested, check the photochemistry of the protein to ensure that the activity is correlated correctly to the photochemical state of the protein.
10. The distance should not be too close, as aliquots are taken from the reaction mixture using a pipette. Thus, a pipette and the workers hand should fit in between the lamp and vial. If illumination for the spectroscopic characterization was done at a very close distance, consider redoing the spectroscopic analysis at a

distance that is comparable to the enzymatic experiments described here. However, distance and light source must be the same for comparing light-dependent photostates with enzymatic behavior.

11. In the method described here, c-di-GMP concentration is measured using HPLC-UV/VIS and a single injection into HPLC machine takes 10 μL sample. One injection is enough as the HPLC-UV measurement is an extremely accurate method and thus does not need technical duplicates (with exception for the calibration curve, *see* also **Note 16**). Still it is necessary to provide more sample volume than 10 μL as in most HPLC machines the injection needle for taking up the sample does not reach the absolute bottom of the vial. A certain dead volume is always present (for the determination of the dead volume of a specific HPLC machine, *see* **Note 17**). Considering the case the dead volume is 8 μL and 10 μL will be injected, a volume of more than 18 μL is recommended to secure the whole 10 μL have been injected (if less than 10 μL or an unknown volume of the samples are injected, the amount of c-di-GMP detected in the measurement cannot be correlated to a specific concentration). Considering pipetting errors, a volume of 25 μL (18 μL + 7 μL = 25 μL) is a suitable amount in an HPLC vial to secure a 10 μL injection. Thus, a 25 μL sample volume is needed for one HPLC measurement. In addition, a threefold volume (75 μL) is suggested to be prepared in case the analytics has to be repeated and because of inaccuracies in pipetting.

12. The enzymatic reaction has to be stopped at a certain time point, e.g., by incubating the reaction solution at 98 °C for 3 min. The heat exposure should denature the protein but not the enzymatic product or substrate. However, it is possible that the enzymatic product or substrate might be degraded, either during illumination or heat inactivation. Therefore, when performing an enzymatic assay, it is important to include controls containing only the substrate/product but lacking the enzyme, with the controls being treated the same way as the samples containing the enzymes. GTP can be dephosphorylated (*see* **Note 4**) and depending on the enzyme, GDP or GMP might have a competitive or inhibitory effect. "Control 1" of the setup (*see* Table 1) contains only GTP without enzyme and will demonstrate whether GTP was dephosphorylated before the reaction was started (sample taken at time point $t = 0$ min from the vial "control 1") and during the assay (sample taken at later time points from the vial "control 1"). "Control 2" contains only c-di-GMP without enzyme and aliquots taken from the vial "control 2" demonstrate whether c-di-GMP is degraded over time. In the third control the

enzyme is incubated in buffer. In some cases, this control has given a hint to unknown signals appearing in the analytics, which turned out to be cofactors, nucleotides, or other molecules bound to the enzyme and released by heat denaturation.

13. Example of calculation: the volume needed for final analysis is 100 μL; samples shall be taken at 5 time points: $t = 0, 1, 2, 10, 20$ min; two different light colors shall be tested; and enzyme solution is needed for two setup lines, for the actual reaction (Rx) and for "control 3" (C3) (*see* also **Note 13** and Table 1):

    100 μL × 5 time points × 2 light colors × 2 vials (Rx & C3) = 2000 μL in total

    Thus, 2000 μL final solution are needed, 500 μL for the "reaction" and 500 μL for the "control 3" per light condition (here two light conditions). Further, the GTP stock solution needs to be added in a 1:100 dilution, see the following pipetting scheme:

$$\frac{\begin{array}{c}1980\ \mu L \text{ protein solution } (1.01\ \mu M)\\ +20\ \mu L \text{ GTP stock solution } (100\text{ mM})\end{array}}{2000\ \mu L \text{ final volume}}$$

    The protein concentration in the final volume will be 1 μM. The same volumes are used for control reactions. Instead of GTP, c-di-GMP or water is used. Depending on the control, the protein solution is replaced by reaction buffer (*see* Table 1).

14. Safe seal vials are recommended. Alternatively, clamps to secure the lid from popping open or a heavy object to keep the lid closed, can be used.

15. If the absorbed wavelength of the photoreceptor is in a range where the human vision is poor and thus, the room is too dark to perform the pipetting steps, an additional light source can be considered. The additional light source should generate a wavelength that cannot be absorbed by the protein (e.g., for red/far-red absorbing phytochromes, green light can be used as "safety light"). To ensure that the wavelength generated by the safety light does not disturb the photocycle, include the respective light source into the spectroscopic characterization of the protein prior to use. Also, during the enzymatic assay, the light can be placed at the far side of the samples. This usually gives enough light for the experimenter to perform experiments.

16. One injection is enough as the HPLC-UV measurement is an extremely accurate method and thus does not need technical duplicates. The error in this enzymatic assay is much higher due to inaccuracy of the determination of protein

concentration and pipetting errors while preparing enzyme dilution and adding GTP.

17. The dead volume of an HPLC machine can be determined prior to sample injection by setting up several HPLC vials in the autosampler providing a decreasing amount of water, e.g., 20 μL, 18 μL, 16 μL, ... 2 μL. The HPLC machine is then programmed to inject 10 μL each. After injection, the water rest remaining in the HPLC vial is measured either by determining the volume by a pipet or by taking the weight of the water spotted onto an ultra-fine balance (1 mg = 1 μL). The difference in the volume before and after injection is determined. The sample where the difference is less than 10 μL indicates that the HPLC needle could not take up the full 10 μL injection volume. This sample indicates the dead volume.
18. The calibration curve should cover the range where c-di-GMP production is obtained from enzymatic reactions. In case of offset, repeat calibration with a different calibration range. For example, if the amount of c-di-GMP obtained from enzymatic reaction is between 0 μmol (for $t = 0$ min) and 70 μmol (for $t = 10$ min), the calibration curve should cover a range of 0–100 μmol c-di-GMP.
19. It can be discussed which value is the correct one for the time point $t = 0$ min. Theoretically, at $t = 0$ min no substrate has been consumed and no product should be detected. Practically, it needs a few seconds to perform Subheading 3.2, **steps 1–17** and another few seconds until the enzyme has been heat inactivated. Therefore, when the sample $t = 0$ min is taken *after* GTP addition, enzymes with a very fast DGC activity will already show some c-di-GMP production at $t = 0$ min. In this case, it can be discussed if one sample is taken *before* GTP addition, and the first time point not earlier than a few seconds (e.g., 10 s) after GTP addition. This needs to be considered for every protein and every application individually.

Acknowledgment

V.A. acknowledges funding by the Jürgen-Manchot-Stiftung.

References

1. Ross P, Weinhouse Y, Aloni Y et al (1987) Regulation of cellulose synthesis in *Acetobacter xylinum* by cyclic diguanylic acid. Nature 325:279–281
2. Ha D, O'Toole G (2015) C-di-GMP and its effects on biofilm formation and dispersion: a *Pseudomonas aeruginosa* review. Microbiol Spectr 3:1–12
3. Savakis P, De Causmaecker S, Angerer V et al (2012) Light-induced alteration of c-di-GMP level controls motility of *Synechocystis* sp. PCC 6803. Mol Microbiol 85:239–251

4. Römling U, Gomelsky M, Galperin M (2005) C-di-GMP: the dawning of a novel bacterial signalling system. Mol Microbiol 57:629–639
5. Hengge R (2009) Principles of c-di-GMP signalling in bacteria. Nat Rev Microbiol 7:263–273
6. Römling U, Galperin M, Gomelsky M (2013) Cyclic di-GMP: the 25 years of a universal bacterial second messenger. MMBR 77:1–52
7. Batschauer A (ed) (2003) Photoreceptors and light signaling. The Royal Society of Chemistry, Cambridge
8. Briggs WR, Spudich JL (eds) (2005) Handbook of photosensory receptors. Wiley-VCH, Weinheim
9. Christie JM, Schwartz T, Bogomolni RA et al (2002) Phototropin LOV domains exhibit distinct roles in regulating photoreceptor function. Plant J 32:205–219
10. Gomelsky M, Klug G (2002) BLUF: a novel FAD-binding domain involved in sensory transduction in microorganisms. Trends Biochem 27:497–500
11. Rockwell N, Su Y, Lagarias J (2006) Phytochrome structure and signaling mechanisms. Annu Rev Plant Biol 57:837–858
12. Ikeuchi M, Ishizuka T (2008) Cyanobacteriochromes: a new superfamily of tetrapyrrole-binding photoreceptors in cyanobacteria. Photochem Photobiol Sci 7:1159–1167
13. Agostoni M, Koestler C, Waters CM et al (2013) Occurrence of cyclic di GMP-modulating output domains in cyanobacteria: an illuminating perspective. mBio 4:1–10
14. Tschowri N, Busse S, Hengge R (2009) The BLUF-EAL protein YcgF acts as a direct anti-repressor in a blue-light response of *Escherichia coli*. Genes Dev 23:522–534
15. Gomelsky M, Hoff WD (2011) Light helps bacteria make important lifestyle decisions. Trends Microbiol 19:441–448
16. Berg JM, Tymoczko JL, Gatto GJ Jr, Stryer L (2015) Biochemistry, 8th edn. Macmillan Education, New York
17. Voet D, Voet JG (2010) Biochemistry, 4th edn. John Wiley & Sons, Inc., New York
18. Strauss HM, Schmieder P, Hughes J (2005) Light-dependent dimerisation in the N-terminal sensory module of cyanobacterial phytochrome 1. FEBS Lett 579:3970–3974
19. Psakis G, Mailliet J, Lang C et al (2011) Signaling kinetics of cyanobacterial phytochrome Cph1, a light regulated histidine kinase. Biochemistry 50:6178–6188
20. Pratt LH (1975) Photochemistry of high molecular weight phytochrome *in vitro*. Photochem Photobiol 22:33–36
21. Pratt LH (1978) Molecular properties of phytochrome. Photochem Photobiol 27:81–105
22. Manchinelli AL (1986) Comparison of spectral properties of phytochromes from different preparations. Plant Physiol 82:956–961
23. Billo EJ (2001) Analysis of spectrophotometric data. In: Excel® for chemists: a comprehensive guide, 2nd edn. John Wiley & Sons, Inc., New York
24. Anders K, von Stetten D, Mailliet J et al (2011) Spectroscopic and photochemical characterization of the red-light sensitive photosensory module of Cph2 from *Synechocystis* PCC 6803. Photochem Photobiol 87:160–173
25. Rockwell NC, Martin SS, Lagarias JC (2015) Identification of DXCF cyanobacteriochrome lineages with predictable photocycles. Photochem Photobiol 14:929–941
26. Schwinte P, Gärtner W, Sharda S (2009) The photoreactions of recombinant phytochrome CphA from the cyanobacterium *Calothrix* PCC7601: a low-temperature UV–Vis and FTIR study. Photochem Photobiol 85:239–249
27. Kunkel T, Tomizama K, Kern R et al (1993) In vitro formation of a photoreversible adduct of phycocyanobilin and tobacco apophytochrome B. Eur J Biochem 215:587–594
28. Paul R, Weiser S, Amiot NC et al (2004) Cell cycle-dependent dynamic localization of a bacterial response regulator with a novel di-guanylate cyclase output domain. Genes Dev 18:715–727
29. Hickman JW, Tifrea DF, Harwood CS (2005) A chemosensory system that regulates biofilm formation through modulation of cyclic diguanylate levels. Proc Natl Acad Sci U S A 102:14422–14427
30. Ryjenkow DA, Tarutina M, Moskvin OV et al (2005) Cyclic diguanylate is a ubiquitous signaling molecule in bacteria: insights into biochemistry of the GGDEF protein domain. J Bacteriol 187:1792–1798

Chapter 16

Probing the Role of Cyclic di-GMP Signaling Systems in Disease Using Chinese Radish

Shi-Qi An, Ji-Liang Tang, and J. Maxwell Dow

Abstract

The determination of the genome sequences of pathogenic bacteria has facilitated functional analyses that aim to understand the molecular basis of virulence. In particular, genome sequence information of the pathogen *Xanthomonas campestris* pathovar *campestris* has allowed researchers to identify and functionally analyze the role of intracellular signaling involving cyclic di-GMP in black rot disease of crucifers. Here, we describe leaf clipping and spraying methods for testing the virulence of wild type and derived mutants of *X. campestris* in Chinese radish. These methods address different facets of the disease cycle, which requires the ability to survive epiphytically before entry into the plant and growth and systemic spread within the xylem.

Key words Virulence testing, Phytopathogenesis, *Xanthomonas campestris*, Leaf clipping, Spray inoculation

1 Introduction

Xanthomonas campestris pv. *campestris* (*Xcc*) is the causal of black rot of crucifers, a disease that is of agronomic importance throughout the world [1–3]. *Xcc* is a vascular pathogen that grows in the xylem of host plants and can become systemic. After a period of epiphytic survival on leaf surfaces, bacteria can enter the xylem of intact plants through hydathodes or through wounds such as those produced by insect damage. Systemic spread can lead to contamination of seeds, which is a major route of disease transmission [1]. The method we have used most widely to testing the role of different gene products in black rot disease*Chinese radish* is to inoculate plants by cutting the top of the leaf with a pair of scissors dipped in a bacterial inoculum of a standard density [4]. Disease symptoms develop over a period of 10–14 days as V-shaped necrotic lesions, whose length can be measured. In this way, the relative virulence of a number of different strains can be compared to the wild-type assessed simultaneously, allowing a demonstration

Karin Sauer (ed.), *c-di-GMP Signaling: Methods and Protocols*, Methods in Molecular Biology, vol. 1657, DOI 10.1007/978-1-4939-7240-1_16, © Springer Science+Business Media LLC 2017

of a role for particular cyclic di-GMP signaling proteins in black rot disease [4]. This method of inoculation introduces the bacteria into the vascular system by-passing the hydathode entry step and any requirement for epiphytic growth and survival. More recently, we have adopted a second assay where intact plants are sprayed with a bacterial inoculum [5]. In this case, the percentage of infected leaves can be determined. One limitation of this assay is that not all leaves sprayed with the wild-type show disease symptoms. Nevertheless, deployment of both the assays has begun to reveal genes that are required for the epiphytic growth and survival but not for growth in the xylem [5].

2 Materials

Prepare all media using deionized water. Autoclave operated at 121 °C, 15 psi for 15 min. Wear appropriate Personal Protective Equipment (PPE) (such as lab coat, gloves, and safety glasses) when performing the procedure. Diligently follow all waste disposal regulations when disposing waste materials.

2.1 Bacterial Growth

1. NYG broth: Oxoid Bacteriological Peptone 5 g/L; Difco Yeast Extract 3 g/L; Glycerol 20 mL/L. To make 400 mL of NYG broth, weigh 2 g peptone and 1.2 g yeast extact, add water to a volume of 200 mL to dissolve the components. Add 8 mL of glycerol, mix and add water to a volume of 400 mL. Autoclave and store at room temperature.
2. NYG agar plate: Add 1% of agar (Agar-agar technical for microbiology, Merck) to the NYG broth described as above. After autoclaving, cool to ~55 °C. Under aseptic conditions add antibiotic (if required), and pour ~20 mL media into each 90 mm sterile petri dish. Let it solidify, then invert and store at +4 °C in the dark.
3. Deionized water: Autoclave and store at room temperature.
4. 20 mL sterile universal bottles.
5. 10 μL sterile inoculation loops.

2.2 Chinese Radish Cultivation

1. Seeds of Chinese radish (*Raphanus sativus* L. var. *radiculus* Pers.) (*see* **Note 1**).
2. Plastic lightweight plant pots (~8 cm diameter top, ~6 cm diameter base, ~10 cm depth).
3. Potting soil: mix seedling substrate (normal potting mix; organic matter: 55%, total nutrient content: 3.5–4%, water: 10%, pH: ~6–7) and soil at 1:1 ratio.

4. Stainless steel trays: choose the size base on the experiment scale and the compatibility with the facility, hoods, and greenhouse.
5. Operating Scissors (14 cm, Straight, Sharp/Sharp) (*see* **Note 2**).
6. Growth conditions for plants: A greenhouse (bacteria free) with a 12-h day and 12-h night cycle of illumination by L.E. D. lamps with temperature maintained at 25–28 °C, humidity 75% unless specified. No specific illumination required until germination. Tap water should be added to the tray daily (enough water must be used to cover the bottom of the tray).

2.3 Inoculation and Infection Trial

1. Operating Scissors (14 cm, Straight, Sharp/Sharp) (*see* **Note 2**).
2. 100 mL plastic trigger sprayer/atomizer bottles (*see* **Note 3**).
3. 20 mL sterile universal bottles.
4. Laminar flow hood. Clean and disinfect the hood before and after the procedure.
5. Separate greenhouse, maintained at the same condition as described previously (*see* **Note 4**).

3 Methods

3.1 Preparation of Plants for Infection Trial

1. Approximately 4 weeks before the infection trial, place the plant pots onto a stainless steel tray. Add ~6–7 cm depth of soil to each pot and plant 3–5 seeds of Chinese radish per pot (~1–1.5 cm deep).
2. Maintain in the greenhouse under the conditions described above.
3. 3–5 days after seeding, the cotyledons should fully expand. These seedlings are thinned to three per pot.
4. The plants are maintained in the greenhouse for another ~20–25 days. Add potting soil to the edge of the pot. Seedlings with 3–4 fully expanded leaves are selected for experiments (Fig. 1).
5. The day before inoculation, transfer the plants for testing onto a clean tray and use sterilized scissors to carefully trim away any unwanted or damaged leaves. Trays are then moved to a separate greenhouse for inoculation and infection trial (*see* **Note 4**).

3.2 Preparation of Bacterial Suspension for Inoculation

1. The day before inoculation, aliquot 5 mL NYG broth into a universal bottle and inoculate with a loop of bacteria from an NYG agar plate and incubate overnight at 28–30 °C with shaking at 200 rpm. Perform the step under aseptic conditions (*see* **Notes 5** and **6**).

Fig. 1 Representative image of a four-week old *Raphanus sativus* seedling which has the mandatory 3–4 fully expanded or unfurled leaves needed for infection tests

2. Transfer 1 mL of the overnight culture to a 1.5 mL eppendorf tube and pellet the bacteria by centrifugation in a microfuge at 10,000 x *g* for 2 min. Perform this step under aseptic conditions.
3. Remove the supernatant and add 1 mL of sterile water to the tube (*see* **Note** 7). Fully resuspend the bacterial pellet by vortexing.
4. Measure the Optical Density (OD) at 600 nm. If the optical density of the sample is greater than 1.0, dilute the sample 1:1 with sterile water and read the optical density again. Record the value and discard this sample.
5. Using sterile water to adjust the optical density of the remaining suspension to 0.001 for leaf clipping or 0.01 for leaf spraying (*see* **Note** 7) by serial dilution. Perform the step under aseptic conditions.

3.3 Virulence Testing: Inoculation by Leaf Clipping

1. Transfer previously prepared testing plants (Subheading 3.1, **step 5**), sterilized scissors and bacterial suspension (Subheading 3.2, **step 5**, $OD_{600} = 0.001$) into a laminar flow hood.
2. Dip a pair of scissors into the bacterial suspension for ~5 s.
3. Slowly remove the scissors from the bacteria suspension to avoid any dripping and immediately gently clip away the tips of the last two completely expanded leaves (~5 s per leaf).

Fig. 2 Image illustrating the clipping of the *Raphanus sativus* leaf using inoculated stainless steel scissors (**a**). Image showing typical symptom development (lesion) 14 days after inoculation (**b**)

Approximately 1 cm from the apex and perpendicular to the leaf midrib (*see* **Notes 8** and **9**, Fig. 2a).

4. Return the tray with inoculated plants to the greenhouse (*see* **Note 4**).
5. Maintain the plants for 24 h at 100% humidity. After this time, maintain the inoculated plants in the growth condition described previously (*see* **Note 10**).
6. Measure the lesion length by ruler at 14 days after inoculation (*see* Fig. 2b, **Note 11**). Data are analyzed by Student's t-test (*see* **Note 12**).

3.4 Virulence Testing: Inoculation by Leaf Spraying

1. Transfer previously prepared test plants (Subheading 3.1, **step 5**) and sprayers that contain bacterial suspension (*see* Subheading 3.2, **step 5**, $OD_{600} = 0.01$) into a laminar hood.
2. Spray sufficient of the 50 mL of the bacterial suspension from a distance of ~20 cm onto the leaves so as to wet them (*see* **Note 13**, Fig. 3a).
3. Rotate the tray in the hood in order to spray all leaves evenly.

Fig. 3 Image illustrating the arrangement for spraying of the *Raphanus sativus* plants (**a**). Image showing typical symptom development 10 days after inoculation (**b**)

4. Allow plants to dry in the hood before return to the greenhouse (*see* **Note 4**).
5. Maintain inoculated plants for 24 h at 100% humidity. After this time maintain inoculated plants in the growth condition described previously (*see* **Note 10**).
6. At 10 days after inoculation, determine the percentage of the total inoculated leaves that show the typical black rot disease symptoms (*see* **Notes 12** and **13**, Fig. 3b)

4 Notes

1. *Xanthomonas campestris* pathovar *campestris* strain used here is strain 8004 [6]. Chinese radish (*Raphanus sativus* L. var. radiculus Pers.) is susceptible to this strain and black rot disease symptoms are seen. However, this cultivar may be resistant to other strains of *Xanthomonas campestris.* Chinese radish may not be a host for other phytopathogenic bacteria.

2. Wrap stainless steel scissors in aluminum foil, sterilize by autoclaving. Make sure the scissors are sufficiently dried before use.
3. Wrap the trigger and the bottle separately in aluminum foil and sterilize by autoclaving. It is important that both parts are free of moisture after autoclaving and before use.
4. Separate greenhouses/growth rooms are required for plant growth with and without infection. One room for plant development and another for inoculation and infection of plants to avoid contamination is crucial.
5. Long-term storage of *Xcc* strains is in 25% (v/v) glycerol at −80 °C. Add 50 mL of water to 50 mL of glycerol to make 50% (v/v) glycerol and sterilize by autoclaving. Inoculate a 5 mL culture of NYG broth with bacteria and incubate overnight at 28–30 °C with shaking. Mix 1 mL of the overnight culture with 1 mL sterile 50% (v/v) glycerol in a 2-mL screw top tube and store at −80 °C.
6. To prepare working cultures, thaw a frozen sample and plate the sample onto an NYG agar plate (supplemented with antibiotics if needed) by gently spreading the bacteria over the plate using a sterile inoculation loop. Incubate at 28–30 °C for 2 days. Store plates at 4 °C for up to a month.
7. 10 mL bacteria suspension per technical replicate is required for leaf clipping assay. Make the bacterial suspension using sterilized universal bottles. 50 mL bacterial suspension per technical replicate is required for leaf spray assay. Make the final bacteria suspension in the sterilized sprayer bottle. The bacteria are pelleted and resuspended in sterile water so as to remove media components. Some strains may be sensitive to resuspension in water, in which case bacteria may be resuspended in 10 mM $MgCl_2$.
8. It is routine to inoculate up to 20 leaves per technical replicate.
9. It is important to change scissors, gloves and disinfect the hood between different inoculums to avoid cross contamination.
10. Seal the trays with inoculated plants using polythene sheeting or equivalent to maintain 100% humidity.
11. Measurement can be taken between day 10 to day 14 depending on the experimental scale and the compatibility with the facility. However, it is important to keep the times consistent for comparative purposes.
12. Three technical replicates are used for each experiment and three independent experiments are carried out.
13. It is routine to inoculate up to 30 seedlings (~100 leaves) per technical replicate. Each leaf is scored independently as either showing no disease symptoms or showing disease symptoms, irrespective of the number or size of the lesions.

References

1. Vicente JG, Holub EB (2013) *Xanthomonas campestris* pv. *campestris* (cause of black rot of crucifers) in the genomic era is still a worldwide threat to brassica crops. Mol Plant Pathol 14:2–18
2. Ryan RP, Vorhölter FJ, Potnis N, Jones JB, Van Sluys MA, Bogdanove AJ, Dow JM (2011) Pathogenomics of *Xanthomonas*: understanding bacterium-plant interactions. Nat Rev Microbiol 9:344–355
3. Büttner D, Bonas U (2010) Regulation and secretion of *Xanthomonas* virulence factors. FEMS Microbiol Rev 34:107–133
4. Ryan RP, Fouhy Y, Lucey JF, Jiang BL, He YQ, Feng JX, Tang JL, Dow JM (2007) Cyclic di-GMP signalling in the virulence and environmental adaptation of *Xanthomonas campestris*. Mol Microbiol 63:429–442
5. An SQ, Allan JH, McCarthy Y, Febrer M, Dow JM, Ryan RP (2014) The PAS domain-containing histidine kinase RpfS is a second sensor for the diffusible signal factor of *Xanthomonas campestris*. Mol Microbiol 92:586–597
6. Qian W, Jia Y, Ren SX, He YQ, Feng JX, Lu LF, Sun Q, Ying G, Tang DJ, Tang H, Wu W, Hao P, Wang L, Jiang BL, Zeng S, Gu WY, Lu G, Rong L, Tian Y, Yao Z, Fu G, Chen B, Fang R, Qiang B, Chen Z, Zhao GP, Tang JL, He C (2005) Comparative and functional genomic analyses of the pathogenicity of phytopathogen *Xanthomonas campestris* pv. *campestris*. Genome Res 15 (6):757–767

Chapter 17

Contribution of Cyclic di-GMP in the Control of Type III and Type VI Secretion in *Pseudomonas aeruginosa*

Ronan R. McCarthy, Martina Valentini, and Alain Filloux

Abstract

Bacteria produce toxins to enhance their competitiveness in the colonization of an environment as well as during an infection. The delivery of toxins into target cells is mediated by several types of secretion systems, among them our focus is Type III and Type VI Secretion Systems (T3SS and T6SS, respectively). A thorough methodology is provided detailing how to identify if cyclic di-GMP signaling plays a role in the *P. aeruginosa* toxin delivery mediated by T3SS or T6SS. This includes in vitro preparation of the samples for Western blot analysis aiming at detecting possible c-di-GMP-dependent T3SS/T6SS switch, as well as in vivo analysis using the model organism *Galleria mellonella* to demonstrate the ecological and pathogenic consequence of the switch between these two secretion systems.

Key words *Pseudomonas aeruginosa*, Cyclic di-GMP, Type III secretion, Type VI secretion, Bacterial toxins, *Galleria mellonella*

1 Introduction

Cyclic di-GMP (CdGMP) is an intracellular eubacterial second messenger that has been shown to regulate a wide variety of phenotypes [1, 2]. In *Pseudomonas aeruginosa* it coordinates the patho-adaptive transition from motile bacteria associated with acute infection to sessile bacteria associated with chronic infections, such as those found in the cystic fibrosis lung [3, 4]. The distinction between these two lifestyles also extends into which secretion systems are active, with low levels of CdGMP promoting activation of the Type III secretion system (T3SS), while high levels of CdGMP are associated with inhibiting T3SS and activating the Type VI secretion system (T6SS) [5]. CdGMP levels within the cell are regulated by two classes of enzymes: diguanylate cyclases (DGCs) that synthesize CdGMP from two GTP molecules and phosphodiesterases (PDEs) that break it down [1, 2]. *P. aeruginosa* species typically contain over 40 different PDEs or DGCs with a wide array of different genes controlling their expression and in turn

Karin Sauer (ed.), *c-di-GMP Signaling: Methods and Protocols*, Methods in Molecular Biology, vol. 1657, DOI 10.1007/978-1-4939-7240-1_17, © Springer Science+Business Media LLC 2017

influencing the levels of CdGMP within the cell [4]. The impact of regulatory events fine-tuning the intracellular CdGMP levels can dramatically alter the behavior of *P. aeruginosa*.

In this section, we outline a concise methodology to establish if a gene of interest, involved in CdGMP signaling, regulates the *P. aeruginosa* switch between secretion systems and, by extension, its virulence. By performing targeted Western blot analysis it is possible to qualitatively assess in vitro the production of T3SS and T6SS components and the secretion of toxins associated with both secretion systems. Furthermore, we outline the use of the model organism *Galleria mellonella* to study in vivo the overall impact of the CdGMP-related gene on *P. aeruginosa* pathogenicity. *G. mellonella* has been established as a robust model organism and has been used to investigate virulence in a wide variety of pathogens (Fig. 1). There are a range of similarities between mammalian and insect innate immunity and the *G. mellonella* model offers a competitive advantage with respect to cost and ease of handling compared to more traditional mammalian models of infection [6–9].

The methods presented in this chapter can be applied to clinical isolates of *P. aerugionosa* as a marker for determining the virulence potential of these strains [10]. Vaccination against a T3SS component ensures protections against *Yersinia* or *P. aeruginosa* infections in animal models [11]. Furthermore, T6SS components, such as Hcp1, are readily recovered in the sputa of CF patients chronically infected with *P. aeruginosa* [12]. Thus, the methodology described here can readily serve for quick phenotyping of clinical strains, diagnostics, and prognosis.

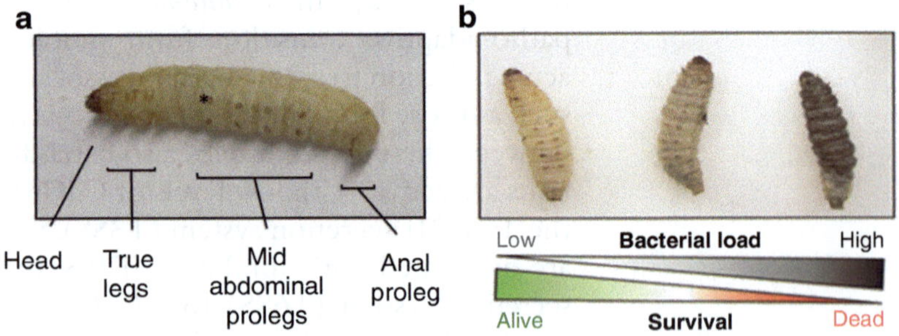

Fig. 1 Visualization of *Galleria mellonella* infection progression. (**a**) *Galleria mellonella* were held on their backs revealing their legs. The top right of the mid abdominal prolegs was injected with 10 μL of the respective bacterial dilution or PBS at an angle parallel to the larvae. Asterisk indicates optimal injection site. (**b**) Disease progression can be visualized by melanization of the organism, with the larvae darkening from prophenoloxidase release, part of the innate immune response to infection

2 Materials

2.1 T3SS vs. T6SS in vitro Assay

1. Lysogeny Broth growth medium (LB): 5 g/L bacto-yeast extract, 10 g/L bacto-tryptone, 5 g/L NaCl; dissolve in distilled water; sterilize by autoclaving.
2. 0.5 M Ethylene-bis(oxyethylenenitrilo)tetraacetic acid (EGTA) solution (pH 8.0): Add 190.1 g of EGTA (FW 380.35) to 800 mL of water. Stir vigorously on a magnetic stirrer. Adjust pH to 8.0, first with NaOH pellets, then with 1 m NaOH solution. Note that EGTA will not go into solution until pH is ~8.0. Adjust final volume to 0.1 L. Filter-sterilize through a 0.22 μm filter.
3. 1 M $MgCl_2$, sterile.
4. 100% (w/v) trichloroacetic acid (TCA).
5. 90% acetone. This solution needs to be ice cold, so store at −20C.
6. SDS-PAGE loading buffer 6×.
7. 15 mL Falcon tubes.
8. Microcentrifuge.
9. Vortex shaker with tube adapter.

2.2 Wax Moth Pathogenicity Assay

1. *Galleria mellonella*. This wax moth species can be ordered from most live food vendors, they should be ordered prior to the experiment and kept for no longer than 14 days at room temperature in the dark.
2. Hamilton 20 Gauge syringe.
3. Cleaning wires for syringe.
4. 85 mm Whattman paper discs.
5. Petri dishes, 90 mm.
6. 50 mL Falcon tubes.
7. 70% ethanol (EtOH).
8. Phosphate-Buffered Saline (PBS) solution: To approximately 800 mL of water add 8.0 g NaCl, 0.2 g KCl, 2.56 g Na2H-PO4•7H2O, and 0.24 g KH2PO4. Mix until all salts are dissolved. Adjust the pH to 7.4 with HCl. Add water to a total volume of 1 L. Autoclave at 120v for 20 min to sterilize and store at 4 °C.
9. Microcentrifuge.
10. Lysogeny Broth growth medium (LB): 5 g/L bacto-yeast extract, 10 g/L bacto-tryptone, 5 g/L NaCl; dissolve in distilled water, sterilize by autoclaving.

11. LB agar medium: 5 g/L bacto-yeast extract, 10 g/L bacto-tryptone, 5 g/L NaCl, 15 g/L agar; dissolve in distilled water, sterilize by autoclaving.

3 Methods

Here below, a typical experiment for *P. aeruginosa* cultures is described.

3.1 T3SS vs. T6SS Assay

1. Inoculate *P. aeruginosa* cultures from single colonies into 5 mL LB and grow them for 16–18 h at 37 °C with shaking at 200 rpm.
2. Subculture into 20 mL of LB supplemented with 5 mM EGTA and 20 mM $MgCl_2$ to a starting OD_{600} (optical density at 600 nm) of 0.1 (*see* **Note 1**).
3. Grow cultures at 37 °C with shaking at 200 rpm for 6–7 h to a final OD_{600} 3.5–4.5.
4. For cell fraction, harvest cells corresponding to 1.5 OD Units (ODU, *see* **Note 2**) in a microcentrifuge (16,000 × *g*) for 5 min at 4 °C.
5. Resuspend pellet in 150 μL 1× SDS loading buffer (i.e., to a final concentration of 0.01 ODU/μL) and store samples at −20 °C.
6. For supernatant fraction, harvest 10 mL culture into a 15 mL Falcon tube and sediment cells by centrifugation (1300 × *g*) for 15 min and 4 °C.
7. Carefully and without disturbing the pellet, transfer 7 mL supernatant to a new 15 mL Falcon tube and repeat the centrifugation for 20 min.
8. If the supernatant is clear, transfer 1.8 mL into a 2 mL Eppendorf tube. If not, repeat the centrifugation step above. These centrifugation steps are required to remove intact cells and cell debris.
9. Precipitate proteins by adding 200 μL of 100% TCA (final concentration 10% w/v) to 1.8 mL of cleared supernatant (Subheading 3.1, **step 8**) and invert each tube 2–3 times to mix.
10. Incubate tube(s) on ice for at least 1 h (max overnight, 16–18 h).
11. Sediment precipitated proteins by centrifugation in a microcentrifuge (16,000 × *g*) for 30 min at 4 °C.
12. Carefully remove the supernatant, first the majority with a 1 mL pipette tip, then the rest with a 200 μL pipette tip. This is to avoid disturbing the pellet.

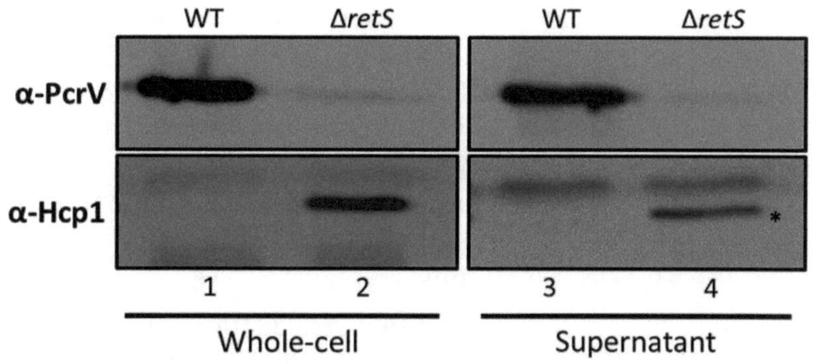

Fig. 2 Visualization of the T3SS/T6SS switch in *P. aeruginosa*. The *P. aeruginosa* PAK wild-type and *retS* mutant strains were grown in LB medium in the presence of EDTA and $MgCl_2$. Whole cell lysates (lanes 1 and 2) and supernatant fractions (lane 3 and 4) were prepared as described in Subheading 3. Samples were analyzed by Western blot using antibodies directed against PcrV (T3SS) and Hcp1 (T6SS). Primary antibodies, anti-Hcp1 and anti-PcrV were used at dilutions of 1:500 and 1:1000 respectively. Secondary antibody (horseradish peroxidase-conjugated goat anti-rabbit IgG) was used at a dilution 1:5000, as reported previously [5]. Visualization was achieved using the SuperSignal West Pico Chemiluminescent Substrate Kit (Thermo) and a LAS3000 Imaging System (Fuji)

13. Wash the pellet with 1 mL of 90% ice cold acetone and place on a vortex shaker with tube adapter for 15 min at max setting.
14. Sediment pellet by centrifugation (16,000 × *g*, 4 °C, 30 min) and carefully remove the supernatant.
15. Dry pellets with open lid on the bench next to the flame (max 30 min).
16. Resuspend pellets in 1× SDS loading buffer to a final concentration of 0.1 ODU/μL (*see* **Note 3**).
17. Samples collected in **steps 5** and **16** can now be subjected to Western blot analysis with specific T3SS/T6SS targeting antibodies (*see* **Note 4**). An example of Western blot analysis is shown in Fig. 2, while others are examples have been previously published [5, 13] (*see* **Note 5**). Western blot analysis is the method that has a high level of sensitivity and specificity for T3SS/T6SS components detection, nonetheless alternatives to this assay are described in Subheading 4 (*see* **Note 6**).

3.2 Wax Moth Pathogenicity Assay

Insects are obtained in advance and stored at room temperature in the dark until use.

1. Inoculate *P. aeruginosa* cultures from single colonies into 5 mL LB and grow them at 37 °C overnight with shaking at 200 rpm.

2. Subculture in 20 mL LB to a starting OD_{600} of 0.01 and proceed to exponential growth to reach OD_{600} 0.7–0.9.
3. Pellet bacteria by centrifugation at room temperature and set to an OD_{600} of 1 in 1 mL of Phosphate-Buffered Saline (PBS).
4. Pellet the bacteria again by centrifugation and resuspend in 1 mL PBS, repeat this **step 2** further times.
5. Following the final wash step, resuspend the pellet in 1 mL PBS.
6. Perform serial dilutions of your bacterial suspension (assuming an OD 1, *see* Subheading 3.2, **step 3**) in PBS to a dilution of 10^{-8}.
7. Select larvae for the pathogenicity assay based on size (between 80 mm and 130 mm) and the absence of melanization (*see* **Note 7**).
8. Prior to injection, fill the Hamilton syringe to capacity with 70% EtOH and leave to sit for minimum 10 min. Between each injection, the syringe must be washed in two baths of 70% EtOH and two baths of sterile PBS (use 50 mL Falcons to store the EtOH and PBS solutions). Washing is done by filling the syringe to capacity and expelling the contents immediately into a liquid waste bin. Between different strains, the syringe must be filled to capacity with 70% EtOH and left to sit for minimum 10 min prior to subsequent washing in the two baths of 70% EtOH and the two baths of sterile PBS.
9. Inject ten individual insects with 10 μL each of the relevant bacterial suspension (*see* **Note 8**) into the first right pro-leg of the second set of pro-legs (Fig. 1) using the Hamilton syringe (*see* **Note 9**).
10. Place all ten insects in a 90 mm Petri dish lined with 85 mm Whatmann paper, then incubate at the appropriate temperature, usually 37 °C in the dark until 100% mortality is achieved, which depending on the dilution injected can take between 10 and 72 h (*see* **Note 8**).
11. Inject the same numbers of insects with PBS alone as control.
12. Examine insects individually for pigment production and time of death on an hourly basis (*see* **Note 10**). A 10% mortality rate is the maximum acceptable for the PBS control inoculations.
13. Plate 100 μL of each of the serial dilutions from **step 6** on LB immediately after *Galleria* inoculation. Incubate the plates at 37 °C overnight and count the number of colony forming units (cfu) for each inoculation to determine the precise number of CFU inoculated per dilution (*see* **Note 11**).

3.3 Data Analysis

The strategy described in this method was tested in *P. aeruginosa* PAK wild-type and *retS* mutant. The RetS sensor represents a well-known example of antagonistic regulator of T3SS and T6SS production [12, 14]. The example involving a *retS* mutant is therefore used as a pilot experiment to illustrate the data that can be obtained following the experimental procedure reported here. The effect of a *retS* mutation on the T3SS/T6SS switch is revealed by the Western blot analysis shown in Fig. 2. In the figure, the production of a structural component of either the T3SS or T6SS machinery is monitored: PcrV, located at the tip of the T3SS needle [15], and Hcp1, the main component of the H1-T6SS tail tube [16]. Western blot analysis shows that the wild-type strain displays PcrV production while low Hcp1 levels. Conversely, in a *retS* mutant Hcp1 is detected but not PcrV.

As a consequence of this regulation, in a *retS* mutant a significant reduction in virulence can be easily observed in the model organism *G. mellonella* when compared to the wild-type strain (Fig. 3). At 30 h post infection, 80% of the larvae are still alive for a *retS* mutant, while 100% of mortality is reached for the wild-type after even shorter incubation times. This difference in virulence is likely to be attributed to the low level of T3SS activity in the *retS* mutant.

The RetS-mediated T3SS/T6SS switch has been demonstrated to be dependent on the high intracellular CdGMP levels that could be observed in a *retS* mutant compared to the parental strain [5]. The protocols outlined here can be therefore adapted for studying the specific impact of any CdGMP-related gene on the T3SS/T6SS switch, as shown with the *retS* mutant [17, 18]. In the case of the *retS* mutation, the elevated levels of CdGMP were directly related to the activity of the SadC diguanylate cyclase [19]. The same

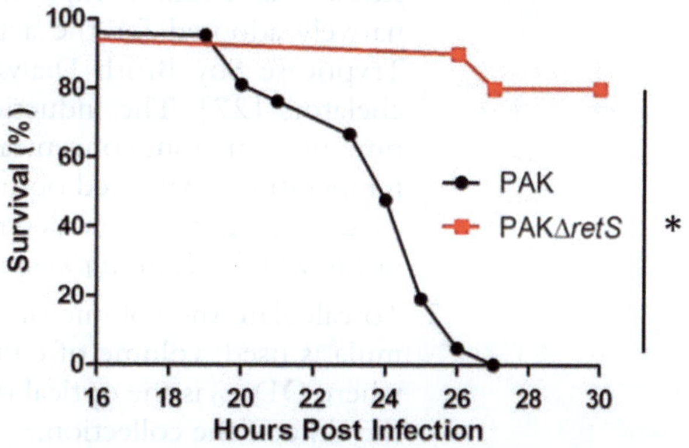

Fig. 3 Impact of *retS* on pathogenicity in *G. mellonella*. Inoculation of *G. mellonella* with 5 × 10^3 *P. aeruginosa* PAK wild-type (*black line*) or *retS* mutant (*red line*) cells was performed as described in Subheading 3. The data obtained are plotted in a Kaplan-Meier curve to estimate the fraction of larvae living after 30 h post-infection, *$p < 0.0001$ (log-rank)

strategy may be applied to any bacteria; however, one should consider that the switch from T3SS to T6SS correlated to CdGMP levels may not necessarily be a hallmark for all bacteria. In this case, other regulatory mechanisms will be revealed. For the feasibility of the *Galleria* toxicity assay, the bacteria of choice must be pathogenic to the insect. This has been shown for several other opportunistic pathogens like *Proteus vulgaris*, *Proteus mirabilis*, and *Serratia marcescens* [6], as well as plant-beneficial bacteria like *P. fluorescens* biocontrol strain CHA0 [20]. Kaplan–Meier Survival analysis can be used to determine differences in pathogenicity using statistical software such as GraphPad Prism. Each death is recorded as an event "1" and each larva left alive at the end of the assay is considered censored and recorded as a "0." The survival probability (S_t) at any given point over the course of the assay can be determined using the following formula

$$S_t = \frac{\text{Number of Galleria alive at the start} - \text{number of Galleria deceased}}{\text{Number of Galleria living at the start}}$$

[21]. The log-rank test for multiple comparisons should be used to determine significance with a p value of <0.05 indicative of a significant difference in virulence [22].

4 Notes

1. Addition of 5 mM EGTA-20 mM $MgCl_2$ to the medium is used to induce the T3SS regulon in *P. aeruginosa*. The expression of T3SS genes in *P. aeruginosa* is regulated by a diversity of mechanisms and it responds to a wide range of environmental conditions as well as to the internal cell metabolism [23–26]. Additional media or experimental conditions have been alternatively adopted for the activation/induction of T3SS, e.g., Trypticase Soy Broth Dialyzate does not involve the use of chelators [27]. The induction parameters (type of medium, time of induction, concentration and type of chelator, growth temperature) may need optimization depending on the strains used. The parameters described in this method have been optimized for *P. aeruginosa* strains PAO1 and PAK [5, 24].
2. To calculate the volume of culture to take, the following formula is used: volume of culture (μL) = $1.5/OD_{600} \times 1000$, where OD_{600} is the optical density at 600 nm of the culture at the time of the collection.
3. The volume of 1× SDS loading buffer to use is calculated as follows: volume of 1× SDS loading buffer (μL) = $(OD_{600} \times 1.8\ \text{mL})/0.1$. The final concentration of the supernatant is 0.1 ODU/μL.

4. For Western blot analysis, whole cell extract and 10× supernatant protein samples (0.1 and 1 ODU, respectively) are separated by SDS-PAGE with PcrV being run on a 10% acrylamide gel and Hcp1 on a 15% acrylamide gel. Prior to loading, samples are boiled for 10 min at 95 °C. Following electrophoresis proteins are transferred to nitrocellulose membranes for immunoblotting.
5. Antibodies can be generated specifically for the required protein by purifying the full length protein, e.g., using the pET System [28], or by using synthetic peptides and using them as antigens [29]. In *P. aeruginosa*, primary anti-Hcp1 antibodies (against the T6SS tail tube) have been raised in rabbits by injecting VKGESKDKTHAEEID and GPVKYGW-NIRQNVQA purified peptides [30]. While Primary anti-PcrV antibodies (against the tip of the T3SS needle) have been raised in rabbits against recombinant His-tagged PcrV [11, 31]. Alternatively, any other antibodies against T3SS/T6SS components or effectors can be used [30, 32–39], although abundant proteins should be preferentially chosen.
6. When generation of specific antibodies is not possible, the protein of interest can be detected by epitope tagging the encoding chromosomal gene (e.g., His6, FLAG, V5 tagging) and using commercial available antibodies against the epitope [40, 41]. In addition, secretion of the ExoS, ExoT, ExoU, and ExoY T3SS effectors can be detected in *P. aeruginosa* by SDS-page and Comassie blue staining [42]; while T6SS-related phenotype can be monitored with a visual competition assay [43].
7. For the ease of handling larvae can be chilled at 4 °C for 20 min prior to injection.
8. Selecting the optimal dilution is dependent on the ability to record death over the full course of the experiment. Typically, it is best to optimize this by inoculating three different dilutions (5000 CFU, 500 CFU, 50 CFU) to assess the timings required to record the full experiment. Alternatively, the full dilution series from neat to 10^{-8} can be inoculated to determine the LD_{50} as described by Giannouli et al. 2014 [44].
9. After the injection of ~20 *G. mellonella* the needle needs to be cleaned using a cleaning wire to prevent the build-up of aggregates that can block the needle.
10. Larvae are considered dead when they no longer react to being poked with a pipette tip.
11. In the case where a difference in *P. aeruginosa* in vivo growth rate between mutant and wild-type strains is suspected, *Galleria* can be macerated immediately after being scored as deceased. This can be performed by first surface sterilizing

the larva using 70% EtOH and then transferring the larva to a 1.5 mL microfuge tube containing 1 mL PBS. Three glass beads should also be added to the microfuge tube. It should then be vortexed for 15 min before serially diluting the macerated larva 1:10 in PBS to a dilution of 10^{-9}. This dilution series should then be plated on *Pseudomonas* Isolation Agar to determine the in vivo CFU at the time of death.

References

1. Jenal U, Malone J (2006) Mechanisms of cyclic-di-GMP signaling in bacteria. Annu Rev Genet 40:385–407. doi:10.1146/annurev.genet.40.110405.090423
2. Romling U, Galperin MY, Gomelsky M (2013) Cyclic di-GMP: the first 25 years of a universal bacterial second messenger. Microbiol Mol Biol Rev 77(1):1–52. doi:10.1128/mmbr.00043-12
3. Ha DG, O'Toole GA (2015) C-di-GMP and its effects on biofilm formation and dispersion: a *Pseudomonas aeruginosa* review. Microbiol Spect 3(2):MB-0003–MB-2014. doi:10.1128/microbiolspec.MB-0003-2014
4. Valentini M, Filloux A (2016) Biofilms and cyclic di-GMP (c-di-GMP) signaling: lessons from *Pseudomonas aeruginosa* and other bacteria. J Biol Chem 291(24):12547–12555. doi:10.1074/jbc.R115.711507
5. Moscoso JA, Mikkelsen H, Heeb S, Williams P, Filloux A (2011) The *Pseudomonas aeruginosa* sensor RetS switches type III and type VI secretion via c-di-GMP signalling. Environ Microbiol 13(12):3128–3138. doi:10.1111/j.1462-2920.2011.02595.x
6. Glavis-Bloom J, Muhammed M, Mylonakis E (2012) Of model hosts and man: using *Caenorhabditis elegans*, *Drosophila melanogaster* and *Galleria mellonella* as model hosts for infectious disease research. Adv Exp Med Biol 710:11–17. doi:10.1007/978-1-4419-5638-5_2
7. Lopez Hernandez Y, Yero D, Pinos-Rodriguez JM, Gibert I (2015) Animals devoid of pulmonary system as infection models in the study of lung bacterial pathogens. Front Microbiol 6:38. doi:10.3389/fmicb.2015.00038
8. Joyce SA, Gahan CG (2010) Molecular pathogenesis of Listeria monocytogenes in the alternative model host Galleria mellonella. Microbiology 156(Pt 11):3456–3468. doi:10.1099/mic.0.040782-0
9. Koch G, Nadal-Jimenez P, Cool RH, Quax WJ (2014) Assessing Pseudomonas virulence with nonmammalian host: Galleria mellonella. Methods Mol Biol 1149:681–688. doi:10.1007/978-1-4939-0473-0_52
10. Sawa T, Shimizu M, Moriyama K, Wiener-Kronish JP (2014) Association between Pseudomonas aeruginosa type III secretion, antibiotic resistance, and clinical outcome: a review. Crit Care (London, England) 18(6):668. doi:10.1186/s13054-014-0668-9
11. Goure J, Pastor A, Faudry E, Chabert J, Dessen A, Attree I (2004) The V antigen of *Pseudomonas aeruginosa* is required for assembly of the functional PopB/PopD translocation pore in host cell membranes. Infect Immun 72 (8):4741–4750. doi:10.1128/iai.72.8.4741-4750.2004
12. Mougous JD, Cuff ME, Raunser S, Shen A, Zhou M, Gifford CA, Goodman AL, Joachimiak G, Ordonez CL, Lory S, Walz T, Joachimiak A, Mekalanos JJ (2006) A virulence locus of Pseudomonas aeruginosa encodes a protein secretion apparatus. Science 312 (5779):1526–1530. doi:10.1126/science.1128393
13. Jones C, Allsopp L, Horlick J, Kulasekara H, Filloux A (2013) Subinhibitory concentration of kanamycin induces the *Pseudomonas aeruginosa* type VI secretion system. PLoS One 8 (11):e81132. doi:10.1371/journal.pone.0081132
14. Goodman AL, Kulasekara B, Rietsch A, Boyd D, Smith RS, Lory S (2004) A signaling network reciprocally regulates genes associated with acute infection and chronic persistence in *Pseudomonas aeruginosa*. Dev Cell 7 (5):745–754. doi:10.1016/j.devcel.2004.08.020
15. Dacheux D, Goure J, Chabert J, Usson Y, Attree I (2001) Pore-forming activity of type III system-secreted proteins leads to oncosis of *Pseudomonas aeruginosa*-infected macrophages. Mol Microbiol 40(1):76–85
16. Ballister ER, Lai AH, Zuckermann RN, Cheng Y, Mougous JD (2008) In vitro self-assembly

of tailorable nanotubes from a simple protein building block. Proc Natl Acad Sci U S A 105 (10):3733–3738. doi:10.1073/pnas.0712247105

17. Portaliou AG, Tsolis KC, Loos MS, Zorzini V, Economou A (2016) Type III secretion: building and operating a remarkable nanomachine. Trends Biochem Sci 41(2):175–189. doi:10.1016/j.tibs.2015.09.005
18. Miyata ST, Bachmann V, Pukatzki S (2013) Type VI secretion system regulation as a consequence of evolutionary pressure. J Med Microbiol 62(Pt 5):663–676. doi:10.1099/jmm.0.053983-0
19. Moscoso JA, Jaeger T, Valentini M, Hui K, Jenal U, Filloux A (2014) The diguanylate cyclase SadC is a central player in Gac/Rsm-mediated biofilm formation in Pseudomonas aeruginosa. J Bacteriol 196(23):4081–4088. doi:10.1128/JB.01850-14
20. Pechy-Tarr M, Bruck DJ, Maurhofer M, Fischer E, Vogne C, Henkels MD, Donahue KM, Grunder J, Loper JE, Keel C (2008) Molecular analysis of a novel gene cluster encoding an insect toxin in plant-associated strains of *Pseudomonas fluorescens*. Environ Microbiol 10(9):2368–2386. doi:10.1111/j.1462-2920.2008.01662.x
21. Goel MK, Khanna P, Kishore J (2010) Understanding survival analysis: Kaplan-Meier estimate. Int J Ayurveda Res 1(4):274–278. doi:10.4103/0974-7788.76794
22. Dudley WN, Wickham R, Coombs N (2016) An introduction to survival statistics: Kaplan-Meier analysis. J Adv Pract Oncol 7(1):91–100
23. Yahr TL, Wolfgang MC (2006) Transcriptional regulation of the *Pseudomonas aeruginosa* type III secretion system. Mol Microbiol 62 (3):631–640. doi:10.1111/j.1365-2958.2006.05412.x
24. Kim J, Ahn K, Min S, Jia J, Ha U, Wu D, Jin S (2005) Factors triggering type III secretion in *Pseudomonas aeruginosa*. Microbiology 151(Pt 11):3575–3587. doi:10.1099/mic.0.28277-0
25. Vallis AJ, Yahr TL, Barbieri JT, Frank DW (1999) Regulation of ExoS production and secretion by *Pseudomonas aeruginosa* in response to tissue culture conditions. Infect Immun 67(2):914–920
26. Rietsch A, Mekalanos JJ (2006) Metabolic regulation of type III secretion gene expression in *Pseudomonas aeruginosa*. Mol Microbiol 59 (3):807–820. doi:10.1111/j.1365-2958.2005.04990.x
27. Iglewski BH, Sadoff J, Bjorn MJ, Maxwell ES (1978) *Pseudomonas aeruginosa* exoenzyme S: an adenosine diphosphate ribosyltransferase distinct from toxin A. Proc Natl Acad Sci U S A 75(7):3211–3215
28. Mierendorf RC, Morris BB, Hammer B, Novy RE (1998) Expression and purification of recombinant proteins using the pET system. Methods Mol Med 13:257–292. doi:10.1385/0-89603-485-2:257
29. Hancock DC, O'Reilly NJ (2005) Synthetic peptides as antigens for antibody production. Methods Mol Biol 295:13–26
30. Hachani A, Lossi NS, Hamilton A, Jones C, Bleves S, Albesa-Jove D, Filloux A (2011) Type VI secretion system in *Pseudomonas aeruginosa*: secretion and multimerization of VgrG proteins. J Biol Chem 286 (14):12317–12327. doi:10.1074/jbc.M110.193045
31. Schoehn G, Di Guilmi AM, Lemaire D, Attree I, Weissenhorn W, Dessen A (2003) Oligomerization of type III secretion proteins PopB and PopD precedes pore formation in Pseudomonas. EMBO J 22(19):4957–4967. doi:10.1093/emboj/cdg499
32. Bleves S, Soscia C, Nogueira-Orlandi P, Lazdunski A, Filloux A (2005) Quorum sensing negatively controls type III secretion regulon expression in *Pseudomonas aeruginosa* PAO1. J Bacteriol 187(11):3898–3902. doi:10.1128/jb.187.11.3898-3902.2005
33. Forster A, Planamente S, Manoli E, Lossi NS, Freemont PS, Filloux A (2014) Coevolution of the ATPase ClpV, the sheath proteins TssB and TssC, and the accessory protein TagJ/HsiE1 distinguishes type VI secretion classes. J Biol Chem 289(47):33032–33043. doi:10.1074/jbc.M114.600510
34. Hachani A, Allsopp LP, Oduko Y, Filloux A (2014) The VgrG proteins are "a la carte" delivery systems for bacterial type VI effectors. J Biol Chem 289(25):17872–17884. doi:10.1074/jbc.M114.563429
35. Lossi NS, Manoli E, Forster A, Dajani R, Pape T, Freemont P, Filloux A (2013) The HsiB1C1 (TssB-TssC) complex of the *Pseudomonas aeruginosa* type VI secretion system forms a bacteriophage tail sheathlike structure. J Biol Chem 288(11):7536–7548. doi:10.1074/jbc.M112.439273
36. Lossi NS, Manoli E, Simpson P, Jones C, Hui K, Dajani R, Coulthurst SJ, Freemont P, Filloux A (2012) The archetype *Pseudomonas aeruginosa* proteins TssB and TagJ form a novel subcomplex in the bacterial type VI secretion system. Mol Microbiol 86(2):437–456. doi:10.1111/j.1365-2958.2012.08204.x
37. Mikkelsen H, McMullan R, Filloux A (2011) The *Pseudomonas aeruginosa* reference strain

PA14 displays increased virulence due to a mutation in *ladS*. PLoS One 6(12):e29113. doi:10.1371/journal.pone.0029113

38. Planamente S, Salih O, Manoli E, Albesa-Jove D, Freemont PS, Filloux A (2016) TssA forms a gp6-like ring attached to the type VI secretion sheath. EMBO J 35(15):1613–1627. doi:10.15252/embj.201694024
39. Soscia C, Hachani A, Bernadac A, Filloux A, Bleves S (2007) Cross talk between type III secretion and flagellar assembly systems in *Pseudomonas aeruginosa*. J Bacteriol 189 (8):3124–3132. doi:10.1128/jb.01677-06
40. Uzzau S, Figueroa-Bossi N, Rubino S, Bossi L (2001) Epitope tagging of chromosomal genes in salmonella. Proc Natl Acad Sci U S A 98 (26):15264–15269. doi:10.1073/pnas.261348198
41. Muhl D, Filloux A (2014) Site-directed mutagenesis and gene deletion using reverse genetics. Methods Mol Biol 1149:521–539. doi:10.1007/978-1-4939-0473-0_40
42. Rolsma SL, Frank DW (2014) In vitro assays to monitor the activity of Pseudomonas aeruginosa type III secreted proteins. Methods Mol Biol 1149:171–184. doi:10.1007/978-1-4939-0473-0_14
43. Hachani A, Lossi NS, Filloux A (2013) A visual assay to monitor T6SS-mediated bacterial competition. J Vis Exp 73:e50103. doi:10.3791/50103
44. Giannouli M, Palatucci AT, Rubino V, Ruggiero G, Romano M, Triassi M, Ricci V, Zarrilli R (2014) Use of larvae of the wax moth Galleria mellonella as an in vivo model to study the virulence of Helicobacter pylori. BMC Microbiol 14:228. doi:10.1186/s12866-014-0228-0

Chapter 18

Semiquantitative Analysis of the Red, Dry, and Rough Colony Morphology of *Salmonella enterica* Serovar Typhimurium and *Escherichia coli* Using Congo Red

Annika Cimdins and Roger Simm

Abstract

The Congo Red (CR) assay is a standard biofilm test assessing the colony morphology of bacteria growing on agar plates supplemented with the diazo dye Congo Red. Biofilm forming *Salmonella enterica* serovar Typhimurium and *Escherichia coli* produce a *r*ed, *d*ry, and *r*ough (rdar) morphotype on CR-plates. The phenotype is characterized by staining of the extracellular matrix components curli (brown color) and cellulose (pink color) by CR. This method allows semiquantitative determination of the expression level of the individual matrix components and dissection of the regulatory networks controlling their production in response to c-di-GMP levels. Here, we describe the CR-assay and its variations and discuss the effect of deletion or overexpression of c-di-GMP turnover proteins on colony morphology.

Key words Biofilm, Rugosity, Rdar morphotype, Congo Red, Calcofluor, C-di-GMP

1 Introduction

Most bacteria naturally exist in the form of biofilms. Per definition, biofilms are bacterial cells growing in a self-produced extracellular matrix attached to biotic or abiotic surfaces [1]. This differs from the planktonic lifestyle occurring under optimal-growth-promoting laboratory conditions that traditionally have been used to characterize properties of most bacterial species. Recognition of the importance of biofilms spurred the development of several complementary biofilm assays that are designed to assess different aspects of biofilm formation under various conditions [2, 3]. Biofilm formation on different materials is for example assayed in liquid growth medium in static or shaken cultures, usually followed by staining of the attached bacteria to facilitate observation and analyses. A very common variant of this assay is the multi-well plate assay in which adherence to a plastic surface can be quantified following staining of attached bacteria with crystal violet [4]. This

Karin Sauer (ed.), *c-di-GMP Signaling: Methods and Protocols*, Methods in Molecular Biology, vol. 1657,
DOI 10.1007/978-1-4939-7240-1_18, © Springer Science+Business Media LLC 2017

assay is useful for high-throughput screening of mutants or compounds influencing biofilm formation, but it does not allow detailed analysis of the biofilm characteristics. The flow cell assay, on the other hand, allows comprehensive analysis of biofilm formation under dynamic conditions by confocal laser scanning microscopy. By fluorescently labeling certain bacterial components, it is possible to characterize the expression of genes and matrix components as the biofilm develops from single cells to a mature three-dimensional structure [5, 6]. Recently, luminescent conjugated oligothiophenes have been used for real-time optotracing of extracellular matrix production in *Salmonella* biofilms [7]. These assays however require special and expensive equipment, are relatively time-consuming (especially in case of the flow cell experiment), and are inconvenient for high-throughput screeningsCongo Red binding assay. Here, we will discuss the Congo Red agar plate assay (CR-assay), which is a fast and straightforward method that allows semiquantitative analysis of biofilm formation on a solid surface. With this method, it is possible to screen genes and compounds that affect biofilm formation and perform detailed analysis of the regulation of certain biofilm matrix components and their contribution to biofilm formation under specific conditions.

The CR-assay has been used extensively to characterize the regulation of biofilm formation in *Escherichia coli* (*E. coli*) and *Salmonella enterica* serovarSalmonella enterica serovar Typhimurium Typhimurium (*S.* Typhimurium). These species can produce a *r*ed, *d*ry, *a*nd *r*ough (rdar) colony morphology when inoculated on LB without salt agar plates supplemented with Congo Red and Coomassie Brilliant Blue (Fig. 1). The color results from binding of the diazodye Congo Red to both the major extracellular matrix components expressed under these conditions; curli and cellulose

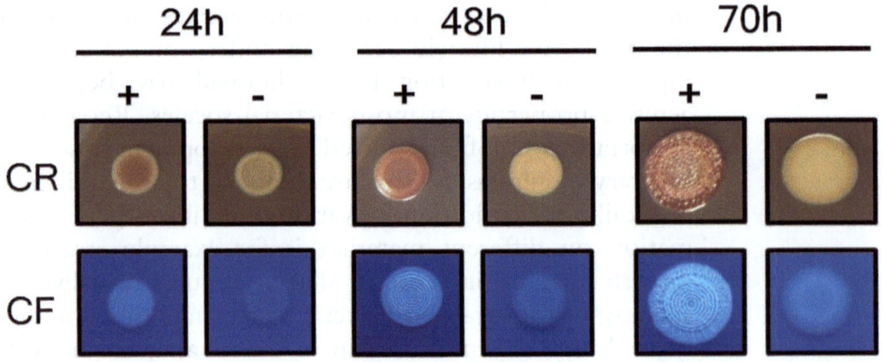

Fig. 1 Colony morphology of model *Salmonella enterica* serovar Typhimurium strains over time on Congo Red (CR) plates (top row) or Calcofluor (CF) plates (lower row). *S.* Typhimurium UMR1 wildtype is the rdar (*r*ed, *d*ry, *a*nd *r*ough) biofilm forming strain (+), UMR1 Δ*csgD* is the non-biofilm forming, saw (*s*mooth *a*nd *w*hite) strain (−). Strains were grown on LB without salt plates supplemented with Congo Red or Calcofluor at 28 °C for up to 70 h. Note the development of rugosity over time for UMR1

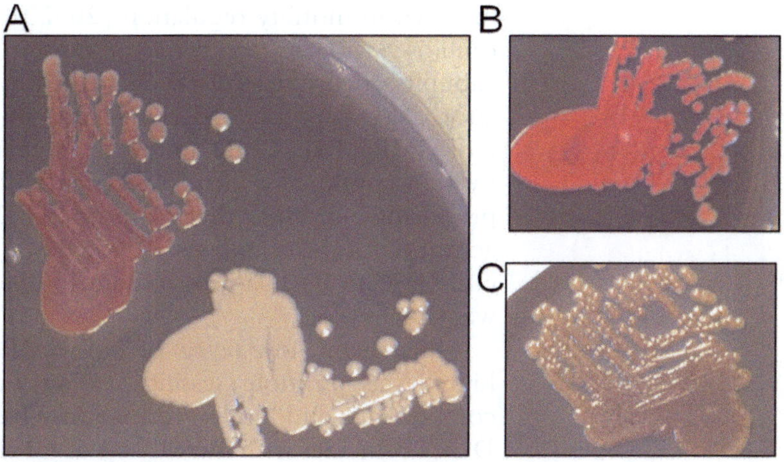

Fig. 2 Streaks of *S.* Typhimurium control strains on CR agar plates. (**a**) From left to right: UMR1 wild type expressing both cellulose and curli (rdar) and the saw, non-biofilm forming colony of UMR1 Δ*csgD* mutant (**b**) pdar (*p*ink, *d*ry, *a*nd *r*ough; cellulose only expression in UMR1Δ*csgBA*), (**c**) bdar (*b*rown, *d*ry, *a*nd *r*ough; curli only expression in UMR1 Δ*bcsA*)

[8]. Curli are amyloids of the major protein subunits CsgA and CsgB [9], whereas cellulose is a polysaccharide consisting of 1,4-β-D-glucan chains [10–12]. Strains lacking the ability to express both curli and cellulose form *s*mooth *a*nd *w*hite (saw) colonies (Fig. 1). Strains expressing curli alone display a *b*rown, *d*ry, *a*nd *r*ough (bdar) colony morphology on CR-plates, whereas curli-deficient strains expressing cellulose only form *p*ink, *d*ry, *a*nd *r*ough (pdar) colonies (Fig. 2). As an alternative, the fluorescent dye Calcofluor White (Fluorescent Brightener 28; CF) can be used in place of CR to detect cellulose production (Fig. 1). When CF binds to cellulose, it can be visualized under UV light at 365 nm. Cellulose positive strains are fluorescent under UV light, while cellulose negative strains are dark. Importantly, in both these cases, dye binding precedes structure development and already indicates low expression levels of biofilm components. The pronounced wrinkled phenotype is mainly associated with high expression of cellulose and can be seen on LB without salt plates even without dye supplementation [13].

The regulatory pathways controlling the rdar morphotype are similar in *E. coli* and *S.* Typhimurium and involve protein regulators, RNA regulators, and integration of extra- and intracellular signals [14–18]. Several stages of biofilm formation in both organisms are influenced by the secondary messenger c-di-GMP. C-di-GMP is synthesized by diguanylate cyclases (DGCs) with a functional GGDEF domain, and degraded by phosphodiesterases (PDEs), harboring functional EAL or HD-GYP domains [19]. Moreover, several c-di-GMP receptors are present, e.g., YcgR, involved in motility regulation [20–22], and proteins involved in cellulose synthesis (*see* below). Expression of *csgBA*, encoding the subunits of curli, is directly controlled by the transcriptional regulator CsgD [23, 24] and *csgD* expression is in turn influenced by c-di-GMP [13]. CsgD indirectly regulates cellulose production by transcriptional control over *adrA* [23, 24]. AdrA is a DGC producing c-di-GMP that promotes cellulose synthesis by allosteric activation of the cellulose synthase BcsA [11]. Recently, binding of c-di-GMP to BcsE, a protein contributing to cellulose production, was shown [25].

In general, low levels of c-di-GMP promote motility, while high levels promote sessility, expression of extracellular matrix components, and biofilm formation. Overexpression of functional DGCs coincides with increased c-di-GMP levels, while overexpression of PDEs is linked to reduced levels of c-di-GMP. Additionally, overexpression of DGCs generally enhances rdar morphotype formation and reduces motility, while overexpression of PDEs results in downregulated rdar morphotype and increased motility. However, chromosomally encoded c-di-GMP turnover proteins usually control specific phenotypes: motility, the rdar morphotype, or even cellulose production alone. For instance, in *S.* Typhimurium, chromosomally encoded AdrA (a DGC) plays a critical role in cellulose production, but has no effect on motility under standard swimming assay conditions (Roger Simm and Ute Römling, unpublished results). Chromosomally encoded phosphodiesterase STM4264 has an important role in controlling rdar morphotype development (Fig. 3b), but has no impact on motility [27]. In contrast, the chromosomally encoded phosphodiesterase YhjH is a key regulator of motility [27, 28], but has only a moderate effect on rdar morphotype development [27]. Consequently, the CR-assay can be used to determine the function of a c-di-GMP turnover protein and elucidate its role in rdar morphotype development by plasmid-mediated overexpression and chromosomal inactivation of the gene, respectively.

Application of a drop of a bacterial suspension onto a CR-assay plate results in a macrocolony that facilitates observation of biofilm growth and development over time. Interestingly, a macrocolony consists of different zones that vary in color and structure (Fig. 3a), reflecting that bacteria are at different growth stages at different locations in the macrocolony. It has been shown that flagellated, dividing cells dominate at the bottom of the rdar macrocolony close to the nutrient source, whereas stationary phase cells producing curli and cellulose are predominantly present at the surface of the colony, further away from the agar [14, 29–31]. The zones can be further divided into subzones, depending on their structure and color [15, 29]. Importantly, the provided picture is an example. The zones may vary in occurrence and size depending on the strain and growth conditions. It is therefore possible that detailed

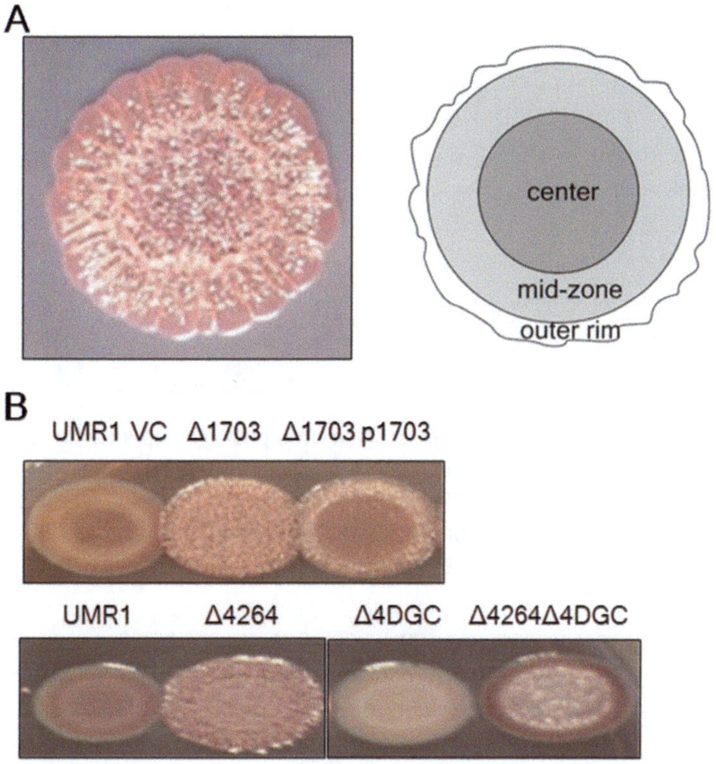

Fig. 3 Zone formation of a bacterial macrocolony. (**a**) Left: Colony morphology of commensal *E. coli* strain Fec51 [26] grown on Congo Red plates at 28 °C for 72 h. Right: Schematic overview of the macrocolony. Zone differentiation can be strain dependent. (**b**) Spotting assay of *S.* Typhimurium UMR1 and respective deletion mutants. Top: Deletion of the active phosphodiesterase STM1703 results in strong upregulation of rdar morphotype formation. Upon overexpression of STM1703 from a plasmid, rdar morphotype is still upregulated in the outer zone of the colony. Bottom: Deletion of the phosphodiesterase STM4264 enhances rdar morphotype, deletion of four diguanylate cyclases abolishes rdar morphotype expression in UMR1. Upon additional deletion of the phosphodiesterase STM4264 in the Δ4DGC background, rdar morphotype formation is restored, but mainly in the inner zone of the colony. VC = vector control pBAD30. Figure 3b is derived from Figs. 4b and 5c of a previous publication by Ahmad et al. [13]. For layout reasons, the panels were relabeled. Only parts of the figures were used. The figure was reprinted with respect to the Creative Commons Attribution 4.0 International License (http:/creativecommons.org/licenses/by/4.0/)

microscopic analysis of the macrocolonies from various strains can facilitate our understanding of even subtle roles of specific proteins in biofilm development. An interesting example of this is shown in Fig. 3b, where a combination of chromosomal deletions and plasmid mediated overexpression of c-di-GMP turnover proteins resulted in phenotypic effects that were limited to distinct zones of the macrocolony (Fig. 3b, [13]).

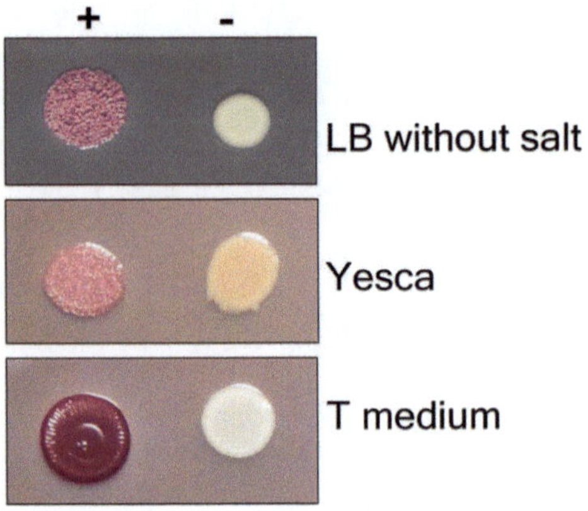

Fig. 4 Colony morphology depends on the chosen growth medium. Colony morphology of *E. coli* commensal strain Tob1 (+) and Tob1 Δ*csgD* (−) [26] on CR-plates prepared with LB without salt (top row), Yesca medium (middle row), or T medium (bottom row) incubated at 28 °C for 48 h. Both structure and color development depend on the chosen medium

Rdar morphotype development is dependent on the growth environment [8, 15, 32]. Besides incubation temperature, the nutrients and osmolarity of the growth medium can affect colony morphology (Fig. 4). Hence, various growth conditions can be chosen to help clarify subtle differences in regulation of rdar morphotype formation among strains.

Congo Red binding is not specific to curli and cellulose *(see also Congo Red binding assay*, but can also bind to other amyloid proteins and polysaccharides. The CR-assay is therefore a useful tool to study regulation of biofilm formation and to dissect the network controlling expression of biofilm matrix components in bacteria other than enterobacteriaceae, as has been demonstrated, e.g., for *Staphylococci* where CR binds to poly-*N*-acetylglucosamine [33–35].

In the following section, we describe the CR-assay and its variations, and discuss its application in the analysis of biofilm formation by *S.* Typhimurium and *E. coli.*

2 Materials

Agar and stock solutions are custom-made. Growth media need to be autoclaved (*see* **Note 1**). Calcofluor stock solution, antibiotics, and inducers for plasmid expression should be sterilized by filtration using a 0.2 μm filter. As the Congo Red stock is prepared in 70% ethanol, no further sterilization is needed. Use a sterile flask for preparation.

2.1 Culture Media and Solutions

1. Lysogeny broth (LB) agar with salt: 10 g/L Tryptone, 5 g/L yeast extract, 10 g/L NaCl, and 16 g/L Agar. Dissolve in deionized water. Prepare 600 mL in a 1 L glass bottle; sterilize by autoclaving (20 min; 121 °C).
2. Lysogeny broth (LB) agar with salt supplemented with ampicillin: 100 μg/mL ampicillin sodium salt; add 600 μL sterilized ampicillin stock solution (100 g/L) to 600 mL autoclaved LB agar with salt after letting the media cool down to 50–60 °C.
3. Lysogeny broth (LB) agar without salt: 10 g/L Tryptone, 5 g/L yeast extract, and 16 g/L Agar. Dissolve in deionized water. Prepare 600 mL in a 1 L glass bottle; sterilize by autoclaving (20 min; 121 °C).
4. Yesca medium: 10 g/L Casamino acids, 1 g/L yeast extract, and 20 g/L Agar. Dissolve in deionized water. Prepare 600 mL in a 1 L glass bottle; sterilize by autoclaving (20 min; 121 °C).
5. T medium: 10 g/L Tryptone and 15 g/L Agar. Dissolve in deionized water. Prepare 600 mL in a 1 L glass bottle; sterilize by autoclaving (20 min; 121 °C).
6. Congo Red stock solution: 2 g/L Congo Red (CR) (*see* **Note 2**) and 1 g/L Coomassie Brilliant Blue G-250. Dissolve in 70% ethanol. Prepare 100 mL in a 100 mL sterilized glass flask. Wrap the lid with parafilm to avoid evaporation. Store the stock solution in the dark (e.g., wrapped with aluminum foil) at room temperature.
   (a) Congo Red (CR) agar plates: Add 12 mL stock solution to 600 mL autoclaved medium after letting the media cool down to 50–60 °C. The final concentration is 40 μg/mL Congo Red and 20 μg/mL Coomassie Brilliant Blue G-250. Pour at least 40 mL of the mixture into standard petri dishes (8.5 cm diameter) and let the plates solidify (*see* **Note 3**). Store the plates in the dark at 4 °C. Before usage, remove residual water on the agar surface by placing the plates without the lid in a laminar flow hood for 30 min or until no moisture is left.
   (b) Congo Red (CR) agar plates supplemented with ampicillin and arabinose: 12 mL CR stock solution, 600 μL sterilized ampicillin stock solution (100 g/L), and 6 mL 10% L-arabinose stock per 600 mL medium (final concentration 100 μg/mL ampicillin, 0.1% L-arabinose (*see* **Notes 4** and **13**)). Proceed as described under 6 *a*.
7. Calcofluor stock solution: 5 g/L Calcofluor (Fluorescent Brightener 28; CF). Dissolve in deionized water and sterilize by filtration (0.2 μm pore size). Prepare 10 mL solution and store in the dark (e.g., wrapped with aluminum foil) at 4 °C in a sterile falcon tube.

8. Calcofluor (CF) agar plates: Add the stock in 1:100 dilution after letting the media cool down to 50–60 °C, i.e., 6 mL CF stock to 600 mL media to a final concentration of 50 μg/mL. Pour the mixture into small petri dishes (diameter of 8.5 cm) and let the plates solidify. Store plates in the dark (e.g., wrapped with aluminum foil) at 4 °C. Before usage, remove residual water on the agar surface by placing the plates without the lid in a laminar flow hood for 30 min or until no moisture is left.
9. Ampicillin stock: 100 g/L ampicillin sodium salt; dissolve in deionized water; sterilize by filtration (0.2 μm pore size). Prepare 5 mL and store at −20 °C in a sterile falcon tube.
10. L-Arabinose stock: 10% L-Arabinose (i.e., 100 g/L) (*see* **Notes 4** and **13**); dissolve in deionized water; sterilize by filtration (0.2 μm pore size); prepare 10 mL and store at 4 °C in sterile falcon tube.

2.2 Bacterial Strains

1. *Salmonella enterica* serovar Typhimurium UMR1 wild type [36].
2. *S.* Typhimurium UMR1 Δ*csgD* (MAE50) [23].
3. *S.* Typhimurium UMR1 Δ*csgBA* (MAE14) [23].
4. *S.* Typhimurium UMR1 Δ*bcsA* (MAE1267) [37].
5. *E. coli* Tob1 wild type [26].
6. *E. coli* Tob1 Δ*csgD* [26].

2.3 General Equipment Needed

1. 1 L Glass bottles for media preparation.
2. Sterilized 100 mL flask for Congo Red stock solution preparation.
3. Sterilized Falcon tubes (15 mL, 50 mL) for preparation of solutions.
4. Sterilized 1.5 mL Eppendorf-tubes.
5. Sterilized tips and inoculation loops.
6. Petri dishes, 8.5 cm diameter.
7. Aluminum foil.
8. Parafilm.
9. Balance suitable to weigh ingredients for media preparation (e.g., Sartorius ED822-CW, 820 g capacity and 0.01 g readability) and for preparation of stock solutions (e.g., Ohaus Explorer, 62 g capacity and 0.1 mg readability).
10. Autoclave.
11. Spectrophotometer for the determination of cell density at 600 nm.
12. Digital camera, e.g., Panasonic DMC TZ-40.
13. Hand-held UV lamp 365 nm.

3 Methods

In principle, the Congo Red assay is suitable for a broad range of species producing an extracellular matrix consisting of amyloid proteins or suitable polysaccharides binding to CR (and/or CF) and has been used to study biofilm formation in a number of enterobacteriaceae in addition to *E. coli* (*see* **Note 5**) and *S. enterica* spp. such as *Hafnia alveri* [38], *Citrobacter* sp. [39], and *Klebsiella pneumoniae* [40]. Similarly, solid agar plates without dye supplementation were used for semiquantitative assessment of *Vibrio fischeri* biofilms over time [41]. The CR-assay has been applied to analyze the extracellular matrix of gram-positives as well, including *Staphylococci* [33–35] and *Listeria monocytogenes* [42].

A positive control (i.e., a biofilm forming strain) and a negative control (i.e., non-biofilm forming strain) should be included when analyzing the biofilm formation of a strain of interest. We frequently use *S.* Typhimurium UMR1, a derivative of ATCC14028s, as a model strain for regulated red, dry, and rough colony morphology, hence as positive control [36]. As a negative control, we use a deletion mutant of *csgD* since *S.* Typhimurium UMR1 Δ*csgD* does not express the extracellular matrix components cellulose and curli, forms smooth and white colonies on Congo Red agar plates (Figs. 1 and 2) and does not fluoresce under UV light when tested on Calcofluor plates [23] (Fig. 1). The controls should be chosen based on the species and strains assayed. If you are, e.g., analyzing mutants in cellulose synthesis genes for colony morphology, a Δ*bcsA* strain, incapable of producing cellulose but able to produce curli, hence producing a brownish colony (Fig. 2), should be included as control. When analyzing the impact of curli fibers on biofilm formation, a Δ*csgBA* mutant can be used, producing cellulose only, and exhibiting a pink colony morphology (Fig. 2). Evaluation of color and roughness is a simple method, but provides a plethora of information. A red, dry, but smooth colony appearance indicates low levels of curli and cellulose production. A brownish colony appearance hints on an impairment of cellulose production, but presence of curli. Bacterial cellulose synthesis operons are diverse [12]. In *E. coli* and *S.* Typhimurium, cellulose production is mediated by two operons, *bcsRQABZC* and *bcsEFG*. The appearance of the morphotype can vary depending on which gene is affected. For instance, in *S.* Typhimurium, deletion of *bcsA* will lead to a bdar morphotype, while deletion of *bcsE* has a less drastic effect, indicating that BcsA is essential for cellulose synthesis, while BcsE is contributing to maximal production [25]. Thus, the CR-assay provides information about the impact of a gene product of interest on rdar morphotype formation. Accordingly, a pink colored colony morphology represents cellulose only expression and thus disturbance of the curli *csgBAC*, *csgDEFG* operon expression, and/or curli assembly. A complementary assay to elucidate the nature of the impairment can be the detection of the curli subunits via Western Blot analysis or electron microscopy to visualize the compounds present in the extracellular matrix [29, 36].

The CR spotting assay*CR-assay* described below uses *S*. Typhimurium UMR1 wild type and Δ*csgD* as example strains. When working with pathogens and genetically modified microorganisms (GMOs), follow the appropriate guidelines and safety rules of your institution. All the steps described below require aseptic techniques. Use sterilized equipment and compounds only. Work under a laminar flow hood approved for microorganisms of biosafety level 2 and genetically modified organisms, if applicable.

3.1 Preparation of Bacterial Samples

1. Inoculate *S*. Typhimurium UMR1 wild type and Δ*csgD* from glycerol stocks kept at −80 °C onto Congo Red agar plates:
   (a) Streak for single colonies.
   (b) Incubate the plates at 28 °C for 1–2 days (*see* **Note 6**).
   (c) Check that they have the correct phenotype: *S*. Typhimurium UMR1 wild type is rdar, UMR1 Δ*csgD* is saw (Figs. 1 and 2).
2. Pick single colonies and re-streak onto LB agar plates with salt. Following incubation at 37 °C, this re-streaked culture can be used as the overnight preculture.
3. The next day, harvest the bacteria from the plate with a sterile loop and resuspend them in 100 μL sterile deionized water in a sterilized 1.5 mL Eppendorf tube (*see* **Note 7**).
4. Measure the cell density (OD_{600}) in a 1:100 dilution in a spectrophotometer at 600 nm.
5. Adjust the bacterial suspension to an OD_{600} of 5 using deionized sterile water.

3.2 Spotting Assay

1. Apply 5 μL drops of *S*. Typhimurium UMR1 wild type and UMR1 Δ*csgD* onto the same CR- or CF-plate (*see* **Note 8**). CR-plates visualize expression of curli and cellulose, while CF-plates can be used to mainly detect the presence of cellulose. When spotting the samples, consider that the bacterial colony will grow and expand. Thus, leave at least 1.5 cm space between the drops.
   (a) To avoid spills, move your hand slowly and choose the most direct way for moving the pipette from the edge over the agar surface.
   (b) While pipetting, steady the pipette with a finger.
   (c) Avoid introducing bubbles in the drop.
2. Let the drops dry completely (for around 10–15 min).

3. After the drops are dry, turn the plates upside down and incubate them at the desired temperature (*see* **Note 9**) for 24 h, 48 h, and longer if necessary, until no further development of the rdar colony can be observed.
4. Document development of colony morphology over time by photographing the plates (*see* **Note 10**). CF-plates have to be illuminated with UV light at 365 nm to allow detection of cellulose positive colonies (*see* **Note 11**).

3.3 Overexpression/Deletion of Diguanylate Cyclases and Phosphodiesterases and Their Effect on rdar Morphotype Development of S. Typhimurium

The CR-assay is an important tool in the analysis of the c-di-GMP signaling network. It is suitable for the investigation of the role of gene products in rdar morphotype expression by comparative analysis of strains after chromosomal deletion of genes (*see* **Note 12**). For example, the central role of the DGC AdrA in cellulose production was elucidated following deletion of *adrA* and comparison of the phenotypes of the mutant and wild type in the CR-assay (bdar and rdar phenotypes respectively on CR-plates at 28 °C) [23]. Since cellulose synthesis is dependent on c-di-GMP for activation, the CR-assay can also be used to determine the function of a specific gene product, such as DGCs and PDEs by overexpression from a plasmid (*see* **Note 13**). For example, YhjH (from *S.* Typhimurium) and HmsT (from *Yersinia pestis*) were shown to function as a PDE and DGC, respectively when overexpressed from a plasmid in *S.* Typhimurium on CR-plates [43, 44]. This demonstrates that the CR-assay can be used to characterize proteins from different species after heterologous expression in a well-characterized rdar morphotype expressing strain such as *S.* Typhimurium UMR1. Furthermore, by site-directed mutagenesis and comparative analysis of wild type and mutated variants after overexpression from a plasmid, gene products such as DGCs and PDEs can be further characterized to identify amino acids important for protein function (*see* **Note 14**). This method was for example used to characterize the structure function relationship of the DGC WspR from *Pseudomonas* spp. in *E. coli* BL21 [45].

1. Prepare bacterial samples according to Subheading 3.1. Make sure to add appropriate antibiotics to the LB plates used to grow the overnight cultures as well as the desired concentration of the compound necessary to induce expression of the plasmid-encoded proteins into the respective CR-assay plates. Do not forget to include an empty vector control strain (*see* **Note 13**).
2. Continue with the steps described under Subheading 3.2.
3. Evaluation of the plates:
   (a) Performing complementation of chromosomal mutations, compare the respective PDE/DGC deletion mutant strain harboring the empty vector control with

the wild-type strain harboring the empty vector control and with the deletion mutant strain harboring the plasmid with encoded protein. While the deletion of a PDE is expected to result in upregulated rdar morphotype, deletion of a DGC is expected to lead to downregulation of rdar morphotype formation. Upon expression of the respective protein from plasmid in the deletion mutant background, the wild-type phenotype will be restored or even hypercomplemented. This can depend on the concentration of inducer added and can be titrated.

(b) When analyzing the sole effect of overexpression, compare wild-type strain harboring empty vector control to wild type harboring the plasmid with the encoded PDE or DGC of interest. Overexpression of a PDE will lead to decreased rdar morphotype, while overexpression of a DGC will provoke enhanced rdar morphotype formation.

(c) Assessing the impact of single amino acids in catalytic motifs of c-di-GMP turnover proteins on rdar morphotype formation, compare the colony morphology upon overexpression of wild-type protein to the colony morphologies upon overexpression of the respective mutated protein variants. If the effect of the mutant protein on rdar morphotype expression in relation to the wild-type protein is reduced, the mutated amino acid could be critical for functionality of the protein in c-di-GMP turnover or protein stability.

3.4 Congo Red Plates with Alternative Growth Media

The choice of growth medium for Congo Red assays can be adjusted to the requirements of the chosen species (i.e., to match the conditions required by a specific species for biofilm formation and expression of the extracellular matrix). For instance, for *Staphylococci*, CR-assays are performed on brain heart infusion agar (BHIA) [33, 35] or Blood Base Agar-2 (BAB-2) [34]. For *L. monocytogenes*, HTM/G agar has been used [42]. Variations of growth medium can affect the colony morphology of a specific strain in both color and roughness. Growth of the commensal *E. coli* strain Tob1 and its Δ*csgD* mutant on CR-plates prepared with LB without salt, Yesca, or T medium, demonstrates how both the rdar phenotype of the wild type and the saw phenotype of the Δ*csgD* mutant vary on the different growth media (Fig. 4). The color contrast between the rdar and saw phenotypes is especially pronounced on T medium, being dark red and bright white, respectively. Thus, comparison of colony morphology on different growth media can be used to study slight modulations of rdar morphotype expression.

4 Notes

1. The agar can be used for pouring plates immediately after autoclaving when it is still in liquid form. However, before use, let it cool down so that you can easily touch it. If you want to use the agar at a later time point, either keep it in a water bath set to 50–60 °C or allow it to solidify at room temperature. If you choose the latter, consider preparing and autoclaving the agar containing medium in a glass bottle. The easiest method of melting the agar is using a microwave oven. Make sure the bottle of choice fits into the microwave oven. If not, reduce the amount of agar and bottle size accordingly.
2. Congo Red is toxic, thus use appropriate safety measures according to the safety description of the compound and the laboratory safety rules of your institution. As the solution is prepared in 70% ethanol, no further sterilization is needed (! never! autoclave ethanol solutions).
3. To minimize differences between plates due to thickness of agar, and for better visualization, add equal volumes of medium to each plate, i.e., 40 mL into a 8.5 cm diameter petri dish. Pouring the plates too thin will hinder the visual examination of the phenotype.
4. The choice of inducer and its concentration depends on the chosen expression vector (*see* **Note 13**). Different inducer concentrations should be applied to identify the optimal expression conditions.
5. *E. coli* K-12 is not capable of producing cellulose due to a stop codon in *bcsQ* [46].
6. To get a first impression of the unknown biofilm forming capacity of the strain of interest and to ensure homogeneity of the culture, a first streak for single colonies on Congo Red plates is recommended. If strains express curli only, brown color development is observed in contrast to the pink colony morphotype of cellulose only strains (Fig. 1). Due to the semi-specificity of Congo Red for amyloids and 1,4- and 1,3- beta glucans, presence of the exopolysaccharide poly-β-1,6-*N*-acetylglucosamine (PNAG) is also detected. Analysis of the genome sequence of the strain in combination with colony morphology and construction of respective deletion mutants will give you the best possible result.
7. Depending on the strain of choice, lysis of bacteria in water might occur. However, this is not an issue for *E. coli* and *Salmonella* spp. For instance, preparation of electrocompetent cells through washing with water is a frequently used method [47]. The preparation of bacterial samples can alternatively be performed with 1× PBS (phosphate-buffered saline) or liquid

growth medium. Another option is applying a drop of liquid culture.

8. As colony morphology development can differ between plates due to various reasons like moisture or oxygen supply, always apply strains to be compared and the controls onto the *same* plate.
9. Usually, we use 28 °C and 37 °C. Our standard model *S.* Typhimurium strain UMR1 forms biofilms at 28 °C but not at 37 °C.
10. A standard, commercially available digital camera like Panasonic Lumix DMC-TZ40 is sufficient for photographing the plates. Avoid reflections. Use a dark, non-patterned surface as background when taking pictures. Take a picture of the entire plate, showing all colonies in the same picture. If you want to take close-up pictures of single colonies, ensure that the same conditions (light, zoom, etc.) are used for all other single colonies to receive comparable pictures. If available, use a scaffold to keep the camera steady.
11. Use a hand-held UV lamp (e.g., UVL-56 from UVP), emitting 365 nm wavelength light to visualize the colonies. Only cellulose positive colonies will be fluorescent. Take pictures in a dark room using a dark, non-patterned background. Make sure to illuminate the plate evenly and keep a constant distance of approximately 30 cm to the plate surface. UV light is harmful, do *not* stare into the lamp and avoid exposing body parts to strong UV light. Wear gloves and a UV protection face shield if applicable.
12. A well-established method for chromosomal gene deletion is the one-step gene inactivation protocol by Datsenko and Wanner, which is based on Lambda red recombination. Thereby, the target gene is disrupted by a PCR synthesized antibiotic cassette with flanking homologous regions [48].
13. We recommend pBAD28 or pBAD30 [49] as expression vectors for c-di-GMP turnover proteins. These vectors confer ampicillin-resistance and their promoters are inducible with L-arabinose. Use various concentrations of L-arabinose to determine the optimal expression conditions, e.g., no inducer, 0.001%, 0.01%, and 0.1% L-arabinose to discriminate subtle effects of the overexpressed protein on colony morphology.
14. To assess the catalytic activity of putative c-di-GMP turnover proteins, perform site-directed mutagenesis to exchange amino acids important for the catalytic activity [19], e.g., by changing GGDEF to GGAAF in DGCs and EAL to AAL in PDEs [13]. Mutagenesis kits like the QuikChange II Site-Directed Mutagenesis Kit from Agilent are commercially available. Perform additional western blot experiments to verify equal expression levels of wild-type and catalytic mutant proteins.

Acknowledgment

A.C. receives funding from the German Research Foundation (DFG; CI 239/1-1, CI 239/2-1). The authors are grateful to Ute Römling for scientific discussions and critical reading of the manuscript.

References

1. Costerton JW, Lewandowski Z, Caldwell DE, Korber DR, Lappin-Scott HM (1995) Microbial biofilms. Annu Rev Microbiol 49:711–745
2. Pantanella F, Valenti P, Natalizi T, Passeri D, Berlutti F (2013) Analytical techniques to study microbial biofilm on abiotic surfaces: pros and cons of the main techniques currently in use. Ann Ig 25(1):31–42
3. Azeredo J, Azevedo NF, Briandet R, Cerca N, Coenye T, Costa AR, Desvaux M, Di Bonaventura G, Hebraud M, Jaglic Z, Kacaniova M, Knochel S, Lourenco A, Mergulhao F, Meyer RL, Nychas G, Simoes M, Tresse O, Sternberg C (2016) Critical review on biofilm methods. Crit Rev Microbiol:1–39
4. O'Toole GA (2011) Microtiter dish biofilm formation assay. J Vis Exp (47):2437
5. Sternberg C, Tolker-Nielsen T (2006) Growing and analyzing biofilms in flow cells. Curr Protoc Microbiol Chapter 1:Unit 1B 2.
6. Tolker-Nielsen T, Sternberg C (2014) Methods for studying biofilm formation: flow cells and confocal laser scanning microscopy. Methods Mol Biol 1149:615–629
7. Choong FX, Bäck M, Fahlén S, Johansson LBG, Melican K, Rhen M, Nilsson KPR, Richter-Dahlfors A (2016) Real-time optotracing of curli and cellulose in live *Salmonella* biofilms using luminescent oligothiophenes. npj Biofilms and Microbiomes 2:16024
8. Römling U (2005) Characterization of the rdar morphotype, a multicellular behaviour in Enterobacteriaceae. Cell Mol Life Sci 62 (11):1234–1246
9. Hammar M, Arnqvist A, Bian Z, Olsen A, Normark S (1995) Expression of two *csg* operons is required for production of fibronectin- and Congo Red-binding curli polymers in *Escherichia coli* K-12. Mol Microbiol 18(4):661–670
10. Ross P, Mayer R, Benziman M (1991) Cellulose biosynthesis and function in bacteria. Microbiol Rev 55(1):35–58
11. Zogaj X, Nimtz M, Rohde M, Bokranz W, Römling U (2001) The multicellular morphotypes of *Salmonella typhimurium* and *Escherichia coli* produce cellulose as the second component of the extracellular matrix. Mol Microbiol 39(6):1452–1463
12. Römling U, Galperin MY (2015) Bacterial cellulose biosynthesis: diversity of operons, subunits, products, and functions. Trends Microbiol 23(9):545–557
13. Ahmad I, Cimdins A, Beske T, Römling U (2017) Detailed analysis of c-di-GMP mediated regulation of *csgD* expression in *Salmonella typhimurium*. BMC Microbiol 17 (1):27
14. Serra DO, Hengge R (2014) Stress responses go three dimensional - the spatial order of physiological differentiation in bacterial macrocolony biofilms. Environ Microbiol 16 (6):1455–1471
15. Simm R, Ahmad I, Rhen M, Le Guyon S, Römling U (2014) Regulation of biofilm formation in *Salmonella enterica* serovar Typhimurium. Future Microbiol 9(11):1261–1282
16. Mika F, Hengge R (2013) Small regulatory RNAs in the control of motility and biofilm formation in *E. coli* and *Salmonella*. Int J Mol Sci 14(3):4560–4579
17. Gerstel U, Römling U (2001) Oxygen tension and nutrient starvation are major signals that regulate *agfD* promoter activity and expression of the multicellular morphotype in *Salmonella typhimurium*. Environ Microbiol 3 (10):638–648
18. Gerstel U, Park C, Römling U (2003) Complex regulation of *csgD* promoter activity by global regulatory proteins. Mol Microbiol 49 (3):639–654
19. Römling U, Galperin MY, Gomelsky M (2013) Cyclic di-GMP: the first 25 years of a universal bacterial second messenger. Microbiol Mol Biol Rev 77(1):1–52
20. Ryjenkov DA, Simm R, Römling U, Gomelsky M (2006) The PilZ domain is a receptor for the second messenger c-di-GMP: the PilZ domain protein YcgR controls motility in enterobacteria. J Biol Chem 281(41):30310–30314

21. Boehm A, Kaiser M, Li H, Spangler C, Kasper CA, Ackermann M, Kaever V, Sourjik V, Roth V, Jenal U (2010) Second messenger-mediated adjustment of bacterial swimming velocity. Cell 141(1):107–116
22. Paul K, Nieto V, Carlquist WC, Blair DF, Harshey RM (2010) The c-di-GMP binding protein YcgR controls flagellar motor direction and speed to affect chemotaxis by a "backstop brake" mechanism. Mol Cell 38 (1):128–139
23. Römling U, Rohde M, Olsen A, Normark S, Reinköster J (2000) AgfD, the checkpoint of multicellular and aggregative behaviour in *Salmonella typhimurium* regulates at least two independent pathways. Mol Microbiol 36 (1):10–23
24. Brombacher E, Dorel C, Zehnder AJ, Landini P (2003) The curli biosynthesis regulator CsgD co-ordinates the expression of both positive and negative determinants for biofilm formation in *Escherichia coli*. Microbiology 149 (Pt 10):2847–2857
25. Fang X, Ahmad I, Blanka A, Schottkowski M, Cimdins A, Galperin MY, Römling U, Gomelsky M (2014) GIL, a new c-di-GMP-binding protein domain involved in regulation of cellulose synthesis in enterobacteria. Mol Microbiol 93(3):439–452
26. Bokranz W, Wang X, Tschäpe H, Römling U (2005) Expression of cellulose and curli fimbriae by *Escherichia coli* isolated from the gastrointestinal tract. J Med Microbiol 54(Pt 12):1171–1182
27. Simm R, Lusch A, Kader A, Andersson M, Römling U (2007) Role of EAL-containing proteins in multicellular behavior of *Salmonella enterica* serovar Typhimurium. J Bacteriol 189 (9):3613–3623
28. Le Guyon S, Simm R, Rehn M, Römling U (2015) Dissecting the cyclic di-guanylate monophosphate signalling network regulating motility in *Salmonella enterica* serovar Typhimurium. Environ Microbiol 17(4):1310–1320
29. Serra DO, Richter AM, Klauck G, Mika F, Hengge R (2013) Microanatomy at cellular resolution and spatial order of physiological differentiation in a bacterial biofilm. MBio 4 (2):e00103–e00113
30. Römling U, Rohde M (1999) Flagella modulate the multicellular behavior of *Salmonella typhimurium* on the community level. FEMS Microbiol Lett 180(1):91–102
31. Hobley L, Harkins C, MacPhee CE, Stanley-Wall NR (2015) Giving structure to the biofilm matrix: an overview of individual strategies and emerging common themes. FEMS Microbiol Rev 39(5):649–669
32. Beloin C, Roux A, Ghigo JM (2008) *Escherichia coli* biofilms. Curr Top Microbiol Immunol 322:249–289
33. Kaiser TD, Pereira EM, Dos Santos KR, Maciel EL, Schuenck RP, Nunes AP (2013) Modification of the Congo Red agar method to detect biofilm production by *Staphylococcus epidermidis*. Diagn Microbiol Infect Dis 75 (3):235–239
34. Mariana NS, Salman SA, Neela V, Zamberi S (2009) Evaluation of modified Congo Red agar for detection of biofilm produced by clinical isolates of methicillin–resistance *Staphylococcus aureus*. Afr J Microbiol Res 3 (6):330–338
35. Brooks JL, Jefferson KK (2014) Phase variation of poly-*N*-acetylglucosamine expression in *Staphylococcus aureus*. PLoS Pathog 10(7): e1004292
36. Römling U, Bian Z, Hammar M, Sierralta WD, Normark S (1998) Curli fibers are highly conserved between *Salmonella typhimurium* and *Escherichia coli* with respect to operon structure and regulation. J Bacteriol 180 (3):722–731
37. Ahmad I, Rouf SF, Sun L, Cimdins A, Shafeeq S, Le Guyon S, Schottkowski M, Rhen M, Römling U (2016) BcsZ inhibits biofilm phenotypes and promotes virulence by blocking cellulose production in *Salmonella enterica* serovar Typhimurium. Microb Cell Factories 15 (1):177
38. Chapartegui-González I, Lázaro-Díez M, Redondo-Salvo S, Amaro-Prelezzo E, Esteban-Rodríguez E, Ramos-Vivas J (2016) Biofilm formation in *Hafnia alvei* HUMV-5920, a human isolate. AIMS Microbiol 2(4):412–421
39. Zogaj X, Bokranz W, Nimtz M, Römling U (2003) Production of cellulose and curli fimbriae by members of the family Enterobacteriaceae isolated from the human gastrointestinal tract. Infect Immun 71(7):4151–4158
40. Barati A, Ghaderpour A, Chew LL, Bong CW, Thong KL, Chong VC, Chai LC (2016) Isolation and characterization of aquatic-borne *Klebsiella pneumoniae* from tropical estuaries in Malaysia. Int J Environ Res Public Health 13(4):426
41. Ray VA, Morris AR, Visick KL (2012) A semi-quantitative approach to assess biofilm formation using wrinkled colony development. J Vis Exp (64): e4035
42. Köseoglu VK, Heiss C, Azadi P, Topchiy E, Guvener ZT, Lehmann TE, Miller KW,

Gomelsky M (2015) *Listeria monocytogenes* exopolysaccharide: origin, structure, biosynthetic machinery and c-di-GMP-dependent regulation. Mol Microbiol 96(4):728–743

43. Simm R, Morr M, Kader A, Nimtz M, Römling U (2004) GGDEF and EAL domains inversely regulate cyclic di-GMP levels and transition from sessility to motility. Mol Microbiol 53 (4):1123–1134
44. Simm R, Fetherston JD, Kader A, Römling U, Perry RD (2005) Phenotypic convergence mediated by GGDEF-domain-containing proteins. J Bacteriol 187(19):6816–6823
45. De N, Pirruccello M, Krasteva PV, Bae N, Raghavan RV, Sondermann H (2008) Phosphorylation-independent regulation of the diguanylate cyclase WspR. PLoS Biol 6(3):e67
46. Serra DO, Richter AM, Hengge R (2013) Cellulose as an architectural element in spatially structured *Escherichia coli* biofilms. J Bacteriol 195(24):5540–5554
47. Pfau J, Youderian P (1990) Transferring plasmid DNA between different bacterial species with electroporation. Nucleic Acids Res 18 (20):6165
48. Datsenko KA, Wanner BL (2000) One-step inactivation of chromosomal genes in *Escherichia coli* K-12 using PCR products. Proc Natl Acad Sci U S A 97(12):6640–6645
49. Guzman LM, Belin D, Carson MJ, Beckwith J (1995) Tight regulation, modulation, and high-level expression by vectors containing the arabinose PBAD promoter. J Bacteriol 177(14):4121–4130

Part VI

Measuring c-di-GMP Modulating Activities

Part IV

Chapter 19

Fluorescent 2-Aminopurine c-di-GMP and GpG Analogs as PDE Probes

Jie Zhou, Clement Opoku-Temeng, and Herman O. Sintim

Abstract

c-di-GMP is widely recognized as an important ubiquitous signaling molecule in bacteria. c-di-GMP phosphodiesterases (PDEs) regulate the intracellular concentration of c-di-GMP and some could be potential drug targets. Here, we describe a class of dinucleotide probes suitable for monitoring the enzymatic activities of c-di-GMP PDEs in real time. Such probes contain fluorescent nucleobases and can be readily cleaved by PDEs, resulting in a change in fluorescence. Fluorescent cyclic and linear dinucleotide probes could be used in diverse applications, such as confirming the activity of an expressed PDE or oligoribonuclease (Orns) or identifying inhibitors of PDEs or Orns using high-throughput screening formats.

Key words c-di-GMP, Fluorescent bases, c-di-GMP synthases, c-di-GMP PDEs, 2-Aminopurine (2AP), Oligoribonuclease (Orn), High-throughput screening, Inhibitor

1 Introduction

c-di-GMP is an important signal molecule found in many Gram-negative bacteria. It is synthesized in response to changes in environment and relays signals to various downstream effector molecules [1]. The intracellular concentration of c-di-GMP is controlled by the generation and/or degradation of c-di-GMP. The signaling molecule is synthesized from two GTPs and cleaved by PDE-A (containing EAL domain) into a linear form, pGpG [2], and then further into two GMP by PDE-B [3] or oligoribonuclease (Orn) [4, 5]. c-di-GMP can also be cleaved directly into GMP by PDEs containing an HD-GYP domain [6]. PDEs of c-di-GMP play important roles in controlling bacterial fitness and virulence as deletion of genes encoding PDEs is often associated with attenuated virulence [7, 8]. Thus, virulence-associated PDEs, which are not involved in biofilm dispersal, could become druggable targets [8].

Karin Sauer (ed.), *c-di-GMP Signaling: Methods and Protocols*, Methods in Molecular Biology, vol. 1657, DOI 10.1007/978-1-4939-7240-1_19, © Springer Science+Business Media LLC 2017

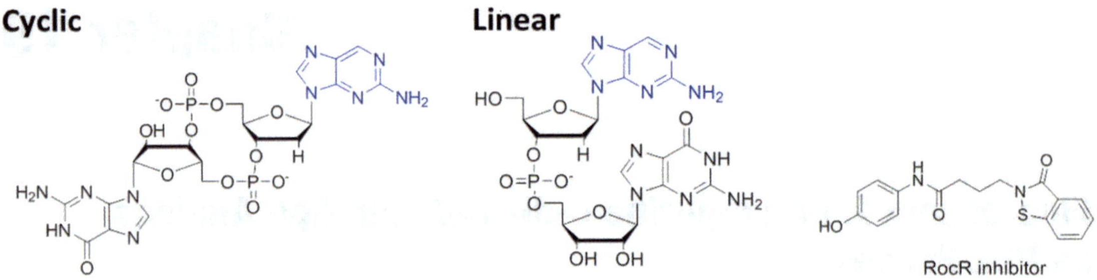

Fig. 1 Structures of fluorescent probes designed and synthesized

There is a paucity of inhibitors targeting c-di-GMP signaling [8–11], probably due to the lack of assays that could be used in a high-throughput manner to identify "hit" molecules. Although some detection methods of c-di-GMP have been reported [12–20], they suffer from drawbacks that prevent them from being used in high-throughput assays and/or real-time monitoring to screen for inhibitors. Therefore, there is an urgent need for assays that could be used in high-throughput screening.

In our design, we replaced one of the guanine moieties of c-di-GMP with a fluorescent 2-aminopurine (2AP) (*see* Fig. 1). These fluorescent probes report the activities of cyclic and linear dinucleotide PDEs. In fact, fluorescent nucleobase analogs have been widely used in the study of nucleic acid dynamics [21] and structure [22]. Precedents have established assays with fluorescent nucleobases incorporated into regular nucleotide or oligonucleotide, in order to monitor activities of nucleic acid [23–26]. Compound **1** has low fluorescence due to quenching (via electron transfer) between guanine (G) and the fluorescent base [27, 28]. Once cleaved by PDE-A, the two nucleobases in the linear form (shown as **2** in Fig. 2) are "loosely" bound (*see* Fig. 2). Since the two nucleobases have the freedom to be orientated in various ways, the distance between them (D2 and D3) could be more, less, or equal to that in the parent cyclic probe (D1)[29]. Therefore, the overall fluorescence change upon cleavage by PDE-A could either increase or decrease. Upon further cleavage by PDE-B or Orn, the mononucleotides generated (shown as **3** in Fig. 2) are considered as far from each other (*see* Fig. 2, D4 >> > D1, D2, and D3), with fluorescence increment expected [29].

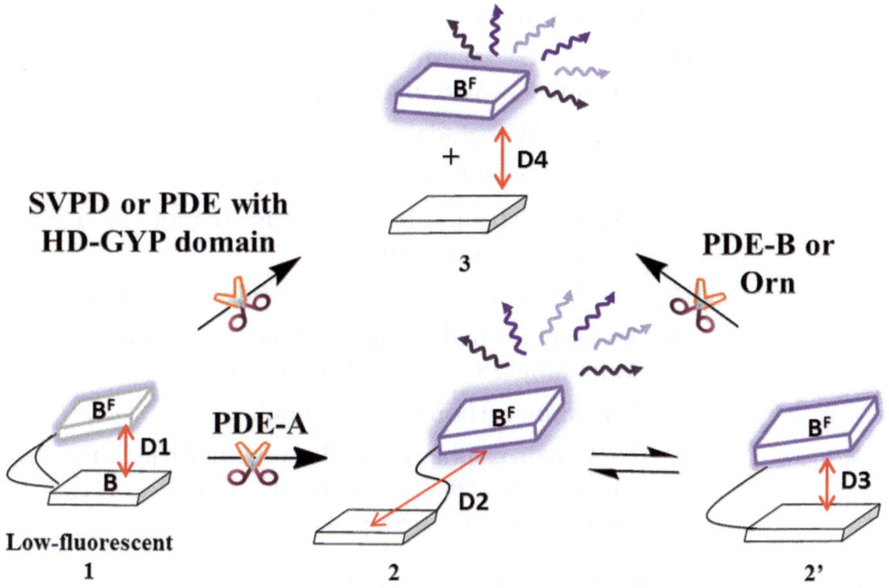

Fig. 2 Our working model of fluorescence changes upon PDE cleavage. A *block* indicates either fluorescent base (2AP) or normal base (G). A *black curved line* indicates sugar and phosphodiester bond between bases. *Purple arrows* represent fluorescence. *Red arrows* represent the distance (D1, D2, D3, and D4) between the fluorescent base and the normal base. Figure adapted from [29]

2 Materials

2.1 Reagents for Probe Synthesis

1. Phosphoramidites or phosphoramidites on controlled-pore glass (CPG) (Chemgenes): 2-Amino-2′-deoxypurine riboside-3′-CEP, 2′-*O*-TBDMS-Guanosine (N-iBu)-3′-CEP, 2′-*O*-TBDMS-Guanosine (N-iBu)-3′-lcaa CPG.
2. Acetonitrile (HPLC grade).
3. Dichloromethane (HPLC grade).
4. Pyridine (HPLC grade).
5. Water (HPLC grade).
6. Acetone (HPLC grade).
7. Diethyl ether (HPLC grade).
8. Ethyl acetate (HPLC grade).
9. Triethylamine (HPLC grade).
10. Pyridinium trifluoroacetate (ACS grade).
11. *tert*–Butyl amine (ACS grade).
12. Dichloroacetic acid (ACS grade).
13. 5.5 M *tert*-butyl hydroperoxide in decane (ACS grade, Sigma-Aldrich, St. Louis, MO).

14. Sodium thiosulfate (ACS grade).
15. 95% 5,5-dimethyl-2-oxo-2-chloro-1,3,2-dioxaphosphinane (DMOCP) (ACS grade, Sigma-Aldrich, St. Louis, MO).
16. Iodine (ACS grade).
17. Sodium bicarbonate (ACS grade).
18. 33% methylamine in anhydrous ethanol (ACS grade, Sigma-Aldrich, St. Louis, MO).
19. Triethylamine trihydrogen fluoride (ACS grade).
20. Ammonium hydroxide (ACS grade, Sigma-Aldrich): solution containing 28.0–30.0% NH_3 basis.
21. Potassium carbonate (ACS grade).
22. *N*-(4-hydroxyphenyl)-4-(3-oxobenzo[d]isothiazol-2(3H)-yl) butanamide (RocR inhibitor) (ACS grade, ChemBridge Chemical Store), *see* Fig. 1.

2.2 Reagents for Protein Purification and Assay Performance

1. BL21(DE3) cell (New England Biolabs).
2. Isopropyl β-D-1-thiogalactopyranoside (IPTG).
3. Ethylenediaminetetraacetic acid (EDTA): 500 mM EDTA stock solution in water (pH 8.0, minimum 500 μL). It should be freshly prepared each time before assay performance.
4. Lysis buffer for RocR (PDE-A): 50 mM Tris–HCl pH 7.5, 250 mM NaCl and 10 mM imidazole.
5. Lysis buffer for Orns (*Escherichia coli*, *Pseudomonas aeruginosa*, and *Mycobacterium smegmatis* Orns): 10 mM Tris–HCl, pH 8.0, 100 mM NaCl.
6. Reaction buffer for RocR (PDE-A): 100 mM Tris–HCl, pH 8.0, 20 mM KCl, 25 mM $MgCl_2$.
7. Reaction buffer for Orns (*E. coli, P. aeruginosa, and M. smegmatis* Orns): 10 mM Tris–HCl, pH 8.0, 100 mM NaCl, 5 mM $MgCl_2$.
8. HPLC buffer A: 0.1 M triethylammonium acetate (TEAA) in water (HPLC grade), pH 7.0, filtered through a solvent filter equipped with 0.22 μm polyethersulfone (PES) membrane.
9. HPLC buffer B: 100% acetonitrile (HPLC grade).
10. Luria Bertani Miller broth (LB).
11. Kanamycin.
12. LB amended with 50 μgm/mL kanamycin.
13. Pierce BCA Protein Assay Kit.
14. 6% dichloroacetic acid in dichloromethane: mix 94 mL of dichloromethane and 6 mL of dichloroacetic acid in a beaker. It should be freshly prepared each time before reaction.
15. HisTrap HP column (GE Healthcare Life Sciences).

2.3 Equipment

1. High performance liquid chromatography (HPLC) instrument equipped with UV detector and column compartment.
2. Nuclear magnetic resonance (NMR) spectrophotometer.
3. Mass spectrophotometer.
4. UV spectrophotometer.
5. Fluorescence spectrophotometer or plate reader covering wavelength (280–600 nm).
6. Ultracentrifuge.
7. Rotatory evaporator.
8. Vacuum pump.
9. Qsonica Q55 Sonicator Ultrasonic Processor.
10. Applied Biosystem 392 DNA/RNA synthesizer.
11. Shaking incubator (250 rpm) for bacterial culture with temperature control.

2.4 Glassware and Accessories

1. 100 mL round-bottom flask.
2. Precision Seal® rubber septa.
3. Drying bag (Chemgenes, catalog number DMT-1973).
4. Disposable syringes and needles.
5. Glass extraction funnel.
6. Conical flask.
7. Transfer pipet and tips.
8. Stir bar.
9. Sub-Micro Quartz Cuvette, 100 μL Cuvette for measuring absorbance and/or fluorescence.
10. Dialysis bag (Molecular Weight Cut Off = 20 K).
11. HPLC column: Nacalai tesque 5C18-PAQ column, 4.6 ID × 250 mm
12. 0.6 mL, 2 mL, 15 mL, and 50 mL plastic tubes for centrifugation
13. Plastic vial (20 mL).
14. Tubes for ultracentrifugation.
15. 500 mL Beaker.
16. Magnetic stirrer hotplate.

3 Methods

3.1 Synthesis of Compound 1 (See Fig. 3)

1. Weigh fluorescent base phosphoramidite (250 mg, 2AP) into a 50 mL round-bottom flask, dissolve in 5 mL of dry acetonitrile (*see* **Note 1**).
2. Evaporate to dryness by rotary evaporator and add 5 mL of dry acetonitrile to repeat the evaporation process for two more times. Instead of evaporating to dryness for the last time, retain around 1 mL. Add one drying bag into the round-bottom flask.
3. Note: How to retain 1 mL in the round-bottom flask? Before adding 5 mL of dry acetonitrile, one can add 1 mL of dry acetonitrile into the round-bottom flask first and mark the solvent level on the outside of round-bottom flask. When evaporating for the last time, one can carefully watch the solvent level and stop evaporation when the solvent level reached the marked position. If one accidentally over dries, add 5 mL of dry acetonitrile and repeat the whole process until 1 mL of solvent is left in the round-bottom flask.
4. Using a separate 100 mL round-bottom flask, weigh guanine phosphoramidite (450 mg).
5. Dissolve guanine phosphoramidite in 5 mL of dry acetonitrile with 100 mg of pyridinium trifluoroacetate and 18 μL of water, and react for 1 min.
6. Add 5 mL of *tert*–butyl amine and react for 10 min.

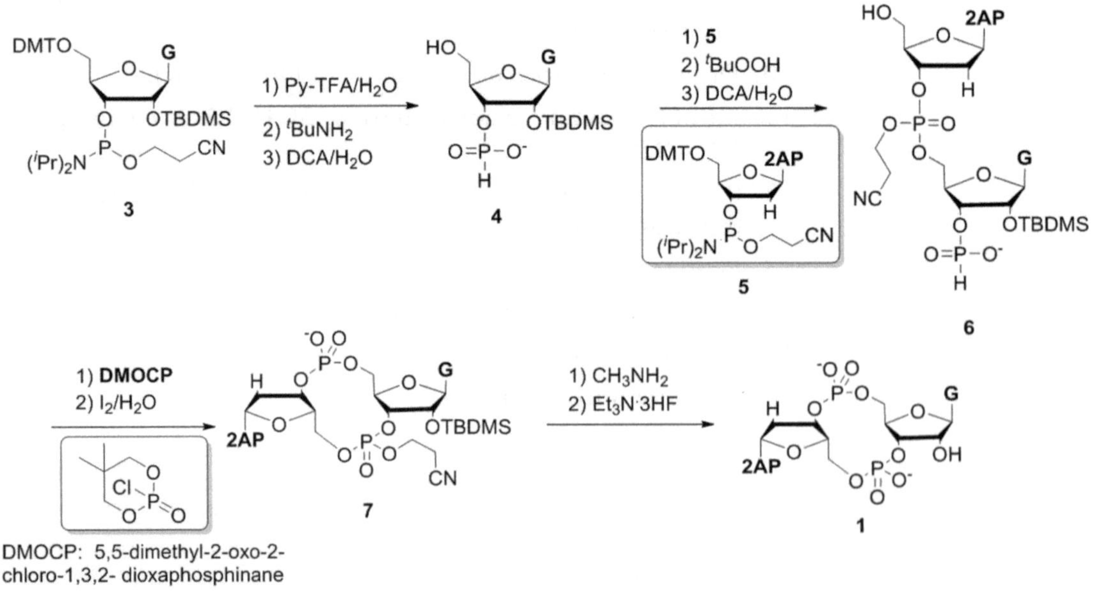

Fig. 3 Synthesis strategy for compound **1**

7. Concentrate the reaction mixture by rotary evaporator and wash three times with 5 mL of dry acetonitrile.
8. Dissolve the concentrated mixture in 6 mL of dichloromethane, 90 μL of water, and 6 mL of 6% dichloroacetic acid in dichloromethane.
9. React for 10 min.
10. Quench the reaction with 0.7 mL of dry pyridine and concentrate into foam by rotary evaporator (*see* **Note 1**).
11. Wash with 5 mL of dry acetonitrile and concentrate to dry.
12. Dry it further on vacuum pump for an additional 30 min.
13. Transfer the reagent prepared in **steps 1** and **2** into the flask from **steps 3–11** and react for 10 min.
14. Add 0.25 mL of 5.5 M *tert*-butyl hydroperoxide in decane and react for an additional 30 min.
15. Quench the reaction by adding 0.1 g of sodium thiosulfate in 0.25 mL of water and stir for 5 min.
16. Concentrate the reaction mixture into foam and wash with 5 mL of dry acetonitrile.
17. Dissolve the mixture in 8 mL of dichloromethane, 90 μL of water, and 8 mL of 6% dichloroacetic acid in dichloromethane and react for 10 min.
18. Quench the reaction with 5 mL of dry pyridine and concentrate into foam by rotary evaporator.
19. Add 20 mL of dry pyridine, evaporate solvent, and stop when around 1 mL of solvent is left in the flask.
20. Repeat **step 18** twice and leave 10 mL after the second time.
21. Weigh 400 mg of 95% 5,5-dimethyl-2-oxo-2-chloro-1,3,2-dioxaphosphinane (DMOCP), add into reaction mixture, and react for 10 min. Watch for color change from light yellow to dark.
22. Immediately add 0.3 mL of water to quench the reaction, followed by addition of 100 mg of iodine. React for additional 5 min.
23. Pour reaction mixture into a beaker containing 100 mg of sodium thiosulfate and 100 mL of water, and observe the formation of a white or light yellow precipitate. Stir vigorously on a magnetic stirrer hotplate for 5 min.
24. Weigh 2 g of sodium bicarbonate and slowly add into reaction mixture.
25. Stir for additional 5 min and watch for bubble formation. Bubble formation is a good indication. Make sure that

magnetic stirrer hotplate is turned to at least 600 stir speed to avoid overflow.

26. Extract with 120 mL of 1:1 ethyl acetate: diethyl ether.
27. Wait for 10 min for the separation of aqueous layer and organic layer. Organic layer is the top layer and aqueous layer is the bottom layer. Open the stopcock and transfer the bottom layer (aqueous layer) into a conical flask. Close the stopcock and transfer the top layer (organic layer) from the top of the funnel into a 250 mL round-bottom flask.
28. Concentrate organic layer to dryness, followed by the addition of 10 mL of 33% methylamine in anhydrous ethanol and stir for 90 min. Save the aqueous layer until compound has been characterized by NMR and mass spectrometer.
29. Concentrate to small volume before transferring into a 20 mL plastic vial.
30. Use a plastic syringe to add 0.5 mL of triethylamine trihydrogen fluoride into the plastic vial and react for 90 min at 55 °C in a magnetic stirrer hotplate.
31. Prepare saturated potassium carbonate solution in water. Put 100 mL water in a beaker, keep adding potassium carbonate and shaking until no potassium carbonate can be dissolved. Immediately rinse plastic syringe used for transferring triethylamine trihydrogen fluoride before discarding it (*see* **Note 2**). Quench the reaction by adding 15 mL of acetone.
32. Centrifuge at 3738 × *g* for 15 min at room temperature.
33. Decant acetone.
34. Dissolve precipitate in 2–5 mL of water and spin down at 3738 × *g* for 15 min. Carefully transfer the supernatant to another tube for HPLC purification.
35. Purify crude product by HPLC (Nacalai tesque 5C18-PAQ column), HPLC condition: 1→13% B, 0→16 min (A: 0.1 M TEAA in water, pH 7.0; B: acetonitrile), flow rate: 1 mL/min, with 1 mL injection loop (max 1 mL as injection volume).
36. Collect each peak shown in HPLC and concentrate each fraction at a reduced pressure.
37. Add acetone (2 mL × 5), vortex and centrifuge. Then carefully remove the supernatant containing excess of TEAA buffer.
38. Confirm product identity by NMR and high-resolution mass spectroscopy.

3.2 Synthesis of Compound 2 (See Fig. 4)

1. Dissolve 2AP phosphoramidite (250 mg) in 2.5 mL distilled acetonitrile.
2. Couple 2AP phosphoramidite into guanosine phosphoramidite on CPG on Applied Biosystem 392 DNA/RNA synthesizer.

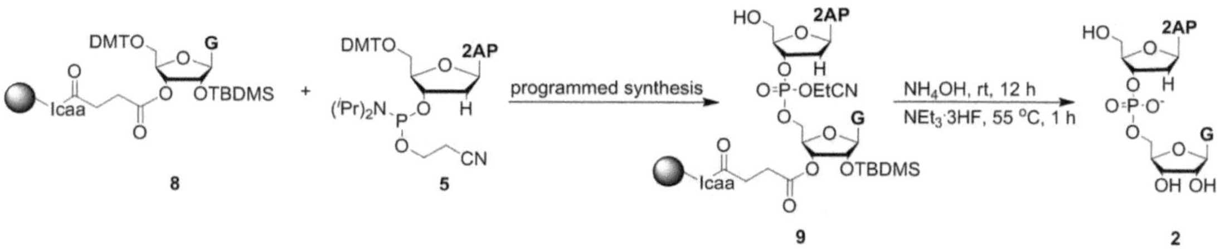

Fig. 4 Synthesis strategy for compound **2**

3. After the reaction, collect all CPG in one round-bottom flask.
4. Transfer 5 mL of ammonium hydroxide (28–30% ammonia in water, commercially available) into the round-bottom flask and allow the reaction to proceed for 12 h at room temperature.
5. Split the sample into each volume and transfer into two 15 mL Eppendorf centrifuge tube and spin at 3738 × *g* for 15 min.
6. Collect the supernatant into a 20 mL plastic vial and evaporate by rotatory evaporator to dry. Alternatively, speed vac can also be used for drying.
7. Transfer 1 mL of triethylamine trihydrogen fluoride using plastic syringe into the plastic vial and allow the reaction at 55 °C for 1 h.
8. Prepare saturated potassium carbonate solution in water. Put 100 mL water in a beaker, keep adding potassium carbonate and shaking until no more potassium carbonate can be dissolved. Immediately rinse plastic syringe used for transferring triethylamine trihydrogen fluoride before discarding it (*see* **Note 2**).
9. Quench the reaction by adding 50 mL acetone.
10. Centrifuge at 3738 × *g* for 15 min at room temperature.
11. Decant acetone.
12. Dissolve precipitate in 2–5 mL of water without air-drying the precipitate pellet and purify crude product by HPLC (Nacalai tesque 5C18-PAQ column), HPLC condition: 1→13% B, 0→16 min flow rate: 1 mL/min, with 1 mL injection loop (max 1 mL as injection volume).
13. Collect each peak shown in HPLC and concentrate each fraction at a reduced pressure.
14. Add acetone (2 mL × 5), vortex and centrifuge. Then carefully remove the supernatant to remove the excess of TEAA buffer.
15. Confirm product identity by NMR and high-resolution mass spectroscopy.

3.3 Protein Purification

1. Transfer the plasmids containing the gene encoding the protein of interest (RocR (PDE-A) or Orns, His-tagged) into *E. coli* BL21 (DE3) cells.
2. Grow *E. coli* BL21(DE3) cells in LB medium amended with 50 μgm/mL kanamycin (or an appropriate antibiotic as specified by plasmid vector) to maintain the plasmid at 37 °C for approximately 4–6 h.
3. Transfer 1 mL of medium into cuvette and measure optical density at 600 nm (OD600).
4. When OD600 reaches 0.6, add IPTG into medium to reach a final concentration of 1 mM.
5. Incubate at 16 °C for 16–18 h, allowing for stable expression of protein. Alternatively, one can also incubate at 37 °C for 3–4 h.
6. Centrifuge the bacterial suspension at 5841 × *g* for 20 min to harvest cells as pellets.
7. Resuspend cells in 20–25 mL of corresponding lysis buffer.
   (a) For RocR (PDE-A), use Lysis buffer for RocR.
   (b) For Orns, use Lysis buffer for Orns.
8. Sonicate on ice to lyse the cells with settings at 40% duty for 1 min, followed by 45 s cooling. Repeat three times.
9. Centrifuge at 113,092 × *g* for 25 min, to remove cell debris.
10. Transfer the supernatant into 10 mL syringe and load onto GE HisTrap HP column for purification. Use another 10 mL syringe to pick up 10 mL lysis buffer, wash through the column, and collect elution into tube A. Then, use the same syringe to pick up 10 mL of 200 mM imidazole in lysis buffer to elute protein and collect elution into tube B.
11. Transfer tube B into dialysis bag and dialyze into corresponding lysis buffer for 24 h at 4 °C.

3.4 Enzymatic Assays

3.4.1 Determine Concentrations and Prepare Stock Solutions

1. Determine concentrations of each probe by measuring UV absorbance at 260 nm and calculate concentration using the corresponding extinction coefficients listed below.
   (a) Formula for calculating the concentration is: $A = \varepsilon bc$, where A is the UV absorbance at 260 nm, ε is the extinction coefficient (M^{-1} cm^{-1}), and b is the path length of the cuvette used (cm) and c is the concentration of probe.
   (b) ε for compounds **1** and **2** (260 nm) = 15,300 M^{-1} cm^{-1}.
2. Prepare probe stock solutions in water (tenfold of final concentration required for the assay, minimum 500 μL for each probe).
   (a) Compound **1 stock solution** = 25 μM.
   (b) Compound **2 stock solution** = 25 μM.

3. Prepare stock solutions of inhibitors or screening compounds (*see* **Note 3**).
   (a) RocR inhibitor stock solution = 1 mM.
   (b) Orn inhibitor stock solution = 1 mM.
4. Determine protein concentrations (*see* Subheading 3.3) by using the BCA assay kit (following the manufacture's protocol).
5. Prepare protein stock solutions (tenfold of final protein concentration required for the assay, minimum 500 μL for each protein) in corresponding reaction buffers (*see* Subheading 2.1, **item** 7 for reaction buffers).
   (a) RocR (PDE-A) stock solution = 2.5 μM, prepare in reaction buffer for RocR.
   (b) *P. aeruginosa* Orn stock solution = 2.5 μM, prepare in reaction buffer for Orns.
   (c) *M. smegmatis* Orn stock solution = 2.5 μM, prepare in reaction buffer for Orns.
   (d) *E. coli* Orn stock solution = 2.5 μM, prepare in reaction buffer for Orns.

3.4.2 Perform RocR Assay

The assays described below for PDEs such as RocR and/or Orns such as *P. aeruginosa*, *M. smegmatis* and *E. coli* Orn can also be applied to other dinucleotide PDEs and Orns. For each enzyme, one should ensure that an appropriate buffer for that enzyme (containing the appropriate divalent cation) is used in determining enzyme activity.

In RocR assay, compound **1** and RocR are mixed together in the absence or presence of RocR inhibitor. In the absence of RocR inhibitor, upon cleavage by RocR, fluorescence decrement (40–60% decrement) would be expected. The decrement in fluorescence usually means that RocR is active. In the presence of a RocR inhibitor, fluorescence decrement would be much lower (15–20%). EDTA is a known inhibitor of RocR as it chelates divalent metal ions in reaction buffers. We use EDTA as a control to verify whether the fluorescence change in the assay is a true reflection of enzyme inhibition. We also used a published inhibitor of RocR [8], known here as RocR inhibitor to test in the assay as an example of a small molecule inhibitor (*see* Figs. 1 and 5).

1. Transfer 10 μL of Compound **1 stock solution** and 80 μL RocR reaction buffer to another microcentrifuge tube and initiate the reaction by adding 10 μL of RocR stock solution. Label as RocR_1_no EDTA.
2. Transfer 10 μL of Compound **1** stock solution, 10 μL of 500 mM EDTA, and 70 μL RocR reaction buffer into one

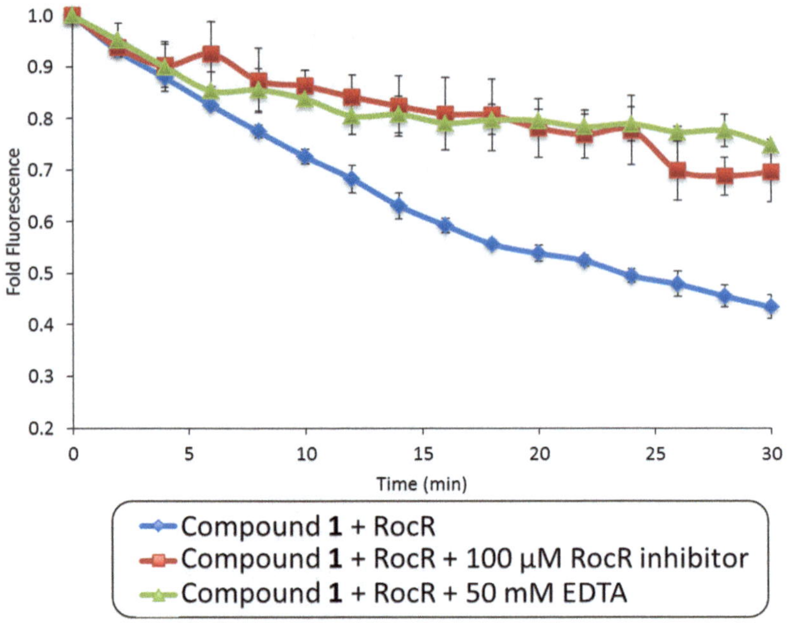

Fig. 5 Fluorescence changes upon cleavage of compound **1** by RocR in the absence or presence of a published RocR inhibitor (100 μM) [8] or 50 mM EDTA at 37 °C for 30 min. $\lambda_{ex} = 310$ nm and $\lambda_{em} = 375$ nm. Final concentrations in assay mixture: [RocR] = 0.25 μM, [Compound **1**] = 2.5 μM. Figure adapted from [29]

microcentrifuge tube and initiate the reaction by adding 10 μL of RocR stock solution. Label as RocR_1_EDTA.

3. Transfer 10 μL of Compound **1 stock solution**, 10 μL of RocR inhibitor **stock solution,** and 70 μL RocR reaction buffer to one microcentrifuge tube and initiate the reaction by adding 10 μL of RocR **stock solution**. Label as RocR_1_inhibitor.
4. Perform each of the above steps in triplicate.
5. Incubate each reaction mixture (**steps 1–3**) at 37 °C for 30 min on a plate reader.
6. Monitor fluorescence by choosing excitation wavelength at 310 nm and emission wavelength at 375 nm over the course of 30 min.
7. Normalize the fluorescence readout to the same initial intensity.
8. Average the data for the triplicate samples and present as average including standard deviation (*see* Fig. 5 as an example).

3.4.3 Perform E. coli Orn Assay

Compound **2** and Orn are mixed together in the absence or presence of Orn inhibitor. In the absence of Orn inhibitor, upon cleavage by Orn, fluorescence increment (350–450%) would be expected. The increment in fluorescence usually means that Orn is active. In the presence of Orn inhibitor, fluorescence would stay as the same. EDTA is a known inhibitor for Orn as it chelates to the divalent metal ion in reaction buffers. We use EDTA as a control to verify whether the fluorescence change in the assay is a true reflection of enzyme inhibition. If the fluorescence stays unchanged in the presence of EDTA, it verifies that Orn is active and is ready to be used for screening purposes. This general concept applies to the assays described below for Orns (*P. aeruginosa*, *M. smegmatis,* and *E. coli* Orn).

1. Transfer 10 μL of Compound **2 stock solution** and 80 μL of corresponding reaction buffer to another microcentrifuge tube and initiate the reaction by adding 10 μL of *E. coli* Orn stock solution. Label as *E. coli* Orn_2_no EDTA.
2. Transfer 10 μL of Compound **2 stock solution**, 10 μL of EDTA stock solution, and 70 μL of the corresponding reaction buffer to one microcentrifuge tube and initiate the reaction by adding 10 μL of *E. coli* Orn stock solution. Label as *E. coli* Orn_2_EDTA.
3. Transfer 10 μL of Compound **2 stock solution**, 10 μL of Orn inhibitor **stock solution,** and 70 μL of corresponding reaction buffer to one microcentrifuge tube and initiate the reaction by adding 10 μL of *E. coli* Orn stock solution. Label as *E. coli* Orn_2_inhibitor.
4. Perform each of the above steps in triplicate.
5. Measure fluorescence by choosing excitation wavelength at 310 nm and emission wavelength at 375 nm at 37 °C for 30 min on a plate reader.
6. Normalize the fluorescence readout to the same initial intensity and present as average (*see* Fig. 6 as an example).

3.4.4 Perform *P. aeruginosa* Orn Assay

1. Transfer 10 μL of Compound **2 stock solution** and 80 μL of corresponding reaction buffer to another microcentrifuge tube and initiate the reaction by adding 10 μL of *P. aeruginosa* Orn stock solution. Label as *P. aeruginosa* Orn_2_no EDTA.
2. Transfer 10 μL of Compound **2 stock solution**, 10 μL of 500 mM EDTA, and 70 μL of corresponding reaction buffer to one microcentrifuge tube and initiate the reaction by adding 10 μL of *P. aeruginosa* Orn stock solution. Label as *P. aeruginosa* Orn_2_EDTA.
3. Transfer 10 μL of Compound **2 stock solution**, 10 μL of Orn inhibitor **stock solution** and 70 μL of corresponding reaction buffer to one microcentrifuge tube and initiate the reaction by adding 10 μL of *P. aeruginosa* Orn stock solution. Label as *P. aeruginosa* Orn_2_inhibitor.
4. Perform each of the above steps in triplicate.

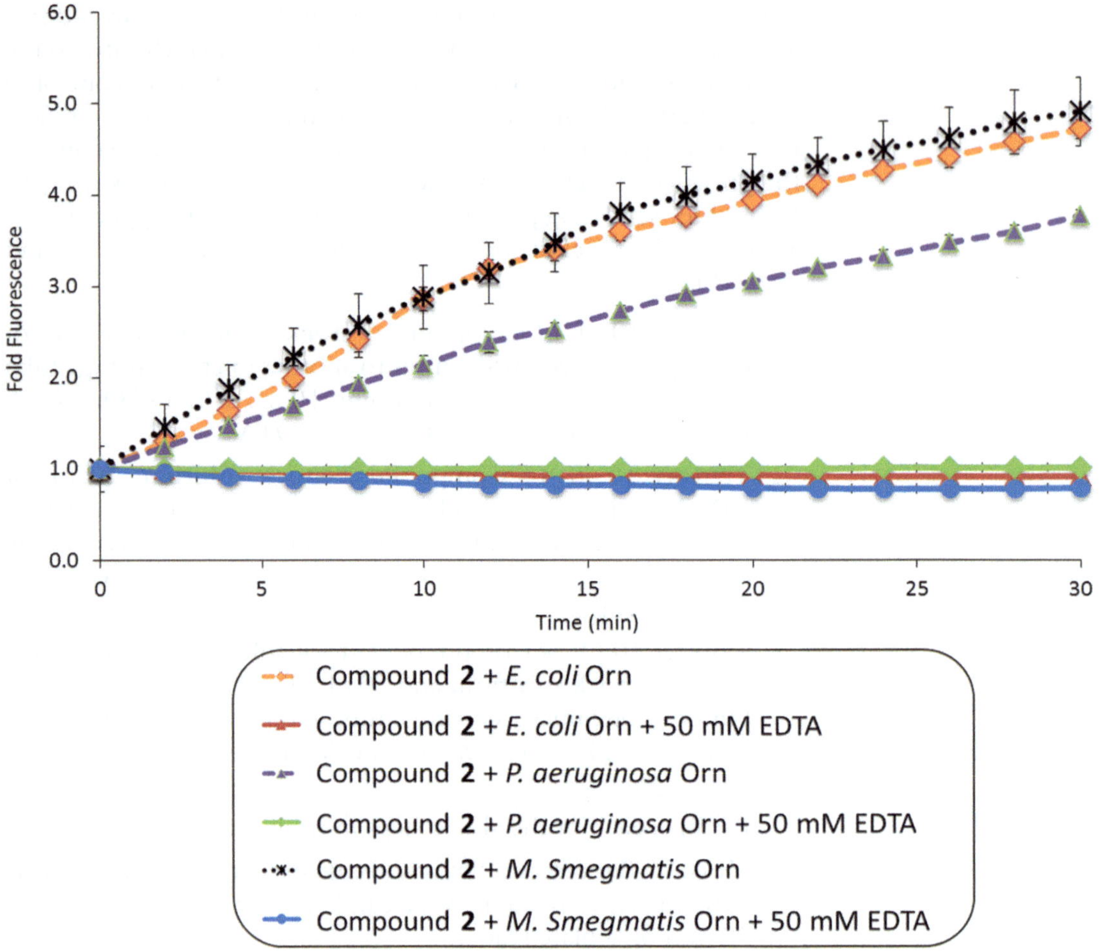

Fig. 6 Fluorescence changes upon cleavage of compound **2** by Orns in the absence or presence of EDTA (50 mM) at 37 °C for 30 min $\lambda_{ex} = 310$ nm and $\lambda_{em} = 375$ nm. [Orn] = 0.25 μM, [Compound **2**] = 2.5 μM. Enzymatic assays were carried out at 37 °C. Figure adapted from [29]

5. Measure fluorescence by choosing excitation wavelength at 310 nm and emission wavelength at 375 nm at 37 °C for 30 min on a plate reader.
6. Normalize the fluorescence readout to the same initial intensity and present as average (*see* Fig. 6 as an example).

3.4.5 Perform M. smegmatis Orn Assay

1. Transfer 10 μL of Compound **2 stock solution** and 80 μL of corresponding reaction buffer to another microcentrifuge tube and initiate the reaction by adding 10 μL of *M. smegmatis* Orn stock solution. Label as *M. smegmatis* Orn_2_no EDTA.
2. Transfer 10 μL of Compound **2 stock solution**, 10 μL of 500 mM EDTA, and 70 μL of corresponding reaction buffer to one microcentrifuge tube and initiate the reaction by adding 10 μL of *M. smegmatis* Orn stock solution. Label as *M. smegmatis* Orn_2_EDTA.

3. Transfer 10 μL of Compound **2 stock solution**, 10 μL of Orn inhibitor **stock solution,** and 70 μL of corresponding reaction buffer to one microcentrifuge tube and initiate the reaction by adding 10 μL of *M. smegmatis* Orn stock solution. Label as *M. smegmatis* Orn_2_inhibitor.
4. Perform each of the above steps in triplicate.
5. Measure fluorescence by choosing excitation wavelength at 310 nm and emission wavelength at 375 nm at 37 °C for 30 min on a plate reader.
6. Normalize the fluorescence readout to the same initial intensity and present as average (*see* Fig. 6 as an example).

4 Notes

1. The synthesis described in Subheadings 3.1 and 3.2 is moisture sensitive. Solvents such as acetonitrile and pyridine should be distilled over CaH_2 and dried overnight with drying trap prior to usage.
2. Hydrogen fluoride component in triethylamine trihydrogen fluoride is extremely dangerous. When using, one should wear heavy gloves to avoid skin contact. Moreover, one should also avoid breathing hydrogen fluoride vapors from the flask. At the completion of the experiment, one should rinse all related experimental accessories such as needles and syringes with freshly prepared saturated potassium carbonate solution before discarding.
3. The stock concentration of inhibitor or screening compounds can be modified to fit the screening requirements of the user. Following the stock concentrations used in this document will result in a final inhibitor concentration of 100 μM.

Acknowledgment

This work was funded by the National Science Foundation (CHE1307218 and CHE1636752), Purdue University. Plasmids were provided by Dr. Zhaoxun Liang (RocR plasmid), Dr. Ehud Banin (*P. aeruginosa* Orn plasmid), and Dr. Nicholas Dixon (*E. coli* and *M. smegmatis* Orn plasmids).

References

1. Kalia D, Merey G, Nakayama S, Zheng Y, Zhou J, Luo Y, Guo M, Roembke BT, Sintim HO (2013) Nucleotide, c-di-GMP, c-di-AMP, cGMP, cAMP, (p) ppGpp signaling in bacteria and implications in pathogenesis. Chem Soc Rev 42(1):305–341. doi:10.1039/c2cs35206k
2. Schmidt AJ, Ryjenkov DA, Gomelsky M (2005) The ubiquitous protein domain EAL is a cyclic diguanylate-specific phosphodiesterase:

enzymatically active and inactive EAL domains. J Bacteriol 187(14):4774–4781. doi:10.1128/JB.187.14.4774-4781.2005

3. Orr MW, Donaldson GP, Severin GB, Wang J, Sintim HO, Waters CM, Lee VT (2015) Oligoribonuclease is the primary degradative enzyme for pGpG in *Pseudomonas aeruginosa* that is required for cyclic-di-GMP turnover. Proc Natl Acad Sci U S A 112(36): E5048–E5057. doi:10.1073/pnas.1507245112
4. Ghosh S, Deutscher MP (1999) Oligoribonuclease is an essential component of the mRNA decay pathway. Proc Natl Acad Sci U S A 96 (8):4372–4377
5. Cohen D, Mechold U, Nevenzal H, Yarmiyhu Y, Randall TE, Bay DC, Rich JD, Parsek MR, Kaever V, Harrison JJ, Banin E (2015) Oligoribonuclease is a central feature of cyclic diguanylate signaling in Pseudomonas Aeruginosa. Proc Natl Acad Sci U S A 112 (36):11359–11364. doi:10.1073/pnas.1421450112
6. Ryan RP, Fouhy Y, Lucey JF, Crossman LC, Spiro S, He YW, Zhang LH, Heeb S, Camara M, Williams P, Dow JM (2006) Cell-cell signaling in Xanthomonas campestris involves an HD-GYP domain protein that functions in cyclic di-GMP turnover. Proc Natl Acad Sci U S A 103(17):6712–6717. doi:10.1073/pnas.0600345103
7. Peng X, Zhang Y, Bai G, Zhou X, Wu H (2016) Cyclic di-AMP mediates biofilm formation. Mol Microbiol 99(5):945–959. doi:10.1111/mmi.13277
8. Zheng Y, Tsuji G, Opoku-Temeng C, Sintim HO (2016) Inhibition of *P. aeruginosa* c-di-GMP phosphodiesterase RocR and swarming motility by a benzoisothiazolinone derivative. Chem Sci 7:6238–6244
9. Nakayama S, Zhou J, Zheng Y, Szmacinski H, Sintim HO (2016) Supramolecular polymer formation by cyclic dinucleotides and intercalators affects dinucleotide enzymatic processing. Future Sci OA 2(1):FSO93. doi:10.4155/fso.4115.4193
10. Opoku-Temeng C, Sintim HO (2016) Inhibition of cyclic diadenylate cyclase, DisA, by polyphenols. Sci Rep 6:25445. doi:10.1038/srep25445
11. Opoku-Temeng C, Sintim HO (2016) Potent inhibition of cyclic diadenylate monophosphate cyclase by the antiparasitic drug, suramin. Chem Commun (Camb) 52 (19):3754–3757. doi:10.1039/c5cc10446g
12. Simm R, Morr M, Rerriminghorst U, Andersson M, Romling U (2009) Quantitative determination of cyclic diguanosine monophosphate concentrations in nucleotide extracts of bacteria by matrix-assisted laser desorption/ionization-time-of-flight mass spectrometry. Anal Biochem 386(1):53–58. doi:10.1016/j.ab.2008.12.013
13. Wang J, Zhou J, Donaldson GP, Nakayama S, Yan L, Lam Y-f, Lee VT, Sintim HS (2011) Conservative change to the phosphate moiety of cyclic diguanylic monophosphate remarkably affects its polymorphism and ability to bind DGC, PDE, and PilZ proteins. J Am Chem Soc 133(24):9320–9330
14. Nakayama S, Roelofs K, Lee VT, Sintim HO (2012) A C-di-GMP-proflavine-hemin supramolecular complex has peroxidase activity-implication for a simple colorimetric detection. Mol Biosyst 8(3):726–729
15. Roembke BT, Zhou J, Zheng Y, Sayre D, Lizardo A, Bernard L, Sintim HO (2014) A cyclic dinucleotide containing 2-aminopurine is a general fluorescent sensor for c-di-GMP and 3',3'-cGAMP. Mol Biosyst 10 (6):1568–1575
16. Zhou J, Sayre DA, Zheng Y, Szmacinski H, Sintim HO (2014) Unexpected complex formation between coralyne and cyclic diadenosine monophosphate providing a simple fluorescent turn-on assay to detect this bacterial second messenger. Anal Chem 86 (5):2412–2420
17. Nakayama S, Kelsey I, Wang JX, Sintim HO (2011) c-di-GMP can form remarkably stable G-quadruplexes at physiological conditions in the presence of some planar intercalators. Chem Comm 47(16):4766–4768
18. Gu H, Furukawa K, Breaker RR (2012) Engineered allosteric ribozymes that sense the bacterial second messenger cyclic diguanosyl 5'-monophosphate. Anal Chem 84 (11):4935–4941
19. Nakayama S, Luo Y, Zhou J, Dayie TK, Sintim HO (2012) Nanomolar fluorescent detection of c-di-GMP using a modular aptamer strategy. Chem Comm 48(72):9059–9061
20. Underwood AJ, Zhang Y, Metzger DW, Bai G (2014) Detection of cyclic di-AMP using a competitive ELISA with a unique pneumococcal cyclic di-AMP binding protein. J Microbiol Methods 107:58–62
21. Venkatesan N, Seo YJ, Kim BH (2008) Quencher-free molecular beacons: a new strategy in fluorescence based nucleic acid analysis. Chem Soc Rev 37(4):648–663. doi:10.1039/b705468h
22. Börjesson K, Preus S, El-Sagheer AH, Brown T, Albinsson B, Wilhelmsson LM (2009)

Nucleic acid base analog FRET-pair facilitating detailed structural measurements in nucleic acid containing systems. J Am Chem Soc 131 (12):4288–4293. doi:10.1021/ja806944w

23. Wilson JN, Cho Y, Tan S, Cuppoletti A, Kool ET (2008) Quenching of fluorescent nucleobases by neighboring DNA: the "insulator" concept. ChemBioChem 9(2):279–285. doi:10.1002/cbic.200700381
24. Wilhelmsson LM (2010) Fluorescent nucleic acid base analogues. Q Rev Biophys 43 (2):159–183. doi:10.1017/S0033583510000090
25. Singleton SF, Roca AI, Lee AM, Xiao J (2007) Probing the structure of RecA-DNA filaments. Advantages of a fluorescent guanine analog. Tetrahedron 63(17):3553–3566. doi:10.1016/j.tet.2006.10.092
26. Sinkeldam RW, Greco NJ, Tor Y (2010) Fluorescent analogs of biomolecular building blocks: design, properties, and applications. Chem Rev 110(5):2579–2619. doi:10.1021/cr900301e
27. Somsen O, Hoek V, Amerongen V (2005) Fluorescence quenching of 2-aminopurine in dinucleotides. Chem Phys Lett 402(1-3):61–65. doi:10.1016/j.cplett.2004.11.122
28. Leonard NJ (1984) Etheno-substituted nucleotides and coenzymes: fluorescence and biological activity. CRC Crit Rev Biochem 15 (2):125–199
29. Zhou J, Zheng Y, Roembke BT, Robinson SM, Opoku-Temeng C, Sayre DA, Sintim HO (2017) Fluorescent analogs of cyclic and linear dinucleotides as phosphodiesterase and oligoribonuclease activity probes. RSC Advances 7:5421–5426

Chapter 20

Measuring Cyclic Diguanylate (c-di-GMP)-Specific Phosphodiesterase Activity Using the MANT-c-di-GMP Assay

Dorit Eli, Trevor E. Randall, Henrik Almblad, Joe J. Harrison, and Ehud Banin

Abstract

The second messenger, cyclic diguanylate (c-di-GMP), regulates a variety of bacterial cellular and social behaviors. A key determinant of c-di-GMP levels in cells is its degradation by c-di-GMP-specific phosphodiesterases (PDEs). Here, we describe an assay to determine c-di-GMP degradation rates in vitro using 2′-O-(N′-methylanthraniloyl)-cyclic diguanylate (MANT-c-di-GMP). Additionally, a protocol for the production and purification of recombinant *Pseudomonas aeruginosa* RocR, a c-di-GMP-specific PDE that may serve as a control in MANT-c-di-GMP assays, is provided. The use of the fluorescent MANT-c-di-GMP analogue can deliver fundamental information about PDE function, and is suitable for identifying and investigating c-di-GMP-specific PDE activators and inhibitors.

Key words Cyclic diguanylate, Phosphodiesterase, MANT-c-di-GMP, RocR, pGpG

1 Introduction

Cyclic diguanylate (c-di-GMP) is a bacterial second messenger that was first identified 30 years ago as an allosteric activator of the *Gluconacetobacter xylinus* cellulose synthetase complex [1]. Since then microbiologists have discovered that c-di-GMP regulates a diverse range of physiological and multicellular behaviors in many bacterial species. These behaviors include, for example, virulence factor expression, biofilm development, flagellar motility, and cell cycle progression [2–6].

C-di-GMP biosynthesis depends on diguanylate cyclases (DGCs) that have a GGDEF domain [7], whereas c-di-GMP degradation is catalyzed by phosphodiesterases (PDEs) that possess either an EAL or HD-GYP domain [8]. Enzymes with an EAL domain hydrolyze c-di-GMP into 5′-phosphoguanylyl-(3′,5′)-

Dorit Eli and Trevor E. Randall contributed equally to this work.

Karin Sauer (ed.), *c-di-GMP Signaling: Methods and Protocols*, Methods in Molecular Biology, vol. 1657,
DOI 10.1007/978-1-4939-7240-1_20, © Springer Science+Business Media LLC 2017

guanosine (pGpG) [9], which is subsequently degraded into two guanosine monophosphate (GMP) molecules by oligoribonuclease (Orn) [10, 11]. By contrast, HD-GYP domains degrade c-di-GMP directly to GMP [12–14].

Although many bacteria possess a multitude of DGC and PDE enzymes, there is remarkable signaling fidelity in c-di-GMP networks, and accumulating evidence indicates that specific DGCs and PDEs regulate discrete aspects of bacterial cellular physiology [15]. Moreover, enzymes with GGDEF, EAL, and HD-GYP domains are absent from mammalian cells. Thus, there is a growing interest in developing small molecule therapeutics that target c-di-GMP signal transduction pathways [16–18]. Methods for quantitative analysis of c-di-GMP, therefore, are quickly becoming important techniques not only for studying bacterial physiology, but also for early-stage drug discovery and development.

Nucleotide extraction and mass spectrometry (MS)-based approaches using MALDI-TOF [19] or LC-MS/MS [20–23] have been used by several groups to directly measure c-di-GMP in vitro and from living cells. MS-based approaches are precise and sensitive, yet are expensive, time consuming, and ill-suited for high-throughput screening. A colorimetric nucleotide intercalator assay based on thiazole orange has also been used to quantify c-di-GMP in vitro [24]; however, this assay requires lengthy incubations and cannot be used to monitor c-di-GMP degradation in real time. Additional approaches to c-di-GMP quantitation have employed circular dichroism [25], a Spinach-based riboswitch [26], a fluorescent 2-aminopurine analogue that complexes c-di-GMP in the presence Mn(II) [27], and the colorimetric peroxidation of 2,2′-azino-bis(3-ethylbenzothiazoline-6-sulfonic acid) (ABTS) by c-di-GMP-hemin complexes [28, 29]. However, these latter approaches have not been widely adopted due to assay complexity, or sensitivity to interfering compounds that directly affect assay chemistry or optical measurements.

A variety of fluorescent c-di-GMP analogues have also been synthesized for the purpose of monitoring c-di-GMP-specific PDE activity [30, 31]. In this chapter, we present a simple protocol for measuring PDE activity using the fluorescent c-di-GMP analogue, 2′-O-(N′-methylanthraniloyl)-cyclic diguanylate (MANT-c-di-GMP) (illustrated in Fig. 1). MANT-c-di-GMP was originally synthesized by Sharma and colleagues [31], and like other fluorescent c-di-GMP analogues, has several advantages: it can be used to measure c-di-GMP degradation rates by PDEs in real time, to evaluate PDE activation or inhibition by soluble small molecules, and to identify modulators of PDE activity in high-throughput screening applications. MANT nucleotide analogues show high quantum yields and resistance to photobleaching [32, 33], and their absorption and emission spectra do not overlap with most other biological molecules [31–33]. Finally, MANT-conjugation of

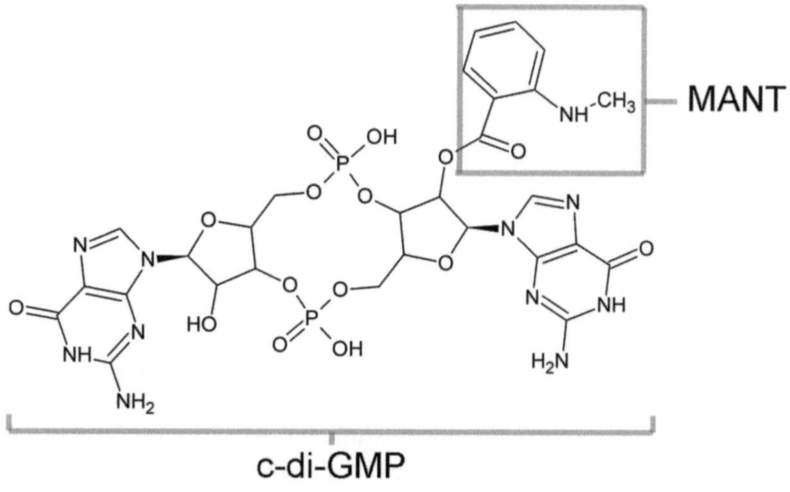

Fig. 1 Structure of MANT-c-di-GMP, a fluorescent cyclic diguanylate orthologue. The N′-methylanthraniloyl (MANT) fluorophore is covalently linked to c-di-GMP via an ether linkage to the 2′ carbon of one ribosyl moiety

c-di-GMP at one of its ribosyl moieties does not seem to affect its abilities to bind to several tested c-di-GMP-binding proteins that are structurally, catalytically, and functionally diverse [31].

In addition to the MANT-c-di-GMP assay, we detail an established protocol [10, 34, 35] for producing and purifying the *Pseudomonas aeruginosa* c-di-GMP-specific phosphodiesterase RocR (Subheadings 2.1–2.5 and 3.1–3.3). The biochemistry of this enzyme is well characterized [34, 35], making RocR an ideal positive control and comparator PDE for use in MANT-c-di-GMP assays. In principle, however, any purified c-di-GMP-specific PDE may be used in these assays. Readers using another c-di-GMP-specific PDE and not wishing to purify RocR can proceed directly to the preparation of reagents and procedures for the MANT-c-di-GMP assay (Subheadings 2.6 and 3.4).

2 Materials

2.1 Buffers, Equipment, and Growth Media for RocR Production

1. Super broth: In 900 mL MilliQ water, add 32 g tryptone, 20 g yeast Extract, 5 g NaCl, and 5 mL 1 N NaOH. Bring to 1 L volume with MilliQ water, subdivide into 400 mL volumes in 2 L baffled flasks, and autoclave to sterilize.
2. Lysogeny broth (LB) agar: In 900 mL MilliQ water, add 10 g tryptone, 5 g yeast extract, 5 g NaCl, and 15 g Bacto agar. Bring to 1 L final volume with MilliQ water. Autoclave to sterilize.
3. 50 mg/mL kanamycin (1000 × working concentration).

4. LB agar containing 50 μg/mL kanamycin: Prepare 1 L of LB agar as described above. Allow sterilized agar to cool to 60 °C. Add 1 mL of 50 mg/mL kanamycin.
5. 1 M Tris–HCl (pH 8.0): Add 157.64 g of Tris–HCl to 950 mL of MilliQ water. Adjust to pH 8.0 with NaOH and bring to 1 L final volume with MilliQ water.
6. 1 M dithiothreitol (DTT): Add 1.54 mg of DTT to 10 mL MilliQ water. Split into aliquots by transferring 1.0 mL aliquots of the 1.0 M DTT solution into 1.5 mL microcentrifuge tubes. Store at −20 °C.
7. RocR lysis/wash buffer: Starting with 800 mL of MilliQ water, add 50 mL 1 M Tris–HCl (pH 8.0), 14.5 g NaCl, 0.7 g imidazole, 50 mL glycerol, 0.5 mL 1 M dithiothreitol, bring pH to 7.5 with HCl. Bring to 1 L final volume with MilliQ water, and filter with a 0.45 μm mixed cellulose esters (MCE) filter (e.g., Millex-HA filter, Millipore) (*see* **Note 1**).
8. RocR elute buffer (containing 250 mM imidazole): Start with 900 mL of MilliQ water, add 50 mL 1 M Tris–HCl (pH 8.0), 14.61 g NaCl, 17.01 g imidazole, 50 mL glycerol, 0.5 mL 1 M dithiothreitol, pH to 7.5 with concentrated HCl, bring to 1 L final volume with MilliQ water, and filter with a 0.45 μm MCE filter (*see* **Note 1**).
9. RocR cleaning buffer: Starting with 800 mL of MilliQ water, add 20 mL 1 M Tris–HCl (pH 8.0) 1.49 g KCl, 50 mL glycerol, 2 mL 1 M dithiothreitol, then pH to 7.5. Bring to 1 L final volume with MilliQ water, and filter with a 0.45 μm MCE filter.
10. RocR storage buffer: Start with 500 mL of water, add 50 mL 1 M Tris–HCl (pH 8.0), 14.5 g NaCl, 1.86 g KCl, 1.02 g MgCl2·6H2O, 200 mL glycerol, 1 mL 1 M dithiothreitol, adjust pH to 8, bring to 1 L final volume with MilliQ water, and filter with a 0.45 μm MCE filter.
11. Baffled flasks, 2 L (sterilized).
12. Incubated and refrigerated shaking incubator (with brackets for 125 mL and 2 L baffled flasks).
13. Analytical balance.
14. Graduated cylinder.
15. Magnetic stir plate.
16. Benchtop centrifuge (with a rotor for 1.5 and 2.0 mL microcentrifuge tubes).
17. Super speed centrifuge with:
    (a) Rotor for 15 and 50 mL conical tubes (fixed angle or swinging bucket).
    (b) Rotor for 50 mL centrifuge tubes (fixed angle).

(c) Rotor for 250 mL bottles (fixed angle).
(d) Rotor for 500 mL bottles (fixed angle).

18. Ultralow temperature freezer (−80 °C).
19. Fast protein liquid chromatography (FPLC) instrumentation.
20. Nickel-affinity column for FPLC (e.g., HisTrap® Columns).

2.2 Reagents for RocR Expression and Production

1. *E. coli* NiCo21 (DE3) pET26b::*rocR*^{WT} cells grown on LB agar containing kanamycin (50 μg/mL) (*see* **Note 2**).
2. 125 mL Erlenmeyer flask.
3. 1.5 mL microcentrifuge tubes.
4. 50 mL polypropylene conical centrifuge tubes.
5. 1 M isopropyl β-D-1-thiogalactopyranoside (IPTG).
6. Trace metal mix (1000 × working concentration): 0.54 g/L $FeCl_3 \cdot 6H_2O$, 1.948 g/L $MnCl_2 \cdot 4H_2O$, 0.0148 g/L $Zn(NO_3)_2 \cdot 6H_2O$, 0.1108 g/L $CaCl_2$, 0.028 g/L $CoSO_4 \cdot 7H_2O$, 0.0240 g/L $NaMoO_4 \cdot H_2O$, 0.4928 g/L $MgSO_4 \cdot 7H_2O$, 0.1296 g/L $NiCl_2$, filter sterilize, aliquot into 50 mL volumes, and freeze at −20 °C.

2.3 Reagents and Equipment for Cell Lysis

1. 5 ¾ inch disposable lime glass Pasteur pipettes.
2. 15 mL polypropylene conical centrifuge tubes (suitable for centrifugation at >20,000 × *g*).
3. 100 μM phenylmethane sulfonyl fluoride (PMSF).
4. EDTA-free protease inhibitor tablets (e.g., cOmplete™ EDTA-free Protease Inhibitor Cocktail, Roche).
5. Probe-tip sonicator (*see* **Note 3**).

2.4 Reagents and Equipment for SDS-PAGE

1. Laemmli Solubilization Buffer (LSB) [36] (2× working solution). Prepare and mix the ingredients for 2× LSB as follows:
   (a) 0.5 M Tris–HCl (pH 6.8): Add 6 g Tris base to 75 mL of MilliQ water. Adjust pH to 6.8 with concentrated HCl. Bring to a final volume of 100 mL with MilliQ water.
   (b) 10% (w/v) sodium dodecyl sulfate (SDS): Weigh 10 g of SDS and quantitatively transfer to a graduated cylinder. Bring volume to 100 mL. Mix ingredients using a magnetic stir plate.
   (c) 1% (w/v) bromophenol blue: Weigh 0.1 g bromophenol blue and quantitatively transfer the dye to a 15 mL conical tube. Bring volume to 10 mL with MilliQ water. Mix with a vortex mixer until the dye is dissolved.
   (d) To make up a 5 mL volume of 2× LSB, combine 500 μL 1 M DTT, 500 μL 0.5 M Tris–HCl (pH 6.8), 1 mL 10% (w/v) sodium dodecyl sulfate, 2 mL 100% glycerol, 250 μL 1% (w/v) bromophenol blue, and 750 μL ddH_2O.

2. Precast SDS-PAGE gel (*see* **Note 4**).
3. Sodium dodecyl sulfate (SDS) running buffer (10×): Starting with 600 mL of MilliQ water, add 30.3 g Tris Base, 144 g glycine, 100 g SDS and bring to 1 L volume. Make up the 1× running buffer by adding 100 mL of 10× SDS running buffer to 900 mL MilliQ water.
4. Protein standard ladder (with recombinant protein standards ranging from approximately 10 to 250 kDa).
5. Fixing solution (1 L): Combine 500 mL methanol, 100 mL glacial acetic acid, and 400 mL MilliQ water.
6. Coomassie Brilliant Blue R250 Solution (1 L): Add 0.1 g Coomassie Brilliant Blue R250 to 500 mL methanol, 100 mL glacial acetic acid, and 400 mL MilliQ water.
7. Protein gel electrophoresis equipment.

2.5 Reagents for RocR Purification

1. 10 kDa molecular weight cut-off centrifugal filters (e.g., Amicon® Ultra, 15 mL volume).
2. Pierce 660 Reagent® (or alternative protein assay reagent, *see* **Note 5**) with Ionic Detergent Compatibility Reagent® for use with protein samples in LSB.
3. 0.45 μm polyethersulfone (PES) syringe filter.

2.6 Buffers, Equipment, and Reagents for MANT-c-di-GMP Assays

1. MANT-c-di-GMP (10 μM) stock solution in MilliQ water (Biolog®).
2. Purified recombinant RocR-His × 6 (5 μM) and or another purified recombinant c-di-GMP-specific phosphodiesterase (5 μM).
3. 1.5 mL black microcentrifuge tubes (*see* **Note 6**).
4. PDE Reaction Buffer (2×): 100 mM Tris–HCl, pH 8.0, 500 mM NaCl, 50 mM KCl, 10 mM $MgCl_2$, and 2 mM DTT.
5. Black, clear-bottomed 96-well microtiter plates.
6. Vortex mixer.
7. Multichannel micropipettes (with 0.5–20 μL and 20–200 μL capacities).
8. Microtiter plate reader capable of fluorescence detection.

3 Methods

A basic knowledge of molecular and microbiological techniques is required to execute protein production (Subheadings 3.1–3.4). Expert users will find that a variety of reagents, enzymes, and kits may be substituted with appropriate alternatives. All the procedures should be carried out at room temperature unless specified otherwise. Steps involving work with live bacteria should be executed using standard aseptic technique and suitable biosafety precautions.

3.1 Recombinant Production of RocR in E. coli

1. Streak out *E. coli* NiCo21 (DE3) pET26b::*rocR*^{WT} on LB agar containing 50 μg/mL kanamycin. Incubate the agar plate overnight at 37 °C.
2. Add 25 μL of the 1000 × kanamycin solution to 25 mL of super broth in a 125 mL Erlenmeyer flask. Pick 2–3 colonies of *E. coli* NiCo21 (DE3) pET26b::*rocR*^{WT} from the agar plate and inoculate the 25 mL of super broth. Incubate this broth starter culture at 37 °C and 250 r.p.m. for 18 h.
3. Add 400 μL of the 1000 × kanamycin solution and 400 μL of the 1000 × trace metal stock to each of the 2 L baffled flasks containing 400 mL super broth.
4. Inoculate each flask with 4 mL of the broth starter culture prepared in Subheading 3.1, **step 2**.
5. Incubate these broth cultures at 37 °C and 250 r.p.m. until the cultures reach an optical density at 600 nm (OD_{600}) of 0.6–0.8.
6. Collect an "uninduced cell pellet" sample for subsequent analysis by SDS PAGE.
   (a) Transfer a 1 mL aliquot of each culture to a 1.5 mL microcentrifuge tube.
   (b) Collect the cells via centrifugation at 15,000 × *g*.
   (c) Remove the supernatant and store the "uninduced cell pellet" at −20 °C for the analysis by SDS-PAGE (*see* Subheading 3.2, **step 9**).
7. Lower the temperature of the incubator to 28 °C and allow the temperature to equilibrate for 15 min.
8. Induce RocR expression by adding 40 μL of 1 M IPTG to the cultures.
9. Incubate these cultures at 28 °C and 250 r.p.m. for 16 h.
10. Collect an "induced cell pellet" sample for subsequent analysis by SDS PAGE
    (a) Transfer a 1 mL aliquot of each induced culture to a 1.5 mL microcentrifuge tube.
    (b) Collect the cells via centrifugation at 15,000 × *g*.
    (c) Remove the supernatant and store the "induced cell pellet" at −20 °C for the analysis by SDS-PAGE (*see* Subheading 3.2, **step 9**).

11. Harvest the remaining *E. coli* cells expressing RocR. For the following steps, work on ice and use rotors and centrifuge bottles that are prechilled to 4 °C:
    (a) Transfer the rest of the cultures to 500 mL centrifugation bottles that are evenly balanced.
    (b) Collect the cells at 3000 × *g* for 20 min using a suitable rotor and super speed centrifuge. Remove the supernatant and discard it.
    (c) Combine the cell pellets together in 250 mL of RocR lysis/wash buffer. Ensure the cells are entirely suspended in the buffer.
    (d) Split the cell suspensions into equal 125 mL aliquots that are evenly balanced, and collect the cells at 3000 × *g* for 20 min using a suitable rotor and super speed centrifuge. Remove the supernatant and discard it.
    (e) Suspend each of the pellets into 50 mL of RocR lysis/wash buffer.
    (f) Transfer each of the two 50 mL aliquots to pre-weighed 50 mL conical tubes.
    (g) Collect the cells at 3000 × *g* for 20 min using a suitable rotor and super speed centrifuge. Remove the supernatant and discard it.
    (h) Weigh the conical tubes, and determine the mass of the cell pellets. Cell pellet mass is used to calculate the amount of RocR lysis/wash buffer and protease inhibitor cocktail used in the subsequent steps. Freeze the cell pellet at −80 °C (*see* **Note 7**).

3.2 Cell Lysis and Verification of RocR Expression

1. Thaw the frozen cell pellets in an ice water bath for 30 min.
2. Working on ice, suspend the cells in RocR lysis/wash buffer:
    (a) Dissolve the EDTA-free protease inhibitor tablets in RocR lysis/wash buffer. Prepare 3.25 mL of RocR lysis/wash buffer for each gram of cell mass harvested in Subheading 3.1, **step 11**. Following the manufacturer's directions, add the required amount of EDTA-free protease inhibitor to the final volume of RocR lysis/wash buffer calculated at this step. Ensure that the protease inhibitor is completely dissolved.
    (b) Transfer the entire volume of RocR lysis/wash buffer containing the protease inhibitor that was prepared in the previous step to the 50 mL conical tube containing the thawed cell pellet. Suspend the bacteria by mixing with a disposable Pasteur pipette.

   (c) Split the resulting cell suspension into 6–9 mL aliquots in 15 mL conical tubes.
   (d) Add 1 μL of 100 μM PMSF per mL of cell suspension to each tube.
   (e) Mix the tubes by inverting them several times, and place the tubes in an ice water bath.
3. Sonicate the cells at mid to high power (50–75% maximum output) using 12 sonication cycles consisting of 45 s sonication followed by at least 1 min of cooling on ice.
4. Centrifuge the cell lysates for 20 min at 7000 × *g* and 4 °C.
5. Transfer the supernatant to a new conical tube. Store the "low-speed" pellet at −80 °C for SDS-PAGE analysis (*see* Subheading 3.2, **step 9**) (*see* **Note 7**).
6. Balance the conical tubes by bringing the volume up to the nearest 10 mL with RocR lysis/wash buffer.
7. Centrifuge the supernatants for 1 h at 20,000 × *g* and 4 °C. Transfer the supernatant, termed the "cell-free extract," to new 15 mL conical tubes.
8. Aliquot 50 μL of the cell-free extract into a 1.5 μL microcentrifuge tube for SDS-PAGE analysis. Store the remaining cell-free extract in conical tubes at −80 °C (*see* **Note 7**). Store the "high-speed" pellet at −80 °C for SDS-PAGE analysis (*see* Subheading 3.2, **step 9**) (*see* **Note 7**).
9. Using SDS-PAGE, verify that RocR was produced in *E. coli* by analyzing the uninduced cell pellet (Subheading 3.1, **step 6**) and induced cell pellet (Subheading 3.1, **step 9**), the low speed pellet (Subheading 3.2, **step 5**), and finally the cell-free extract (Subheading 3.2, **step 8**) via standard methods for SDS-PAGE. The samples can be prepared for SDS-PAGE and analyzed as follows:
   (a) Add 150 μL of LSB to the "uninduced" and "induced" cell pellets. Suspend the bacteria by pipetting up and down with a P1000 micropipette.
   (b) Add 5 μL of the semisolid low speed pellet to 150 μL LSB.
   (c) Add 5 μL of the semisolid high speed pellet to 150 μL LSB.
   (d) Add 50 μL of 2 × LSB to 50 μL of the cell-free extract.
   (e) Boil all of the samples for 15 min at 100 °C. Allow samples to cool to approximately 50 °C (requires approximately 15–20 min).
   (f) Centrifuge the samples for 30 s at 5000 × *g*, and then place the samples on ice.
   (g) Pour and load the SDS-PAGE gel according to standard methods. Load 5–10 μL of a suitable protein standard

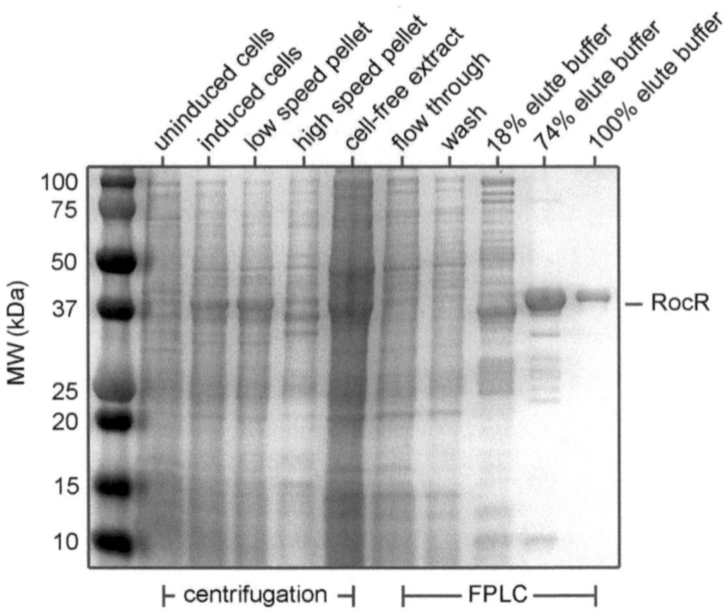

Fig. 2 Production and purification of *Pseudomonas aeruginosa* RocR, a c-di-GMP-specific phosphodiesterase for use in MANT-c-di-GMP assays. Qualitative SDS-PAGE analysis was performed with uninduced and induced cells, as well as with low speed and high speed pellets after *E. coli* cell lysis. Differential centrifugation is used to remove insoluble protein and cellular debris, yielding high quantities of soluble, recombinant RocR in the cell-free extract. Subsequently, RocR is purified via FPLC using a block elution protocol, with the vast majority of RocR eluting in 74% RocR elution buffer with >95% purity

ladder, and 2–5 μL of each sample, into adjacent lanes of the polyacrylamide gel.

(h) Run the SDS-PAGE gel.

(i) After the gel has run, rinse the gel in MilliQ water for 5 min, and then soak in fixing solution for 10 min.

(j) Subsequently, rinse the gel twice in MilliQ water for 5 min.

(k) Stain with a Coomassie Brilliant Blue R250 solution for 1 h, and then soak in MilliQ water for at least 30 min to destain.

(l) Image the gel with a gel documentation system (Fig. 2).

3.3 Purification of Recombinant RocR

1. Thaw the cell-free extract (from Subheading 3.2, **step 8**) in an ice water bath. Once thawed, filter the cell-free extract with a 0.45 μm PES membrane filter.
2. Using FPLC instrumentation fitted with a 5 mL nickel affinity column, fractionate the cell-free extract and purify RocR with a block elution protocol as follows:

(a) Equilibrate the column by washing it with 2 column volumes (10 mL for a 5 mL column) of RocR lysis/wash buffer.

(b) Load one column volume of the sample (5 mL for a 5 mL column) onto the column (*see* **Note 8** for recommended protein concentration).

(c) Wash the column with 5 column volumes (25 mL for a 5 mL column) of RocR lysis/wash buffer.

(d) Subsequently, fractionate the protein using elution steps with 16%, 74%, and 100% RocR elution buffer, washing with 5 column volumes at each step (*see* **Note 1**).

(e) Collect the eluate in 2 mL fractions, and monitor protein elution in each fraction using ultraviolet (UV) light (280 nm) detection. It is expected that RocR will elute in 74% RocR elute buffer.

3. Identify fractions containing protein using the UV chromatogram generated by the FPLC instrument.
4. Analyze these fractions by standard methods for SDS-PAGE to identify fractions with RocR and estimate its purity (Fig. 2). We recommend loading 5–10 μL of each fraction per SDS-PAGE gel lane.
5. Guided by the SDS-PAGE results, pool the fractions containing purified RocR.
6. Transfer the pooled fractions containing RocR to a 15 mL centrifugal filter apparatus with a 10 kDa molecular weight cutoff (MWCO).
7. Centrifuge the apparatus at 4000 × *g* at 4 °C for 1 h.
8. To verify that the protein has been concentrated and retained correctly during centrifugation, test the protein concentration in the retentate using the Pierce 660® reagent (or another suitable protein assay).
9. To reduce the imidazole concentration, add 14 mL of the RocR cleaning buffer to the retentate.
10. Transfer the diluted RocR protein sample to a new 15 mL centrifugal filter apparatus with a 10 kDa MWCO.
11. Centrifuge the apparatus at 4000 × *g* at 4 °C for 1 h to obtain 1– 2 mL of concentrated RocR.
12. Bring the volume of the retentate to 15 mL with RocR storage buffer.
13. Transfer the diluted RocR protein sample to a new centrifugal filter apparatus with a 10 kDa MWCO.
14. Centrifuge the apparatus at 4000 × *g* at 4 °C for 1 h to obtain ~2–4 mL of concentrated, purified RocR.

15. Determine the concentration of purified RocR using the Pierce 660® reagent (or another suitable protein assay). A total yield of ≥20 mg purified, recombinant RocR is expected with an estimated purity of 95–99%.
16. Take an aliquot of purified RocR and prepare a 50 μM stock solution for use in MANT-c-di-GMP assays (Subheading 3.4).
17. Store purified RocR at −20 °C (*see* **Note 9**).

3.4 MANT-c-di-GMP Assay

1. Working on ice and following the order specified in Table 1, aliquot the reagents for the MANT-c-di-GMP assay into the wells of a black, clear-bottom, 96-well microtiter plate. Each control and test condition should be set up in technical triplicate. Multichannel pipettes may be used to dispense reagents quickly. Do not add the PDE RocR yet.
2. Add PDE RocR.
3. After the addition of the PDE*Phosphodiesterase)*, immediately move the microtiter plate into the microtiter plate reader.
4. Monitor fluorescence (excitation at 355 nm, emission at 440 nm) using time-dependent kinetics. An assay temperature in the range of 25–37 °C is recommended, depending on the temperature-dependent activity of the PDE. Shake for 5 s before each read, and measure fluorescence every 2 min for 30 min.
5. The reduction in c-di-GMP concentration over time is calculated as percentage of the concentration at time zero and plotted (Fig. 3).

Table 1
Reagent volumes (μL) for the MANT-c-di-GMP assay (per well, *see* Note 10)

	No PDE control	PDE only	PDE + agent	Agent only control[c]
PDE Reaction buffer (2 ×)	50	50	50	50
MANT-c-di-GMP (10 μM)	5	5	5	5
Test agent[a]	–	–	10	10
ultrapure water	45	43	33	35
PDE[b] (50 μM)	–	2	2	–

[a]Concentration of working solution depends on the activity of the agonist or antagonist
[b]Any c-di-GMP-specific PDE may be used (e.g., purified RocR from Subheading 3.3, **step 16**)
[c]Control to identify direct interactions between a test agent and MANT-c-di-GMP (optional, depends on application and test agent)

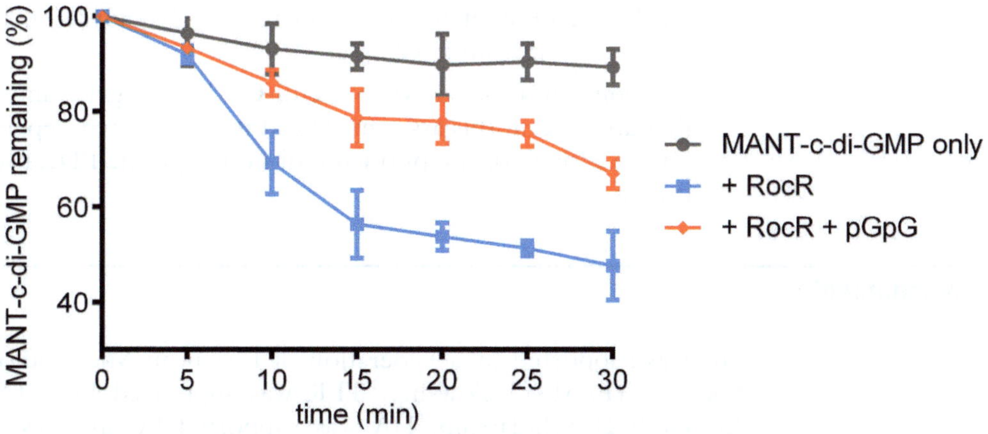

Fig. 3 Quantifying c-di-GMP-specific phosphodiesterase (PDE) activity and inhibition using MANT-c-di-GMP. EAL domain PDEs, such as RocR, break a single phosphodiester bond in c-di-GMP to produce the linear nucleotide pGpG, which exerts product inhibition on the PDE when added exogenously [10, 11]. Here, 0.4 mM pGpG was added to the assays to inhibit RocR

4 Notes

1. Two buffers are used to establish an imidazole gradient on the FPLC instrument. Here, RocR lysis/wash buffer and RocR elute buffer are used for this purpose, the latter of which contains a high concentration of imidazole.
2. All strains and plasmids used in this protocol are available by request to Joe J. Harrison (jjharris@ucalgary.ca) or Ehud Banin (ehud.banin@biu.ac.il).
3. Although a protocol for using a probe-tip sonicator to lyse *E. coli* has been provided here, alternative means of physical cell lysis—including, for example, the use of a French press—can be carried out at the reader's preference.
4. Gels may be poured in-house according to standard methods for SDS-PAGE [37], or purchased from various suppliers for the ease of use and consistency.
5. Alternative compatible protein assay reagents may be used at the reader's discretion.
6. Microcentrifuge tubes may also be wrapped in aluminum foil to protect light-sensitive reagents.
7. All cell pellets and cell-free extract can be stored at −80 °C for at least 4 months prior to processing or analysis.
8. We recommend diluting the cell-free extract to a protein concentration of ~10 mg/mL prior to loading the FPLC column.

9. Purified RocR contains approximately 35–40% glycerol, which will prevent it from freezing at −20 °C.
10. The concentrations of MANT-c-di-GMP, test agent, and PDE that are used in this assay may need to be adjusted depending on the activity of the purified c-di-GMP specific PDE used in the assay.

Acknowledgments

DC was supported by a Federation of European Microbiological Societies (FEMS) fellowship. TER was supported by a Queen Elizabeth II Scholarship. HA was supported by an Eyes High Postdoctoral Fellowship. JJH has been supported by a Canada Research Chair from the Canadian Institutes for Health Research (CIHR), and a Discovery Grant from the Natural Sciences and Engineering Research Council of Canada (NSERC). EB was supported by the Israel Science Foundation Grant 1124/12 and the Dyna and Fala Weinstock Foundation.

References

1. Ross P, Weinhouse H, Aloni Y, Michaeli D, Weinberger-Ohana P, Mayer R, Braun S, de Vroom E, van der Marel GA, van Boom JH, Benziman M (1987) Regulation of cellulose synthesis in *Acetobacter xylinum* by cyclic diguanylic acid. Nature 325:279–281
2. Romling U, Galperin MY, Gomelsky M (2013) Cyclic di-GMP: the first 25 years of a universal bacterial second messenger. Microbiol Mol Biol Rev 77:1–52
3. Hengge R (2009) Principles of c-di-GMP signalling in bacteria. Nat Rev Microbiol 7:263–273
4. Mills E, Pultz IS, Kulasekara HD, Miller SI (2011) The bacterial second messenger c-di-GMP: mechanisms of signalling. Cell Microbiol 13:1122–1129
5. Hengge R, Grundling A, Jenal U, Ryan R, Yildiz F (2016) Bacterial signal transduction by cyclic Di-GMP and other nucleotide second messengers. J Bacteriol 198:15–26
6. Almblad H, Harrison JJ, Rybtke M, Groizeleau J, Givskov M, Parsek MR, Tolker-Nielsen T (2015) The cyclic AMP-Vfr signaling pathway in *Pseudomonas aeruginosa* is inhibited by cyclic di-GMP. J Bacteriol 197:2190–2200
7. Paul R, Weiser S, Amiot NC, Chan C, Schirmer T, Giese B, Jenal U (2004) Cell cycle-dependent dynamic localization of a bacterial response regulator with a novel di-guanylate cyclase output domain. Genes Dev 18:715–727
8. Schirmer T, Jenal U (2009) Structural and mechanistic determinants of c-di-GMP signalling. Nat Rev Microbiol 7:724–735
9. Christen M, Christen B, Folcher M, Schauerte A, Jenal U (2005) Identification and characterization of a cyclic di-GMP-specific phosphodiesterase and its allosteric control by GTP. J Biol Chem 280:30829–30837
10. Cohen D, Mechold U, Nevenzal H, Yarmiyhu Y, Randall TE, Bay DC, Rich JD, Parsek MR, Kaever V, Harrison JJ, Banin E (2015) Oligoribonuclease is a central feature of cyclic diguanylate signaling in *Pseudomonas aeruginosa*. Proc Natl Acad Sci U S A 112:11359–11364
11. Orr MW, Donaldson GP, Severin GB, Wang J, Sintim HO, Waters CM, Lee VT (2015) Oligoribonuclease is the primary degradative enzyme for pGpG in *Pseudomonas aeruginosa* that is required for cyclic-di-GMP turnover. Proc Natl Acad Sci U S A 112:E5048–E5057
12. Ryan RP, Fouhy Y, Lucey JF, Crossman LC, Spiro S, He YW, Zhang LH, Heeb S, Camara M, Williams P, Dow JM (2006) Cell-cell signaling in *Xanthomonas campestris* involves an HD-GYP domain protein that functions in cyclic di-GMP turnover. Proc Natl Acad Sci U S A 103:6712–6717
13. Stelitano V, Giardina G, Paiardini A, Castiglione N, Cutruzzola F, Rinaldo S (2013) C-

di-GMP hydrolysis by *Pseudomonas aeruginosa* HD-GYP phosphodiesterases: analysis of the reaction mechanism and novel roles for pGpG. PLoS One 8:e74920

14. Bellini D, Caly DL, McCarthy Y, Bumann M, An SQ, Dow JM, Ryan RP, Walsh MA (2013) Crystal structure of an HD-GYP domain cyclic-di-GMP phosphodiesterase reveals an enzyme with a novel trinuclear catalytic iron center. Mol Microbiol 91:26–38
15. Valentini M, Filloux A (2016) Biofilms and cyclic di-GMP (c-di-GMP) signaling: lessons from *Pseudomonas aeruginosa* and other bacteria. J Biol Chem 291:12547–12555
16. Groizeleau J, Rybtke M, Andersen JB, Berthelsen J, Liu Y, Yang L, Nielsen TE, Kaever V, Givskov M, Tolker-Nielsen T (2016) The anticancerous drug doxorubicin decreases the c-di-GMP content in *Pseudomonas aeruginosa* but promotes biofilm formation. Microbiology 162(10):1797–1807
17. Kim HS, Cha E, Kim Y, Jeon YH, Olson BH, Byun Y, Park HD (2016) Raffinose, a plant galactoside, inhibits *Pseudomonas aeruginosa* biofilm formation via binding to LccA and decreasing cellular cyclic diguanylate levels. Sci Rep 6:25318
18. Antoniani D, Bocci P, Maciag A, Raffaelli N, Landini P (2010) Monitoring of diguanylate cyclase activity and of cyclic-di-GMP biosynthesis by whole-cell assays suitable for high-throughput screening of biofilm inhibitors. Appl Microbiol Biotechnol 85:1095–1104
19. Simm R, Morr M, Remminghorst U, Andersson M, Romling U (2009) Quantitative determination of cyclic diguanosine monophosphate concentrations in nucleotide extracts of bacteria by matrix-assisted laser desorption/ionization-time-of-flight mass spectrometry. Anal Biochem 386:53–58
20. Irie Y, Parsek MR (2014) LC/MS/MS-based quantitative assay for the secondary messenger molecule, c-di-GMP. Methods Mol Biol 1149:271–279
21. Spangler C, Bohm A, Jenal U, Seifert R, Kaever V (2010) A liquid chromatography-coupled tandem mass spectrometry method for quantitation of cyclic di-guanosine monophosphate. J Microbiol Methods 81:226–231
22. Waters CM, Lu W, Rabinowitz JD, Bassler BL (2008) Quorum sensing controls biofilm formation in *Vibrio cholerae* through modulation of cyclic di-GMP levels and repression of vpsT. J Bacteriol 190:2527–2536
23. Hickman JW, Harwood CS (2008) Identification of FleQ from *Pseudomonas aeruginosa* as a c-di-GMP-responsive transcription factor. Mol Microbiol 69:376–389
24. Nakayama S, Kelsey I, Wang J, Roelofs K, Stefane B, Luo Y, Lee VT, Sintim HO (2011) Thiazole orange-induced c-di-GMP quadruplex formation facilitates a simple fluorescent detection of this ubiquitous biofilm regulating molecule. J Am Chem Soc 133:4856–4864
25. Stelitano V, Brandt A, Fernicola S, Franceschini S, Giardina G, Pica A, Rinaldo S, Sica F, Cutruzzola F (2013) Probing the activity of diguanylate cyclases and c-di-GMP phosphodiesterases in real-time by CD spectroscopy. Nucleic Acids Res 41:e79
26. Nakayama S, Luo Y, Zhou J, Dayie TK, Sintim HO (2012) Nanomolar fluorescent detection of c-di-GMP using a modular aptamer strategy. Chem Commun 48:9059–9061
27. Roembke BT, Zhou J, Zheng Y, Sayre D, Lizardo A, Bernard L, Sintim HO (2014) A cyclic dinucleotide containing 2-aminopurine is a general fluorescent sensor for c-di-GMP and 3′,3′-cGAMP. Mol BioSyst 10:1568–1575
28. Nakayama S, Roelofs K, Lee VT, Sintim HO (2012) A C-di-GMP-proflavine-hemin supramolecular complex has peroxidase activity–implication for a simple colorimetric detection. Mol BioSyst 8:726–729
29. Roembke BT, Wang J, Nakayama S, Zhou J, Sintim HO (2013) Octameric G8 c-di-GMP is an efficient peroxidase and this suggests that an open G-tetrad site can effectively enhance hemin peroxidation reactions. RSC Adv 3:6305–6310
30. Zhou J, Zheng Y, Roembke BT, Robinson S, Opoku-Temeng C, Sayre DA, Sintim HO (2017) Fluorescent analogs of cyclic and linear dinucleotides as phosphodiesterase and oligoribonuclease activity probes. RSC Adv 7:5421–5426
31. Sharma IM, Dhanaraman T, Mathew R, Chatterji D (2012) Synthesis and characterization of a fluorescent analogue of cyclic di-GMP. Biochemistry 51:5443–5453
32. Hiratsuka T (1982) New fluorescent analogs of cAMP and cGMP available as substrates for cyclic nucleotide phosphodiesterase. J Biol Chem 257:13354–13358
33. Ren J, Goss DJ (1996) Synthesis of a fluorescent 7-methylguanosine analog and a fluorescence spectroscopic study of its reaction with wheatgerm cap binding proteins. Nucleic Acids Res 24:3629–3634
34. Kotaka M, Dutta S, Lee HC, Lim MJ, Wong Y, Rao F, Mitchell EP, Liang ZX, Lescar J (2009) Expression, purification and preliminary crystallographic analysis of *Pseudomonas aeruginosa*

RocR protein. Acta Crystallogr Sect F Struct Biol Cryst Commun 65:1035–1038

35. Rao F, Yang Y, Qi Y, Liang ZX (2008) Catalytic mechanism of cyclic-di-GMP-specific phosphodiesterase: a study of the EAL domain-containing RocR from *Pseudomonas aeruginosa*. J Bacteriol 190:3622–3631

36. Laemmli UK (1970) Cleavage of structural proteins during the assembly of the head of bacteriophage T4. Nature 227:680–685

37. Green MR, Sambrook J (2012) Molecular cloning: a laboratory manual. Cold Spring Harbor Laboratory Press, Cold Spring Harbor, NY

Chapter 21

Determining Phosphodiesterase Activity (Radioactive Assay)

Barbara I. Kazmierczak

Abstract

Cyclic-di-GMP phosphodiesterases (PDEs) catalyze the hydrolysis of the bacterial second messenger c-di-GMP. This protocol describes a sensitive radioactive assay for PDE activity in which substrate and product can be quickly and easily separated by thin-layer chromatography.

Key words Phosphodiesterase, Cyclic-di-GMP, Thin-layer chromatography

1 Introduction

Two distinct protein domains found in bacterial phosphodiesterases (PDEs) can catalyze the hydrolysis of c-di-GMP, leading to two different products: EAL domains result in the production of a linear 5′-pGpG molecule, while HD-GYP domain proteins hydrolyze c-di-GMP to GMP [1]. Phosphodiesterase activity can be assayed using enzymatically synthesized [^{32}P]-c-di-GMP (*see* Chap. 3) and thin-layer chromatography (TLC) (*see* Chap. 22). Although c-di-GMP can now be commercially purchased, the use of a radioactive substrate allows for highly sensitive assays of phosphodiesterase activity. Thin-layer chromatography (TLC) is used for the detection of radiolabeled nucleotides, a methodology that may be more accessible to some labs than other techniques that are used for the qualitative and quantitative detection of these molecules, such as liquid chromatography–tandem mass spectrometry [2]. Furthermore, the ability to separate and accurately quantify the amount of radioactive substrate and product also allows for quantitative characterization of phosphodiesterase activity, including determination of V_{max} and K_m.

Karin Sauer (ed.), *c-di-GMP Signaling: Methods and Protocols*, Methods in Molecular Biology, vol. 1657,
DOI 10.1007/978-1-4939-7240-1_21, © Springer Science+Business Media LLC 2017

2 Materials

2.1 Radioactive Phosphodiesterase Activity Assay

1. [^{32}P]-c-di-GMP as prepared in Chap. 3
2. Phosphodiesterase (PDE), purified to near homogeneity.
3. Buffer A: 0.05 M Tris–HCl pH 8.0, 0.25 M NaCl, 0.005 M β-mercaptoethanol (BME).
4. GTP stock solution: 0.001 M in Buffer A.
5. $MgCl_2$: 0.1 M stock solution.
6. EDTA, 0.5 M stock solution (pH 8.0).

2.2 TLC Assay for Nucleotides

1. Saturated $(NH_4)_2SO_4$: Dissolve 766.8 g in 1 L water.
2. 1.5 M KH_2PO_4 buffer: Dissolve 204.2 g in ca. 800 mL of water. Bring to pH 3.65 with phosphoric acid, then adjust final volume to 1000 mL with water.
3. PEI cellulose TLC plate (20 × 20 cm), stored at 4 °C.
4. Glass tank with lid, large enough to accommodate the TLC plate.
5. Whatman filter paper, cut to 15 × 15 cm.
6. [α^{32}P]-GTP (3000 Ci/mmol, 10 mCi/mL).

3 Methods

3.1 Radioactive Phosphodiesterase Activity Assay

All personnel must be trained in and follow appropriate safety precautions when working with ^{32}P. The high-energy beta emissions from ^{32}P can present a substantial skin and eye dose hazard. Appropriate shielding, personal protective equipment and monitoring must be used when working with samples containing ^{32}P. All materials must be disposed of in accordance with your institution's radiation safety protocols.

1. Known or putative PDE proteins of interest should be purified to near homogeneity and dialyzed against Buffer A (*see* **Note 1**).
2. The presence of PDE activity can be assayed by mixing the PDE protein preparation (8 μL in Buffer A) with 1 μL of 0.1 M $MgCl_2$ and 1 μL of the heat-inactivated [^{32}P]-c-di-GMP containing reaction mixture prepared in Chap. 3.
3. Incubate the reaction at room temperature for 30 min.
4. Stop the reaction by adding 2 μL of 0.5 M EDTA and mixing (*see* **Note 2**).
5. A "no enzyme" control reaction should be set up in parallel to provide a c-di-GMP migration control for TLC. This also provides a control for non-enzymatic hydrolysis of c-di-GMP, which is characteristically quite minimal.

3.2 TLC Assay for Nucleotides

1. Mix 100 mL saturated $(NH_4)_2SO_4$ and 150 mL 1.5 M KH_2PO_4, pH 3.65 (ratio 2:3 v/v).
2. Pour the resulting solution into the bottom of a glass tank ("development chamber") that will accommodate the TLC plate. The level of the solvent solution must be below 1 cm. A Whatman paper wick (cut smaller than the TLC plate) should be placed in the tank against the back wall.
3. Cover the tank to allow the atmosphere to become saturated with solvent vapor (*see* **Note 3**).
4. Allow a PEI cellulose plate to warm up to RT (*see* **Note 4**).
5. Handling the plate by the edge, place it on a clean surface, powdered side up.
6. Using a ruler, lightly draw a pencil line 1.5 cm from the end of the plate—do not dig into and scratch the surface of the plate, as this will interfere with the migration of solvent. Mark 1 cm intervals along the line to guide spotting of samples.
7. Samples are loaded by spotting 2 μL along the pencil line at 1 cm intervals using a micropipetor. Pipet slowly and allow the spots to dry completely before developing the plate.
8. Include the "no enzyme" control reaction to provide a c-di-GMP migration control for TLC.
9. Place the TLC plate in the equilibrated development chamber. Make sure that the level of solvent is below the level of the spotted samples, and the chamber is level (*see* **Note 5**). The filter paper wick should not make contact with the sides of the plate. Cover the development chamber and allow the solvent front to advance almost to the top of the plate (ca. 2h).
10. Upon completion, remove the TLC plate from the development chamber and mark the position of the solvent front before it dries.
11. Cover the dry TLC plate with Saran Wrap and expose to autoradiography film. The pGpG product is more polar than c-di-GMP and will migrate further in the solvent system (*see* Fig. 1) (*see* **Note 6**).

4 Notes

1. Buffer exchange can be accomplished by placing a protein-containing solution into hydrated dialysis tubing that is tied off or clamped at both ends, and incubating the dialysis bag in a large volume (~ 1 L) of the final buffer. This is usually done at 4 °C overnight, and the buffer should be changed at least once after 4–6 h of equilibration. Alternatives to dialysis tubing include Slide-A-Lyzer™ Dialysis Cassettes (Thermo Fisher

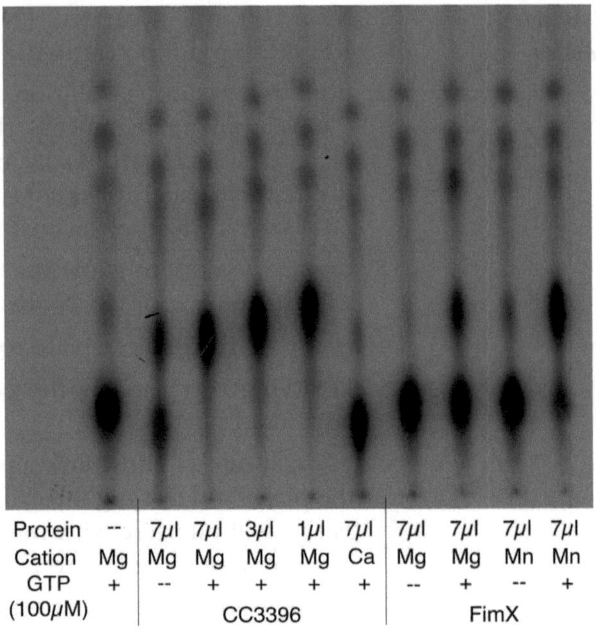

Fig. 1 Thin layer chromatography of phosphodiesterase reaction products. Purified *Caulobacter crescentus* CC3396 phosphodiesterase [3] or *Pseudomonas aeruginosa* FimX were incubated with synthesized [^{32}P]-c-di-GMP in the presence of the indicated divalent cation (0.01 M final concentration). In some reactions, 0.1 M GTP was present as an allosteric activator of PDE activity. The "no protein" control reaction shows the migration of c-di-GMP, while the faster migrating spot corresponding to pGpG is produced by CC3396 in the presence of Mg^{2+} plus GTP. The PDE activity of FimX is less than that of CC3396 (note the presence of unhydrolyzed substrate), but activation by GTP is clearly demonstrated

Scientific) and ultrafiltration devices such as Amicon® Ultra centrifugal devices (EMD Millipore). These are more expensive, but also more convenient to use.

2. Some PDEs may require allosteric activation by GTP to show activity [3, 4]. This is the case for CC3396 and FimX (Fig. 1).
3. We keep the covered tank in a chemical hood.
4. The plate can be incubated briefly in an incubator or oven to speed up warming (37–55 °C).
5. If the solvent is above the level of the spots, the spotted material will dissolve in the development solvent. If the cellulose makes contact with the filter paper wick, the solvent will move the spots sideways. If the plate is not level, the compounds will not migrate perpendicular to the bottom of the plate. If the front migrates to the top of the plate, the Rf cannot be calculated.

6. The TLC plate can also be exposed to a phosphor storage screen, which will record a latent image produced by ionizing radiation. This image can then be read by laser scanning using a PhosphorImager (e.g., GE Amersham Molecular Dynamics Typhoon) and saved in a digital format. The response of a phosphor storage screen is linear, unlike that of autoradiography film, aiding in quantification of substrate and product "spots" on the TLC plate.

References

1. Romling U, Galperin MY, Gomelsky M (2013) Cyclic di-GMP: the first 25 years of a universal bacterial second messenger. Microbiol Mol Biol Rev 77(1):1–52. doi:10.1128/mmbr.00043-12
2. Jain R, Behrens AJ, Kaever V, Kazmiercak BI (2012) Type IV pilus assembly in *Pseudomonas aeruginosa* over a broad range of cyclic-di-GMP concentrations. J Bacteriol 194:4285–4294. doi:10.1128/JB.00803-12
3. Christen M, Christen B, Folcher M, Schauerte A, Jenal U (2005) Identification and characterization of a cyclic di-GMP specific phosphodiesterase and its allosteric control by GTP. J Biol Chem 280:30829–30837
4. Kazmierczak BI, Lebron MB, Murray TS (2006) Analysis of FimX, a phosphodiesterase that governs twitching motility in *Pseudomonas aeruginosa*. Mol Microbiol 60(4):1026–1043

Chapter 22

Determining Diguanylate Cyclase Activity (Radioactive Assay)

Barbara I. Kazmierczak

Abstract

Bioinformatics approaches can identify sequence motifs associated with diguanylate cyclases (DGCs), but experimental demonstration of DGC enzymatic activity is often desired. This protocol describes a sensitive radioactive assay for DGC activity in which substrate and product are quickly and easily separated by thin-layer chromatography.

Key words Diguanylate cyclase, Cyclic-di-GMP, Thin-layer chromatography

1 Introduction

Cyclic-di-GMP is an intracellular signaling molecule that is synthesized from two molecules of GTP by enzymes called diguanylate cyclases (DGCs) [1]. In this protocol, a known or putative diguanylate cyclase (DGC) is incubated with [α^{32}P]-GTP in the presence of Mg^{2+}, and the production of c-di-GMP is detected using thin layer chromatography (TLC). TLC, in which compounds within a mixture are separated based on their partitioning between a stationary phase and mobile (solvent) phase [2], may be more accessible to some labs than other techniques that are used for the qualitative and quantitative detection of these molecules, such as liquid chromatography-tandem mass spectrometry [3]. The ability to quickly separate and accurately quantify the amount of radioactive substrate and product also allows for quantitative characterization of diguanylate cyclase activity, including determination of V_{max} and K_m. Standard protein purification techniques should be employed to purify the DGC to near homogeneity.

Karin Sauer (ed.), *c-di-GMP Signaling: Methods and Protocols*, Methods in Molecular Biology, vol. 1657,
DOI 10.1007/978-1-4939-7240-1_22, © Springer Science+Business Media LLC 2017

2 Materials

2.1 Radioactive Diguanylate Cyclase Activity Assay

1. Diguanylate cyclase (DGC), purified to near homogeneity (*see* **Note 1**).
2. [$\alpha^{32}P$]-GTP (3000 Ci/mmol, 10 mCi/mL).
3. $MgCl_2$: 0.1 M stock solution.
4. EDTA, 0.5 M stock solution (pH 8.0).
5. Buffer A: 0.05 M Tris–HCl pH 8.0, 0.25 M NaCl, 0.005 M β-mercaptoethanol (BME).

2.2 TLC Assay for Nucleotides

1. Saturated $(NH_4)_2SO_4$: Dissolve 766.8 g in 1 L water.
2. 1.5 M KH_2PO_4 buffer: Dissolve 204.2 g in ca. 800 mL of water. Bring to pH 3.65 with phosphoric acid, then adjust final volume to 1000 mL with water.
3. PEI cellulose TLC plate (20 × 20 cm), stored at 4 °C.
4. Glass tank with lid, large enough to accommodate the TLC plate.
5. Whatman filter paper, cut to 15 × 15 cm.
6. [$\alpha^{32}P$]-GTP (3000 Ci/mmol, 10 mCi/mL).

3 Methods

3.1 Radioactive Diguanylate Cyclase Activity Assay

All personnel must be trained in and follow appropriate safety precautions when working with ^{32}P. The high-energy beta emissions from ^{32}P can present a substantial skin and eye dose hazard. Appropriate shielding, personal protective equipment and monitoring must be used when working with samples containing ^{32}P. All materials must be disposed of in accordance with your institution's radiation safety protocols.

1. Purify known or putative diguanylate cyclase to near homogeneity using established techniques (*see* **Note 1**). The protein should be dialyzed against a buffer such as Buffer A after purification, and EDTA (an inhibitor of DGC activity) should be avoided. Repeated freeze–thaw cycles may denature protein and destroy enzymatic activity; thus, it is recommended that freshly purified protein be stored on ice at 4 °C and assayed as soon as possible for activity. For convenience, some investigators may prefer to freeze purified protein in single use aliquots (stabilized with 5–10% glycerol or sucrose); however, the effects of such manuevers on enzymatic activity must be experimentally determined.
2. Combine 40 μL of purified DGC with 5 μL of 0.1 M $MgCl_2$ (final concentration, 0.01 M) (*see* **Note 2**).

3. Start the enzymatic reaction by adding 5 μL [α^{32}P]-GTP (3000 Ci/mmol, 10 mCi/mL) and mix by gently pipetting (*see* **Note 3**). Incubate at RT for up to 1–2 h.
4. At time intervals, sample 5 μL of reaction into a tube containing an equal volume (5 μL) of 0.5 M EDTA and vortex to stop reaction.
5. Analyze reaction products by TLC, as described below; remainder of sample can be stored at −20 °C.

3.2 TLC Assay for Nucleotides

1. Mix 100 mL saturated $(NH_4)_2SO_4$ and 150 mL 1.5 M KH_2PO_4, pH 3.65 (ratio 2:3 v/v).
2. Pour the resulting solution into the bottom of a glass tank ("development chamber") that will accommodate the TLC plate. The level of the solvent solution must be below 1 cm. A Whatman paper wick (cut smaller than the TLC plate) should be placed in the tank against the back wall.
3. Cover the tank to allow the atmosphere to become saturated with solvent vapor (*see* **Note 4**).
4. Allow a PEI cellulose plate to warm up to RT (*see* **Note 5**).
5. Handling the plate by the edge, place it on a clean surface, powdered side up.
6. Using a ruler, lightly draw a pencil line 1.5 cm from the end of the plate—do not dig into and scratch the surface of the plate, as this will interfere with the migration of solvent. Mark 1 cm intervals along the line to guide spotting of samples.
7. Samples are loaded by spotting 2 μL along the pencil line at 1 cm intervals using a micropipetor. Pipet slowly and allow the spots to dry completely before developing the plate.
8. Prepare a GTP standard to be run in parallel with the samples by diluting [α^{32}P]-GTP (3000 Ci/mmol, 10 mCi/mL). To dilute [α^{32}P]-GTP, add 0.5 μL of nucleotide into 5 μL 0.5 M EDTA, 4 μL Buffer A and 1 μL of 0.1 M $MgCl_2$. Spot 2 μL of diluted nucleotide using a micropipetor and allow the spot to dry completely.
9. Place the TLC plate in the equilibrated development chamber. Make sure that the level of solvent is below the level of the spotted samples, and the chamber is level (*see* **Note 6**). The filter paper wick should not make contact with the sides of the plate. Cover the development chamber and allow the solvent front to advance almost to the top of the plate (ca. 2h).
10. Upon completion, remove the TLC plate from the development chamber and mark the position of the solvent front before it dries.
11. Cover the dry TLC plate with Saran Wrap and expose to autoradiography film. An image of a typical separation by TLC is shown in Fig. 1 (*see* **Note 7**).

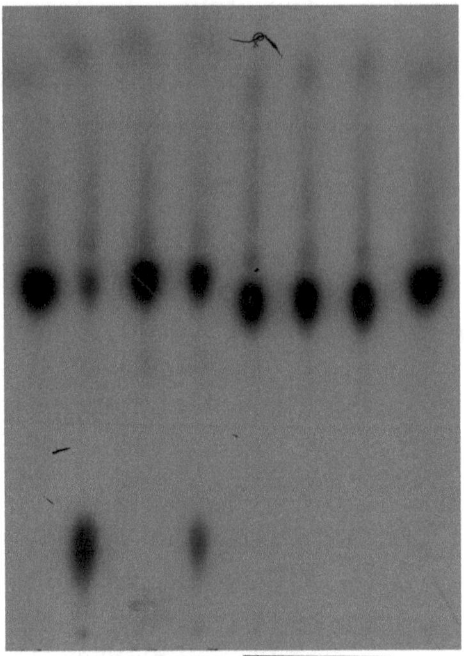

Fig. 1 Thin-layer chromatography of diguanylate cyclase reaction products. Purified PleD*-His_6 or *Pseudomonas aeruginosa* His_6-FimX [4] were incubated with [α^{32}P]-GTP in the presence of the indicated divalent cation (0.01 M) for 30 min at 25 °C. Reactions were stopped by the addition of 0.5 M EDTA, and 2 μL of each reaction was spotted to PEI cellulose, as indicated. The flanking lanes are "no protein" controls. The nonpolar cyclic dinucleotide migrates much more slowly than GTP. Note that Ca^{2+} inhibits the DGC activity of PleD*, and that FimX shows no DGC activity despite possessing a degenerate DGC domain

12. Rf is the ratio of distance traveled by nucleotide divided by the distance traveled by the solvent front, and will be between 0 and 1. Rf reflects the degree to which the nucleotide interacts with the solid phase, and is least for cyclic-di-GMP, more for linearized pGpG and greatest for GTP.

4 Notes

1. Purification conditions must be experimentally determined with the goal of maintaining protein activity. Native conditions are favored, although it may be possible to purify a protein under denaturing conditions and allow it to refold.
2. If the goal is to determine whether a purified protein has any DGC activity, I recommend using a molar excess of protein to

substrate. For further characterization of enzymatic activity, the substrate should be present in excess.

3. This technique can be used to characterize the K_m and V_{max} of a DGC. In this case, a known, constant amount of enzyme is added to varying concentrations of substrate (nonradioactive GTP spiked with a small amount of [α^{32}P]-GTP). A "blank" reaction containing no enzyme should be prepared in parallel at each concentration of substrate. At measured time intervals the reaction is sampled to 0.5 M EDTA to stop enzymatic activity, and the amount of substrate and product at each time point is determined by quantifying the amount of GTP and c-di-GMP using TLC followed by PhosphoImager analysis. This allows the initial velocity to be determined as a function of substrate concentration. The Michaelis–Menten equation can then be solved for K_m and V_{max}.
4. We keep the covered tank in a chemical hood.
5. The plate can be incubated briefly in an incubator or oven to speed up warming (37–55 °C).
6. If the solvent is above the level of the spots, the spotted material will dissolve in the development solvent. If the cellulose makes contact with the filter paper wick, the solvent will move the spots sideways. If the plate is not level, the compounds will not migrate perpendicular to the bottom of the plate. If the front migrates to the top of the plate, the Rf cannot be calculated.
7. The TLC plate can also be exposed to a phosphor storage screen, which will record a latent image produced by ionizing radiation. This image can then be read by laser scanning using a PhosphorImager (e.g., GE Amersham Molecular Dynamics Typhoon) and saved in a digital format. The response of a phosphor storage screen is linear, unlike that of autoradiography film, aiding in quantification of substrate and product "spots" on the TLC plate.

References

1. Jenal U, Malone J (2006) Mechanisms of cyclic-di-GMP signaling in bacteria. Annu Rev Genet 40:385–407
2. Cai L (2014) Thin layer chromatography. In: Current protocols essential laboratory techniques. John Wiley & Sons, Inc, Hoboken, NJ. doi:10.1002/9780470089941.et0603s08
3. Jain R, Behrens AJ, Kaever V, Kazmiercak BI (2012) Type IV pilus assembly in *Pseudomonas aeruginosa* over a broad range of cyclic-di-GMP concentrations. J Bacteriol 194:4285–4294. doi:10.1128/JB.00803-12
4. Kazmierczak BI, Lebron MB, Murray TS (2006) Analysis of FimX, a phosphodiesterase that governs twitching motility in *Pseudomonas aeruginosa*. Mol Microbiol 60(4):1026–1043

Part VII

c-di-GMP Binding Proteins

Chapter 23

Detection of c-di-GMP-Responsive DNA Binding

Jacob R. Chambers and Karin Sauer

Abstract

Modulation of signal transduction via binding of the secondary messenger molecule cyclic di-GMP to effector proteins is a near universal regulatory schema in bacteria. In particular, direct binding of c-di-GMP to transcriptional regulators has been shown to alter gene expression of a variety of processes. Here, we illustrate a pull-down-based DNA:protein binding reaction to determine the relative importance of c-di-GMP in the binding affinity of a target protein to specific DNA sequences. Specifically, the pull-down-based assay enables DNA binding to be analyzed with differing concentrations of c-di-GMP in the absence/presence of specific and nonspecific competitors.

Key words Cyclic di-GMP, Transcriptional regulators, Pull-down assay, Immunoblotting, DNA binding

1 Introduction

Cyclic di-GMP is a secondary messenger molecule that plays a role in a wide array of regulatory processes. In the past few years, the turnover mechanisms for c-di-GMP, whereby synthesis occurs through diguanylate cyclases and hydrolysis through phosphodiesterases, have been well studied and characterized [1, 2]. As attention has turned away from these mechanisms and toward uncovering downstream signal transduction pathways, it has become clear that binding of c-di-GMP to both proteins and RNA-based effectors is capable of modulating transcriptional, posttranscriptional, and posttranslational processes [3]. This is apparent by the recent identification of several transcriptional regulators being able to directly bind c-di-GMP, with c-di-GMP binding coinciding with differential expression of the respective target genes. In *Pseudomonas aeruginosa*, binding of c-di-GMP to the transcriptional regulator FleQ has been shown to reduce expression of genes associated with flagellar motility and relieve FleQ-mediated repression of exopolysaccharide biosynthesis genes [4]. Additionally, in *P. aeruginosa*, the transcriptional regulator BrlR

Karin Sauer (ed.), *c-di-GMP Signaling: Methods and Protocols*, Methods in Molecular Biology, vol. 1657,
DOI 10.1007/978-1-4939-7240-1_23, © Springer Science+Business Media LLC 2017

was found to directly bind c-di-GMP resulting in altered binding affinity for downstream DNA binding targets [5]. The Clp protein in *Xanthomonas campestris* has also been shown to bind c-di-GMP and alter expression of genes linked to *Xanthomonas* virulence and motility [6, 7]. As more c-di-GMP-responsive transcriptional regulators are uncovered, it will be beneficial to have a variety of tools and procedures to verify the role of c-di-GMP in direct binding of the regulators to their DNA targets.

Here we describe a streptavidin-mediated pull-down assays to determine the role c-di-GMP in DNA binding of a protein of interest. As the assay is performed in the presence of increasing concentrations of c-di-GMP, the c-di-GMP concentration at which optimal DNA binding occurs, can be determined. Moreover, the assay can be performed in the presence of competitor DNA and other mononucleotide and dinucleotide monophosphates, thus assisting in verifying the role of c-di-GMP in DNA binding of a particular protein. The streptavidin-mediated pull-down assay has been used by us to verify the importance of c-di-GMP in the binding of DNA by the transcriptional regulator BrlR in *P. aeruginosa* to several promoters of interest [5]. The procedure itself is relatively straightforward, involving a DNA–protein binding reaction in the presence or absence of c-di-GMP, followed by a pull-down assay, SDS-PAGE, and immunoblotting. However, the method requires the use of a specific antibody be available for the protein of interest. In this procedure, we make use of a protein engineered to harbor a V5-tag and the commercially available anti-V5 antibody.

2 Materials

All solutions are to be prepared in*Diguanylate cyclase* ultrapure water (purified deionized water to 18 MΩ-cm) and analytical grade reagents (unless otherwise indicated). All solutions are prepared and stored at room temperature (23 °C) unless otherwise indicated.

2.1 Pull-Down Assays

1. Cyclic-di-GMP (c-di-GMP). Dilute with water to a final concentration of 200 pmol/μl. Store at −80 °C.
2. Streptavidin (SA) magnetic beads (Thermo Scientific, 100 μg). Store at 4 °C.
3. Magnetic stand suitable to hold microfuge tubes.
4. 10× Reaction buffer: 100 mM Tris, 500 mM KCl, 10 mM DTT, pH 7.5. Store at −20 °C.
5. EDTA: 200 mM stock solution, pH 8.0.
6. Protein: Cell-free protein extract harboring protein (*see* **Note 1**).

7. Antibody specific to protein of interest (*see* **Note 1**).
8. Biotinylated promoter DNA: 0.5 pmol of DNA of interest coupled to a biotin tag (*see* **Note 2**).
9. Competitor DNA: Use increasing concentrations (anywhere from 1× to 50×) of nonbiotinylated promoter DNA (*see* **Note 3**).
10. Nonspecific competitors (optional): Other mononucleotide or dinucleotide monophosphates such as cAMP or GTP can also be used in similar concentration ranges as c-di-GMP to verify specificity of binding in the presence of c-di-GMP. Store solutions at −80 °C.
11. Tris-buffered saline (TBS): 0.1 M Tris–HCl, 0.9% w/v NaCl, pH 7.0. Store at 4 °C.
12. TBS containing 0.1% Tween 20 (TTBS). Store at 4 °C.
13. Thermomixer set to room temperature.

2.2 Sodium Dodecyl Sulfate–Polyacrylamide Gel Electrophoresis (SDS-PAGE)

1. 30% acrylamide–bisacrylamide (29:1). Store at 4 °C.
2. Resolving gel buffer: 1.5 M Tris–HCl, pH 8.8. Weigh 90.85 g Tris–HCl and dissolve in approximately 300 ml water in a 500 ml glass bottle. Mix and adjust pH with concentrated (12 M) HCl (*see* **Note 4**). Bring up volume to 500 ml with water and store at 4 °C.
3. Stacking buffer: 0.5 M Tris–HCl, pH 6.8. Weigh 6.06 g Tris–HCl and dissolved in approximately 80 ml water in a 250 ml glass bottle. Mix and adjust pH with concentrated (12 M) HCl (*see* **Note 4**) before bringing up final volume to 100 ml with water. Store at 4 °C.
4. SDS: 10% solution in water (*see* **Note 5**).
5. Ammonium persulfate: 10% solution in water (*see* **Note 6**).
6. N,N,N′,N′-tetramethylethane-1,2-diamine: Store at 4 °C.
7. SDS-PAGE running buffer: 0.025 M Tris–HCl, 0.192 M glycine, 0.1% SDS.
8. 4× SDS-PAGE sample buffer: 0.048 M Tris–HCl, 8% SDS, 40% glycerine, 0.4 ml 2-mercaptoethanol, 0.1% bromophenol blue. Aliquot and store at −20 °C.

2.3 Immunoblotting

1. PVDF Membranes.
2. Stacking Paper: Whatman filter paper cut to size of one gel. 10 pieces per gel.
3. Western blot transfer buffer: 100 ml SDS-PAGE running buffer, 100 ml ethanol, 800 ml water.
4. 95% ethanol.

5. Tris-buffered saline (TBS): 0.1 M Tris–HCl, 0.9% w/v NaCl, pH 7.0. Store at 4 °C.
6. TBS containing 0.1% Tween 20 (TTBS). Store at 4 °C.
7. Blocking solution: 1% BSA in TTBS. Store at 4 °C.
8. Diluent Solution: 1% BSA in TTBS. Store at 4 °C.
9. Antibody for pull-down: Anti-V5-HRP (Invitrogen) (*see* **Note 1**).
10. Clarity™ Western ECL Substrate (Bio-Rad).
11. Developer Solution: Kodak Professional developer. Prepare per manufacturer's instructions.
12. Fixer Solution: Kodak Professional fixer. Prepare per manufacturer's instructions.
13. X-ray film: CL-XPosure™ Film.
14. Three plastic containers.
15. Trans-Blot® Turbo™ Transfer system (Bio-Rad).

3 Methods

Carry out all procedures at room temperature unless otherwise indicated.

3.1 Pull-Down Assays

1. In a 1.5 ml microcentrifuge reaction, set up the binding reaction. Per 20 μl reaction, add increasing concentrations of c-di-GMP (1–50 pmol), 2 μl 10× reaction buffer, 0.2 μl of 200 mM EDTA, biotinylated promoter DNA (0.5 pmol), and cell extract (25 μg) (*see* **Note 7**). Bring up the reaction to a final volume of 20 μl with water.
   (a) The pull-down assay described above can be varied to include competitor DNA. To do so, set up the pull-down binding reaction (20 μl) using the optimal c-di-GMP concentration, 2 μl 10× reaction buffer, 0.2 μl of 200 mM EDTA, cell extract (25 μg), biotinylated promoter DNA (0.5 pmol), and increasing concentration of non-biotinylated competitor DNA (e.g., 0.1–5 pmol; *see* **Note 8**).
   (b) The pull-down assay described above can be varied to determine specificity of c-di-GMP binding, by include mononucleotide and/or dinucleotide monophosphates such as cAMP or GTP. These can also be used as competitors for c-di-GMP. To test for specificity, replace c-di-GMP with mononucleotide and/or dinucleotide monophosphates as follows: Per 20 μl reaction, add increasing concentrations of mononucleotide and/or dinucleotide monophosphates (1–50 pmol), 2 μl 10× reaction buffer, 0.2 μl of 200 mM EDTA, biotinylated promoter DNA (0.5 pmol), and cell extract (25 μg).

2. Incubate the reaction mixture for 30 min at room temperature.
3. Binding reactions containing competitor DNA can be performed on reaction mixtures containing an optimal concentration of c-di-GMP.
4. Incubate the reaction mixture for 30 min at room temperature.
5. While waiting, prepare the SA magnetic beads by vortexing for 30 s.
6. Immediately after vortexing, aliquot 10 μl of beads into a fresh 1.5 ml tube. Prepare 1 microfuge tube per pull-down reaction (*see* **Note 9**).
7. Wash the beads twice: Add 200 μl TTBS, vortex briefly, and collect the beads using a magnetic tube stand (~3–5 min) and remove the supernatant while the tube remains in the magnetic stand. Avoid pipetting up the magnetic beads (*see* **Note 10**).
8. Combine the washed beads from **step 5**, the protein/biotinylated c-di-GMP reaction mixture from **steps 1–2**, and 250 μl TTBS. Mix by pipetting.
9. Collect the beads using the magnetic stand by letting the tubes sit for ~3–5 min. Then, remove the supernatant.
10. Wash the beads by adding 1 ml TTBS, incubating for 2 min at room temperature at 184 x *g* (thermomixer), collecting the beads using the magnetic stand (~3–5 min), and removing the supernatant. Repeat this step three more times.
11. Once the beads have been washed for times, beads can be stored at −20 °C until further use. Otherwise, proceed to sample preparation for SDS-PAGE and immunoblotting.

3.2 Preparing 10% SDS-PAGE Gel

1. In a small beaker, mix 2.5 ml 30% acrylamide–bisacrylamide, 1.88 ml resolving buffer, 75 μl 10% SDS, and 3.5 ml water. Mix well by swirling or pipetting.
2. Add 50 μl 10% APS and 5 μl TEMED, then gently swirl to mix (*see* **Note 11**).
3. Cast the resolving gel within cassette. Allow space for the stacking gel and gently overlay the resolving gel solution with water. Water can be added using a pipette or squirt bottle (*see* **Note 12**).
4. Allow the gel to polymerize for 45 min (*see* **Note 13**).
5. Once the resolving gel has polymerized, remove overlaid water from resolving gel in the casting cassette, and blot-dry (*see* **Note 14**).
6. Prepare the stacking gel by mixing 660 μl 30% acrylamide–bisacrylamide, 1.25 ml stacking buffer, 50 μl 10% SDS, and 3.0 ml water.

7. Add 35 μl 10% APS and 10 μl TEMED, then gently swirl to mix. Cast stacking gel on top of resolving gel and gently insert 10-well comb without introducing air bubbles. Let the gel sit for 60 min until fully polymerized.
8. Before loading the gel, gently remove the 10-well comb, and rinse out wells with water. Rinsing can be done using a pipette or squirt bottle.

3.3 Running SDS-PAGE Gel

1. Resuspend the beads in 15 μl of water.
2. Add 5 μl 4× SDS-PAGE sample buffer. Mix by vortexing.
3. Boil the sample for 10 min at 100 °C (*see* **Note 15**).
4. Remove samples and allow to cool to room temperature. Centrifuge at max speed for 1 min to collect liquid.
5. Collect the magnetic beads using the magnetic stand (~3–5 min).
6. Keeping the samples in the magnetic stand, use a pipette to carefully remove the entire supernatant. Avoid pipetting up the magnetic beads. Transfer the supernatant to a well of the prepared SDS-PAGE gel (Subheading 3.2). Avoid pipetting up the magnetic beads (*see* **Note 16**).
7. Run the SDS-PAGE gels at 100–150 V in SDS-PAGE running buffer for approximately 1.5 h until the dye front has reached the bottom of the gel.

3.4 Immunoblotting

1. Prepare three containers containing 95% ethanol, water, and blotting buffer, respectively.
2. Immediately following SDS-PAGE, turn off the power supply, and remove the gel from the unit.
3. Separate the glass plates holding the gel using a spatula.
4. Remove the stacking gel by separating it from the resolving gel with a plastic spatula and discarding it.
5. Carefully transfer the resolving gel to a plastic container containing Western blot transfer buffer and incubate for a minimum of 2 min at room temperature. Make sure that the SDS-gel is completely submerged.
6. Meanwhile, cut PVDF membrane to approximate size of the resolving gel.
7. Activate the membrane by submerging in 95% ethanol for 5 s, then transferring to water for 10 s, followed by complete submersion in Western blot transfer buffer for at least 1 min.
8. While gel and membrane incubate in Western blot transfer buffer, organize stacking paper into two stacks of approximately five pieces each.

9. Briefly submerge one group of stacking paper in Western blot transfer buffer until wet throughout.
10. Place in transfer apparatus (Trans-Blot® Turbo™ Transfer system).
11. Gently lay PVDF membrane on top of stacking paper.
12. Place the resolving gel on top of the PVDF membrane.
13. Place the second stack of stacking paper, on top of the resolving gel.
14. Using a rolling pin, glass tube or similar device, gently roll out any air bubbles from between the layers. Start rolling from the center to the outside.
15. Place lid on transfer apparatus and run at 1.3 A and 25 V for 7 min. For two gels run together in the same transfer apparatus, use 2.5 A.
16. Prepare a fresh plastic container.
17. Following transfer, remove PFDV membrane and place the membrane in the fresh plastic container.
18. Block the membrane by adding ~10 ml blocking solution until membrane is completely submerged.
19. Incubate the membrane in the blocking solution at room temperature for 2 h at 70 rpm.
20. Prepare 1:10,000 dilution of Anti-V5-HRP antibody in 10 ml diluent solution.
21. Following blocking step, pour off blocking solution.
22. Add antibody-containing diluent solution to the membrane.
23. Incubate at room temperature for 2 h at 70 rpm.
24. Pour off diluent solution.
25. Wash membrane three times by adding ~10 ml TTBS, incubating for 10 min at room temperature at 70 rpm followed by pouring off of the TTBS.
26. Prepare Clarity™ Western ECL Substrate per manufacturer's instructions. Add to membrane and incubate covered from light for 3–5 min at room temperature.
27. Remove membrane from Western substrate, and place in plastic back. Roll out any air bubbles introduced.
28. Place the sealed membrane into a developing cassette.
29. Place X-ray on top of sealed membrane within the developing cassette and exposure X-ray for 1–5 min.
30. Briefly submerge X-ray film in developer solution and wait for bands to appear followed by quick rinse with water and complete submersion in fixer solution for a minimum of 2 min. Allow X-ray film to air-dry.

4 Notes

1. To ensure that the protein of interest is detected in a specific manner by immunoblot analysis, it is preferable to engineer the protein of interest to harbor an antibody-specific tag. Based on our experience, a V5- or HA-tag is preferable, as anti-V5 (Life Technologies) and anti-HA (Covance) antibodies produce little to no background/cross-reactivity. These constructs are then placed into an appropriate expression vector and transformed into *E. coli*. Cells were grown to mid-exponential phase under inducing conditions, lysed via sonication, and subsequent debris removed following centrifugation at 21,000 × *g* for 10 min at 4 °C. Concentration of protein was determined by a Modified Lowry Assay. Concentration of protein of interest was assumed to be ~1% of total protein. Store lysates in aliquots at −20 °C and avoid multiple freeze–thaw cycles.
2. DNA sequence for a five prime biotin label was added to PCR primers used to amplify the promoter region of interest. Optimal length of resulting PCR products was between 100 and 400 nucleotides. The resulting PCR products were then cleaned up with a PCR cleanup kit and quantitated.
3. Competitor DNA should be the same sequence as target DNA, just lacking the biotin label in the primer sequence. PCR, cleanup, and quantitation are performed as above. Additionally, nonspecific competitor DNA sequences can be used to verify specificity.
4. 12 M HCl can be used for initial pH adjustment to narrow large gaps between initial and target pH while lower ionic strength HCl can be used to adjust pH once a smaller gap has been obtained. This will help avoid sudden pH drops below the target pH value.
5. SDS will precipitate out of solution if stored at 4 °C.
6. Prepare a fresh solution each time when casting a SDS-PAGE gel.
7. All components should be thawed on ice prior to use. The concentration of protein lysate (or purified protein) required to obtain efficient DNA binding needs to be determined. This is best achieved by testing increasing amounts of protein. Each protein concentration will require a separate binding reaction to be set up.
8. Competitor reactions allow for determination of specificity in the DNA binding reaction. A range of concentrations of competitor DNA should be utilized and an inverse correlation should be observed between concentration of competitor DNA and efficiency of DNA:protein binding.

9. SA magnetic beads will quickly settle to the bottom of tubes. Make sure to vortex vigorously to ensure proper mixing.
10. When placed in magnetic rack, SA magnetic beads should visually accumulate on the side of the tube closest to magnet. During wash steps, place pipette tip away from the beads to ensure minimal loss of beads while supernatant is removed.
11. SDS-PAGE gels will begin to polymerize once APS and TEMED are added to the solution. Have all pipets and material ready before adding. Once added, quickly cast gels.
12. Spacer plate dimensions (W × L) are 10.1 × 8.2 cm. Short plate dimensions are 10.1 × 7.3 cm. Final gel dimensions are 8.6 × 6.7 cm.
13. Final concentration of acrylamide in resolving gel solution can be adjusted to obtain better separation of proteins depending on the size of the target protein. A 10% SDS-PAGE gel is suitable to resolve proteins having a molecular mass of 50–70 kDa proteins. A lower percentage acrylamide gel (8%) is preferable for larger molecular weight proteins (>70 kDa) while a higher percentage SDS-PAGE gel (12%) may better separate proteins of lower molecular weight proteins (<50 kDa).
14. Gel polymerization can be determined by gently tipping gel cassette to one side. The water should clearly separate from a solid gel mass. Excess water should then be poured off and any excess water removed with a paper towel.
15. Boiling samples for 10 min in 1.5 ml tubes can result in the caps popping open, and subsequent loss of sample volume. Secure the caps with tape or a holder to prevent the cap from opening during boiling.
16. A protein marker added to the SDS-PAGE gel should transfer to the membrane. If the marker contains visible dye, the prestained marker bands should be visible on the membrane following transfer. Detecting prestained marker bands on the membrane is usually a good initial indication of a successful transfer. If the ladder bands are not visible or distorted, an issue may have occurred during the transfer. Consult the manufacturer of the transfer apparatus for aid if needed.

Acknowledgments

This work was supported by a grant from the National Institute of Health (R01AI080710).

References

1. Chan C, Paul R, Samoray D, Amiot NC, Giese B, Jenal U, Schirmer T (2004) Structural basis of activity and allosteric control of diguanylate cyclase. Proc Natl Acad Sci U S A 101:17084–17089
2. Chou S-H, Galperin MY (2016) Diversity of cyclic di-GMP-binding proteins and mechanisms. J Bacteriol 198:32–46
3. Mills E, Pultz IS, Kulasekara HD, Miller SI (2011) The bacterial second messenger c-di-GMP: mechanisms of signalling. Cell Microbiol 13:1122–1129
4. Hickman JW, Harwood CS (2008) Identification of FleQ from *Pseudomonas aeruginosa* as a c-di-GMP-responsive transcription factor. Mol Microbiol 69:376–389
5. Chambers J, Liao J, Schurr MJ, Sauer K (2014) BrlR from *Pseudomonas aeruginosa* is a c-di-GMP-responsive transcription factor. Mol Microbiol 92:471–487
6. Chin KH, Lee YC, Tu ZL, Chen CH, Tseng YH, Yang JM, Ryan RP, McCarthy Y, Dow JM, Wang AH, Chou SH (2010) The cAMP receptor-like protein CLP is a novel c-di-GMP receptor linking cell-cell signaling to virulence gene expression in *Xanthomonas campestris*. J Mol Biol 396:646–662
7. Ryan RP, Fouhy Y, Lucey JF, Crossman LC, Spiro S, He Y-W, Zhang L-H, Heeb S, Camara M, Williams P, Dow JM (2006) Cell-cell signaling in Xanthomonas campestris involves an HD-GYP domain protein that functions in cyclic di-GMP turnover. Proc Natl Acad Sci U S A 103:6712–6717

Chapter 24

Use of Nonradiochemical DNAse Footprinting to Analyze c-di-GMP Modulation of DNA-Binding Proteins

Claudine Baraquet and Caroline S. Harwood

Abstract

The transition of bacteria from a planktonic lifestyle to a collaborative, sessile biofilm lifestyle is a regulated process orchestrated by the intracellular second-messenger c-di-GMP (bis-(3′–5′)-cyclic dimeric guanosine monophosphate). To modulate this transition, c-di-GMP acts at the transcriptional, posttranscriptional, and posttranslational levels. In this chapter, we describe a method to study of how a transcriptional regulator modulates gene expression in response to c-di-GMP binding. DNase I footprinting is a valuable tool for use in analyzing how regulatory proteins bind to DNA, the location of their binding sites or how c-di-GMP affects their binding to DNA. This chapter describes a protocol for nonradiochemical DNase I footprinting experiments using a capillary electrophoresis method based on the interaction of the *Pseudomonas aeruginosa* FleQ protein with the promoter regions of biofilm-related genes.

Key words Nonradiochemical footprint, DNase I, Capillary electrophoresis, c-di-GMP, Transcription regulator, c-di-GMP binding protein, Automated capillary DNA sequencer

1 Introduction

Bis-(3′–5′)-cyclic dimeric guanosine monophosphate (c-di-GMP) is an intracellular secondary messenger that modulates the transition between planktonic and biofilm lifestyles in many gram-negative bacteria [1–3]. High intracellular c-di-GMP is often associated with biofilm formation and exopolysaccharide production, whereas reduced intracellular c-di-GMP is generally correlated with increased motility and virulence. To modulate cell physiology, c-di-GMP has been shown to act at the transcriptional, posttranscriptional, and posttranslational levels [4].

Our focus for this chapter is on regulatory proteins that respond to c-di-GMP to control gene transcription. Typically, c-di-GMP modulates the ability of transcription factors to bind to the promoter region of the genes they regulate. But the way that c-di-GMP modulates the action of transcription factors is highly variable. For example, c-di-GMP can act as a negative regulator, as

Karin Sauer (ed.), *c-di-GMP Signaling: Methods and Protocols*, Methods in Molecular Biology, vol. 1657,
DOI 10.1007/978-1-4939-7240-1_24, © Springer Science+Business Media LLC 2017

shown for the *Xanthomonas campestris* Clp proteins where c-di-GMP impairs its binding to DNA and the induction of genes important for virulence [5]. By contrast, c-di-GMP enhances the binding of MrkH from *Klebsiella pneumoniae*, Bcam1349 from *Burkholderia cenocepacia* and LtmA from *Mycobacterium smegmatis* to the promoter regions of type 3 fimbriae genes, cellulose synthase genes, and lipid transport or metabolism genes, respectively leading to an increase expression of these genes [6–8]. In *Vibrio cholerae*, two c-di-GMP binding transcriptional regulators, VpsR and VpsT, transcribe genes encoding VPS (*Vibrio* polysaccharide) and biofilm matrix*DNA binding* proteins when the concentration of intracellular c-di-GMP is elevated [9–11]. VpsR binds to the regulatory region of its targets in vitro in the presence as well as in the absence of c-di-GMP, whereas the c-di-GMP-dependent interaction between two VpsT monomers is required for DNA recognition and transcriptional regulation. Finally, BrlR, a MerR family member from *Pseudomonas aeruginosa*, was recently shown to bind c-di-GMP to control gene expression [12]. While BrlR–DNA binding was detected in the absence of c-di-GMP, addition of c-di-GMP induces BrlR dimerization leading to an increased binding of BrlR to DNA. We showed that FleQ from *P. aeruginosa*, modulates flagella- and biofilm-related gene expression in response to c-di-GMP. C-di-GMP binding to FleQ does not affect the affinity of FleQ for either flagella or biofilm gene promoters [13–15]. However, a conformational change of FleQ that is induced by c-di-GMP appears to cause DNA distortions at certain biofilm gene promoters [14, 16].

DNase footprinting is a valuable tool for use in studying DNA–protein interactions, identifying DNA binding sites and for examining the role of protein ligands in modulating DNA/protein interactions. Usually this technique involves labeling DNA encompassing the promoter region of a target gene with radioactive phosphate, incubating the labeled DNA with the protein of interest, and then partially hydrolyzing the DNA, typically with DNAse I. The hydrolyzed DNA fragments are separated on polyacrylamide sequencing gels. After autoradiography, fragmentation patterns are manually analyzed by comparing samples with and without the protein of interest. Regions of DNA that are bound by protein are protected from DNAse I cleavage and this is visualized as a "footprint" that appears on the gel and is characterized by a lack of DNA fragments where the protein binds. Footprinting experiments also reveal hypersensitive sites, which are often indicative of flexible DNA regions, able to bend or to loop. This DNA distortion, often induced by protein binding, facilitates DNAse I access to DNA and induces the appearance of bands of increased intensity on gels.

Although traditional footprinting experiments are still widely used, the development of a new method using fluorescent labeled DNA and capillary electrophoresis provides an opportunity for safer and easier footprinting with the big advantage of avoiding handling large slab gels and radioactive compounds [17–20]. This chapter describes a protocol for DNase I footprinting experiments using a nonradiochemical capillary electrophoresis method based on the interaction of the *Pseudomonas aeruginosa* FleQ protein with the promoter regions of biofilm-related genes [13, 14].

2 Materials

Prepare all solutions using nanopure water (18 MΩ-cm ionic purity at 25 °C). For PCR reactions, use ultrapure distilled water.

2.1 Equipment Needed for Generation of Fluorescently Labeled DNA Fragments

1. Micropipettes (2 μL–1 mL).
2. 0.2 mL PCR tubes.
3. NanoDrop spectrophotometer.
4. Thermal cycler polymerase chain reaction (PCR) machine.
5. Apparatus for agarose gel electrophoresis of DNA.

2.2 Reagents Needed for Generating Labeled DNA Fragments

1. 6-FAM labeled primers can be purchased from almost any company able to synthesize custom DNA oligos.
2. Chromosome or plasmid purification kits.
3. Purified genomic DNA or plasmid to be used as template for PCR.
4. PCR is performed with High Fidelity DNA polymerase and standard dNTPs.
5. PCR purification kit.
6. Genetic material may be visualized on agarose gels made using TAE buffer. Standard DNA visualization protocols and agarose gel conditions (1–1.5% agarose) should be employed.

2.3 DNase I Treatment and Phenol–Chloroform Extraction Equipment

1. 1.5 mL microfuge tubes.
2. Vortex mixer.
3. Refrigerated microcentrifuge.
4. Ice.
5. −80 °C freezer.
6. Timer.

2.4 DNase I Treatment and Phenol–Chloroform Extraction Reagents

2.4.1 Products and Stock Solutions Required

1. 1 M Tris: dilute 12.11 g Tris base (MW = 121.14) in 100 mL of water, adjust to pH 7.8 with HCl.
2. 2 M magnesium acetate: dilute 4.3 g magnesium acetate ($(CH_3COO)_2Mg \cdot 4H_2O$, MW = 214.45) in 10 mL of water.
3. 1 M KCl: dilute 7.45 g KCl (MW = 74.56) in 100 mL of water.
4. 0.5 M EDTA: dilute 14.6 g ethylenediaminetetraacetic acid (MW = 292.25) in 100 mL of water.

5. Ethanol (EtOH): 100 and 70%, ice cold. Store at −20 °C.
6. TE buffer: 10 mM Tris–HCl pH 7.8, 1 mM EDTA. Prepare by adding 100 μL of 1 M Tris–HCl pH 7.8 and 20 μL of 0.5 M EDTA to 10 mL of water.
7. 0.1 M $MgCl_2$: dilute 2.03 g $MgCl_2$ ($MgCl_2$. $6H_20$, MW = 203.31) in 100 mL of water.
8. 3 M Sodium acetate:dilute 6.12 g sodium acetate ($CH_3COO^-Na^+ \cdot 3H_2O$, MW = 136.08) in 15 mL of water, adjust pH to 5.2 with HCl.
9. 80% glycerol.
10. UltraPure phenol–chloroform–isoamyl alcohol (Phenol-CIAA) (25:24:1, v/v).
11. DNase RQ1 1000 U (Promega).
12. DNA carrier such as GlycoBlue from Ambion (15 mg/mL).
13. Binding buffer: The binding buffer is composed of 10 mM Tris–HCl, pH 7.8; 50 mM KCl, 8 mM magnesium acetate, 250 ng/μL bovine serum albumin (BSA) and 5% glycerol. Prepare 10 mL of a 5× concentration of the buffer by adding 500 μL of 1 M Tris–HCl pH 7.8, 2.5 mL of 1 M KCl, 200 μL of 2 M magnesium acetate and 3.125 mL of 80% glycerol. Add 12.5 mg of bovine serum albumin (BSA) to the binding buffer and make it up to 10 mL. The 5× binding buffer can be prepared in advance and kept at 4 °C.

2.5 Purification of Protein of Interest

Purify the protein of interest with an appropriate protocol. If the protein has to be tagged, it is usually best to use an N terminus tag because DNA binding domains are usually located at the C terminus of transcription factors. In general, it is better to purify the protein in the absence of any tag as tags might affect protein function. The authors used the IMPACT (Intein Mediated Purification with an Affinity Chitin-binding Tag) system from NEB which is based on the inducible (by thiol reagents) self-cleavage of inteins to separate the target protein from the affinity tag. Each intein tag harbors a chitin binding domain required for the affinity purification of the fusion protein on a chitin resin. Cleavage of the fusion protein is induced by adding thiol reagents such as dithiothreitol, leading to the release of the target protein from the intein tag.

In addition to the purified protein, a small amount of the appropriate protein buffer is required for use as a control for footprinting experiments.

2.6 Equipment and Reagents for Footprint Analysis

1. A capillary electrophoresis DNA analyzer instrument such as the Applied Biosystems 3730xl DNA Analyzer. We use a company (such as Genewiz) or DNA analysis facility that offers DNA fragment analysis services to run our samples.

2. Fragment analysis size standard. Can be purchased or can be included and added to the samples by the DNA fragment-analysis service.
3. ABI Peak Scanner or GeneMapper software to analyze the electropherograms.

3 Methods

3.1 Generation of DNA Fragments Labeled at One End with 6-FAM (6-Carboxyfluorescein Phosphoramidate)

The first step is to generate a fragment of DNA duplex containing the promoter region of interest labeled at one end of one strand with a covalently attached fluorescent dye. To do this, the fragment of interest is amplified by PCR using one 5′ fluorescently labeled primer and one unlabeled primer. The PCR product is then checked for purity, purified, and quantified as follows.

3.1.1 Primer Design

The easiest way to generate the labeled fragment is by ordering the oligonucleotide that is used to generate the amplicon (probe) with the appropriate dye attached. 6-FAM (6-carboxyfluorescein phosphoramidate) is the most commonly used fluorescent dye attachment for oligonucleotides and is compatible with most fluorescence detection equipment. It does not matter which primer of the two primers set is labeled. It is usually a good idea to confirm the data obtained with one labeled primer by using a different primer annealing to the other DNA strand. The labeled DNA fragment must include at least 50 bases flanking the expected protein binding sites. This is due to the loss of short fragments that might occur during analysis. The size of the DNA probe for footprinting experiments and general rules for primers design are discussed in **Notes 1** and **2,** respectively. One particularity of 5′ 6-FAM primers is the necessity to avoid 5′6-FAM dye next to a guanine base, which may result in decreased fluorescence.

3.1.2 Template DNA for PCR

For footprinting experiments, the fragment of DNA generated by PCR needs to be as pure as possible. Two options are available. The first is to use purified chromosome as template for the PCR. This option is suitable only if the PCR amplification gives only one clear product without any contamination. If a contamination occurs, the band corresponding to the expected fragment can be extracted from a gel but this method usually gives low yield. A second option is to first clone the fragment of interest in a plasmid and then use this plasmid as template. This usually avoids contamination.

1. Chromosomal DNA will need to be purified from the target strain. This can be achieved by using a genomic DNA purification kit. For successful DNA purification, follow the protocol for bacterial cell lysate, included with the kit. Usually 1 mL of an overnight culture is sufficient for successful purification, and elution of DNA in 100 μL of water is recommended.

2. Following purification, check for the presence of DNA by running a sample on an agarose gel and measure its concentration using a NanoDrop spectrophotometer. A genomic DNA concentration of 50–100 ng/μL will be sufficient.
3. Optional. First clone the fragment of interest in a plasmid using a classical cloning protocol and then extract the plasmid with a standard plasmid extraction kit.
4. Check for the presence of plasmid DNA by running an agarose gel and measure the DNA concentration using a Nanodrop spectrophotometer. A plasmid DNA concentration of 10–50 ng/μL will be sufficient.

3.1.3 PCR Reaction

1. For general PCR reactions first prepare a PCR reaction mix (200 μL) in a microcentrifuge tube according to the protocol provided by the Taq polymerase manufacturer. A high fidelity polymerase is suitable. The authors used the Phusion® High-Fidelity DNA Polymerase from NEB and set up the PCR reaction mix as follows:
   - 40 μL of Phusion Buffer 5×.
   - 4 μL of 10 mM dNTPs (final concentration 200 μM).
   - 10 μL of 5′FAM primer at 10 μM (final concentration 0.5 μM).
   - 10 μL of unlabeled primer at 10 μM (final concentration 0.5 μM).
   - 2 μL of Phusion DNA polymerase (final concentration 1.0 units/50 μL PCR).
   - Variable concentration of template DNA (usually around 100 ng/50 μL of reaction for chromosome template or 10 ng/50 μL of reaction for plasmid template).
   - Variable amount of water to reach a final volume of 200 μL.
2. Gently mix the reaction and split the 200 μL PCR reaction mixture into four 0.2 mL PCR tubes (50 μL each).
3. Set up the PCR machine with the appropriate program as recommended by the manufacturer. Annealing temperature and extension time may need to be optimized for each reaction. Example PCR reaction conditions for the Phusion DNA polymerase are:
   - Initial denaturation at 98 °C: 30 s.
   - 30 cycles as follows:
     - Denaturation at 98 °C: 15 s.
     - Annealing from 50 to 72 °C: 30 s.
     - Elongation at 72 °C: 30 s per kb.
   - Final extension at 72 °C for 10 min.

3.1.4 Purification and Quantification

Following PCR, mix 10 μL of the reaction mixture with 2 μL of DNA loading dye and visualize PCR products using an agarose gel. Run the PCR product on the gel with a suitable molecular weight marker, check that the product has the correct molecular weight, and is free from nonspecific PCR products.

1. If the PCR product is free of nonspecific PCR products, purify the PCR product using a standard PCR purification kit. In case of PCR products contamination, purify the correct PCR product using a gel extraction kit.
2. Measure DNA concentration using a NanoDrop spectrophotometer.
3. Evaluate the molecular weight (MW in Da) of the double-stranded DNA fragment using a free online software or approximate the size of the fragment using the formula:

   MW of dsDNA (Da) $= x \times 607.4 + 157.9$,
   where x is the size of the fragment in bases (in other words, the number of nucleotides).
4. Calculate the concentration of the fragment in μM (or pmol/μL) using the formula:
   Concentration (pmol/μL or μM) $= x \times 1000/$MW of dsDNA,
   where x is the concentration of the PCR fragment in μg/mL.

3.2 Footprinting Reactions, DNA Extraction and Sample Preparation

The next step is to incubate labeled DNA with the protein of interest to allow the binding of the protein to the target DNA. If the protein binds to a ligand such as c-di-GMP or ATP, as in the case of FleQ, proteins and ligands can be incubated together prior to the addition of DNA. This incubation period is followed by a partial hydrolysis of the DNA fragment by DNAse I. The rationale is that protein binding offers a protection from DNAse I cleavage and/or induces DNA conformational changes that may result in hypersensitive sites at which enzyme access is facilitated leading to enhanced DNAse I cleavage.

3.2.1 Binding Experiments

It is recommended to perform binding experiments on ice, but depending on the interaction between protein and DNA the experiment can be performed at room temperature. The authors suggest using three concentrations of protein as a start: 7, 14, and 28 pmol (*see* **Note 3**).

1. Dilute the DNA probe to a concentration of 0.15 pmol/μL using water.
2. Dilute the protein of interest to a concentration of 7 pmol/μL using the appropriate protein buffer.
3. In a microcentrifuge tube mix, water (standard NanoPure water) with binding buffer and different concentrations of protein (1, 2, or 4 μL leading to 7, 14, and 28 pmol of protein

Table 1
Typical binding experiment setup reaction

Samples	1 no protein	2 Protein 7 pmol	3 Protein 14 pmol	4 Protein 28 pmol
DNA probe (0.15 pmol/μL)	3 μL	3 μL	3 μL	3 μL
Purified protein (7 pmol/μL)	–	1 μL	2 μL	4 μL
Protein buffer (1×)	4 μL	3 μL	2 μL	–
Binding buffer (5×)	10 μL	10 μL	10 μL	10 μL
Water	33 μL	33 μL	33 μL	33 μL
Total	50 μL	50 μL	50 μL	50 μL

per reaction). An example reaction setup is shown in Table 1. If the protein interacts with a ligand (such as c-di-GMP for FleQ) or needs a particular nucleotide to become active, add this compound and incubate the mixture for 10 min (*see* **Note 4**).

4. Do a control experiment with water, binding buffer and protein buffer only.
5. Add the DNA probe (0.45 pmol) to each test tube.
6. Incubate on ice for 30 min and then put the tubes at room temperature for few minutes to warm up.

3.2.2 DNase I Cleavage

Optimization of the DNAse I reaction might be required (*see* **Note 5**).

1. Dilute the DNase RQ1 1000 U (Promega) with standard nanopure water (DNase 3 μL + H_2O 47 μL) and keep the diluted enzyme at room temperature. Diluted DNAse I is discarded at the end of the experiment.
2. Add 2.5 μL of 100 mM $MgCl_2$ to the reaction tubes.
3. Add 5 μL of diluted DNase I to the reaction tubes (0.3 U final).
4. Incubate at room temperature for 2 min. (*see* **Note 6**).
5. Stop the reaction by adding 58 μL of Phenol-CIAA and vortex for 5–10 s.

3.2.3 DNA Extraction Protocol and Sample Preparation

1. Centrifuge the tubes at 16,000 rcf for 15 min and at room temperature. Centrifugation of the mixture yields two phases: the lower organic phase and the upper aqueous phase.
2. Collect 25 μL of the upper aqueous phase containing nucleic acids into a new microcentrifuge tube.
3. Add 3 μL of 3 M sodium acetate.

4. Optional: add DNA carrier (usually 1 μL—GlycoBlue Ambion 15 mg/mL). (*see* **Note 7**).
5. Add 60 μL of 100% cold EtOH.
6. Incubate overnight at −80 °C.
7. Centrifuge the tubes at 16,000 rcf for 15 min and at 4 °C.
8. Remove the supernatant by pipetting. If a DNA carrier is used a colored pellet must be visible (blue with the Glycoblue DNA carrier). The pellet might be stable at the bottom of the tube or totally unstable, floating in the liquid. The pellet may even be dispersed in few pieces. In these two last cases, remove most of the supernatant but do not try to remove it all, the risk of losing the pellet or few pieces of the pellet is too high, which will result in a decreased intensity of fluorescence on the electropherogram.
9. Leave open tubes on ice and allow the ethanol to evaporate for few minutes.
10. Add 200 μL of 70% cold EtOH, do not try to resuspend the pellet.
11. Centrifuge the tubes at 16,000 rcf for 10 min and at 4 °C.
12. Remove the supernatant by pipetting (same comments as for Subheading 3.2.3, **step 8**).
13. Let ethanol evaporate by putting the tubes open at 42 °C or under a fume hood for few minutes, no liquid can remain.
14. Resuspend the pellet with TE or water according to the DNA service facility's recommendation (2 μL for Genewiz, to give an example) by pipeting or vortexing and spin down several times.
15. Samples can be stored at −20 °C for a long period until used.
16. If the samples have to be shipped, wrap the tubes with parafilm to prevent the caps from opening during transit. Shipment at ambient temperature was perfectly successful.

3.3 Capillary Electrophoresis

Samples are ready to be analyzed by capillary electrophoresis, on an Applied Biosystems 3730 DNA Analyzer, for example. Usually, the fragment analysis service adds a size standard (such as LIZ-500 standard) to the samples and runs them together in the same capillary. The size standard enables a standard curve to be generated for each sample and thus allows determination of the relative size of each unknown fragment. It is recommended to use the same size standard and instrument conditions for all samples.

3.4 Electropherogram Analysis

The electropherograms files produced by the 3730 DNA Analyzer (*.fsa) have a proprietary format. GeneMapper or PeakScanner software programs are required to view the files. The authors used the PeakScanner software. PeakScanner does not allow electropherograms to be superimposed, while Genemapper does. PeakScanner allows one to download as many files as needed, and up to 12 digestion patterns can be visualized at once. It is possible to zoom in on a particular region or obtain a sizing table view which details the position and height of each peak (*see* **Note 8**).

Figure 1 gives an example of electropherograms obtained with a 5′FAM labeled fragment encompassing the *pelA* promoter region incubated in the absence of protein (upper digestion panel) and in the presence of 14 pmol of FleQ (lower digestion panel). In general, the first 40 bases of the fragment do not resolve well on the DNA analyzer, and cannot be interpreted for protection sites. Electropherograms obtained in the absence of protein should have peaks with a relatively uniform intensity of fluorescence throughout the sequence. However, a few regions might be more resistant or favorable to cleavage. When the height of a peak decreases in a protein concentration-dependent manner, this would indicate a protected region and thus a region of protein binding (Fig. 1). Inversely, when the height of a peak increases with protein concentration, this would indicate a hypersensitive site (see examples of hypersensitive sites in references [13, 14]).

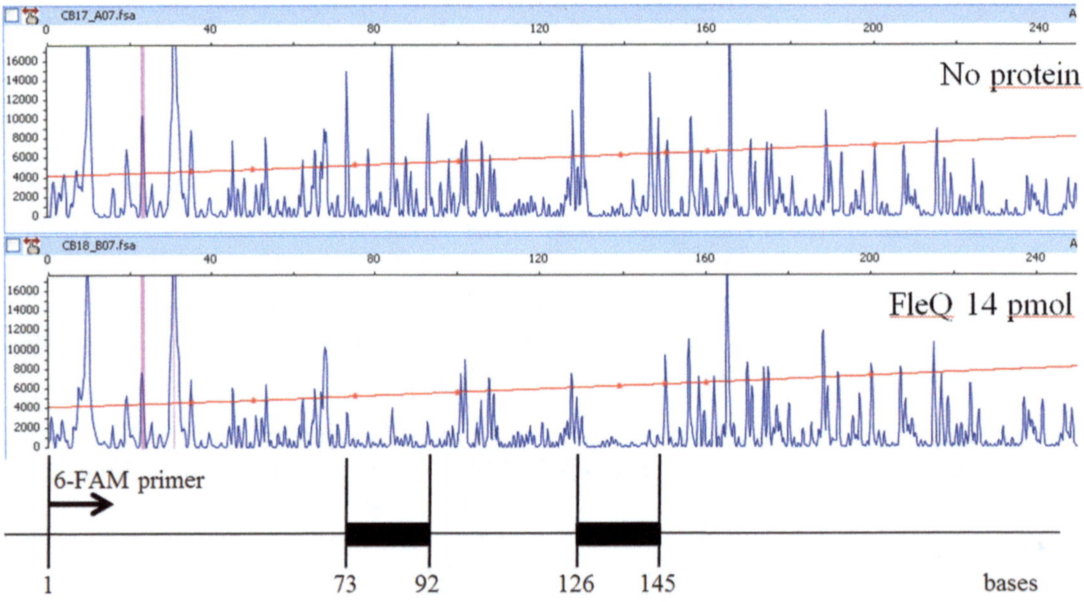

Fig. 1 Example of an electropherogram from a footprinting experiment. This example shows binding of FleQ to *pel* promoter DNA. A 5′FAM-labeled *pel* DNA fragment (0.45 pmol) was incubated without (*upper panel*) or with 14 pmol of FleQ (lower panel), submitted to DNAse I cleavage (0.3 U) and then analyzed by capillary electrophoresis. The fluorescence intensity of the FAM-labeled DNA fragment (in arbitrary units) is plotted against the sequence length of the fragment (in bases), relative to the first base of the 5′FAM labeled primer. The horizontal *red line* indicates the internal standard. Two protected regions are indicated by *rectangles* and the position of the FleQ binding sites are indicated in bases relative to the first base of the 5′FAM-labeled primer

By comparing the position of the region of interest with the position of the labeled primers, the sequence of the protein binding site can be determined (Fig. 1) (*see* **Note 9**).

4 Notes

1. The DNA probe can be from 100 to 600 bases long. Long fragments (>450 bases in length) can be more challenging due to the digestion requirements. However, we performed footprinting experiments with fragments from 438 to 594 bases length without any trouble, using the exact same protocol [13, 14].
2. General rules for primers design apply: primer length is usually between 18 and 24 bases; primer melting (and annealing) temperature is set between 50 and 72 °C; the GC content of primers is between 40 and 60% with a balanced distribution of GC-rich and AT-rich domains; and primer secondary structures or primer homology should be avoided.
3. In most cases, the ability of a protein to bind DNA is first measured using EMSA (electrophoretic mobility shift assay). Then, use the same conditions as used for protein/DNA incubation (temperature, binding buffer, and time of incubation) with EMSA for footprinting experiments.
4. If the protein interacts with a ligand, add this compound to the protein mix and incubate for 10 min before adding the DNA probe. The suggested protein concentration for footprinting experiments ranged from 0.14 to 0.56 μM, then adding 10 μM of ligand is usually largely enough to saturate the protein. In our experiments with FleQ and c-di-GMP, we varied the c-di-GMP concentration from 1 to 100 μM.
5. Digestion with DNAse I might require optimization and an optimal protocol must be empirically determined. The digestion conditions can be adjusted by varying two parameters: DNAse I incubation time or concentration. According to Yindeeyoungyeon and Schell, digestion conditions giving, on a gel, a moderately smeared pattern with about 60% of undigested fragment gives optimum results [18]. The digestion protocol given in this chapter is recommended for DNA fragments from 350 to 600 bases in length. For fragments smaller than 350 bases, the authors recommended decreasing the incubation time with DNase I to 1 min as was done by Hirakawa et al. with a similar protocol [21].
6. To obtain reproducible experiments it is crucial to incubate all samples with DNAse I for the exact same amount of time. It is therefore recommended to measure precisely the 2 min of incubation using a timer. DNAse I is inactivated when Phenol-CIAA [chloroform–isoamyl alcohol (24:1)] is added.

7. The authors strongly recommended the use of a DNA carrier as DNA pellets can be easily lost at the DNA extraction step.
8. A quantitative analysis of the electropherogram is possible without the need of densitometric analysis as it is the case with traditional footprinting experiments using gels and radiolabeled DNA fragments. The PeakScanner software gives, in the sizing table view, data about the height or the area under a peak, which can be valuable for quantitative footprinting experiments. Footprinting experiments performed on a large spectrum of protein concentrations will allow for determining the dissociation constants of each individual protein binding site on DNA with this information.
9. Alternatively, to obtain a perfectly accurate determination of the protein binding site, dideoxynucleotide sequencing reactions using the same fragment and the same primer can be performed [17, 19].

Acknowledgments

We gratefully thank Hidetada Hirakawa, Keiji Murakami, and Gael Panis for useful discussions and technical advice. The National Institute of General Medical Sciences (NIGMS) provided funding to Caroline S. Harwood, under grant number GM56665.

References

1. Romling U, Galperin MY, Gomelsky M (2013) Cyclic di-GMP: the first 25 years of a universal bacterial second messenger. Microbiol Mol Biol Rev 77(1):1–52. doi:10.1128/MMBR.00043-12
2. Ha DG, O'Toole GA (2015) c-di-GMP and its effects on biofilm formation and dispersion: a *Pseudomonas aeruginosa* review. Microbiol Spectr 3(2):MB-0003-2014. doi:10.1128/microbiolspec.MB-0003-2014
3. Teschler JK, Zamorano-Sanchez D, Utada AS, Warner CJ, Wong GC, Linington RG, Yildiz FH (2015) Living in the matrix: assembly and control of *Vibrio cholerae* biofilms. Nat Rev Microbiol 13(5):255–268. doi:10.1038/nrmicro3433
4. Sondermann H, Shikuma NJ, Yildiz FH (2012) You've come a long way: c-di-GMP signaling. Curr Opin Microbiol 15(2):140–146. doi:10.1016/j.mib.2011.12.008
5. Tao F, He YW, Wu DH, Swarup S, Zhang LH (2010) The cyclic nucleotide monophosphate domain of *Xanthomonas campestris* global regulator Clp defines a new class of cyclic di-GMP effectors. J Bacteriol 192(4):1020–1029. doi:10.1128/JB.01253-09
6. Fazli M, O'Connell A, Nilsson M, Niehaus K, Dow JM, Givskov M, Ryan RP, Tolker-Nielsen T (2011) The CRP/FNR family protein Bcam1349 is a c-di-GMP effector that regulates biofilm formation in the respiratory pathogen *Burkholderia cenocepacia*. Mol Microbiol 82(2):327–341. doi:10.1111/j.1365-2958.2011.07814.x
7. Wilksch JJ, Yang J, Clements A, Gabbe JL, Short KR, Cao H, Cavaliere R, James CE, Whitchurch CB, Schembri MA, Chuah ML, Liang ZX, Wijburg OL, Jenney AW, Lithgow T, Strugnell RA (2011) MrkH, a novel c-di-GMP-dependent transcriptional activator, controls *Klebsiella pneumoniae* biofilm formation by regulating type 3 fimbriae expression. PLoS Pathog 7(8):e1002204. doi:10.1371/journal.ppat.1002204
8. Li W, He ZG (2012) LtmA, a novel cyclic di-GMP-responsive activator, broadly regulates the expression of lipid transport and

metabolism genes in *Mycobacterium smegmatis*. Nucleic Acids Res 40(22):11292–11307. doi:10.1093/nar/gks923

9. Krasteva PV, Fong JC, Shikuma NJ, Beyhan S, Navarro MV, Yildiz FH, Sondermann H (2010) *Vibrio cholerae* VpsT regulates matrix production and motility by directly sensing cyclic di-GMP. Science 327(5967):866–868. doi:10.1126/science.1181185
10. Srivastava D, Harris RC, Waters CM (2011) Integration of cyclic di-GMP and quorum sensing in the control of *vpsT* and *aphA* in *Vibrio cholerae*. J Bacteriol 193 (22):6331–6341. doi:10.1128/JB.05167-11
11. Zamorano-Sanchez D, Fong JC, Kilic S, Erill I, Yildiz FH (2015) Identification and characterization of VpsR and VpsT binding sites in *Vibrio cholerae*. J Bacteriol 197(7):1221–1235. doi:10.1128/JB.02439-14
12. Chambers JR, Liao J, Schurr MJ, Sauer K (2014) BrlR from *Pseudomonas aeruginosa* is a c-di-GMP-responsive transcription factor. Mol Microbiol 92(3):471–487. doi:10.1111/mmi.12562
13. Baraquet C, Harwood CS (2015) FleQ DNA binding consensus sequence revealed by studies of FleQ-dependent regulation of biofilm gene expression in *Pseudomonas aeruginosa*. J Bacteriol 198(1):178–186. doi:10.1128/JB.00539-15
14. Baraquet C, Murakami K, Parsek MR, Harwood CS (2012) The FleQ protein from *Pseudomonas aeruginosa* functions as both a repressor and an activator to control gene expression from the pel operon promoter in response to c-di-GMP. Nucleic Acids Res 40 (15):7207–7218. doi:10.1093/nar/gks384
15. Hickman JW, Harwood CS (2008) Identification of FleQ from *Pseudomonas aeruginosa* as a c-di-GMP-responsive transcription factor. Mol Microbiol 69(2):376–389. doi:10.1111/j.1365-2958.2008.06281.x
16. Matsuyama BY, Krasteva PV, Baraquet C, Harwood CS, Sondermann H, Navarro MV (2016) Mechanistic insights into c-di-GMP-dependent control of the biofilm regulator FleQ from *Pseudomonas aeruginosa*. Proc Natl Acad Sci U S A 113(2):E209–E218. doi:10.1073/pnas.1523148113
17. Sivapragasam S, Pande A, Grove A (2015) A recommended workflow for DNase I footprinting using a capillary electrophoresis genetic analyzer. Anal Biochem 481:1–3. doi:10.1016/j.ab.2015.04.013
18. Yindeeyoungyeon W, Schell MA (2000) Footprinting with an automated capillary DNA sequencer. Biotechniques 29(5):1034–1036. 1038, 1040-1031
19. Zianni M, Tessanne K, Merighi M, Laguna R, Tabita FR (2006) Identification of the DNA bases of a DNase I footprint by the use of dye primer sequencing on an automated capillary DNA analysis instrument. J Biomol Tech 17 (2):103–113
20. Wilson DO, Johnson P, McCord BR (2001) Nonradiochemical DNase I footprinting by capillary electrophoresis. Electrophoresis 22 (10):1979–1986
21. Hirakawa H, Oda Y, Phattarasukol S, Armour CD, Castle JC, Raymond CK, Lappala CR, Schaefer AL, Harwood CS, Greenberg EP (2011) Activity of the *Rhodopseudomonas palustris* p-coumaroyl-homoserine lactone-responsive transcription factor RpaR. J Bacteriol 193(10):2598–2607. doi:10.1128/JB.01479-10

Chapter 25

Detection of Cyclic di-GMP Binding Proteins Utilizing a Biotinylated Cyclic di-GMP Pull-Down Assay

Jacob R. Chambers and Karin Sauer

Abstract

Cyclic di-GMP is an important regulatory messenger molecule that often directly interacts with proteins to alter function. It is therefore important to find potential c-di-GMP binding proteins and verify a direct interaction between them. Here, we describe a pull-down assay using biotinylated-c-di-GMP to capture a specific protein of interest followed by immunoblot analysis to determine relative protein abundance. This method also allows for addition of both specific and nonspecific competitors to determine specificity of c-di-GMP–protein binding. We also discuss using densitometry analysis on resulting immunoblots to calculate the dissociation constant (K_D) of the binding reaction, allowing for a determination of binding affinity.

Key words Cyclic di-GMP, Pull-down assay, Western blot, Immunoblot, Biotinylated-c-di-GMP

1 Introduction

The secondary messenger molecule cyclic diguanosine monophosphate (c-di-GMP) regulates a wide variety of processes in response to both environmental and cellular signals. Modulation of c-di-GMP levels has been shown to play a vital role in the switch between the motile and sessile modes of growth, as well as regulating cell motility and antimicrobial tolerance [1–3]. The near universal presence of c-di-GMP in bacteria underscores the importance of understanding its regulatory functions, in particular its downstream binding targets. Numerous approaches have been described to identify proteins capable of binding c-di-GMP, ranging from in silico bioinformatics approaches to affinity-binding pull-down assays [4–6]. To date, many of these approaches have led to the identification of proteins involved with c-di-GMP turnover, including numerous diguanylate cyclases, involved in the synthesis of c-di-GMP, and phosphodiesterases, which play a role in its degradation [7, 8]. However, identification and verification of downstream binding targets has remained challenging and hampered by predictive methods based on known motifs often failing to identify novel

Karin Sauer (ed.), *c-di-GMP Signaling: Methods and Protocols*, Methods in Molecular Biology, vol. 1657,
DOI 10.1007/978-1-4939-7240-1_25, © Springer Science+Business Media LLC 2017

or previously uncharacterized c-di-GMP binding proteins. For instance, c-di-GMP responsive transcriptional regulators include the *P. aeruginosa* flagella biosynthesis gene activator FleQ [9] and the *Mycobacterium smegmatis* LtmA that broadly regulates the expression of lipid transport and metabolism genes [10]. Likewise, the *Vibrio cholerae* MshE protein [11] and the *P. aeruginosa* transcriptional regulator BrlR [12] have been demonstrated to bind c-di-GMP. While capable of c-di-GMP binding, these proteins lack regions that have a predicted secondary structure resembling those of known c-di-GMP binding regions, including the *P. aeruginosa* PelD protein, PilZ domains or the I-sites of diguanylate cyclases [4, 13–15]. It is therefore important to have techniques at hand capable of easily verifying any potential binding targets.

Here, we describe a method using biotinylated c-di-GMP to identify potential target proteins. The method is based on pull-downs using streptavidin-coated magnetic beads. This approach is not designed as a broad screen to identify c-di-GMP binding proteins, as detection of c-di-GMP binding to the target protein is based on immunoblot analysis using antibodies specific to the target protein of interest. Instead, the pull-down approach makes use of an engineered/tagged candidate protein, allowing for verification of c-di-GMP binding to a protein of interest using pull-down assays and immunoblotting. In this way, potential c-di-GMP binding proteins can be easily and quickly verified. In addition, densitometry analysis using the western blots can be used to calculate binding kinetics, allowing for the comparison of binding affinities to know c-di-GMP binding proteins. Using this method, we have confirmed binding by c-di-GMP to the *P. aeruginosa* transcriptional regulator BrlR [12].

2 Materials

All solutions are to be prepared using ultrapure water (18 MΩ-cm) and analytical grade reagents (unless otherwise indicated). All solutions are prepared and stored at room temperature (22 °C) unless otherwise indicated.

2.1 Pull-Down Assays

1. Biotinylated cyclic-di-GMP (c-di-GMP): Dilute with water to a final concentration of 200 pmol/μL. Store at −80 °C.
2. Streptavidin (SA) magnetic beads (Thermo Scientific, 100 μg). Store at 4 °C.
3. Magnetic stand suitable to hold microfuge tubes.
4. 10× Reaction buffer: 100 mM Tris, 500 mM KCl, 10 mM DTT, pH 7.5. Store at −20 °C.
5. EDTA: 200 mM stock solution, pH 8.0.

6. Protein: cell-free protein extract containing protein of interest or purified protein of interest (*see* **Note 1**).
7. Unmarked c-di-GMP (Competitor): Prepare increasing concentrations (anywhere from 1× (200 pmol) to 50× (10,000 pmol)) of nonbiotinylated c-di-GMP in water. Store solution at −80 °C.
8. Nonspecific competitors (optional): In addition to nonbiotinylated c-di-GMP, other mononucleotide or dinucleotide monophosphates such as cAMP or GTP can also be used in similar concentration ranges to verify specificity of binding. Store solution at −80 °C.
9. Tris-buffered saline (TBS): 0.1 M Tris–HCl, 0.9% w/v NaCl, pH 7.0. Store at 4 °C.
10. TBS containing 0.1% Tween 20 (TTBS). Store at 4 °C.
11. Thermomixer set to room temperature.

2.2 Sodium Dodecyl Sulfate-Polyacrylamide Gel Electrophoresis (SDS-PAGE)

1. 30% acrylamide–bisacrylamide (29:1). Store at 4 °C.
2. Resolving gel buffer: 1.5 M Tris–HCl, pH 8.8. Weigh 90.85 g Tris–HCl and dissolve in approximately 300 mL water in a 500 mL glass bottle. Mix and adjust pH with concentrated (12 M) HCl (*see* **Note 2**). Bring up the volume to 500 mL with water and store at 4 °C.
3. Stacking buffer: 0.5 M Tris–HCL, pH 6.8. Weigh 6.06 g Tris–HCl and dissolve in approximately 80 mL water in a 250 mL glass bottle. Mix and adjust pH with concentrated (12 M) HCl (*see* **Note 2**) before bringing up final volume to 100 mL with water. Store at 4 °C.
4. SDS: 10% solution in water (*see* **Note 3**).
5. Ammonium persulfate: 10% solution in water (*see* **Note 4**).
6. N,N,N′,N′-tetramethylethane-1,2-diamine: Store at 4 °C.
7. SDS-PAGE running buffer: 0.025 M Tris–HCl, 0.192 M glycine, 0.1% SDS.
8. 4× SDS-PAGE sample buffer: 0.048 M Tris–HCl, 8% SDS, 40% glycerine, 0.4 mL 2-Mercaptoethanol, 0.1% bromophenol blue. Aliquot and store at −20 °C.

2.3 Immunoblotting

1. PVDF membranes.
2. Stacking Paper: Whatman filter paper cut to size of one gel. Ten pieces per gel.
3. Western blot transfer buffer: 100 mL SDS-PAGE running buffer, 100 mL ethanol, 800 mL water.
4. 95% ethanol.

5. Tris-buffered saline (TBS): 0.1 M Tris–HCl, 0.9% w/v NaCl, pH 7.0. Store at 4 °C.
6. TBS containing 0.1% Tween 20 (TTBS). Store at 4 °C.
7. Blocking solution: 1% BSA in TTBS. Store at 4 °C.
8. Diluent Solution: 1% BSA in TTBS. Store at 4 °C.
9. Antibody for pull-down: Anti-V5-HRP (Invitrogen).
10. Clarity™ Western ECL Substrate (Bio-Rad).
11. Developer Solution: Kodak Professional developer. Prepare per manufacturer's instructions.
12. Fixer Solution: Kodak Professional fixer. Prepare per manufacturer's instructions.
13. X-ray film: CL-XPosure™ Film.
14. Three plastic containers.
15. Trans-Blot® Turbo™ Transfer system (Bio-Rad).

2.4 Immunoblot Analysis

1. Scanner.
2. ImageJ or other image analysis software for densitometry determinations.

3 Methods

Carry out all procedures at room temperature unless otherwise indicated.

3.1 Pull-Down Assays

The assay is based on protein–ligand interactions, with immobilized biotinylated c-di-GMP being used to capture a protein of interest. The resulting bound protein is then detected by immunoblot analysis. In general, the protein concentration is increased while the concentration of c-di-GMP is being kept constant. A strategy for the setup for this type of experiment and expected outcome is shown in Fig. 1a. Alternatively, the assay can be carried out at a fixed protein concentration but increasing c-di-GMP concentrations (Fig. 1b).

1. In a 1.5 mL microcentrifuge reaction, set up the protein/biotinylated c-di-GMP reaction. Per 20 μL reaction, add 1 μL biotinylated c-di-GMP (200 pmol), 2 μL 10× reaction buffer, 0.2 μL of 200 mM EDTA, and increasing concentrations of cell extract (1–10 μg). Bring up the reaction to a final volume of 20 μL with water.
2. For pull-down assays in the presence of competitors, *see* Subheading 3.2. To determine the dissociation constant K_D, refer to pull-down assays indicated in Subheading 3.3. Incubate the reaction mixture for 30 min at room temperature.

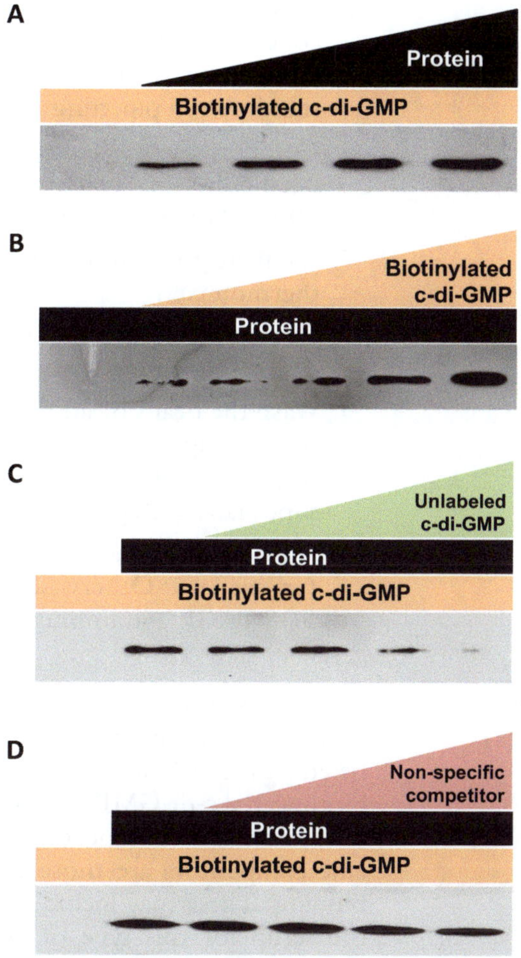

Fig. 1 Schematic overview of pull-down assays. (**a**) Pull-down using biotinylated c-di-GMP (400 pmol) immobilized to streptavidin magnetic beads demonstrating that BrlR binds c-di-GMP. Increasing concentrations of V5-tagged BrlR were used (0, 1.25, 2.5, 5, and 10 μg). (**b**) Pull-down using increasing concentrations of c-di-GMP (0, 5, 10, 25, 50 pmol) and a constant concentration of BrlR (5 μg). (**c**) Pull-downs using a constant concentration of biotinylated-c-di-GMP (400 pmol) and a constant concentration of BrlR (5 μg). An increasing amount of unlabeled competitor c-di-GMP was added (0, 400, 800, 2000, and 4000 pmol). (**d**) Pull-downs using a constant concentration of biotinylated-c-di-GMP (400 pmol) and a constant concentration of BrlR (5 μg). An increasing amount of unlabeled nonspecific competitor (GTP) was added (0, 400, 800, 2000, and 4000 pmol). All figures adapted from [12]

3. While waiting, prepare the SA magnetic beads by vortexing for 30 s.
4. Immediately after vortexing, aliquot 10 μL of beads into a fresh 1.5 mL microcentrifuge tube. Prepare 1 microfuge tube per pull-down reaction (*see* **Note 5**).

5. Wash the beads twice: Add 200 μL TTBS, vortex briefly, and collect the beads using a magnetic tube stand (~3–5 min) and remove the supernatant while the tube remains in the magnetic stand. Avoid pipetting up the magnetic beads (*see* **Note 6**).
6. Combine the washed beads from **step 5**, the protein/biotinylated c-di-GMP reaction mixture from **steps 1–2**, and 250 μL TTBS. Mix by pipetting.
7. Incubate for 1 h at room temperature at 1400 rpm using a thermomixer.
8. Collect the beads using the magnetic stand by letting the tubes sit for ~3–5 min. Then, remove the supernatant.
9. Wash the beads by adding 1 mL TTBS, incubating for 2 min at room temperature at 1400 rpm (thermomixer), collecting the beads using the magnetic stand (~3–5 min), and removing the supernatant. Repeat this step three more times.
10. After the four washes, the beads can be stored at −20 °C until further use. Otherwise, proceed to sample preparation for SDS-PAGE and immunoblotting.

3.2 Pull-Down Assays Using Competitors

Two kinds of competitors can be used, specific competitors and nonspecific competitors. For c-di-GMP binding proteins, unlabeled c-di-GMP can be used as a specific competitor to outcompete biotinylated c-di-GMP. A general strategy for the setup for this type of experiment and expected outcome is shown in Fig. 1c. Nonspecific competitors are those that do not compete with c-di-GMP binding. These may include cAMP or GTP. A general strategy for the setup for this type of experiment and expected outcome is shown in Fig. 1d.

1. Set up the protein/biotinylated c-di-GMP reaction as described in Subheading 3.1 using a single protein concentration but increasing concentrations of competitor.
2. Per 20 μL reaction, add 1 μL biotinylated c-di-GMP (200 pmol), protein extract (e.g., 5 μg), 2 μL 10× reaction buffer, 0.2 μL of 200 mM EDTA, and increasing concentrations of competitor (e.g., 0.4 - 10× relative to biotinylated c-di-GMP, 50–2000 pmol). Bring up the reaction to a final volume of 20 μL with water (*see* **Note 7**).
3. Continue as outlined in Subheading 3.1, **steps 2–10**.

3.3 Pull-Down Assays for K_D Determination

The dissociation constant K_D is commonly used to describe the affinity between a ligand (such as a c-di-GMP) and a protein, and more specifically, how tightly a ligand binds to a particular protein. Experimentally, assays aimed at determining K_D for c-di-GMP binding generally make use of increasing amounts ligand (c-di-GMP) that is titrated against a fixed amount of protein to determine the concentration at which an equilibrium in protein–ligand binding is reached. To do so, set up the protein/biotinylated c-di-GMP reaction as described in Subheading 3.1 using a single protein concentration but increasing concentrations of biotinylated c-di-GMP.

1. A general strategy for the setup of this experiment is shown in Fig. 1b.
2. Per 20 μL reaction, add protein extract (e.g., 5 μg), 2 μL 10× reaction buffer, 0.2 μL of 200 mM EDTA, and increasing concentrations of biotinylated c-di-GMP (e.g., 50–1000 pmol). Bring up the reaction to a final volume of 20 μL with water (*see* **Note 7**).
3. Continue as outlined in Subheading 3.1, **steps 2–10**.

3.4 Preparing a 10% SDS-PAGE Gel

1. In a small beaker, mix 2.5 mL 30% acrylamide–bisacrylamide, 1.88 mL resolving buffer, 75 μL 10% SDS, and 3.5 mL water. Mix well by swirling or pipetting.
2. Add 50 μL 10% APS and 5 μL TEMED, then gently swirl to mix (*see* **Note 8**).
3. Cast the resolving gel within cassette. Allow space for the stacking gel and gently overlay the resolving gel solution with water. Water can be added using a pipette or squirt bottle (*see* **Note 9**).
4. Allow the gel to polymerize for 45 min (*see* **Note 10**).
5. Once the resolving gel has polymerized, remove overlaid water from resolving gel in the casting cassette, and blot dry (*see* **Note 11**).
6. Prepare the stacking gel by mixing 660 μL 30% acrylamide–bisacrylamide, 1.25 mL stacking buffer, 50 μL 10% SDS, and 3.0 mL water.
7. Add 35 μL 10% APS and 10 μL TEMED, then gently swirl to mix. Cast stacking gel on top of resolving gel and gently insert 10-well comb without introducing air bubbles. Let the gel sit for 60 min until fully polymerized.
8. Before loading the gel, gently remove the 10-well comb, and rinse out wells with water. Rinsing can be done using a pipette or squirt bottle.

3.5 Running SDS-PAGE Gel

1. Resuspend the beads in 15 μL of water.
2. Add 5 μL 4× SDS-PAGE sample buffer. Mix by vortexing.
3. Boil the sample for 10 min at 100 °C (*see* **Note 12**).
4. Remove samples and allow to cool to room temperature. Centrifuge at max speed for 1 min to collect liquid.

5. Collect the magnetic beads using the magnetic stand (~3–5 min).
6. Keeping the samples in the magnetic stand, use a pipette to carefully remove the entire supernatant. Avoid pipetting up the magnetic beads. Transfer the supernatant to a well of the prepared SDS-PAGE gel (Subheading 3.2). Avoid pipetting up the magnetic *Western blot analysis* beads.
7. Run the SDS-PAGE gels at 100–150 V in SDS-PAGE running buffer for approximately 1.5 h until the dye front has reached the bottom of the gel.

3.6 Immunoblotting

1. Prepare three containers containing 95% ethanol, water, and blotting buffer, respectively.
2. Immediately following SDS-PAGE, turn off the power supply, and remove the gel from the unit.
3. Separate the glass plates holding the gel using a spatula.
4. Remove the stacking gel by separating it from the resolving gel with a plastic spatula and discarding it.
5. Carefully transfer the resolving gel to a plastic container containing Western blot transfer buffer and incubate for a minimum of 2 min at room temperature. Make sure that the SDS-gel is completely submerged.
6. Meanwhile, cut PVDF membrane to approximate size of the resolving gel.
7. Activate the membrane by submerging in 95% ethanol for 5 s, then transferring to water for 10 s, followed by complete submersion in Western blot transfer buffer for at least 1 min.
8. While gel and membrane incubate in Western blot transfer buffer, organize stacking paper into two stacks of approximately five pieces each.
9. Briefly submerge one group of stacking paper in Western blot transfer buffer until wet throughout.
10. Place in transfer apparatus (Trans-Blot® Turbo™ Transfer system).
11. Gently lay PVDF membrane on top of stacking paper.
12. Place the resolving gel on top of the PVDF membrane.
13. Place the second stack of stacking paper, on top of the resolving gel.
14. Use a rolling pin, glass tube or similar device to gently roll out any air bubbles from between the layers. Start rolling from the center to the outside.

15. Place lid on transfer apparatus and run at 1.3 A and 25 V for 7 min. For two gels run together in the same transfer apparatus, use 2.5 A.
16. Prepare a fresh plastic container.
17. Following transfer, remove PFDV membrane and place the membrane in the fresh plastic container (*see* **Note 13**).
18. Block the membrane by adding ~10 mL blocking solution until membrane is completely submerged.
19. Incubate the membrane in the blocking solution at room temperature for 2 h at 70 rpm.
20. Prepare 1:10,000 dilution of Anti-V5-HRP antibody in 10 mL diluent solution.
21. Following blocking step, pour off blocking solution.
22. Add antibody-containing diluent solution to the membrane.
23. Incubate at room temperature for 2 h at 70 rpm.
24. Pour off diluent solution.
25. Wash membrane three times by adding ~10 mL TTBS, incubating for 10 min at room temperature at 70 rpm followed by pouring off of the TTBS.
26. Prepare Clarity™ Western ECL Substrate per manufacturer's instructions. Add to membrane and incubate covered from light for 3–5 min at room temperature.
27. Remove membrane from Western substrate, and place in plastic back. Roll out any air bubbles introduced.
28. Place the sealed membrane into a developing cassette.
29. Place X-ray on top of sealed membrane within the developing cassette and exposure X-ray for 1–5 min.
30. Briefly submerge X-ray film in developer solution and wait for bands to appear followed by quick rinse with water and complete submersion in fixer solution for a minimum of 2 min. Allow X-ray film to air dry.

3.7 Densitometry of Protein Bands Present on Immunoblots

1. Scan the X-ray film. Save the image, preferably as a TIFF.
2. Use ImageJ software (or similar analysis tools) to measure the intensity of individual protein bands. Please note that the band intensity is a measure of the fraction of protein bound to c-di-GMP. An example of a scanned blot is given in Fig. 2a (*see* **Note 14**).
3. In addition to determining the intensity of bands, measure the intensity of the background of your scan. This value will be used to normalize your intensity measurements.
4. Subtract background intensity from band intensities.

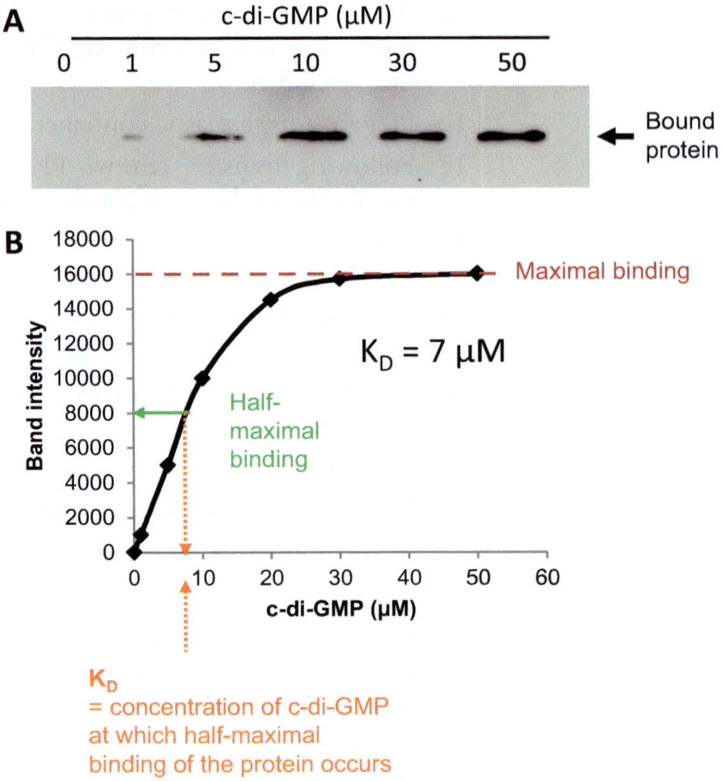

Fig. 2 Overview of K_D determination. (**a**) Image of immunoblot showing increased presence of protein of interest following pull-down assays in the presence of increasing concentrations of c-di-GMP. (**b**) Graph showing the intensity of bound protein relative to c-di-GMP concentrations used in pull-down assays. Band intensities were obtained by analyzing the immunoblot shown in (**a**) by ImageJ. Maximal and half-maximal binding of protein to c-di-GMP is indicated by *arrows*. The K_D can be determined at the c-di-GMP concentration at which half-maximal binding occurs

3.8 Calculating the Dissociation Constant K_D

The formula for determining K_D is as follows: K_D = ([ligand]* [protein])/[protein–ligand complex], with units for [ligand] and [protein] being concentrations (M, mM, mM, etc.). The units for K_D are concentrations (M, mM, mM, etc.). To resolve the equation, the concentrations of ligand, protein, and protein–ligand complex have to be known. However, considering that half-maximal binding of protein occurs when the free ligand concentration is equal to K_D, it is possible to estimate the K_D by inspection of the binding curve. The instructions on how to determine half-maximal binding are given below.

1. Determine the intensity of individual bands obtained from procedure Subheading 3.3.

2. Determine the concentration of c-di-GMP present in the pull-down assay. Convert pmol c-di-GMP used per pull-down assay (20 μL reaction volume) to nmol/L (nM). For example, if you used a total of 200 pmol per 20 μL reaction, first determine how many pmol are present in 1 L. To do so, use the following formula: (200 pmol*50) = 1000 pmol/L.
3. Convert pmol/L to nM. 1000 pmol/L = 1 nM (*see* **Note 15**).
4. Graph data. There are several ways to plot equilibrium data of this kind. The easiest way is to plot the c-di-GMP concentration versus band intensity (obtained in Subheading 3.7, band intensity is a measure of bound protein) using excel. This will generate a simple binding curve in which the fraction of bound protein is plotted vs. the concentration of c-di-GMP. If the reaction is a bimolecular equilibrium reaction, the binding curve will be similar to the hyperbolic binding curve shown in Fig. 2b.
5. From the graph, determine the band intensity at which an equilibrium in protein–ligand binding (maximal binding) has been reached (Fig. 2b, red dashed line).
6. From the graph, determine the band intensity at which half-maximal binding occurred (Fig. 2b, green dashed line). For instance, if maximal binding was noted at a band intensity of 16,000, half-maximal binding will be at a band intensity of 8000 (Fig. 2b, green arrow).
7. Next, determine the concentration of c-di-GMP at which half-maximal binding of the protein is apparent (Fig. 2b, orange arrow). In the given example, the concentration is 7 μM. Thus, the K_D is 7 μM.

4 Notes

1. Cells were grown to mid-exponential phase under inducing conditions, lysed via sonication, and subsequent debris removed following centrifugation at 21,000 × *g* for 10 min at 4 °C. Concentration of protein was determined by a Modified Lowry Assay. Concentration of protein of interest was assumed to be ~1% of total protein. Store lysates in aliquots at −20 °C and avoid multiple freeze/thaw cycles.
2. 12 M HCl can be used for initial pH adjustment to narrow large gaps between initial and target pH, while lower ionic strength HCl can be used to adjust pH once a smaller gap has been obtained. This will help avoid sudden pH drops below the target pH value.
3. SDS will precipitate out of solution if stored at 4 °C.
4. Make this solution fresh each time SDS-PAGE gels are cast.

5. SA magnetic beads will quickly settle to the bottom of tubes. Make sure to vortex vigorously to ensure proper mixing.
6. When placed in magnetic tube holder, SA magnetic beads should visually accumulate on the side of the tube closest to magnet. During wash steps, place pipette tip away from the beads to ensure minimal loss of beads while supernatant is removed.
7. This reaction can also be set up using a fixed protein concentration and varying concentration of biotinylated c-di-GMP.
8. SDS-PAGE gels will begin to polymerize once APS and TEMED are added to the solution. Have all pipets and material ready before adding. Once added, quickly cast gels.
9. Spacer plate dimensions (W × L) are 10.1 × 8.2 cm. Short plate dimensions are 10.1 × 7.3 cm. Final gel dimensions are 8.6 × 6.7 cm.
10. Final concentration of acrylamide in resolving gel solution can be adjusted to obtain better separation of proteins depending on size of target protein. A lower percentage acrylamide gel (8%) can be used for larger molecular weight proteins while a higher percentage gel (12%) may better separate out lower molecular weight proteins.
11. Gel polymerization can be determined by gently tipping gel cassette to one side. The water should clearly separate from a solid gel mass. Excess water should then be poured off and any excess water removed with a paper towel.
12. Boiling samples for 10 min in 1.5 mL tubes can result in the caps popping open, and subsequent loss of sample volume. Secure the caps with tape or a holder to prevent the cap from opening during boiling.
13. A protein marker added to the SDS-PAGE gel should transfer to the membrane. If the marker contains visible dye, this should be visible on the membrane following transfer and can be a good initial indication of a successful transfer. If the ladder bands are not visible or distorted, an issue may have occurred during the transfer. Consult the manufacturer of transfer apparatus for aid if needed.
14. ImageJ is a free software and can be easily installed. Several websites are available containing step-by-step instructions on how to use ImageJ for densitometry. Two useful links are:
    (a) http://lukemiller.org/index.php/2010/11/analyzing-gels-and-western-blots-with-image-j/
    (b) http://www.yorku.ca/yisheng/Internal/Protocols/ImageJ.pdf
15. There are several programs available online to calculate the conversion from pmol/L to nmol/L.

Acknowledgments

This work was supported by a grant from the National Institute of Health (R01AI080710).

References

1. D'Argenio DA, Miller SI (2004) Cyclic di-GMP as a bacterial second messenger. Microbiology 150:2497–2502
2. Mills E, Pultz IS, Kulasekara HD, Miller SI (2011) The bacterial second messenger c-di-GMP: mechanisms of signalling. Cell Microbiol 13:1122–1129
3. Boyd CD, O'Toole GA (2012) Second messenger regulation of Biofilm formation: breakthroughs in understanding c-di-GMP effector systems. Annu Rev Cell Dev Biol 28:439–462
4. Amikam D, Galperin MY (2006) PilZ domain is part of the bacterial c-di-GMP binding protein. Bioinformatics 22:3–6
5. Nesper J, Reinders A, Glatter T, Schmidt A, Jenal U (2012) A novel capture compound for the identification and analysis of cyclic di-GMP binding proteins. J Proteome 75:4874–4878
6. An SQ, Caly DL, McCarthy Y, Murdoch SL, Ward J, Febrer M, Dow JM, Ryan RP (2014) Novel cyclic di-GMP effectors of the YajQ protein family control bacterial virulence. PLoS Pathog 10:e1004429
7. Chou S-H, Galperin MY (2016) Diversity of cyclic di-GMP-binding proteins and mechanisms. J Bacteriol 198:32–46
8. Krasteva PV, Giglio KM, Sondermann H (2012) Sensing the messenger: the diverse was that bacteria signal through c-di-GMP. Protein Sci 21:929–948
9. Hickman JW, Harwood CS (2008) Identification of FleQ from *Pseudomonas aeruginosa* as a c-di-GMP-responsive transcription factor. Mol Microbiol 69:376–389
10. Li W, He Z-G (2012) LtmA, a novel cyclic di-GMP-responsive activator, broadly regulates the expression of lipid transport and metabolism genes in *Mycobacterium smegmatis*. Nucleic Acids Res 40:11292–11307
11. Roelofs KG, Jones CL, Helman SR, Shang X, Orr MW, Goodson JR, Galperin MY, Yildiz FH, Lee VT (2015) Systematic identification of cyclic-di-GMP binding proteins in *Vibrio cholerae* reveals a novel class of cyclic-di-GMP-binding ATPases associated with type II secretion systems. PLoS Pathog 11:e1005232
12. Chambers JR, Liao J, Schurr MJ, Sauer K (2014) BrlR from *Pseudomonas aeruginosa* is a c-di-GMP-responsive transcription factor. Mol Microbiol 92:471–487
13. Chan C, Paul R, Samoray D, Amiot NC, Giese B, Jenal U, Schirmer T (2004) Structural basis of activity and allosteric control of diguanylate cyclase. Proc Natl Acad Sci U S A 101:17084–17089
14. Wassmann P, Chan C, Paul R, Beck A, Heerklotz H, Jenal U, Schirmer T (2007) Structure of BeF3-modified response regulator PleD: implications for diguanylate cyclase activation, catalysis, and feedback inhibition. Structure 15:915–927
15. De N, Pirruccello M, Krasteva PV, Bae N, Raghavan RV, Sondermann H (2008) Phosphorylation-independent regulation of the diguanylate cyclase WspR. PLoS Biol 6:e67

Chapter 26

Probing Protein–Protein Interactions with Genetically Encoded Photoactivatable Cross-Linkers

Richard B. Cooley and Holger Sondermann

Abstract

Fundamental to all living organisms is the ability of proteins to interact with other biological molecules at the right time and location, with the proper affinity, and to do so reversibly. One well-established technique to study protein interactions is chemical cross-linking, a process in which proteins in close spatial proximity are covalently tethered together. An emerging technology that overcomes many limitations of traditional cross-linking methods is one in which photoactivatable cross-linking noncanonical amino acids are genetically encoded into a protein of interest using the cell's native translational machinery. These proteins can then be used to trap interacting biomolecules upon UV illumination. Here, we describe a method for the site-specific incorporation of photoactivatable cross-linking amino acids into fluorescently tagged proteins of interest in *E. coli*. Photo-cross-linking and analysis by SDS-PAGE using in-gel fluorescence detection, which provides rapid, highly sensitive, and specific detection of cross-linked adducts even in impure systems, are also described. An example expression and cross-linking experiment involving transmembrane signaling of a bacterial second messenger receptor system that controls biofilm formation is shown. All reagents needed to carry out these experiments are commercially available, and do not require special or unique technology to perform, making this method tractable to a broad community studying protein structure and function.

Key words Photo-cross-linking, UV cross-linking, *Para*-azidophenylalanine, *Para*-benzoylphenylalanine, Genetic code expansion, Noncanonical amino acid

1 Introduction

Complex networks of protein–protein interactions are the foundation of a cell's ability to send, receive, and respond to signals. Numerous strategies exist to delineate the language of these signaling networks. Cross-linking—the process of forming covalent bonds between two biological molecules, such as proteins and DNA—is a time-honored tool that can yield atomic resolution information about the interactions of biomolecules without the need for specialized equipment or technology [1]. Depending on the particular method, cross-linking*crosslinkerphoto-activatable crosslinkers* can be used to probe both specific and nonspecific

Karin Sauer (ed.), *c-di-GMP Signaling: Methods and Protocols*, Methods in Molecular Biology, vol. 1657,
DOI 10.1007/978-1-4939-7240-1_26, © Springer Science+Business Media LLC 2017

interactions. Cross-linking is commonly performed by introducing small synthetic molecules that contain two functional groups separated by a linker of variable length and that are able to covalently connect amine groups (such as the side chain of lysines), sulfhydryl groups (cysteines), and carboxyl groups (aspartate and glutamate) [1]. Cross-linkers in which the functional cross-linking group is activated with light are also attractive, as it gives the researcher an additional level of temporal control over the cross-linking reaction [2–5].

Here, we describe an increasingly popular cross-linking technique in which a UV photoactivatable cross-linker is introduced site-specifically into a protein in vivo via genetic code expansion [6, 7]. To do this, the protein of interest is coexpressed with an aminoacyl-tRNA syntetase that has been engineered to recognize a cross-linking noncanonical amino acid (ncAA), such as *para*-azidophenylalanine (pAzF) [8–10] or *para*-benzoylphenylalanine (Bpa) [11] (Fig. 1a). This synthetase then charges its cognate tRNA, which itself has been engineered to recognize the amber stop codon (TAG) so that when an amber stop codon is introduced at the desired site, the ncAA is incorporated at that site of the peptide chain during translation. Full-length protein, generated only if the amber stop codon is successfully supressed, can be purified using a C-terminally expressed affinity tag for in vitro cross-linking reactions. Alternatively, cross-linking reactions can be carried out in whole cells or cell lysates.

The methods below outline a general strategy for the incorporation for pAzF or Bpa into recombinantly expressed fluorescent proteins in *E. coli* (*see* **Note 1**), for subsequent cross-linking reactions, and for analysis by SDS-PAGE and detection by in-gel fluorescence. Though this chapter largely describes the use of pAzF as a cross-linker, Bpa is also widely used, and the choice of which to incorporate depends on the system being studied (*see* **Note 2**). As shown in Fig. 1a, pAzF is activated by UV light irreversibly, and will only cross-link if it is in the immediate proximity of a binding partner. Thus pAzF cross-linking efficiency, though variable (depending on the site of cross-linkage) and not stoichiometric, can be used as an analytical tool to reflect the amount of protein--protein binding at a particular point in time [12]. UV activation of Bpa, on the other hand, is reversible and so it can be continuously irradiated until a binding parter engages and thus Bpa-mediated cross-links in principle will continue to accumulate over time with increased UV flux. While both photo-cross-linkers can be used to assess inter-protein interactions, important to note is that intra-protein interactions and dynamics can be evaluated as well with photo-cross-linking (e.g., to analyze conformational states of multidomain proteins) (Fig. 2).

Here, we describe a general protocol for the incorporation of pAzF and Bpa into proteins using the popular T7-based expression

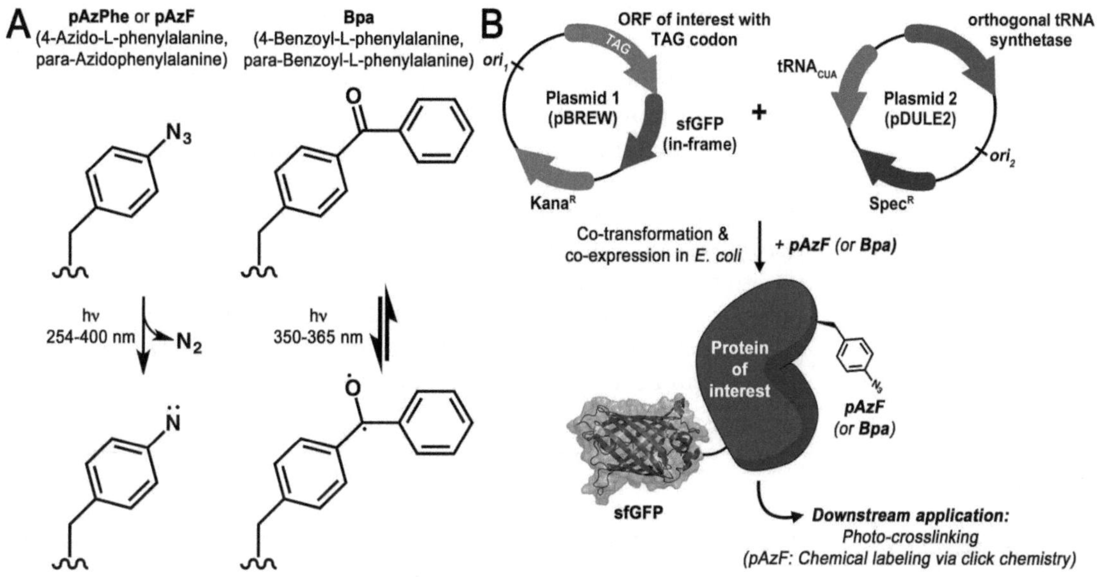

Fig. 1 UV photo-cross-linking using genetic code expansion. (**a**) Structures and mechanism of photoactivation of pAzF and Bpa noncanonical amino acids. (**b**) Expression system and plasmids for the incorporation of noncanonical amino acids into a protein of interest. This sytem requires two plasmids: the expression plasmid (pBrew) in which the protein of interest containing the TAG amber stop codon is C-terminally fused to superfolder GFP, and the machinery plasmid (pDule2) that expresses the orthogonal tRNA synthetase that recognizes and charges its cognate amber-suppressing tRNA (also expressed on this plasmid) with the noncanonical amino acid. Only full-length protein containing the noncanonical amino acid is fluorescent

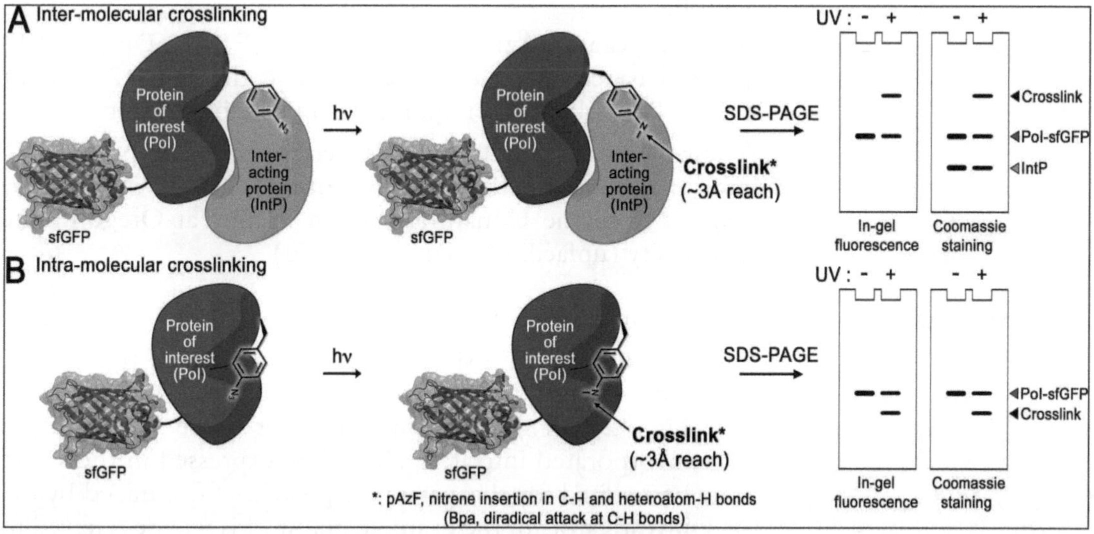

Fig. 2 Applications and analysis of photo-cross-linking: inter-peptide vs. intra-peptide adducts. (**a**) Upon UV illumination, the cross-linker amino acid can form a covalent bond with an interacting protein when placed at a protein–protein interface (*top*), or to an amino acid in the same polypeptide chain distant in primary sequence but close in 3-dimensional space ($< \sim 4$ Å) (*bottom*). (**b**) SDS-PAGE analysis can be used to differentiate these two cross-linking phenomena. Inter-protein interactions will typically migrate slower as the cross-linked species is of larger moleculear weight (*top*), while intra-protein cross-linked species commonly migrate faster due to the formation of a more compact, circular polypeptide (*bottom*)

system in BL21-type cells, as well as a protocol for the ensuing cross-linking reactions. The reader should be aware that many viable variations on the methods described here exist [13], and these protocols will almost always require optimization for the particular system being investigated. In addition, we describe the application of superfolder green fluorescence protein (sfGFP) for the specific and sensitive detection of cross-linked products by in gel-fluorescence imaging. The examples shown here demonstrate its utility for analyzing binding of the bacterial calcium-dependent periplasmic LapG protease (*see* **Note 3**) to its substrate LapA, an outer membrane anchored adhesin protein, a well-studied system involved in bacterial biofilm formation [12, 14–16].

2 Materials

2.1 The Machinery Plasmid

The appropriate genetic machinery plasmid(s) must first be obtained for ncAA incorporation.

1. The "pDule2" machinery plasmid conferring spectinomycin resistance expresses an engineered tRNA synthetase and its cognate amber-suppressing tRNA under constitutive promoters (the "pDule" plasmid is functionally identical but confers tetracycline resistance).
2. To incorporate a particular ncAA of interest, the correct synthetase must be expressed from this pDule2 (or pDule) plasmid.
3. To incorporate pAzF, the pDule2-pCNF (or pDule-pCNF) plasmid is required (*see* **Note 4**) while the pDule2-Bpa (or pDule-Bpa) plasmid is required for Bpa incorporation.
4. These plasmids (as well as others recommended for performing proper control experiments [*see* **Note 5**]) can be obtained by request from the Unnatural Protein Facility at Oregon State University (upfacility@oregonstate.edu).

2.2 The Expression Plasmids

1. At a minimum, two expression plasmids are required:
   (a) The first plasmid should express the protein that will be cross-linked, that is, the "Interacting protein" (IntP, Fig. 2). This protein will not have any ncAA's incorporated into it. It should be expressed in the sytem best suited for this particular protein as optimized by the investigator. In the example below, this will be referred to as the pBAD-LapA plasmid, which confers ampicillin resistance.
   (b) The second plasmid should express the protein in which pAzF (or Bpa) is to be incorporated site-specifically (Fig. 1).

2. The choice of plasmid for this expression system is important and it is essential that it not have a p15a origin of replication (use of such a vector will result in plasmid incompatbility issues as the pDule/pDule2 plasmids also have a p15a origin). Traditional pET (e.g., pET28a) and pBAD plasmids containing the pBR322 origin are suitable expression vectors for this purpose.
3. Note that if the protein is to be purified, it is adventageous that the expression construct contains a C-terminal affinity tag so that only full-length protein is isolated. Here, we describe methods for using the pET28a-based "pBrew" plasmid in which the protein of interest is expressed with a C-terminally fused monomeric superfolder green fluorescence protein (sfGFP) followed by a hexahistidine affinity tag (Fig. 1b) [12]. The sfGFP-His_6 fusion protein is linked to the protein of interest by a TEV protease recognition sequence for convenient removal, if desired. We have observed several cases in which protein expression, as well as cross-linking analysis, was significantly facilitated by the use of the pBrew plasmid (*see* **Note 6**).
4. An amber stop codon (TAG) should be incorporated in the gene at the desired position of ncAA incorporation via standard mutagenesis protocols, while this codon shall not be used as a terminal stop codon to avoid secondary ncAA incorporation at the C-terminus of the protein. The example plasmid here will be referred to here as pBrew-LapG^{TAG}, which has a TAG site located at position Y^{108} (*see* **Note 7**).

2.3 DNA Transformation into BL21 T7-Based Expression *E. coli* Cells

1. *Escherichia coli* BL21ai One Shot electrocompetent cells (*see* **Note 8**).
2. Electroporator.
3. 0.1 cm electroporation cuvettes.
4. SOC medium: 0.5% (w/v) yeast extract, 2% (w/v) tryptone, 10 mM NaCl, 2.5 mM KCl, 20 mM $MgSO_4$, 20 mM glucose (*see* **Note 9**). Prepare 10 mL. Autoclave.
5. LB agar plates: 1% (w/v) yeast extract, 0.5% (w/v) tryptone, 86 mM NaCl, 1.5% (w/v) Bacto agar with appropriate antibiotics. For the examples described here, one LB/agar plate should have 50 μg/mL kanamycin +100 μg/mL spectinomycin for the pBrew-LapG^{TAG}/pDule2-pCNF double transformation, and the other plate should have 100 μg/mL ampicillin for the pBAD-LapA single transformation (*see* **Note 10**).

2.4 Protein Expression

1. LB medium: 1% (w/v) yeast extract, 0.5% (w/v) tryptone, 86 mM NaCl. Prepare 50 mL. Autoclave.
2. Component solutions for making autoinduction media (adapted from [17]).

(a) ZY solution: 1% (w/v) yeast extract, 0.5% (w/v) tryptone. Prepare 500 mL. Autoclave.

(b) 1 M $MgSO_4$: Prepare 50 mL. Autoclave.

(c) 25× M-salts: 0.625 M Na_2HPO_4, 0.625 M KH_2PO_4, 1 M NH_4Cl, 125 mM $(NH_4)_2SO_4$ (do not adjust pH). Prepare 500 mL. Autoclave.

(d) 5052 solution: 12.5 g α-D-glucose, 50 g α-lactose, 125 mL glycerol. Autoclave. Prepare 500 mL (*see* **Note 11**).

(e) 20% (w/v) L-(+)-arabinose. Prepare 50 mL. Sterile filter.

(f) 5000× trace metal solution (*see* **Note 12**): 20 mM $CaCl_2$, 10 mM $MnCl_2$, 10 mM $ZnSO_4$, 2 mM $CoCl_2$, 2 mM $CuCl_2$, 2 mM $NiCl_2$, 2 mM $NaMoO_4$, 2 mM Na_2SeO_3, 2 mM H_3BO_3, 5 mM $FeCl_3$. Prepare 100 mL. Sterile filter.

3. 1000× antibiotic stock solutions. Dissolve with sterile water. Freeze at −20 °C until use. For the examples described here, prepare 1 mL of each of the following:

   (a) 50 mg/mL kanamycin.

   (b) 100 mg/mL spectinomycin.

   (c) 100 mg/mL ampicillin.

4. 100 mM *para*-azido-phenylalanine (pAzF) or *para*-benzoyl-phenylalanine (Bpa) stock in water (*see* **Note 13**).
5. 15 mL culture tubes, sterile.
6. 500 mL baffled culture flasks, sterile.

2.5 Protein Purification

1. 2 mL Ni-NTA resin (Qiagen).
2. PD-10 disposable desalting columns, bed dimensions 14.5 × 50 mm, bed volume 8.3 mL of sephadex G-25 resin (GE Healthcare, product number 17-0851-01).
3. Disposable columns that hold up to 2 mL of resin and 10 mL of sample.
4. Sonicator with 0.5 in. wide probe, or French Press cell disruptor.
5. UV-Visible spectrophotometer.
6. Quartz cuvettes.
7. Wash buffer: 25 mM HEPES, 500 mM NaCl, 20 mM imidazole pH 7.5. Prepare 1 L.
8. Elution buffer: 25 mM HEPES, 500 mM NaCl, 300 mM imidazole pH 7.5. Prepare 50 mL.
9. Reaction buffer: 25 mM HEPES, 250 mM NaCl pH 7.5. Prepare 500 mL.

2.6 Cross-Linking and SDS-PAGE Analysis

1. Clear 96-well microplate.
2. Handheld UV-lamp with short and long wave UV-light illumination capability.
3. Power supply for SDS-PAGE.
4. SDS-PAGE gel running setup (e.g., Bio-Rad Mini-PROTEAN Tetra-cell).
5. 0.25 M Tris base, pH to 6.8 with 6 M HCl.
6. 5× SDS-sample buffer (50 mL): 0.25% (w/v) bromophenol blue, 0.25 M Tris base, pH 6.8, 10% (w/v) SDS, 50% (v/v) glycerol.
7. 5× SDS-sample reducing buffer To make the 5× SDS sample buffer a sample buffer reducing, add 100 μL of β-mercaptoethanol (*see* **Note 14**) to 400 μL of 5× SDS sample buffer.

   10× SDS-PAGE running buffer: 0.25 M Tris base, 1.92 M glycine, 1% (w/v) SDS. Do not adjust pH. Dilute to 1× concentration before use
8. Precast acrylamide SDS-PAGE gel (*see* **Note 15**).
9. In-gel fluorescence imager (e.g., Bio-Rad GelDoc system) with blue light LED excitation and emission filter able to detect ~510–530 nm fluorescence. If not available, or if not using fluorescently tagged proteins like those expressed in the pBrew plasmid, proteins may be visualized via alternative methods including Coomassie Blue stain, silver stain, SYPRO ruby protein stain, or by Western blot.

3 Methods

The methods below discuss the incorporation of pAzF into a protein of interest and subsequent cross-linking to a binding partner protein. The incorporation of Bpa follows the identical method except where noted.

3.1 Plasmid Transformation

We recommend performing a fresh transformation before each expression rather than growing from strains frozen in glycerol stocks (*see* **Note 16**).

1. Transfer 50 μL of freshly thawed electrocompetent BL21ai cells to a prechilled 0.1 cm electroporation cuvette. Add ~1–10 ng of the pBrew-LapG^{TAG} plasmid and ~1–10 ng of the pDule2-pCNF plasmid (*see* **Note 17**).
2. Pulse the cells at 1.8 kV (200 Ω, 25 μF). If the electrocuvette generates an arc, redo the transformation with fresh cells using 1/10th the amount of DNA.

3. Immediately resuspend cells with 1 mL of room temperature SOC media and transfer suspension to a 1.5 mL eppendorf tube.
4. Perform a second, separate transformation with only the pBAD-LapA plasmid by repeating above **steps 1–3**.
5. Incubate cells at 37 °C for 1 h with shaking at 200–250 rpm.
6. Plate ~200 μL of recovered cells onto LB/agar plates with the appropriate antibiotics (50 μg/mL kanamycin + 100 μg/mL spectinomycin for the pBrew-LapG^{TAG}/pDule2-pCNF double transformation, 100 μg/mL ampicillin for the pBAD-LapA single transformation). Discard remaining cells.
7. Incubate plates at 37 °C overnight (14–20 h).

3.2 Protein Expression

Outlined here is a general strategy for expressing proteins with autoinduction media at 37 °C as a reasonable starting point since these are the conditions where ncAA incorporation is most efficient. The ideal expression conditions of the reader's protein may require optimization or alterations.

1. Scrape several colonies of the pBrew-LapG^{TAG}/pDule2-pCNF strain and use them to inoculate a 3 mL LB culture containing 50 μg/mL kanamycin and 100 μg/mL spectinomycin.
2. Repeat for pBAD-LapA except media should contain 100 μg/mL ampicillin.
3. Grow these two starter cultures at 37 °C for 3–4 h with shaking at 200–250 rpm or until visibly turbid (OD_{600} > ~1).
4. While cells are growing, prepare 200 mL of fresh ZY5052 autoinduction media by adding the following stock solutions together (*see* **Note 18**):
   (a) 187 mL ZY media, 0.4 mL 1 M $MgSO_4$, 8 mL 25× M-salts, 40 μL 5000× trace metals, 4 mL 5052 solution, 0.5 mL 20% arabinose (*see* **Note 19**).
5. Divide autoinduction medium into two 500 mL culture flasks each containing 100 mL of medium.
6. Add appropriate antibiotics to each flask. For pBrew-LapG^{TAG}/pDule2-pCNF, use 50 μg/mL kanamycin and 100 μg/mL spectinomycin. For pBAD-LapA use 100 μg/mL ampicillin.
7. Once starter cultures are visually turbid (prepared in Subheading 3.2, **step 2**), add 1 mL of starter culture to each flask containing autoinduction media prepared in Subheading 3.2, **step 6**.
8. Grow at 37 °C for 1–2 h until cells are visible (OD ~0.1–0.3) to ensure cells are growing properly before addition of ncAA.

9. Once cell growth is confirmed (~1–2 h after inoculation), to the pBrew-LapG^{TAG}/pDule2-pCNF culture, add 1 mL of 100 mM pAzF stock for a final concentration of 1 mM in the medium (*see* **Note 20**).
10. Continue shaking at 37 °C for 16–24 h (*see* **Note 21**).

3.3 Preparation of Cell-Free Soluble Lysates

1. Centrifuge cells at 4000 × *g* for 15 min at 4 °C.
2. Pour off supernatant and gently resuspend pelleted cells with 30 mL of wash buffer (*see* **Note 22**). At this point, cells may be frozen by rapid immersion in liquid nitrogen and stored at −80 °C.
3. Lyse each set of cells via sonication for 2 × 30 s intervals at 30–80% power. Appropriate power settings and length of sonication will vary with sonicator manufacturer, probe size, sample size and density of cell suspension (*see* **Note 23**).
4. Centrifuge lysed cells at 20,000 × *g* for 45 min at 4 °C to pellet membranes and insoluble debris.
5. After centrifugation, collect supernatant of each sample. The supernatant contains the soluble fraction of cell lysate with the expressed protein.

3.4 Protein Purification

These steps should all be performed at 4 °C. When working with-proteins containing pAzF, light exposure should be minimized (*see* **Note 24**).

1. Equilibrate ~2 mL bed volume of Ni-NTA resin with 10 column volumes (i.e., 20 mL) of wash buffer. This can be done by removing 4 mL of a 50% slurry suspesion and pipetting into column. Allow the storage solution (usually 20% ethanol) to flow through completely. Discard this flow through. Then gently add 10 mL of wash buffer without disturbing resin bed, and allow all buffer to flow through.
2. Resuspend resin with 2 mL of wash buffer by gentle pipetting. The final volume of this resin solution will be 4 mL.
3. Add 2 mL of this 50% slurry to each freshly prepared cell lysate.
4. Gently rotate (or stir) cell-free lysate with resin for 30 min at 4 °C.
5. Collect resin by pouring lysate–resin mixture through column (*see* **Note 25**).
6. Once lysate has flowed through, wash resin with 20 column volumes (40 mL) of wash buffer.
7. Elute protein from resin by adding 2.5 mL of elution buffer.
8. Collect all 2.5 mL of eluted protein in a single 15-mL conical tube.

9. Equilibrate two PD-10 desalting columns each with 30 mL of reaction buffer.
10. Add the protein (2.5 mL, obtained in Subheading 3.4, **step** 7) that is currently in elution buffer to the PD-10 column. Discard eluate.
11. Add 3.5 mL of reaction buffer, and collect eluate. This eluate contains the purified protein, now in reaction buffer.
12. Quantitate the amount of purified protein.
    (a) For the sfGFP fusion proteins, measure the absorbance at 488 nm and calculate protein concentration using Beer's law (sfGFP extinction coefficient at 488 nm is 83,300/M/cm).
    (b) For nonfluorescent proteins, this can be performed by measuring the absorbance at 280 nm and using the primary sequence to calculate molar extinction coefficients (for LapG, use 11,000/M/cm) (*see* **Note 26**).

3.5 Protein Photo-Cross-Linking

The concentrations, additives, and conditions in which the cross-linking reactions occur will vary and will require optimizing for each individual application.

1. In 25 μL total volume, dilute into reaction buffer the sfGFP fusion protein of interest containing the pAzF (or Bpa) to 2 μM along with 10–100 μM interacting partner (obtained upon completion of the purification outlined in Subheading 3.4). In the example described here, we diluted LapG Y^{108}pAzF-sfGFP (expressed from the pBrew-LapG^{TAG}/pDule2-pCNF expression system) to 2 μM with 50 μM LapA (expressed from the pBAD-LapA system) in the presence of calcium or the calcium chelator EGTA (*see* **Note 3**), allowing us to ascertain the effect calcium has on the binding of LapG to its substrate, LapA (Fig. 3).
2. Allow protein interactions to equilibrate for 20 min at room temperature.
3. Take 12 μL out of each reaction, and place into a well of a clear 96-well microplate. Save the remainder of each reaction on ice.
4. Once all samples are in the 96-well plate, place a dual wavelength handheld UV lamp directly on top of the plate so that the lamp is as close to the samples as possible. Ensure that during the course of the cross-linking process, UV light is evenly distributed across all samples. Wear protective eyewear and/or cover samples with aluminum foil to avoid UV damage to eyes.
    (a) For pAzF cross-linking, illuminate samples with short wave (~254 nm) UV light for 5 min or less (*see* **Note 27**).

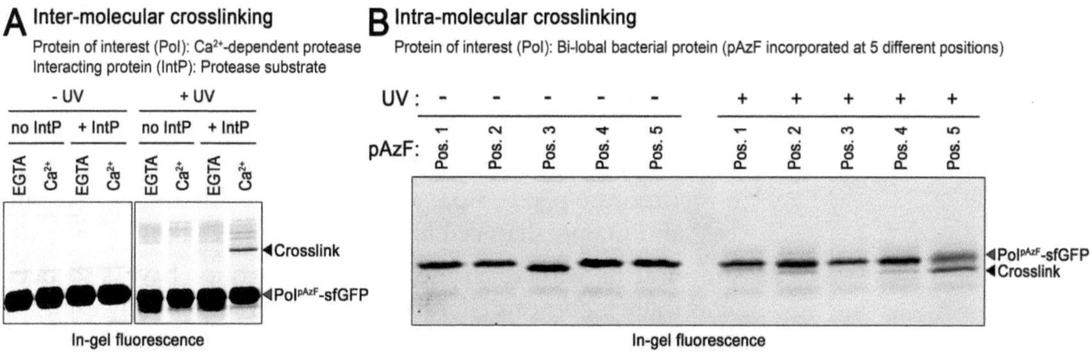

Fig. 3 SDS-PAGE analysis of LapG photo-cross-linking. (**a**) Inter-peptide cross-linking: a fluorescently labeled, covalently cross-linked adduct with slower electrophoretic mobility is observed only when the LapG protease-sfGFP containing pAzF is illuminated with UV light in the presence of its substrate LapA (i.e., "IntP") and calcium. This observation demonstrates calcium is required for protease–substrate interaction. (**b**) Intra-peptide cross-linking: pAzF was incorporated at 5 different positions into a protein containing two structural lobes. Only when AzF was located as "position 5" (Pos. 5) was a new intra-peptide cross-linked species with faster electrophoretic mobility observed upon UV light illumination, indicating this residue (and not the others) is at the interface of two folded sub-domains

   (b) For Bpa cross-linking, use longer wavelength light (~360 nm), and shine light for 10 min–2 h (time may need to be optimized).

5. Once cross-linking reaction is complete, remove 8 μL from each UV-exposed well to a fresh microcentrifuge tube.
6. Add 2 μL of 5× reducing SDS sample buffer and mix well.
7. In a similar manner, remove 8 μL of each non-UV-exposed sample (saved in **step 3**, Subheading 3.5) and mix with 2 μL of 5× reducing SDS sample buffer. This will act as a negative control in which no cross-linking should occur.
8. To maintain sfGFP fluorescence, do not boil these samples (*see* **Note 28**).
9. Run all 10 μL on a 12% SDS-PAGE according to standard protocols.
10. Image gels by fluorescence using a Bio-Rad GelDoc system using blue light emission and a 525/10 (or equivalent) filter in order to visualize fluorescent bands from the sfGFP fusion proteins (Fig. 3). When two proteins are inter-molecularly cross-linked, one should observe a new band in the +UV sample with an apparent molecular weight corresponding to roughly the sum of both protein proteins (Fig. 3a). On the other hand, intra-molecularly cross-linked peptides will commonly migrate with faster electrophoretic mobility than their non-cross-linked counterparts due to increased compactness (Fig. 3b).

4 Notes

1. The synthetase/tRNA system described here, which was derived from the archaeon *Methanocaldococcus janasheii*, works only in *E. coli*. However, systems have been developed to incorporate pAzF, Bpa, and other cross-linking ncAAs into yeast and mammalian cell lines.
2. Amino acids with diazirine functional groups have also been genetically incorporated into proteins and successfully used as a photo-cross-linker. To our knowledge, these reagents are not commercially available and therefore their use is less accessible to the broader community than pAzF or Bpa.
3. In its native host, the LapG protease is targeted to the periplasm, however here we recombinantly express the LapG gene without its leader periplasmic signal sequence so that LapG is expressed in the cytoplasm. Calcium is required for LapG proteolytic activity, and when this protein is incubated with calcium chelators (such as EGTA) the LapG protease is unable to proteolyze substrates [18].
4. Though tRNA synthetases have been engineered to specifically recognize pAzF, reports have serendipitously shown a different synthetase originally selected for para-cyanophenylalanine (pCNF) actually works better for pAzF incorporation. This is why the pDule-pCNF plasmid should be used for pAzF incorporation.
5. Also worthwhile obtaining are control plasmids that express wild-type superfolder GFP (e.g., pET28a-sfGFP) and sfGFP that has a TAG site incorporated (e.g., pET28-sfGFP 150TAG). These allow the researcher to ensure that the ncAA incorporation technology is working in their hands. Perform these controls as outlined for the system described here by substituting the pBrew-LapG^{TAG} plasmid for pET28-sfGFP150TAG plasmid. In these controls, expressions with for example pET28a-sfGFP150TAG/pDule2-pCNF in the presence of pAzF in the media should produce full-length protein and the cells should be notably fluorescent to the eye. In the absence of ncAA, the cells should not be colored.
6. Ideal expression construct for ncAA incorporation will vary depending on the protein being expressed.
7. Expressing the wild-type protein (i.e., protein with no ncAA incorporated) is also an important negative control to include, but omitted here in the methods for brevity. This can be particularly useful to assess the affect of UV light on the protein of interest as some proteins (those with redox cofactors, for example) may have unexpected responses to UV light exposure.

8. Chemically competent cells may also be used for double plasmid transformations provided they are of sufficient competency (>10^8 cfu/μg pUC19 DNA). Also, other BL21 T7-based expressing strains may be used as long as they are compatible with autoinduction media (e.g., BL21(DE3)).
9. Autoclave 2 M glucose and 2 M $MgSO_4$ stocks separately from other components and add to proper concentration once all solutions have cooled to room temperature.
10. After the LB–agar mix is prepared and poured into an autoclave compatible bottle, a magnetic stir bar should be added to the bottle before autoclaving. Once sterilization is complete, the LB/agar solution will be a clear liquid. Do not add antibiotics to the hot medium. Gently stir hot medium until cooled to <65 °C (comfortable to touch). Antibiotics can be added to the cooled medium at this point and the medium poured onto sterile plates to solidify.
11. To get the lactose to dissolve, heat gently (microwave works well). Once dissolved, the lactose will stay in solution indefinitely.
12. When using the described autoinduction media containing rich ZY media, trace metals may not be necessary but they can facilitate cellular growth to higher densities and expression of metal-containing proteins. When making the 5000× trace metal solution, it may be advantageous to first make individual concentrated stocks of each metal, and then mix these stocks together with the appropriate dilution to the desired 5000× concentration.
13. Make fresh immediately before use. To make, weigh appropriate amount of amino acid (~200 mg/L of culture for pAzF). Add sterile water to make a 100 mM solution. At this point the amino acid will not dissolve. To dissolve amino acid, add NaOH to 100–200 mM final concentration from an 8 M stock. The deprotonated amino acid will go readily into solution.
14. Mercaptoethanol must be handled with care and in a fume hood at all times.
15. Alternatively, acrylamide gels for SDS-PAGE may be hand-cast rather than purchasing commercially precast gels.
16. BL21 cells containing expression plasmids often do not express protein as well or at all after being frozen even though they grow normally and are resistant to the correct antibiotic(s). Reasons for this are unclear, and some strains may be more tolerant than others depending on the expression plasmid and protein of interest, however the only way to guarantee reliable expression from BL21 cells is fresh transformation for each expression.

17. If Bpa is to be incorporated, use the pDule2-Bpa machinery plasmid.
18. For proteins that do not express very well, or if more protein is required, the volume of culture can be scaled accordingly. Ensure autoinduction cultures are well aerated. Insufficient aeration will significantly compromise expression.
19. When using BL21ai cells, arabinose is required even when using non-arabinose inducible protein expression vectors because the T7 polymerase was inserted into the genome under the control of an arabinose promoter. When using other BL21 cells lines (e.g., BL21(DE3)) and expressing from a non-arabinose inducible vector (e.g., pET28a), arabinose may be omitted from the media.
20. If Bpa is to be incorporated, add 1 mL of 100 mM Bpa stock instead of pAzF.
21. For proteins that require lower temperatures for expression, grow cultures at 37 °C until OD600 ~ 1–1.5, at which point expression will begin and the temperature should be reduced for the duration of the expression. Lower temperature expressions may require incubation times longer than 24 h.
22. Because pAzF can readily react with primary amines upon activation with UV light, Tris-based buffers should be avoided.
23. Care should be taken to ensure cells are immersed in an ice-water bath during sonication to prevent overheating. Keep sonication probe immersed in sample to avoid excessive frothing of the cell lysate.
24. If using an FPLC to purify the protein containing pAzF, turn off the UV detector as this will inactivate the cross-linker.
25. Alternatively, the resin–lysate suspension can be centrifuged at 500 × *g* for 5 min to sediment the resin. Aspirate the supernatant, with care taken not to evacuate the resin beads. The resin can be resuspended in slurry with buffer and applied to column.
26. If no or low (<0.1 mg/mL) protein concentrations are observed, expression may not be optimal. Try different expression constructs and conditions, or a different site of ncAA incorporation.
27. Placing the 96-well plate on a bed of ice may help prevent samples from becoming too warm during exposures to UV light.
28. When the sample is not boiled, the sfGFP remains folded in SDS-PAGE. As a result, these protein fusions commonly have faster electrophoretic mobility than expected, however it can also result in unusual and unpredictable electrophoretic mobility. If problematic, gel visualization techniques other than in-

gel fluorescence may be used, including Western blot, Coomassie Blue staining, silver staining, or Sypro Ruby staining. We prefer in-gel fluorescence when applicable due to its high sensitivity and specificity even in non-purified systems.

Acknowledgments

This work was supported by the NIH under grants R01-AI097307 (H.S.) and F32-GM108440 (R.B.C.)

References

1. Melcher K, Chen HT (2007) Identification and analysis of multiprotein complexes through chemical crosslinking. Curr Protoc Cell Biol Chapter 17:Unit 17 10
2. Lowder MA, Appelbaum JS, Hobert EM, Schepartz A (2011) Visualizing protein partnerships in living cells and organisms. Curr Opin Chem Biol 15(6):781–788
3. Tanaka Y, Bond MR, Kohler JJ (2008) Photocrosslinkers illuminate interactions in living cells. Mol BioSyst 4(6):473–480
4. Mohibullah N, Hahn S (2008) Site-specific cross-linking of TBP in vivo and in vitro reveals a direct functional interaction with the SAGA subunit Spt3. Genes Dev 22(21):2994–3006
5. Rannversson H, Andersen J, Sorensen L, Bang-Andersen B, Park M, Huber T, Sakmar TP, Stromgaard K (2016) Genetically encoded photocrosslinkers locate the high-affinity binding site of antidepressant drugs in the human serotonin transporter. Nat Commun 7:11261
6. Wang L, Brock A, Herberich B, Schultz PG (2001) Expanding the genetic code of *Escherichia coli*. Science 292(5516):498–500
7. Dumas A, Lercher L, Spicer CD, Davis BG (2015) Designing logical codon reassignment–expanding the chemistry in biology. Chem Sci 6(1):50–69
8. Chin JW, Santoro SW, Martin AB, King DS, Wang L, Schultz PG (2002) Addition of *p*-azido-L-phenylalanine to the genetic code of *Escherichia coli*. J Am Chem Soc 124 (31):9026–9027
9. Young DD, Young TS, Jahnz M, Ahmad I, Spraggon G, Schultz PG (2011) An evolved aminoacyl-tRNA synthetase with atypical polysubstrate specificity. Biochemistry 50 (11):1894–1900
10. Miyake-Stoner SJ, Miller AM, Hammill JT, Peeler JC, Hess KR, Mehl RA, Brewer SH (2009) Probing protein folding using site-specifically encoded unnatural amino acids as FRET donors with tryptophan. Biochemistry 48(25):5953–5962
11. Chin JW, Martin AB, King DS, Wang L, Schultz PG (2002) Addition of a photocrosslinking amino acid to the genetic code of *Escherichia coli*. Proc Natl Acad Sci U S A 99 (17):11020–11024
12. Chatterjee D, Cooley RB, Boyd CD, Mehl RA, O'Toole GA, Sondermann H (2014) Mechanistic insight into the conserved allosteric regulation of periplasmic proteolysis by the signaling molecule cyclic-di-GMP. eLlife 3: e03650
13. Peeler JC, Mehl RA (2012) Site-specific incorporation of unnatural amino acids as probes for protein conformational changes. Methods Mol Biol 794:125–134
14. Cooley RB, Smith TJ, Leung W, Tierney V, Borlee BR, O'Toole GA, Sondermann H (2016) Cyclic di-GMP-regulated Periplasmic proteolysis of a *Pseudomonas aeruginosa* type Vb secretion system substrate. J Bacteriol 198 (1):66–76
15. Navarro MV, Newell PD, Krasteva PV, Chatterjee D, Madden DR, O'Toole GA, Sondermann H (2011) Structural basis for c-di-GMP-mediated inside-out signaling controlling periplasmic proteolysis. PLoS Biol 9(2):e1000588
16. Newell PD, Boyd CD, Sondermann H, O'Toole GA (2011) A c-di-GMP effector system controls cell adhesion by inside-out signaling and surface protein cleavage. PLoS Biol 9 (2):e1000587
17. Studier FW (2005) Protein production by auto-induction in high density shaking cultures. Protein Expr Purif 41(1):207–234
18. Boyd CD, Chatterjee D, Sondermann H, O'Toole GA (2012) LapG, required for modulating biofilm formation by *Pseudomonas fluorescens* Pf0-1, is a calcium-dependent protease. J Bacteriol 194(16):4406–4414

Chapter 27

Identification of c-di-AMP-Binding Proteins Using Magnetic Beads

Jan Kampf, Jan Gundlach, Christina Herzberg, Katrin Treffon, and Jörg Stülke

Abstract

To identify cytosolic proteins that bind to cyclic di-AMP, a biotinylated analog of the nucleotide is used for protein pull-down experiments. In this approach, biotinylated c-di-AMP is coupled to *Strep*tactin-covered beads. After protein separation using standard SDS-PAGE, the protein(s) of interest are identified by mass spectrometric analyses.

Key words Protein pull-down, Biotinylated c-di-AMP, Magnetic beads, Bacteria

1 Introduction

Nucleotide second messengers are employed throughout all kingdoms of life to adapt and to transduce information on changing environmental conditions. Bacteria typically employ second messengers such as cyclic AMP (cAMP)*c-di-GMP*, ppGpp, cyclic di-AMP, and cyclic di-GMP. These molecules are involved in a variety of functions such as carbon catabolite repression, survival of starvation, lifestyle selection, and cell wall homeostasis [1–4]. To understand the physiological functions of these second messengers, the identification of their target molecules is of crucial importance. There are three principal approaches to identify second messenger targets. These are (a) genetic screens based on phenotypes caused by the loss of the signaling nucleotide, (b) screening for binding partners using radioactively labeled nucleotides and analysis of their binding to cell lysates, and (c) the isolation of binding partners by direct purification based on the high affinity interaction to second messengers. Of the two biochemical methods, the screen with labeled effector molecules called differential radial capillary action of ligand assay (DRaCALA) [5, 6] has been applied to identify proteins binding to ppGpp, c-di-GMP and c-di-AMP. A major

Karin Sauer (ed.), *c-di-GMP Signaling: Methods and Protocols*, Methods in Molecular Biology, vol. 1657,
DOI 10.1007/978-1-4939-7240-1_27, © Springer Science+Business Media LLC 2017

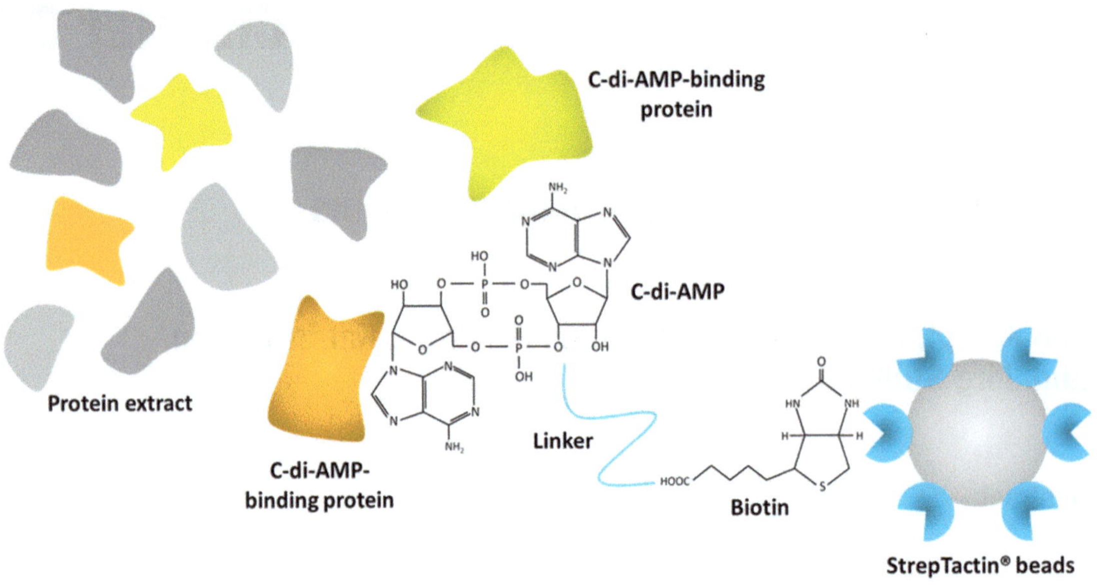

Fig. 1 Schematic overview of pull-down experiments via c-di-AMP coupled beads

advantage of this method is the possibility to screen both large expression libraries and individual proteins for ligand binding [5, 7, 8]. To directly isolate nucleotide-binding proteins from complex mixtures such as cell lysates, nucleotide analogs that carry functional groups allowing for direct binding to a matrix such as affinity beads have been developed [9]. Such labeled nucleotides have been used to identify cyclic nucleotide-binding proteins from cell lysates as well as from peptide libraries [7, 10–12]. In particular, cyclic dinucleotides that have been connected to biotin via a linker molecule can be bound to *Strep*tactin-covered beads in order to isolate proteins that bind to the molecule.

Here, we describe the use of 2′-[biotin]-AHC-c-di-AMP which is an analog of the second messenger c-di-AMP to isolate c-di-AMP-binding proteins from protein extracts. This analog has a hexyl spacer with a terminal amino group attached to one of the two ribose 2′-hydroxy groups by a carbamate bond. Beads incubated with biotinylated c-di-AMP will specifically pull-down binding partners of the nucleotide, while beads which were not preincubated with biotinylated c-di-AMP will not. The principle of the procedure is outlined in Fig. 1. Biotinylated c-di-AMP can be bound both to magnetic beads and to *Strep*tactin columns, giving thus two principal approaches for the purification.

2 Materials

Prepare all solutions using ultrapure water and analytical grade reagents. Prepare and store all reagents at room temperature (unless indicated otherwise). Diligently follow all waste disposal regulations when disposing of waste materials.

2.1 Chemicals and Enzymes

1. 0.5 M EDTA pH 8.0.
2. 1 nM 2′-[biotin]-AHC-c-di-AMP (2′-O-(6[biotinyl]amino-hexylcarbamoyl)-cyclic diadenosine monophosphate, sodium salt; BIOLOG, Germany), dissolved in water.
3. 1 nM 8′-[biotin]-AET-c-di-AMP (8′-(2[biotinyl]amino-ethylthio)-cyclic diadenosine monophosphate, sodium salt; BIOLOG, Germany), dissolved in water.
4. 10× protease inhibitor cocktail without EDTA (Roche, Germany).
5. Acetone (100%, cold).
6. Biotin.
7. BSA (10 mg/ml).
8. CHAPS (5%).
9. D-Desthiobiotin.
10. 1 M Dithiothreitol (DTT) (in water).
11. DNase I powder (Roche, Germany).
12. Glycerol (50%).
13. 1 M Isopropyl-β-D-thiogalactopyranosid (IPTG) (in water).
14. Lysozyme powder.

2.2 Further Materials and Methods

1. 50% *Strep*tactin Sepharose.
2. Magnetic rack for twelve 2 ml tubes.
3. Poly-Prep® Chromatography Columns (2 ml bead volume, 9 cm length, empty polypropylene column, 10 ml reservoir).
4. Equipment and material to carry out SDS-PAGE [13].
5. Silver staining [13].
6. *Strep*tactin-covered Mag*Strep* "type2HC" beads.
7. Equipment and material to carry out Western blot [13] (*see* **Note 1**).
8. Vortex Genie 2 with an adapter for 1,5 and 2 ml tubes.
9. Cell lysis: sonicator, French pressure cell.
10. Ultracentrifuge, refridgerated centrifuge.

2.3 Biological Materials

Depending on the experimental setup, we use different sources of proteins. To identify c-di-AMP-binding membrane proteins, strains expressing FLAG-tagged proteins of interest are used to facilitate detection of the target protein by Western blotting. If c-di-AMP binding was to be verified, we use His-tagged proteins, as they can also be used for subsequent in vitro work.

1. Bacterial strains with protein of interest fused to FLAG tag
2. Protein of interest (POI) fused to His_6-tag

2.4 Buffers

1. ***Buffer CP1:*** *5 mM Tris–HCl, pH 7.5, 230 mM NaCl, 1 mM $MgCl_2$, 0.5 mM DTT, 1× Protease inhibitor (see* ***Note 2****); ad 10 ml H_2O.*
2. ***Buffer CP2:*** *5 mM Tris–HCl, pH 7.5, 230 mM NaCl, 1 mM $MgCl_2$, 0.5 mM DTT, 1× Protease inhibitor (see* ***Note 2****);10% Glycerol, 50 μg/ml BSA, 4 mM EDTA, pH 8.0; ad 10 ml H_2O.*
3. ***Buffer CP3:*** *100 mM Tris–HCl, pH 8.0, 150 mM NaCl, 1 mM EDTA, pH 8.0, 5 mM Biotin.*
4. ***Buffer W:*** *100 mM Tris–HCl, pH 8.0, 150 mM NaCl, 1 mM EDTA, pH 8.0.*
5. ***Buffer MP1:*** *5 mM Tris–HCl, pH 7.5, 230 mM NaCl, 1 mM $MgCl_2$, 0.5 mM DTT, 1× protease inhibitor (see* ***Note 2****); ad 3 ml H_20.*
6. ***Buffer MP2:*** *5 mM Tris–HCl, pH 7.5, 230 mM NaCl, 1 mM $MgCl_2$, 0.5 mM DTT, 1× protease inhibitor (see* ***Note 2****), 50 μg/ml BSA, 0.5% CHAPS (see* ***Note 3****); ad 5 ml H_20.*
7. ***Buffer MP3:*** *920 mM NaCl, 4 mM $MgCl_2$, 2 mM DTT, 1× protease inhibitor, 40% Glycerol, 200 μg BSA, 0.5% CHAPS; ad 1 ml H_2O.*
8. ***Buffer MP4:*** *5 mM Tris–HCl, pH 7.5, 230 mM NaCl, 1 mM $MgCl_2$, 0.5 mM DTT 10% Glycerol, 0.5% CHAPS; ad 10 ml H_2O.*
9. ***Buffer MP5:*** *100 mM Tris–HCl, pH 8.0, 150 mM NaCl, 1 mM EDTA, pH 8.0, 0.5% CHAPS, 5 mM Biotin.*

3 Methods

3.1 Preparation/ Equilibration of Beads

Perform all steps in a magnetic rack.

1. For later protein pull down, magnetic beads coupled to c-di-AMP (+) and magnetic beads without c-di-AMP (−) are required. Per strain, the + and − beads are needed.
2. Add 800 μl buffer CP2 to 40 μl *Strep*tactin covered beads in a 2 ml Eppendorf reaction tube.
3. Mix at medium speed by vortexing and spin down shortly at 2000 × *g*.
4. Place the tubes into a magnetic rack and let the beads sink to the bottom for 30 s. Discard the supernatant using a micropipette.
5. Repeat **steps 2–4** once.
6. Now, the magnetic beads are equilibrated.
7. Add 40 μl biotinylated c-di-AMP only to the + tubes and incubate the mixture for 15 min with gentle agitation to bind the biotinylated c-di-AMP to the magnetic beads.
8. Wash the "+" tubes twice (compare **steps 2–4**) to get rid of unbound biotinylated c-di-AMP.

3.2 Pull-Down of Cytosolic Proteins via c-di-AMP Coupled Beads

1. Grow the bacteria in 50 ml of the required medium until the desired OD_{600} (*see* **Note 4**).
2. Centrifugation for 10 min, 4 °C, 10,000 × *g*. Discard the supernatant. Cell pellets can be stored at −20 °C.
3. Resuspend the cells in 2 ml buffer CP1 (*see* **Buffer CP1**).
4. Disrupt the cell pellet(s) by using an appropriate method (e.g., French press, *see* **Note 5**). Spin down the cell debris for 10 min, 4 °C, 10,300 × *g*. Retain the supernatant for subsequent steps.
5. Separate the cytosolic cell fraction from the membrane fraction by ultracentrifugation for 30 min, 4 °C, 80,000 × *g*.
6. Meanwhile prepare the *Strep*tactin covered beads (with or without bound biotinylated c-di-AMP) (*see* Subheading 3.1).
7. After ultracentrifugation (**step 5**) save the supernatant and determine the protein concentration (*see* **Note 6**). The pellet (membrane fraction) can be discarded. For an analysis of c-di-AMP-binding membrane proteins, proceed to Subheading 3.3.
8. Prepare two assay mixtures (1.2 mg cytosolic protein fraction; 0.2 ml glycerol (50%); 8 μl 0.5 M EDTA, pH 8.0; 50 μl BSA (10 mg/ml), ad 1 ml Buffer CP1) per strain. One assay mixture is used as negative control (− tubes), the other mixture will serve for the actual protein pull-down (+ tubes).
9. Add the complete assay mixtures to the equilibrated magnetic beads (one sample to c-di-AMP-bound beads, one sample to free beads).
10. Incubate the samples for 30 min with gentle agitation at room temperature (*see* **Note 7**).
11. Wash all samples up to seven times with 800 μl buffer W (*see* **Buffer W**) to remove unbound proteins (*see* Subheading 3.1, **steps 2–4** for washing). Keep at least the last washing step for later analyses to make sure that no further proteins are eluted nonspecifically.
12. For elution of proteins bound by c-di-AMP use 50 μl buffer CP3 (*see* **Buffer CP3**) and incubate the samples for 15 min with gentle agitation.
13. Spin down briefly at 2000 × *g* and save the elution fraction (supernatant).
14. Analyze the eluted proteins by SDS-PAGE, silver staining or Western blotting. Prominent bands that appear in the + fraction of the samples after silver staining can be cut out and analyzed via mass spectrometry (Fig. 2).

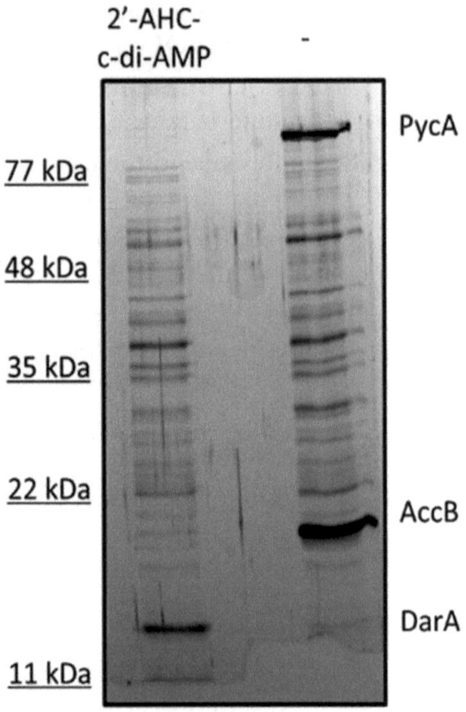

Fig. 2 Identification of the c-di-AMP receptor DarA in *B. subtilis* [11]. A pull-down assay was performed with *B. subtilis* crude extract to identify c-di-AMP binding proteins. After incubation of *Strep*tactin-covered beads with 2′-[biotin]-AHC-c-di-AMP and a *B. subtilis* crude extract, the samples were washed to eliminate nonspecifically bound proteins. The elution was performed with 5 mM biotin. Samples were analyzed by SDS-PAGE with subsequent silver staining. As the negative control, the crude extract was incubated with *Strep*tactin-covered beads without biotinylated c-di-AMP

3.3 Separation and Analysis of c-di-AMP-Binding Membrane Proteins

3.3.1 Preparation of Cytosolic and Membrane Fractions [14]

The following samples have to be prepared and saved for subsequent analysis:

- Crude extract reference: Crude extract prior to ultracentrifugation.
- Cytosol 1 reference: Protein extract after the first ultracentrifugation.
- Cytosol 2 reference: Protein extract after the second ultracentrifugation.
- Membrane reference: The membrane fraction prior to solubilization.
- Solubilized membrane fraction: The membrane fraction after solubilization.
- Flow through 1.
- Flow through 2.

1. Use strains that express the protein of interest attached to a C-terminal FLAG tag for later immunological detection (*see* **Note 8**). Grow the bacterial strains (500 ml) in the required medium to an OD_{600} of ~0.8. Cell pellets can be stored at −20 °C.
2. Resuspend the cell-pellet(s) in 19 ml 20 mM Tris–HCl pH 8.0, 1× protease inhibitor. Add 10 mg lysozyme in 800 μl 20 mM Tris–HCl pH 8.0/400 μl EDTA, pH 8.0 (0.5 M) solution.
3. Incubate the cell suspension 30 min on ice. Centrifuge for 30 min, 4 °C, 3000 × *g*.
4. Discard the supernatant and incubate the pellet for 30 min at −20 °C.
5. Resuspend the pellet in 10 ml 20 mM Tris–HCl pH 8.0, 2 mg DNase I and 1× protease inhibitor.
6. Split each of the cultures into two 15 ml tubes for better handling.
7. Expose each sample four times to ultrasonication (1 min, 0.8 cycle/80% amplitude, 4 °C).
8. Centrifuge each sample 15 min, 4 °C, 10,000 × *g*.
9. Pool the supernatant from the split cultures. Take 100 μl of the sample(s) (*crude extract reference*).
10. Transfer the remaining samples (30 ml) into an ultracentrifuge tube and add 10 ml Tris–HCl pH 8.0 + 1× protease inhibitor. The final volume should be 40 ml.
11. Ultracentrifugation 1 h, 4 °C, 230,000 × *g*. Take 500 μl supernatant of each sample (*cytosol 1 reference*).
12. Discard the remaining supernatant.
13. Wash the cell pellet(s) twice with 2.5 ml 20 mM Tris–HCl pH 8.0 and 1× protease inhibitor. Finally, resuspend the samples in 1 ml 20 mM Tris–HCl pH 8.0.
14. Centrifuge shortly at full speed, 4 °C. Transfer the supernatant into an ultracentrifuge tube.
15. Repeat **step 11** at 100,000 × *g*. Take 500 μl supernatant of each sample (*cytosol 2 reference*).
16. Resuspend cell pellet(s) in 400 μl 20 mM Tris–HCl pH 8.0. Add 5% CHAPS and 1× protease inhibitor to a final volume of 500 μl (*see* **Note 2**).
17. Incubate the samples with gentle agitation overnight at 4 °C to ensure complete solubilization of the membrane protein(s).
18. Determine the protein concentration of the *crude extract*, *cytosol 1* and *cytosol 2* samples. Store the reference samples at 4 °C.

3.3.2 Pull-Down of Membrane Proteins Binding c-di-AMP

1. Take a 25 μl aliquot of the solubilized membrane samples for later SDS-PAGE analysis (*membrane reference*). With the remaining solubilized membrane fraction perform ultracentrifugation 1 h at 4 °C, 100,000 × *g*.
2. In the meantime, equilibrate the *Strep*tactin-covered beads (*see* Subheading 3.1).
3. Take the supernatants following ultracentrifugation and add to a fresh microfuge tube.
4. Use 450 μl solubilized membrane (save 10 μl for later SDS-PAGE, *solubilized membrane fraction*) and add 150 μl buffer MP3 and mix carefully.
5. Add the entire mixture to the "−" beads.
6. Incubate for 20 min with gentle agitation (save 10 μl as *flow through 1*).
7. Centrifuge shortly at 2000 × *g* and separate the beads from the sample using the magnetic rack.
8. Apply the supernatant to the "+" beads.
9. Incubate for 20 min at room temperature with gentle agitation (save 10 μl as *flow through 2*).
10. Centrifuge briefly at 2000 × *g* and separate the beads from the sample via the magnetic rack.
11. Wash the "+" and "−" beads five times with 800 μl buffer MP4. Keep the last washing fraction for SDS-PAGE.
12. Elute the bound proteins, by using 50 μl buffer MP5. Incubate the samples for 20 min with gentle agitation.
13. Add 200 μl ice-cold acetone to precipitate proteins and store the samples at −20 °C over night.

3.3.3 Detection of Membrane Proteins Binding c-di-AMP

1. Centrifuge the samples 30 min at 4 °C, 11,000 × *g*.
2. Discard the supernatant and dry the samples for about 2 h at room temperature to get rid of any residual acetone (no acetone smell detectable any longer).
3. Resuspend the very thin pellet in 10 μl buffer W.
4. To detect any c-di-AMP-bound proteins run a standard SDS-PAGE with all the collected fractions, followed by standard western blotting techniques.

3.4 Verification Using an Alternatively Labeled Variant of c-di-AMP

The results can be verified by using 8′-[biotin]-AET-c-di-AMP instead of 2′-[biotin]-AHC-c-di-AMP. For this purpose, the procedure is done as described above.

3.5 Assay for c-di-AMP-Binding of Tagged Proteins Purified from E. coli

This procedure serves to verify c-di-AMP binding for previously identified candidate proteins.

1. *E. coli* has to be transformed with the plasmid expressing the inducible POI (*see* **Note 9**).
2. Inoculate 250 ml of the bacteria expressing the POI in the required medium. Cultivate the cells to the exponential growth phase (OD_{600} of 0.5–0.8).
3. Add 1 mM IPTG for induction of the overexpression and incubate the culture for additional 3 h at 37 °C and 200 rpm (*see* **Note 10**).
4. Harvest the culture by centrifugation for 10 min, 4 °C, 5000 × *g*. Discard the supernatant. The cell pellet(s) can be stored at −20 °C.
5. Resuspend the pellet in 2 ml cold buffer W (with 1× complete protease inhibitor (without EDTA)).
6. Disrupt the cell pellet(s) by using an appropriate method (i.e., French press, *see* **Note 5**). Centrifuge the lysate for 10 min, 4 °C, 11,000 × *g* to remove cell debris.
7. Separate the cytosolic cell fraction by ultracentrifugation for 30 min, 4 °C, 72,500 × *g*. Save the supernatant and determine the protein concentration (*see* **Note 6**).
8. Incubate 1450 μl of the cytoplasmic cell extracts with 40 μl biotinylated c-di-AMP in a 2 ml Eppendorf reaction tube for 30 min at room temperature with gentle agitation. Do not forget a negative control without biotinylated c-di-AMP.
9. Purify the samples according to standard protocol for *Strep*-tag purification but wash the samples at least seven times (*see* **Note 11**)
10. Load the cytosolic cell extracts with/−out biotinylated c-di-AMP onto the Poly-Prep columns and keep 50–100 μl for later analysis (*crude extract*). Also, collect the flow through.
11. Wash the column by adding at least seven times 2.5 column bed volumes of buffer W. Collect all fractions (W1-W7) to ensure that all nonspecifically binding proteins have been eliminated.
12. To elute specifically bound proteins, add up to four times 0.5 column bed volumes of elution buffer CP3 (*see* **Buffer CP3**) and collect all fractions (E1–E4).
13. Run an SDS-PAGE with 25 μl of the elution and wash fractions, then silver stain the gel (Fig. 3).

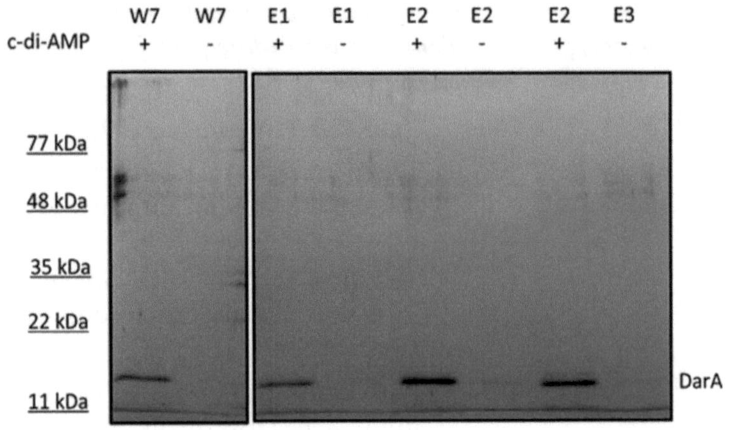

Fig. 3 Identification of overexpressed *B. subtilis* DarA as c-di-AMP-binding protein in *E. coli*. DarA was overexpressed in *E. coli* and subsequently incubated with 2′-[biotin]-AHC-c-di-AMP. After incubation, the sample was purified using a *Strep*tactin column, and bound protein was eluted with 2.5 mM D-desthiobiotin. As a negative control, the overexpressed protein was purified without addition of 2′-[biotin]-AHC-c-di-AMP (−). The last wash fraction (W7) and elution fractions (E1–E3) were analyzed by SDS-PAGE with subsequent silver staining [11]

4 Notes

1. **Western blot.**

   For Western blot, proteins are transferred to a PVDF membrane using a semidry blotting apparatus (80 mA, 1 h). The incubation with primary and secondary antibodies is performed using standard procedures [13].

2. **Quick dissolving of 10× protease inhibitor cocktail without EDTA tablet.**

   Crushing of the protease inhibitor tablet before adding it to the buffers CP1, CP2, MP1, and MP2 results in faster dissolution of the tablet.

3. **Reduction of foam in buffers with CHAPS.**

   Add CHAPS after resuspension. This will reduce the formation of foam.

4. **Mediums and conditions for cultivation.**

   Here, LB (1% trypton, 0.5% yeast extract, 1% NaCl) or CSE-Glucose medium [11] are normally used for the experiments. The bacteria were cultured at 37 °C and 200 rpm.

5. **Cell disruption using the French press.**

Materials	
(a) French press	
(b) Disruption cell	Precooled in a water bath or in the fridge
(c) Resuspension buffer	Precooled in the fridge or on ice
(d) Cell pellets	Frozen cells within a 50 ml falcon tube

Procedure.

All steps must be carried out on ice and the French press has to be precooled!

(a) Resuspend the cell pellet in resuspension buffer (e.g., CP1) by vortexing or pipetting up and down. Usually, for the big French press a pellet from 500 ml of cells with an OD_{600} of 0.8–1.2 is resuspended in 15 ml buffer, for the small French press a pellet from 50 ml of cells with an OD_{600} of 0.8–1.2 is resuspended in 1–2 ml buffer.

(b) Turn on the French Press and select the pressure for cell disruption (18,000 psi for *B. subtilis* and *E. coli*).

(c) Fill the resuspended cells in the disruption cell and close it tightly.

(d) Put the closed disruption cell into the machine so that the piston is on the top. Make sure that the handles of the piston are turned in your direction so they cannot come in contact with the mounting while the cell gets compressed.

(e) Press start and follow the instructions on the display of the machine.

(f) Open the valve slowly while you put a 50 ml falcon under it.

(g) Collect the opened cells that flow out of the valve. Watch out that the cell suspension flows out slowly and the pressure remains almost constant until the whole suspension has flown out.

(h) Reload the disruption cell. For gram-positive bacteria, open up the cells three times. For gram-negative cells one run is sufficient.

(i) Clean the disruption cell with ethanol (70%) and distilled water.

6. **Bradford assay with Roti®-Quant.**

   For all approaches in this chapter, a Bradford assay with Roti®-Quant reagent (Roth, Germany) was used for the protein concentration determination [13].

7. **A Vortex Genie 2 with an adapter for 1,5 and 2 ml tubes was used for agitation of the reaction tubes.**

8. **Adding a C-terminal FLAG-tag to the POI.**

   Here: Vector pGP1087 was used [15] (*see* http://subtiwiki.uni-goettingen.de/wiki/index.php/PGP1087). This vector attaches a C-terminal triple FLAG-tag sequence (DYKDDDDK) to the POI. The modified gene can be introduced into the *Bacillus subtilis* chromosome by transformation. Note that customized plasmids must be used for other species.

9. **Vector for N-terminal His_6-tag to the POI.**

   Here: Vector pWH844 was used [16] (*see* http://subtiwiki.uni-goettingen.de/wiki/index.php/PWH844). This vector attaches an N-terminal His_6-tag to the POI. A vector attaching a *Strep*-tag to the protein must not be used, this would automatically cover the entire beads. Alternatively, a vector that does not attach any affinity tag can be used.

10. **Culturing bacteria at 18 °C to increase the expression rate of POI.**

    Growing the cultures overnight at 18 °C can increase the solubility of the POI. Inoculate the culture to an OD_{600} of 0,1 and culture them until an OD_{600} of 0,3 at 37 °C and 200 rpm. Then change the temperature to 18 °C and keep culturing for 1 h for cooling down the culture. Then, add the IPTG to induce the overexpression. Harvest the cells the next day (OD_{600} of 2,5–3,0).

11. ***Strep*-tag purification.**

    For all steps, allow the solutions on the column to drain off by gravity before adding the next one. All steps are performed at 4 °C. Use 1 ml of the 50% suspension per 1 l of cell culture, column bed volume = 0.5. Remove the bottom cap from the column and apply the appropriate amount of 125 μl of *Strep*-tactin® matrix (since samples were split for the negative control). Equilibrate the column by adding 10 column volume of buffer W (*see* Buffer W).

Acknowledgments

We are grateful to Fabian Commichau for helpful discussions. Frank Schwede, Annette Garbe, and Volkhard Kaever are acknowledged for long-standing support and collaboration in the analysis of c-di-AMP signaling. Work in the authors' lab is supported by the Deutsche Forschungsgemeinschaft through SPP1879.

References

1. Gomelsky M (2011) cAMP, c-di-GMP, c-di-AMP and now cGMP: bacteria use them all! Mol Microbiol 79:562–565
2. Görke B, Stülke J (2008) Carbon catabolite repression in bacteria: many ways to make most out of nutrients. Nat Rev Microbiol 6:613–624

3. Liu K, Bittner AN, Wang JD (2015) Diversity in (p)ppGpp metabolism and effectors. Curr Opin Microbiol 24:72–79
4. Commichau FM, Dickmanns A, Gundlach J, Ficner R, Stülke J (2015) A jack of all trades: the multiple roles of the unique essential second messenger cyclic di-AMP. Mol Microbiol 97:189–204
5. Roelofs KG, Wang J, Sintim HO, Lee VT (2011) Differential radial capillary action of ligand assay for high-throughput detection of protein-metabolite interactions. Proc Natl Acad Sci U S A 108:15528–15533
6. Lee VT (2016) Discovering protein receptors for signaling nucleotides. PLoS Pathog 12: e1005569
7. Corrigan RM, Campeotto I, Jeganathan T, Roelofs KG, Lee VT, Gründling A (2013) Systematic identification of conserved bacterial c-di-AMP receptor proteins. Proc Natl Acad Sci U S A 110:9084–9089
8. Syal K, Chatterji D (2015) Differential binding of ppGpp and pppGpp to *E. coli* RNA polymerase: photo-labeling and mass spectral studies. Genes Cells 20:1006–1016
9. Poppe H, Rybalkin SD, Rehmann H, Hinds TR, Tang XB, Christensen AE, Schwede F, Genieser HG, Bos JL, Doskeland SO, Beavo JA, Butt E (2008) Cyclic nucleotide analogs as probes of signaling pathways. Nat Methods 5:277–278
10. Sureka K, Choi PH, Precit M, Delince M, Pensinger DA, Huynh TN, Jurado AR, Goo YA, Sadilek M, Iavarone AT, Sauer JD, Tong L, Woodward JJ (2014) The cyclic dinucleotide c-di-AMP is an allosteric regulator of metabolic enzyme function. Cell 158:1389–1401
11. Gundlach J, Dickmanns A, Schröder-Tittmann K, Neumann P, Kaesler J, Kampf J, Herzberg C, Hammer E, Schwede F, Kaever V, Tittmann K, Stülke J, Ficner R (2015) Identification, characterization and structure analysis of the cyclic di-AMP-binding PII-like signal transduction protein DarA. J Biol Chem 290:3069–3080
12. Düvel J, Bense S, Möller S, Bertinetti D, Schwede F, Morr M, Eckweiler D, Genieser HG, Jänsch L, Herberg FW, Frank R, Häussler S (2015) Application of synthetic peptide arrays to uncover cyclic di-GMP binding motifs. J Bacteriol 198:138–146
13. Sambrook J, Russell D (2001) Molecular cloning: a laboratory manual. Cold Spring Harbor Laboratory, Cold Spring Harbor, NY
14. Müller VS, Tschauner K, Hunke S (2013) Membrane-SPINE: a biochemical tool to identify protein-protein interactions of membrane proteins *in vivo*. J Vis Exp 81:e50810
15. Diethmaier C, Pietack N, Gunka K, Wrede C, Lehnik-Habrink M, Herzberg C, Hübner S, Stülke J (2011) A novel factor controlling bistability in *Bacillus subtilis*: the YmdB protein affects flagellin expression and biofilm formation. J Bacteriol 193:5997–6007
16. Schirmer F, Ehrt S, Hillen W (1997) Expression, inducer spectrum, domain structure, and function of MopR, the regulator of phenol degradation in *Acinetobacter calcoaceticus* NCIB8250. J Bacteriol 179:1329–1336

Chapter 28

Pull-Down with a c-di-GMP-Specific Capture Compound Coupled to Mass Spectrometry as a Powerful Tool to Identify Novel Effector Proteins

Benoît-Joseph Laventie, Timo Glatter, and Urs Jenal

Abstract

Capture compound technology coupled to mass spectrometry (CCMS) allows to biochemically identify ligand receptors. Using a c-di-GMP-specific Capture Compound, we adapted this method for the identification and characterization of c-di-GMP binding proteins in any bacterial species. Because in silico analysis often fails to predict novel c-di-GMP effectors, this universal method aims at better defining the cellular c-di-GMP network in a wide range of bacteria. CCMS was successfully applied in several bacterial species (Nesper et al., J Proteom 75:4874–4878, 2012; Steiner et al., EMBO J 32:354–368, 2013; Tschowri et al., Cell 158:1136–1147, 2014; Trampari et al., J Biol Chem 290:24470–24483, 2015; Rotem et al., J Bacteriol 198:127–137, 2015). To outline the detailed protocol and to illustrate its power, we use *Pseudomonas aeruginosa*, an opportunistic pathogen in which c-di-GMP plays a critical role in virulence and biofilm control, as an example. CCMS identified 74% (38/51) of the known or predicted components of the c-di-GMP network.

Key words Capture compound, Photoactivable crosslinker, Mass spectrometry, c-di-GMP effector, EAL, GGDEF, PilZ, *Pseudomonas aeruginosa*

1 Introduction

The ubiquitous second messenger c-di-GMP controls growth and behavior of many bacteria. C-di-GMP notably regulates cell cycle progression, planktonic and surface motility, and the expression of adhesins and exopolysaccharides [6–8]. By modulating cellular levels of c-di-GMP, this signaling network coordinates the arrest of flagellar motility, promotes surface motility and adhesion, and ultimately orchestrates biofilm formation. C-di-GMP is synthetized by enzymes called diguanylate cyclases (DGCs) containing a catalytic GGDEF domain [8], and is hydrolyzed by two classes of c-di-GMP-specific phosphodiersterases (PDEs) containing catalytic EAL and HD-GYP domains, respectively [9, 10].

Karin Sauer (ed.), *c-di-GMP Signaling: Methods and Protocols*, Methods in Molecular Biology, vol. 1657,
DOI 10.1007/978-1-4939-7240-1_28,

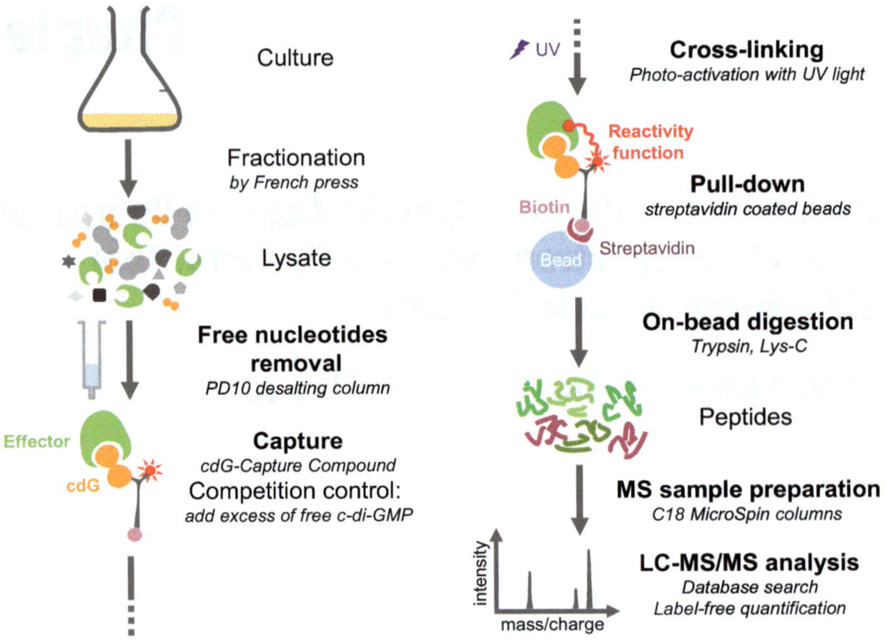

Fig. 1 Schematic overview of the CCMS workflow

C-di-GMP-dependent cellular processes are controlled by a diverse range of c-di-GMP receptors, also called "effectors". Many of the protein-based c-di-GMP effectors characterized so far contain catalytically inactive GGDEF or EAL domains, or harbor a PilZ domain [11–14]. Protein domains belonging to any of these three classes can be predicted in silico relatively safely. In contrast, predicting new c-di-GMP binding domains or motifs remains challenging [15]. A new journey has begun for the identification of new classes of c-di-GMP effectors, and effectors with novel c-di-GMP binding sites have been recently described, such as BcsE, MshE, or BldA [3, 16, 17]. In search for alternative strategies, we designed a biochemical approach to screen for new c-di-GMP effectors.

This approach is called capture compound coupled mass spectrometry (CCMS) and takes advantage of a capture compound to specifically pull down putative effectors directly from a complex mixture of proteins, followed by their identification with mass spectrometry (method overview: *see* Fig. 1) [1, 18, 19]. Capture compound and chemical proteomics-based approaches have been used successfully in the past in several organisms, such as *P. aeruginosa*, *Caulobacter crescentus*, and *Salmonella enterica serovar typhimurium* [1–5, 20]. The trivalent c-di-GMP-specific capture compound (cdG-CC, Fig. 2) is composed of: (1) a c-di-GMP molecule, linked via the 2'OH group of one ribose to the capture compound backbone, which serves as bait to capture c-di-GMP binding proteins, (2) a photoreactive aryl azide group to covalently

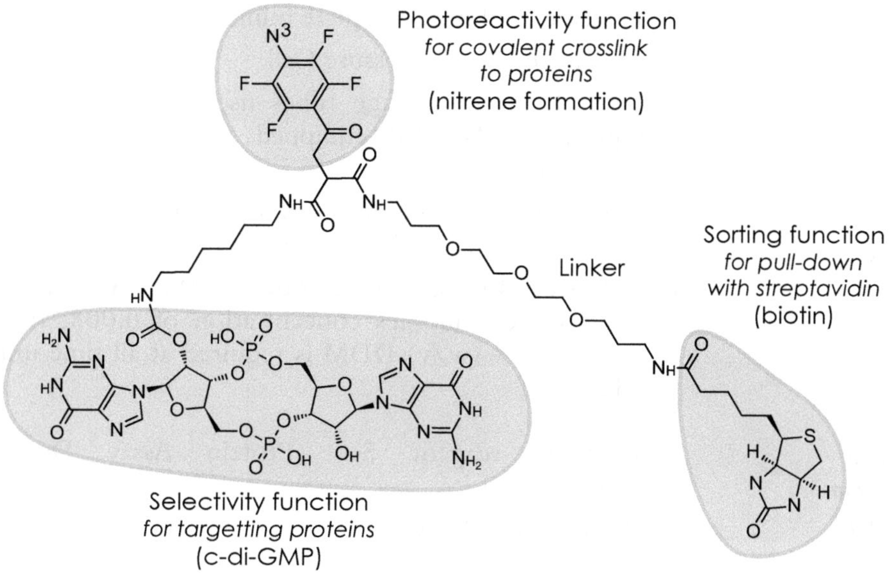

Fig. 2 Molecular structure of the c-di-GMP capture compound

cross-link target proteins to the cdG-CC, and (3) a biotin to purify the protein–CC complex using streptavidin-coated magnetic beads. The cdG-CC specifically interacts and covalently crosslinks c-di-GMP effectors in complex mixtures such as cell lysates. The covalent linkage with the effectors allows the use of stringent washing conditions, thereby reducing false-positive hits significantly. The target proteins are then identified by LC-MS-MS after tryptic digestion. The procedure is designed for soluble proteins, but with a few modifications indicated in the method below, can also be used to enrich for membrane-bound c-di-GMP effectors. For a visual aid to this detailed protocol, note that the method has been published in a video format in the Journal of Visualized Experiments [21]. The CCMS method is illustrated here using the example of *Pseudomonas aeruginosa*.

2 Materials

All the buffers used for the MS sample preparation (starting from **step 6**) should be handled with care and using gloves to prevent keratin contamination. HPLC grade H_2O should be used for buffer preparation.

2.1 Lysate Preparation

1. Lysis buffer: 6.7 mM MES, 6.7 mM HEPES, 200 mM NaCl, 6.7 mM KAc (Potassium Acetate), 1 mM DDT, pH 7.5. Note that although the lysis buffer can be prepared in advance, DTT should be added freshly on the same day.

2. Protease inhibitor: complete mini, EDTA-free (Roche).
3. Miniature French Pressure Cell.
4. 1.5 mL ultracentrifuge tubes used with an ultracentrifuge Optima™ MAX-XP equipped with a TLA-55 rotor or equivalent.
5. PD10 desalting column (GE Healthcare), to remove free c-di-GMP and other nucleotides, from lysates.
6. n-dodecyl-β-D-maltopyranoside (DDM) (Affymetrix). DDM has a critical micelles concentration of 0.008; therefore, at least 0.008% (w/v) DDM is required at all time until urea is added.

2.2 Protein Concentration Measurement

1. Bradford reagent 5× (Protein Assay Dye Reagent Concentrate).
2. BCA Protein Assay kit.

2.3 Capture

1. Nucleotide mix: 10 mM GDP, 10 mM GTP, 10 mM ATP, 10 mM CTP.
2. 5× capture buffer: 100 mM HEPES, 250 mM KAc, 50 mM MgAc, 5% glycerol, pH 7.5.
3. 5× washing buffer: 5 M NaCl, 250 mM Tris, pH 7.5.
4. UV lamp (*e.g.*, CaproBox, $\lambda = 310$ nm, Irradiance ≥ 10 mW/cm^2, distance from the source = 2 cm, Caprotec Bioanalytics, or any UV lamp with similar properties).
5. 12-tube PCR strips.
6. Magnet for 12-tube PCR strips.c-di-GMP-capture compound (Caprotec Bioanalytics).
7. c-di-GMP (Biolog).
8. Streptavidin-coated magnetic beads.
9. UV cross-linker.

2.4 Washing

1. 1 M ABC (ammonium bicarbonate, NH_4CO_3). Prepare extemporaneously in HPLC grade H_2O or store aliquots at −20 °C.
2. 100 mM ABC, 8 M urea: mix 1 mL 1 M ABC, 4 mL HPLC grade H_2O, and 4.8 g urea, solubilize on a rotating wheel for 1 h. at RT, then adjust to 10 mL with HPLC grade H_2O. Can be diluted in 100 mM ABC to prepare the 100 mM ABC, 2 M urea washing buffer for **step 3.6**.

2.5 MS Sample Preparation

1. For an efficient and simple control of temperature and mixing conditions, we recommend using an Eppendorf ThermoMixer or equivalent.
2. 200 mM Tris(2-Carboxyethyl)Phosphine (TCEP) stock solution. Store aliquots at −20 °C.

3. 400 mM iodacetamide stock solution. Prepare freshly and protect from light.
4. 0.5 M N-acetyl-cysteine stock solution. Store aliquots at −20 °C.
5. 0.2 μg/μL Lys-C Endopeptidase (Wako). The lyophilized, mass spectrometry grade, lysyl endopeptidase is solubilized with 100 mM ABC.
6. 0.4 μg/μL Modified Porcine Trypsin (Promega). This trypsin is modified by reductive methylation to improve its resistance to autoproteolysis, as proteolysis of trypsin can interfere with the HPLC peptide analysis. Note that the urea concentration has to be below 1.6 M for trypsin to be active.
7. 5% (v/v) TFA (trifluoroacetic acid).
8. 2 M HCl.
9. C18 MicroSpin columns, 5–60 μg capacity, 5–200 μL loading (The Nest Group, MA USA, ref. SS18V).
10. HPLC LC-MS grade 99.95% Acetonitrile (Reuss Chemie).
11. UIS250v sonicator with VialTweeter (Hielscher Ultrasound Technology).

2.6 LC-MS/MS Analysis

1. For label-free quantitative analysis we present here one of different possible set ups to perform efficient liquid chromatography–mass spectrometry (LC-MS) analysis. We perform LC-MS analysis on an Easy nLC 1000 (Thermo Fisher Scientific) online connected to a Nanospray Flex Ion Scource (Thermo Fisher Scientific) and an Orbitrap Elite hybrid mass spectrometer (Thermo Fisher Scientific).
2. C18 beads (ReproSil-Pur C18-AQ, 1.9 μm; Dr. Maisch).
3. PicoFrit fused silica column (height: 37 cm, 75 μm inner diameter, 10 μm emitter; New Objectives).
4. Pressure cell.
5. LC Solvent A: 98% HPLC grade H_2O, 0.15% formic acid, 2% acetonitrile.
6. LC Solvent B: 98% acetonitrile, 0.15% formic acid.
7. Progenesis QIP software (Nonlinear Dynamics, Version 2.0), for label-free quantification.

3 Methods

3.1 Lysate Preparation

1. Grow bacteria in the appropriate medium to the desired density (*see* **Note 1**). All further steps are carried out at 4 °C unless otherwise noted (*see* **Note 2**).

2. Pellet the bacteria by centrifugation for 20 min at 5000 × *g*.
3. Resuspend 0.5–1 g of pellet in 1 mL lysis buffer supplemented with complete mini protease inhibitor (1 tab/10 mL) and 10 μg/mL DNaseI (*see* **Note 3**).
4. Lyse the cells by three passages through a French pressure cell, at 20′000 psi.
5. Ultra-centrifuge the cell lysate at 100,000 × *g* for 1 h at 4 °C.
6. For the capture of cytoplasmic proteins, continue with the supernatant containing the soluble proteins, as described in Subheading 3.2.
7. For the capture of membrane proteins, wash the pellet with 1 mL of 1× lysis buffer by pipetting up and down.
8. Ultracentrifuge at 100,000 × *g* for 1 h at 4 °C and discard the supernatant. Continue with the pellet as described in Subheading 3.3. At this step, it is possible to flash-freeze the membrane fraction in liquid nitrogen and store it at −80 °C until used.

3.2 Removal of Free Nucleotides (Soluble Fraction Only)

1. Prepare the PD10 desalting columns by washing with 10 mL of cold lysis buffer. Use one column per sample.
2. Load the PD10 columns with 1 mL of the supernatant from Subheading 3.1, **step 6** (*see* **Note 4**).
3. Elute in 8 fractions of 500 μL each, with cold lysis buffer.
4. Perform a Bradford assay to select the most concentrated fractions. Pool the most concentrated fractions. A quick method to assess which fractions to pool is by mixing 1 μL of the fraction with 200 μL of 1× Bradford reagent in a 96-well flat bottom plate, shake and then estimate optically the bluest fractions. Proteins typically elute from fractions 3–8, with a peak around fraction 6.
5. Continue with steps listed in Subheading 3.4.

3.3 Membrane Resuspension and Solubilization (Membrane Fraction Only)

1. Resuspend the membrane pellet (from Subheading 3.1, **step 8**) in 500 μL of 1× capture buffer by pipetting up and down. Then aspirate several times with a 1 mL syringe equipped with a 27G needle until a homogenous suspension is obtained.
2. Add 1% (w/v) DDM to solubilize membrane proteins (*see* **Note 5**) and incubate at 4 °C for at least 2 h (or overnight) on a rotating wheel.
3. Ultracentrifuge at 100,000 × *g* for 1 h at 4 °C and transfer the supernatant containing the solubilized membrane proteins in a fresh tube.
4. Continue with steps listed in Subheading 3.4.

3.4 Protein Concentration Measurement

1. Measure the protein concentration of the soluble fraction by Bradford assay. Measure the protein concentration of the membrane fraction by BCA assay, as the Bradford reagent is incompatible with the concentration of detergent used here.
2. Mix Bradford (or BCA) reagent and the samples in a 1 mL plastic cuvette, incubate 30 min at RT (or 37 °C for the BCA assay).
3. Read OD at $\lambda = 595$ nm (562 nm for the BCA assay) with a spectrophotometer and calculate the protein concentration according to the manufacturer protocol.
4. Perform a standard curve using BSA as a protein standard (0–20 μg/mL).
5. Adjust the protein concentration to 10 mg/mL.

3.5 Capture

1. Mix 30 μL of proteins (300 μg) with 10 μL of nucleotide mix (*see* **Note 6**) and 20 μL of 5× capture buffer in 200 μL 12-tube PCR strips.
2. Adjust the reaction volume to 100 μL with ddH_2O. The total volume includes the volume of the cdG-CC and free c-di-GMP for the competition control (*see* **step 3.5.3**).
3. A typical reaction setup is shown in Table 1.
4. Incubate at 4 °C for 30 min on a rotating wheel.
5. Add 10 μM cdG-CC (final concentration, *see* **Notes 7–9**). For each sample, perform in parallel a "bead control" (sample without cdG-CC), and a "competition control" (sample supplemented with 1 mM free c-di-GMP) (*see* Table 1).
6. Incubate at 4 °C for at least 2 h for the soluble fraction, and overnight for the membrane fraction, on a rotating wheel, protected from light (*see* **Note 8**).
7. Remove the lid of the 12-tube PCR strips (which would otherwise shield UVs), and cross-link by activation of the reactive moiety of the cdG-CC by UV light for 4 min (*see* **Note 10**).
8. Add 25 μL 5× wash buffer.
9. Resuspend the stock of streptavidin magnetic beads by shaking the tube and add 50 μL of beads to the reaction mix.
10. Homogenize and incubate at 4 °C for 1 h on a rotating wheel. Although prewashing the Streptavidin magnetic beads prior to use is normally recommended by the manufacturer, this step can skipped without affecting the quality of the data.

3.6 Washing Steps

1. To wash the magnetic beads, the 12-tube PCR strips are placed on the magnet to separate the beads from the washing buffer. After discarding the supernatant, remove the tubes from the magnet and proceed with the next washing step (*see* **Note 11**).

Table 1
Typical reaction setup. The bead control correspond to a negative control to identify false positive fished in absence of cdG-CC. The capture experiment is the conditions aiming at capturing c-di-GMP effectors. The competition control is a negative control containing a large excess of free c-di-GMP to compete the c-di-GMP moiety of the cdG-CC, to decrease the rate a false positive hits. This example uses 10 μM cdG-CC (*see* Note 7)

	Bead control	Capture experiment	Competition control
Protein extract (10 mg/mL)	30 μL	30 μL	30 μL
c-di-GMP (10 mM)	0 μL	0 μL	10 μL
Nucleotides mix (10 mM of each nucleotide)	10 μL	10 μL	10 μL
Capture buffer 5×	20 μL	20 μL	20 μL
ddH_2O	40 μL	30 μL	20 μL
30 min incubation			
c-di-GMP-CC (stock 100 μM)	0 μL	10 μL	10 μL
Final concentrations:	Bead control	Capture experiment	Competition control
c-di-GMP-CC	0 μM	10 μM	10 μM
free c-di-GMP for competition	0 μM	0 μM	1000 μM

2. For soluble fraction only:
   (a) Wash six times in 200 μL 1× washing buffer.
   (b) Wash once in 200 μL HPLC grade H_2O.
   (c) Wash six times in 200 μL 80% acetonitrile.
   (d) Wash twice in 200 μL HPLC grade H_2O.
   (e) Remove supernatant.
3. Membrane fraction only:
   (a) Wash five times in 200 μL 1× washing buffer +0.1% (w/v) DDM.
   (b) Wash twice in 200 μL 1× washing buffer +0.05% (w/v) DDM.
   (c) Wash once in 200 μL 1× washing buffer containing 0.025% (w/v) DDM.
   (d) Wash once in 200 μL 1× washing buffer containing 0.0125% (w/v) DDM.
   (e) Wash three times in 200 μL 100 mM ABC containing 2 M urea.
   (f) Remove supernatant.

3.7 MS Sample Preparation

1. Resuspend the beads in 20 μL 100 mM ABC (100 mM ABC + 8 M urea for the membrane fraction) and transfer in fresh 1.5 mL tubes.
2. Membrane fraction only: additionally incubate at 60 °C for 5 min with shaking at 500 rpm.
3. Add 0.5 μL 200 mM TCEP and incubate at 60 °C for 1 h, shaking at 500 rpm.
4. Cool down to 25 °C (let sit for approximately 5 min on the bench).
5. Add 0.5 μL 400 mM iodacetamide. Incubate at 25 °C for 30 min, shaking at 500 rpm, protected from light.
6. Add 0.5 μL 0.5 M N-acetyl-cysteine and incubate at 25 °C for 10 min, shaking at 500 rpm.
7. Membrane fraction only: add 1 μL Lys-C and incubate at 37 °C, overnight. Then adjust the urea concentration to <1.6 M by adding 100 μL ABC 100 mM before addition of trypsin.
8. Add 2 μg trypsin and seal the tubes with parafilm to prevent evaporation.
9. Once trypsin has been added, incubate overnight at 37 °C, shaking at 500 rpm. After this step, samples can optionally be stored at −20 °C if desired.
10. Short spin to collect the liquid at the bottom of the tube.
11. Separate the beads with a magnet.
12. Transfer the supernatant into a new 1.5 mL tube (repeat this step if beads are left).
13. For the soluble fraction, add 5 μL 5% (v/v) TFA and 1 μL 2 M HCl. For the membrane fraction, add 15% (v/v) TFA and 1 μL 5 M HCl.
14. Condition C18 MicroSpin columns (one per sample) with 150 μL acetonitrile.
15. Spin the column for 20 s at 550 × *g* using a 2 mL tube as a collection tube.
16. Equilibrate the C18 columns with twice 150 μL 0.1% (v/v) TFA (spin 20 s, 550 × *g*).
17. Load the sample and spin 2 min, 400 × *g*. If the sample is not eluting properly, increase the centrifugation speed up to 1200 × *g*.
18. Reload the flow-through onto the column and repeat the spinning step (2 min, 400 × *g*).
19. Wash the column three times with 150 μL 0.1% TFA, 5% acetonitrile (20 s, 550 × *g*).

20. Elute twice with 150 μL 0.1% (v/v) TFA, 50% (v/v) acetonitrile and collect eluate in a fresh tube (2 min, 400 × *g*).
21. Dry the peptides in a speed-vac for 2 h at 60 °C using the aqueous solution settings.
22. Resuspend the peptides in 40 μL 98% HPLC grade H_2O, 2% (v/v) acetonitrile, 0.15% (v/v) formic acid. Complete resuspension is achieved by 20 s sonication (pulse cycle 0.5, amplitude 100%), followed by 10 s vortexing at max speed.
23. Transfer the peptides to a HPLC vial for LC-MS/MS analysis. Optionally, the samples can be stored at −20 °C until further use.

3.8 LC-MS/MS Analysis

1. Make a slurry of C18 beads with methanol and load a PicoFrit fused silica column using a pressure cell. This setting allows flexible column lengths that can be adapted to the sample complexity. For this study we used a 37 cm column. Backpressure was reduced to around 300 bar using an in-house designed column heater. Although we prefer to self-pack our separating columns for an optionally flexible column length setup, it is possible to purchase prepacked fused silica capillary columns (*e.g.*, via New Objectives) or UPLC columns that have to be connected to a nano spray emitter (*e.g.*, via New Objectives or Thermo Scientific). Please note that, for a flexible column set up, ion sources that allow a more versatile set up are required (*e.g.*, Nanospray Flex Ion source from Thermo Scientific for Orbitrap instruments).
2. Before starting acquisition make sure that the mass spectrometer is properly calibrated.
3. Peptides are separated on the self-packed C18 column using a linear gradient from 95% LC solvent A and 5% LC solvent B to 35% LC solvent B (and 65% solvent A) over the course of 60 min at a flow rate of 200 nL/min. (*see* **Note 12**). Depending on the peptide sample composition the earliest peptides elute at around 8–10% LC solvent B.
4. Mass spectra of ionized peptides are acquired on an Orbitrap Elite. Full MS scans are acquired at 240,000 resolution (at 400 *m/z*) in the Orbitrap part of the instrument and the 20 most intense ions are selected for collusion induced fragmentation (CID) in the linear ion trap. The basic acquisition parameters are: Spray voltage of 2.5 kV, heating capillary temperature 300 °C, MS1 scan range of 375–1600 *m/z*, automatic gain control (AGC) is 1×10^6 for MS survey scans and 1×10^5 for MS/MS scans, ion accumulation time is set to 300 ms (MS) and 50 ms (MS/MS), normalized collusion energy of 35% and dynamic exclusion duration was set to 30 s.

3.9 Database Search

1. Download the *P. aeruginosa* uniprot-database (or equivalent organism-specific databases) via the Uniprot homepage (http://www.uniprot.org/).
2. Use the MassMatrix conversion tool (http://www.massmatrix.net/mm-cgi/downloads.py) to convert the MS raw spectra into mascot generic files (mgf).
3. Search this mgf file using MASCOT version 2.3 (www.matrixscience.com) against the *P. aeruginosa* uniprot-database containing forward and reverse-decoy protein entries.
4. Perform an in silico tryptic digestion after lysine and arginine (unless followed by proline) tolerating two missed cleavages in fully tryptic peptides.
5. Set the database search parameters to allow oxidized methionines (+15.99491 Da) as variable modifications and carboxyamidomethylation (+57.021464 Da) of cysteine residues as fixed modification.
6. For MASCOT searches using high-resolution scans, set the precursor mass tolerance to 10 p.p.m. and set the fragment mass tolerance to 0.6 Da.
7. Import the Mascot searches of *P. aeruginosa* CCMS experiments into Scaffold (Proteomesoftware, Version 3 or higher), set the parameters to obtain a protein FDR close to 1%, and extract the total spectral counts or quantitative value.
8. Data can be exported in any spreadsheet software for further analysis.

3.10 Label-Free Quantification

1. Import the raw files into Progenesis QIP software. Perform LC-MS alignment and feature detection (default settings). Export the data in mgf format.
2. Search the MS/MS spectra using the MASCOT against the uniprot *P. aeruginosa* database containing forward and reverse-decoy protein entries. Reverse-decoy hits are required to calculate identification false-discovery rates.
3. Import the database search results into Progenesis QIP and map the peptides identifications to MS1 features.
4. For data evaluation (calculation of significance levels, fold-change ratios) the R-based SafeQuant script should be used (https://github.com/eahrne/SafeQuant) (*see* **Note 13**). We recommend annotating the known or predicted c-di-GMP effectors and determine their specificity and significance to decide a threshold to refine the list of potent candidates (Fig. 3).

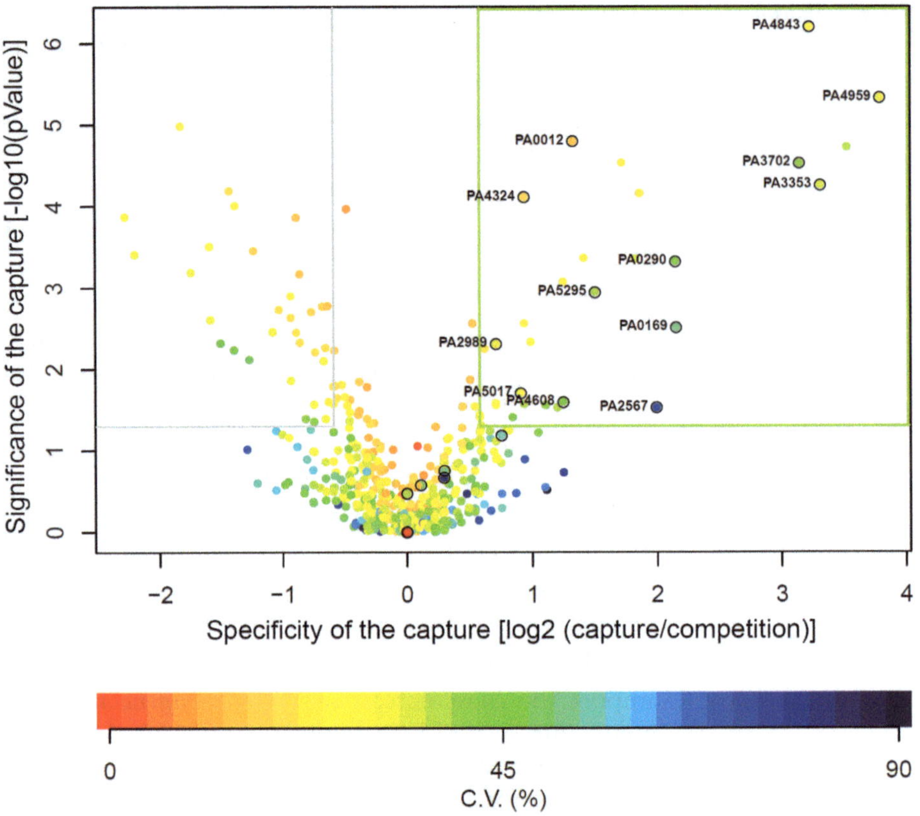

Fig. 3 Example of result obtained with *Pseudomonas aeruginosa* PA01. CCMS was performed for the soluble protein fraction in the presence of 5 μM cdG-CC ($n = 4$). The volcanoplot represents proteins identified by mass spectrometry which are enriched in the capture experiment compared to the competition control (1 mM free c-di-GMP). The enrichment is represented as a Log2-intensity ratio of detected peptides, and plotted against a significance analysis (modified *t*-statistic, empirical Bayes method [25]). Previously reported c-di-GMP binding proteins serve as positive control and are circled in *black*. The *green box* represents a threshold of significance for proteins with intensity ratios >1.5 fold and *p*-values <0.05

4 Notes

1. In this example we used Lysogeny broth (LB) medium for bacterial growth (10 g/L Bacto tryptone, 10 g/L NaCl, 5 g/L yeast extract, autoclaved), which should be replaced by any suitable media adapted to the organism used. The procedure can easily be adapted for other bacteria and has been successfully used for three different bacterial species with minor modifications [1]. For guidance, we used 100 mL culture/sample for stationary phase cultures, and 500 mL culture/sample for log phase cultures ($OD_{600nm} = 0.5$). It is recommended to perform three to four independent biological replicas in parallel.

2. Keeping samples at 4 °C, besides decreasing proteases activity, will also reduce the risk of degradation of the c-di-GMP moiety of the capture compound by phosphodiersterases from the cell lysate.
3. The protein concentration is a critical parameter. An ideal concentration should be in the range of 10 mg/mL of protein. This concentration can be challenging to reach when cells are grown under specific conditions (*e.g.*, biofilms or small colony variants). Therefore, the pellet resuspension should be performed in a small volume of lysis buffer (or alternatively the culture volume increased). The protein concentrations can be decreased to 8 mg/mL with satisfying results.
4. A desalting step allows removing free nucleotides, especially free c-di-GMP, which could compete with the cdG-CC and decrease the capture efficiency. Dialysis is a less preferred option as it requires longer times.
5. The choice of an optimal detergent may need to be optimized for specific bacteria. We found that DDM gave the best result for solubilizing membrane proteins of *Pseudomonas aeruginosa*. Note that the detergent should also be MS compatible, or otherwise washed away before the MS analysis.
6. Compared to the method published by Nesper *et al.* [1], we added various nucleotides to the capture reaction to minimize nonspecific capturing of nucleotide binding proteins. Although the addition of nucleotides improved the specificity, it may at the same time prevent the capture of proteins that bind different nucleotides at the same site. For example the effector FleQ, which was recently shown to bind ATP and c-di-GMP [22] was fished specifically in the absence of ATP in our previous experiment, but not in the experiments outlined here, which were carried out in the presence of an excess of ATP. Despite this optimization, many nucleotide-binding proteins were captured with a relatively low specificity and, to a large degree, may represent false-positive hits.
7. The cdG-CC concentration can be optimized depending on the organisms analyzed. The cdG-CC concentration can be adjusted from 1 to 10 μM.
8. The cdG-CC should be carefully protected from light. Although ambient light contains only a small fraction of UV, it is recommended to keep the capture compound stock wrapped in aluminum foil, as well as the capture mix previous to the activation by UV irradiation.
9. Other capture compounds are commercially available, or can be designed and synthetized for tailored pull downs (*e.g.*, cAMP-CC, *S*-adenosyl-L-homocysteine-CC), and could be used with minor modifications of this method [19, 23, 24]. One example

relevant for the c-di-GMP field is the immobilization c-di-GMP on Sepharose 6B resin by Düvel and colleagues, which was used successfully to identify new c-di-GMP effectors [20]. Moreover, the cdG-CC harbors a chemically modified c-di-GMP molecule, as c-di-GMP is linked via the 2'OH group of one ribose to the Capture Compound backbone. This chemical modification may alter the interaction with certain c-di-GMP effectors, and may therefore generate false-negative results. Using different capture compounds in parallel, or using alternative pull down strategies, may prove useful to improve the identification of bona fide effectors.

10. It is recommended to adjust the reactive group activation by testing various UVs exposure time.
11. As the cdG-CC is covalently bound to target proteins, and as the streptavidin-biotin interaction is strong, stringent washing conditions can be applied (*e.g.*, 1 M NaCl, high detergent concentration, 2 M urea in case of membrane proteins, 80% acetonitrile). This allows reducing the rate of false-positives and limiting the capture of protein complexes, thereby saving time on tedious candidate validation.
12. Depending on the sample complexity different length of analytical columns and gradient lengths can be chosen. Increasing backpressure can be lowered by using a columns heater, which in turn will at least partially compensate for slower acquisition speed on older instrument versions.
13. Although we use Progenesis QIP for label-free quantification followed by SafeQuant evaluation of the data, several other noncommercial solutions can be alternatively be implemented (*e.g.*, MaxQuant for LFQ and Perseus for data evaluation). In addition, Progenesis also supports other search engines/environments such as SEQUEST, Phenyx, and Proteome Discoverer. Depending on the strategy, the calculation of the false-discovery rate is embedded within the search environment, and exported searches can be performed using a standard forward protein database without including the reverse-decoy hits.

Acknowledgments

We thank Jutta Nesper who established the protocol and Alberto Reinders for his work in adapting the CCMS conditions for *P. aeruginosa*. We also thank Pablo Manfredi for the annotation of the *P. aeruginosa* proteins and Erik Ahrné for his help with the MS data evaluation. This work was supported by the Swiss National Science Foundation (SNF) Sinergia grant CRSII3_127433.

References

1. Nesper J, Reinders A, Glatter T, Schmidt A, Jenal U (2012) A novel capture compound for the identification and analysis of cyclic di-GMP binding proteins. J Proteom 75 (15):4874–4878. doi:10.1016/j.jprot.2012.05.033
2. Steiner S, Lori C, Boehm A, Jenal U (2013) Allosteric activation of exopolysaccharide synthesis through cyclic di-GMP-stimulated protein-protein interaction. EMBO J 32 (3):354–368. doi:10.1038/emboj.2012.315
3. Tschowri N, Schumacher MA, Schlimpert S, Chinnam NB, Findlay KC, Brennan RG, Buttner MJ (2014) Tetrameric c-di-GMP mediates effective transcription factor dimerization to control Streptomyces development. Cell 158 (5):1136–1147. doi:10.1016/j.cell.2014.07.022
4. Trampari E, Stevenson CE, Little RH, Wilhelm T, Lawson DM, Malone JG (2015) Bacterial rotary export ATPases are allosterically regulated by the nucleotide second messenger cyclic di GMP. J Biol Chem 290 (40):24470–24483. doi:10.1074/jbc.M115.661439
5. Rotem O, Nesper J, Borovok I, Gorovits R, Kolot M, Pasternak Z, Shin I, Glatter T, Pietrokovski S, Jenal U, Jurkevitch E (2015) An extended cyclic di-GMP network in the predatory bacterium Bdellovibrio bacteriovorus. J Bacteriol 198(1):127–137. doi:10.1128/JB.00422-15
6. Hengge R (2009) Principles of c-di-GMP signalling in bacteria. Nat Rev Microbiol 7 (4):263–273. doi:10.1038/nrmicro2109
7. Sondermann H, Shikuma NJ, Yildiz FH (2012) You've come a long way: c-di-GMP signaling. Curr Opin Microbiol 15 (2):140–146. doi:10.1016/j.mib.2011.12.008
8. Schirmer T, Jenal U (2009) Structural and mechanistic determinants of c-di-GMP signalling. Nat Rev Microbiol 7(10):724–735. doi:10.1038/nrmicro2203
9. Christen M, Christen B, Folcher M, Schauerte A, Jenal U (2005) Identification and characterization of a cyclic di-GMP-specific phosphodiesterase and its allosteric control by GTP. J Biol Chem 280(35):30829–30837. doi:10.1074/jbc.M504429200
10. Ryan RP, Fouhy Y, Lucey JF, Dow JM (2006) Cyclic di-GMP signaling in bacteria: recent advances and new puzzles. J Bacteriol 188 (24):8327–8334. doi:10.1128/JB.01079-06
11. Duerig A, Abel S, Folcher M, Nicollier M, Schwede T, Amiot N, Giese B, Jenal U (2009) Second messenger-mediated spatiotemporal control of protein degradation regulates bacterial cell cycle progression. Genes Dev 23(1):93–104. doi:10.1101/gad.502409
12. Newell PD, Monds RD, O'Toole GA (2009) LapD is a bis-(3′,5′)-cyclic dimeric GMP-binding protein that regulates surface attachment by Pseudomonas fluorescens Pf0–1. Proc Natl Acad Sci U S A 106(9):3461–3466. doi:10.1073/pnas.0808933106
13. Amikam D, Galperin MY (2006) PilZ domain is part of the bacterial c-di-GMP binding protein. Bioinformatics 22(1):3–6. doi:10.1093/bioinformatics/bti739
14. Benach J, Swaminathan SS, Tamayo R, Handelman SK, Folta-Stogniew E, Ramos JE, Forouhar F, Neely H, Seetharaman J, Camilli A, Hunt JF (2007) The structural basis of cyclic diguanylate signal transduction by PilZ domains. EMBO J 26(24):5153–5166. doi:10.1038/sj.emboj.7601918
15. Li W, Cui T, Hu L, Wang Z, Li Z, He ZG (2015) Cyclic diguanylate monophosphate directly binds to human siderocalin and inhibits its antibacterial activity. Nat Commun 6:8330. doi:10.1038/ncomms9330
16. Fang X, Ahmad I, Blanka A, Schottkowski M, Cimdins A, Galperin MY, Romling U, Gomelsky M (2014) GIL, a new c-di-GMP-binding protein domain involved in regulation of cellulose synthesis in enterobacteria. Mol Microbiol 93(3):439–452. doi:10.1111/mmi.12672
17. Wang YC, Chin KH, Tu ZL, He J, Jones CJ, Sanchez DZ, Yildiz FH, Galperin MY, Chou SH (2016) Nucleotide binding by the widespread high-affinity cyclic di-GMP receptor MshEN domain. Nat Commun 7:12481. doi:10.1038/ncomms12481
18. Lenz T, Poot P, Grabner O, Glinski M, Weinhold E, Dreger M, Koster H (2010) Profiling of methyltransferases and other S-adenosyl-L-homocysteine-binding proteins by capture compound mass spectrometry (CCMS). J Vis Exp: JoVE 46. doi:10.3791/2264
19. Köster H, Little DP, Luan P, Muller R, Siddiqi SM, Marappan S, Yip P (2007) Capture compound mass spectrometry: a technology for the investigation of small molecule protein interactions. Assay Drug Dev Technol 5(3):381–390. doi:10.1089/adt.2006.039
20. Düvel J, Bertinetti D, Möller S, Schwede F, Morr M, Wissing J, Radamm L, Zimmermann

B, Genieser H, Jänsch L, Herberg F, Häussler S (2012) A chemical proteomics approach to identify c-di-GMP binding proteins in Pseudomonas aeruginosa. J Microbiol Methods 88 (2):229–236. doi:10.1016/j.mimet.2011.11.015

21. Laventie BJ, Nesper J, Ahrne E, Glatter T, Schmidt A, Jenal U (2015) Capture compound mass spectrometry–a powerful tool to identify novel c-di-GMP effector proteins. J Vis Exp: JoVE 97. doi:10.3791/51404

22. Baraquet C, Harwood CS (2013) Cyclic diguanosine monophosphate represses bacterial flagella synthesis by interacting with the Walker a motif of the enhancer-binding protein FleQ. Proc Natl Acad Sci U S A 110 (46):18478–18483. doi:10.1073/pnas.1318972110

23. Bertinetti D, Schweinsberg S, Hanke SE, Schwede F, Bertinetti O, Drewianka S, Genieser HG, Herberg FW (2009) Chemical tools selectively target components of the PKA system. BMC Chem Biol 9:3. doi:10.1186/1472-6769-9-3

24. Dalhoff C, Huben M, Lenz T, Poot P, Nordhoff E, Koster H, Weinhold E (2010) Synthesis of S-adenosyl-L-homocysteine capture compounds for selective photoinduced isolation of methyltransferases. Chembiochem: a Eur J Chem Biol 11(2):256–265. doi:10.1002/cbic.200900349

25. Smyth GK (2004) Linear models and empirical bayes methods for assessing differential expression in microarray experiments. Stat Appl Genet Mol Biol 3. Article3. doi:10.2202/1544-6115.1027

Chapter 29

Identification of c-di-GMP-Responsive Riboswitches

Johann Peltier and Olga Soutourina

Abstract

Bis-(3′-5′)-cyclic dimeric guanosine monophosphate (c-di-GMP) is an important signaling molecule for community behavior control, cell morphogenesis, and virulence in bacteria. In addition to protein effectors, this second messenger binds RNA molecules that act as riboswitches to control target gene expression. In this chapter, we describe a method for experimental validation of the functionality of c-di-GMP-responsive riboswitches and the analysis of c-di-GMP control of target gene expression by qRT-PCR and Northern blot. This procedure can be used for the studies of in silico-predicted riboswitch candidates, as well as a targeted experimental approach for exploring the data from next-generation sequencing. The examples on the analysis of type I and type II c-di-GMP-responsive riboswitches in *Clostridium difficile* are provided to illustrate the application of the method.

Key words Riboswitches, c-di-GMP, Premature termination of transcription, Target gene, Northern blot, qRT-PCR

1 Introduction

The mRNA leaders, the so-called "riboswitches," affect gene expression *in cis* by direct binding of effector molecules [1]. "Riboswitches" are part of the mRNAs they regulate and usually control vitamin, amino acid, and base biosynthesis and uptake in bacteria [2]. These leader sequences fold into structures subjected to conformational changes upon the binding of small molecules leading to positive or negative effects on transcription termination or translation and also RNA cleavage or transcript processing [2].

These genetic switches are composed of two modular domains consisting of an aptamer and an expression platform. The aptamer is involved in the specific recognition of the metabolite and the expression platform is used to control gene expression by altering the structure of mRNA. The molecular mechanism of riboswitch function includes a ligand-dependent conformational switch between two mutually exclusive structures (Fig. 1). Metabolite binding to an aptamer domain induces an allosteric conformational switch of an expression platform, leading to activation or repression

Karin Sauer (ed.), *c-di-GMP Signaling: Methods and Protocols*, Methods in Molecular Biology, vol. 1657,
DOI 10.1007/978-1-4939-7240-1_29,

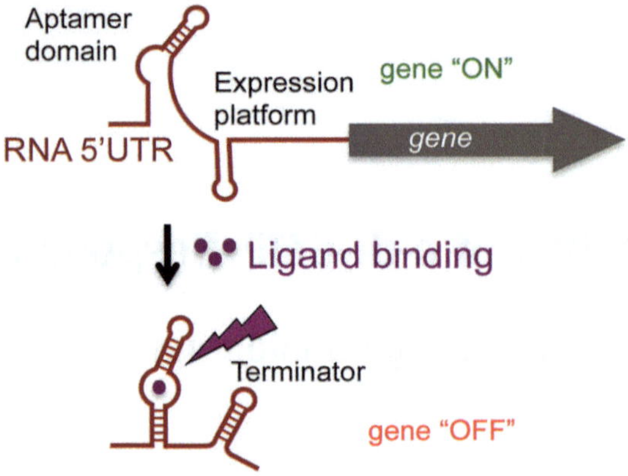

Fig. 1 Ligand-dependent conformational switch for premature termination of transcription

of metabolite-related gene expression. In the case of transcription, the structures serve as a Rho-independent or Rho-dependent terminator and anti-terminator hairpins. In controlling translation, the structures sequester or release ribosome-binding sites. Unconventional mechanisms associated with riboswitch action have been recently discovered including coupling of metabolite sensing with mRNA cleavage by the *glmS* riboswitch-ribozyme, the action of SAM-responsive riboswitch *in trans* as independent sRNA, and the induction of alternative self-splicing and translation activation by the type II c-di-GMP-dependent riboswitch [2].

Bis-(3′-5′)-cyclic dimeric guanosine monophosphate (c-di-GMP) is an important secondary messenger in bacteria that controls lifestyle switch from free-living motile state to attached communities in a biofilm, and also virulence of several pathogens [3]. High intracellular c-di-GMP levels are associated with nonflagellated forms and low intracellular levels with flagellated planktonic forms. C-di-GMP is synthesized from two GTP molecules by enzymes named diguanylate cyclases or c-di-GMP synthetases and it is degraded into pGpG or two GMP by phosphodiesterases in response to a variety of stimuli including light, oxidative conditions, cell density, and antibiotics. Effector molecules bound by this messenger include transcriptional factors, other protein effectors, and RNA*Reporter* molecules that act as riboswitches [4].

Two types of c-di-GMP-responsive riboswitches were discovered that differ in structure and the recognition of ligand [5–7]. Type I c-di-GMP riboswitch (c-di-GMP-I) carries a phylogenetically widespread "GEMM" RNA domain (Fig. 2a). There are over 500 examples of the GEMM RNA domain in many diverse bacterial species that folds into a complex secondary and tertiary structure forming a selective pocket for c-di-GMP binding [8]. The affinity of this binding is very high with Kd of 10 pM representing one of the highest affinity RNA–ligand interactions known [7]. C-di-GMP-I

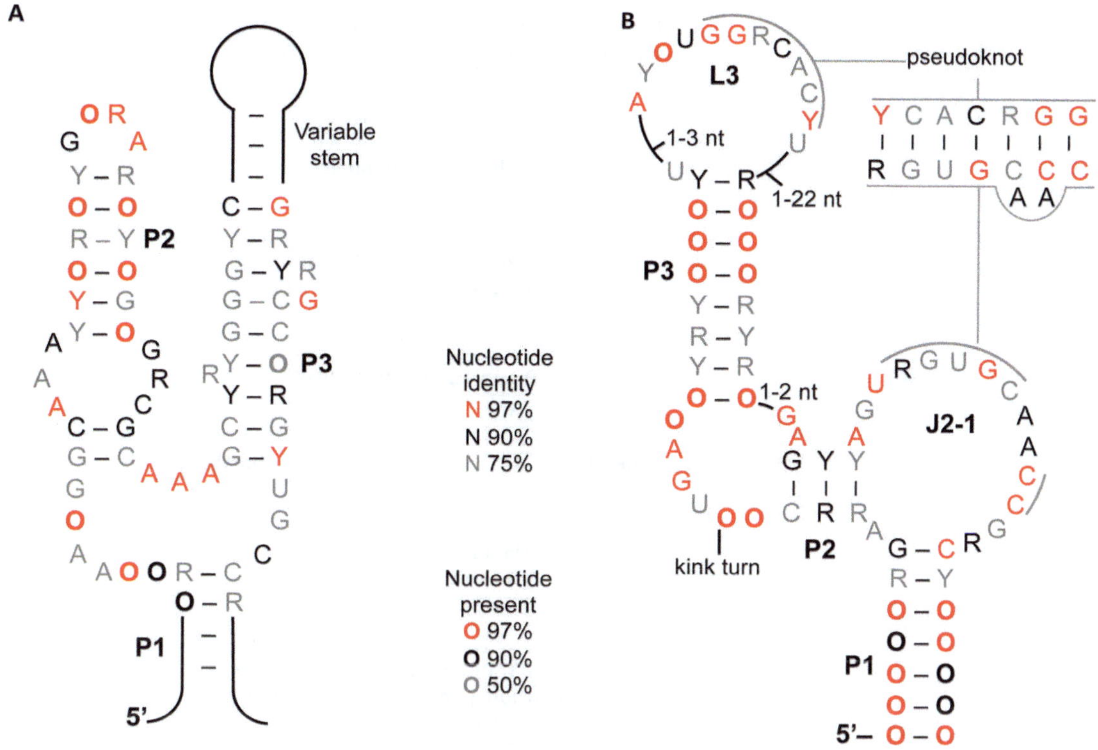

Fig. 2 Consensus sequences of GEMM (**a**) and c-di-GMP-II (**b**) aptamers. The riboswitch structures are based on [5, 20]

riboswitches positively or negatively control their target gene expression through premature termination of transcription or anti-termination [8]. The second type of c-di-GMP binding riboswitches (c-di-GMP-II) was detected in a bioinformatics search in anaerobic Gram-positive members of the Clostridiales family (Fig. 2b) [5]. C-di-GMP-II riboswitches positively affect target gene expression through modulation of target gene translation in association with group I intron self-splicing or through anti-termination of transcription [5, 9, 10].

Here, we provide a comprehensive protocol to experimentally validate the functionality of c-di-GMP-responsive riboswitch candidates (Fig. 3). A necessary prerequisite to this protocol is the prediction of riboswitches responding to c-di-GMP. If such data are not yet available for the microorganism to be studied, genome-wide analysis for systematic search of possible riboswitch activity will need to be conducted. This is made possible by using next-generation sequencing techniques such as the recently published combination of RNA-seq and transcriptional start site mapping [11] or the Term-seq method [12]. These powerful approaches combined with an in silico search for c-di-GMP-specific aptamers included in the Rfam database [13] (Fig. 2) will provide a robust

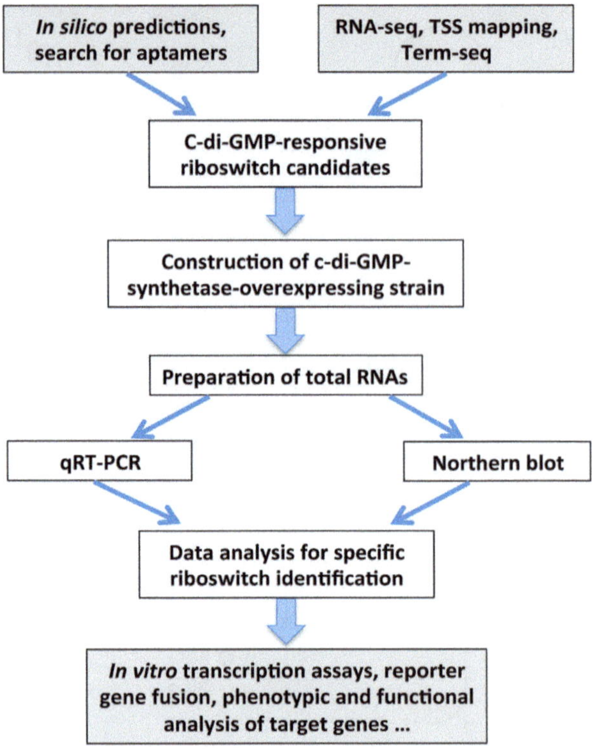

Fig. 3 Schematic overview of the experimental procedure

basis for the prediction of c-di-GMP-dependent riboswitches acting through premature termination of transcription [8, 12, 14]. Other mechanisms have also been revealed for c-di-GMP signaling, for example a particular case of alternative splicing controlled by c-di-GMP riboswitch establishing an accessible RBS for translation of target gene [5].

The focus of this protocol is on analyzing the effects of c-di-GMP on target gene expression. It requires the use of a strain producing high c-di-GMP levels. To artificially elevate intracellular c-di-GMP concentration, the easiest approach consists of cloning a gene encoding a diguanylate cyclase in an appropriate vector for the overexpression of the corresponding protein in the bacterial strain of interest. It is strongly recommended to confirm the overexpression of the cloned gene by comparing its transcript levels with that of a wild type control strain carrying an empty vector. Protein levels can also be compared by Western blot if specific antibodies are available. In addition, measurement of the intracellular c-di-GMP concentration by LC-MS/MS [15] provides valuable data to unambiguously demonstrate the presence of elevated levels of c-di-GMP in the strain overexpressing the diguanylate cyclase in comparison to the control strain carrying an empty plasmid. Phenotypic analysis of diguanylate cyclase-overexpressing strain could also help to find appropriate experimental conditions. Indeed, elevated c-di-GMP levels are usually associated with cell aggregation phenotype, loss of motility and increase in biofilm formation [3].

Two experimental approaches are developed in this chapter to analyze target gene expression and validate relevance of c-di-GMP signaling: quantitative real-time reverse transcription polymerase chain reaction (qRT-PCR) and Northern blotting (Fig. 3). qRT-PCR is the method of choice to determine whether the transcript abundance of a gene adjacent to a predicted c-di-GMP riboswitch is actually affected by an elevation of the c-di-GMP levels. Using target gene-specific primers (and not riboswitch-specific primers) (Fig. 4), expression of the target gene in the strain overexpressing the diguanylate cyclase can be compared to the control strain carrying an empty plasmid. Any significant change in the transcript levels will confirm the presence of a c-di-GMP riboswitch controlling the expression of the target gene. Northern blotting will provide complementary information to that obtained by qRT-PCR. This method presents the compelling advantage of visualization of all detectable transcripts produced and the estimation of their size. Using riboswitch-specific probes (Fig. 4), two major RNA species are typically observed in the wild type strain by Northern blot if the riboswith action is associated with premature termination of transcription. The shorter RNA species corresponds to premature termination of transcription while the longer RNA species corresponds to a transcript of the target gene whose expression is controlled by the riboswitch. In the presence of high c-di-GMP, changes in the abundance of the full-length and short terminated RNA species are classically observed. The functionality of c-di-GMP-riboswitches can be further evaluated by in vitro ligand binding and transcription assays, as well as regulatory region fusion with reporter gene (not covered in the present protocol) (Fig. 3) [8, 9].

2 Materials

All aqueous solutions and reagents are prepared with DEPC-treated or RNase-free water (*see* **Note 1**).

2.1 Total RNA Extraction

1. Diguanylate cyclase-overexpressing strain and corresponding control strain.
2. 96% ethanol.
3. Resuspension buffer: 10% glucose, 12.5 mM Tris pH 7.6, 65 mM ethylenediaminetetraacetic acid (EDTA).
4. Acid phenol solution, pH 4.5 (Sigma-Aldrich).
5. TRI reagent (Sigma-Aldrich).
6. Chloroform–isoamyl alcohol (24:1) (v/v).

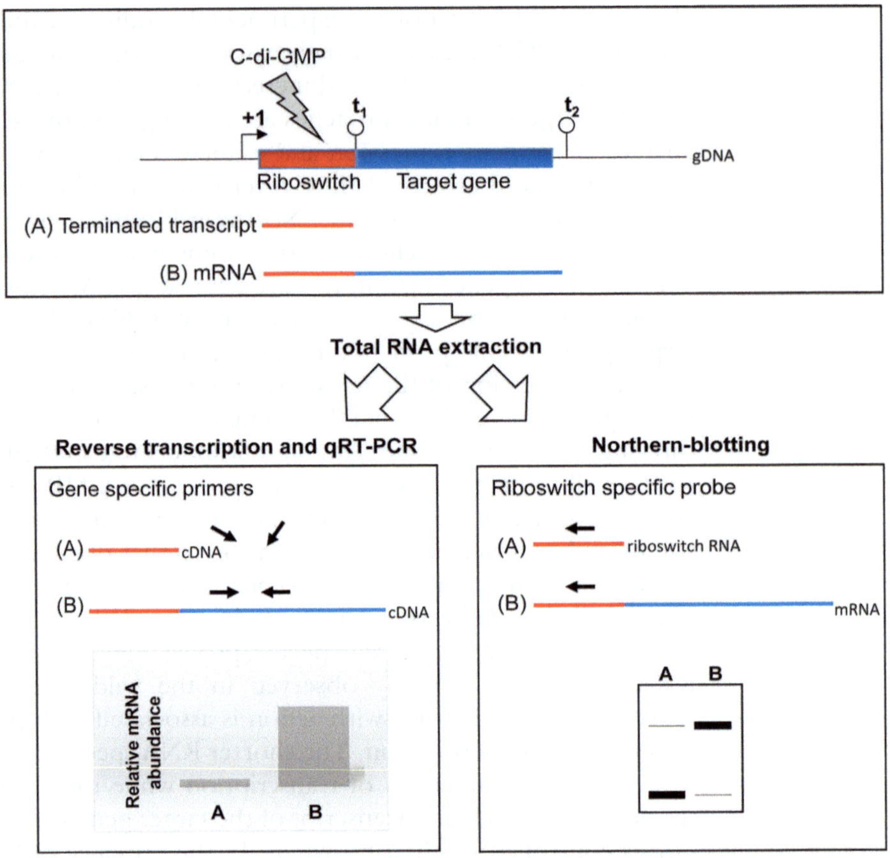

Fig. 4 Schematic overview of the experimental approaches developed in this chapter. Upon binding of c-di-GMP to riboswitches, target gene expression is positively or negatively controlled through premature termination of transcription or antitermination, leading to terminated transcript (**a**) or mRNA (**b**). QRT-PCR with gene specific primers can be performed to compare the transcript levels of a target gene between strains with low or high levels of c-di-GMP. Northern blotting with riboswitch-specific probes can be performed to determine the size and the abundance of all different detectable transcripts in the presence of low or high levels of c-di-GMP. The transcriptional initiation site (+1) and the transcriptional terminators (t_1 and t_2) are represented as well as the position of the gene specific primers or riboswitch specific probe (*black arrows*)

7. Isopropanol.
8. Ice-cold 70% ethanol.
9. RNA buffer: 10 mM Tris–HCl pH 7.6, 0.1 mM EDTA.
10. Screw cap 2 ml tubes (Sarstedt) with acid-washed glass beads 200–300 μm (Sigma-Aldrich) or lysing matrix B 2 ml tubes (MP Biomedicals).
11. TURBO DNA-free kit (Ambion). The contents of TURBO DNA-free kit are as follows:
    - 10× TURBO DNase buffer.
    - TURBO DNase.
    - DNase inactivation reagent.

12. FastPrep homogenizer (MP Biomedicals).
13. Refrigerated centrifuge.
14. Microcentrifuge.
15. Water bath and/or thermal block.
16. NanoDrop spectrophotometer or equivalent.

2.2 Reverse Transcription and Quantitative Real-Time PCR (qRT-PCR)

1. Avian Myeloblastosis Virus (AMV) reverse transcriptase (RT) enzyme and AMV RT 5× buffer (Promega).
2. dNTPs (25 mM).
3. RNasin (Promega).
4. DEPC-treated water.
5. FastStart Universal SYBR Green Master (Rox) (Roche) or equivalent.
6. Forward and reverse gene-specific oligonucleotide primers (10 μM).
7. PCR thermal cycler.
8. Real-time PCR instrument.

2.3 Denaturing Urea Polyacrylamide Gel Electrophoresis for Northern Blot

1. Urea.
2. 10× TBE: 0.89 M Tris base, 0.89 M boric acid, 20 mM EDTA.
3. 40% acrylamide–bis-acrylamide (19: 1) solution.
4. DEPC-treated water or RNase-free water.
5. *N*,*N*,*N*,*N′*-tetramethylethylenediamine (TEMED).
6. Ammonium persulfate 10% (APS).
7. Gel loading buffer II (Ambion) or equivalent.
8. 0.1–2 kb RNA ladder (Life Technologies) or equivalent.
9. 1× TBE buffer.
10. Ethidium bromide (0.7 mg/ml).
11. Vacuum driven disposable filtration system (0.22 μm pore size) (Millipore) or equivalent.
12. Gel electrophoresis system.
13. Saran wrap.

2.4 Electrotransfer of RNAs for Northern Blot

1. Positively charged nylon membrane: Hybond N+ (Amersham) or equivalent.
2. Whatman 3MM paper or Precut blot filter paper for use with the Criterion blotter (Bio-Rad).
3. 1× TBE buffer prepared with DEPC-treated water from RNAse-free 10× stock.
4. DEPC-treated water.

5. Electrotransfer blotting system.
6. UV cross-linker.

2.5 Radioactive Labeling of Probes for Northern Blot

2.5.1 Radioactive Labeling of Oligonucleotides Probes

1. T4 polynucleotide kinase and 10× reaction buffer A (Fermentas).
2. [γ-^{32}P]ATP (800 Ci/mmol, 10 mCi/ml EasyTide) (PerkinElmer).
3. Reverse riboswitch-specific oligonucleotide (10 μM) (*see* **Note 2**).
4. DEPC-treated water.
5. Micro Bio-Spin P-30 gel column, Tris buffer, RNase-free (Bio-Rad) or equivalent.

2.5.2 Radioactive Labeling of Double-Stranded DNA Probes

1. T4 polynucleotide kinase and 10× reaction buffer A (Fermentas).
2. [γ-^{32}P]ATP (800 Ci/mmol, 10 mCi/ml EasyTide) (PerkinElmer).
3. Reverse riboswitch-specific oligonucleotide (10 μM) (*see* **Note 2**).
4. DEPC-treated water.
5. Taq DNA polymerase, 10× Taq buffer with $(NH_4)_2SO_4$ and 25 mM $MgCl_2$ (Thermo Scientific).
6. Forward riboswitch-specific oligonucleotide (10 μM) (*see* **Note 2**).
7. dNTPs (10 mM).
8. Genomic DNA (100 ng/μl).
9. PCR cleanup system (Nucleospin extract II kit from Macherey-Nagel or equivalent).

2.5.3 Radioactive Labeling of RNA Probes

1. Phusion DNA polymerase and 5× Phusion HF buffer.
2. dNTPs (10 mM).
3. Forward riboswitch-specific oligonucleotide (10 μM) (*see* **Note 2**).
4. Reverse riboswitch-specific oligonucleotide with a floating tail containing the T7 polymerase promoter sequence (10 μM) (*see* **Note 2**).
5. Genomic DNA (100 ng/μl).
6. DEPC-treated water.
7. MAXIscript T7 (Ambion). The contents of MAXIscript T7 kit are as follows:
   - 10× transcription buffer.
   - ATP (10 mM).

- CTP (10 mM).
- GTP (10 mM).
- T7 enzyme mix.
- TURBO DNase.

8. [α-^{32} P]UTP (800 Ci/mmol, 10 mCi/ml EasyTide) (PerkinElmer).
9. 0.5 M EDTA pH 8.0.
10. Micro Bio-Spin P-30 gel column, Tris buffer, RNase-free (Bio-Rad) or equivalent.
11. PCR cleanup system (Nucleospin extract II kit from Macherey-Nagel or equivalent).

2.6 Membrane Hybridization for Northern Blot

1. 1× ULTRAhyb buffer (Ambion).
2. Washing buffer I: 2× saline sodium citrate (SSC), 0.1% sodium dodecyl sulfate (SDS).
3. Washing buffer II: 0.1× SSC, 0.1% SDS.
4. Hybridization oven at 42 or 65 °C depending on probes.
5. Typhoon imager or equivalent.

2.7 Membrane Stripping

1. DEPC-treated water containing 0.5% SDS.

3 Methods

All procedures described in the methods section should be carried out in an RNase-free environment (*see* **Note 1**).

3.1 Total RNA Extraction

The RNA extraction method described below is routinely used in the laboratory. Alternative protocols are also available using for example FastRNA PRO™ BLUE kit (MP Biomedicals).

1. Grow 10–15 ml of the diguanylate cyclase-overexpressing strain and the corresponding control strain to the late exponential growth phase.
2. Harvest the bacteria by centrifugation at 4500 × *g* for 10 min at 4 °C.
3. Discard the supernatants and invert tubes briefly over absorbent paper. The cell pellets can be used fresh or immediately frozen in dry ice/96% ethanol and stored at −80 °C.
4. Resuspend the bacterial pellet in 460 μl of resuspension buffer and transfer the suspension into a screw cap 2-ml microtube containing 500 μl of acid phenol and 0.4 g of acid-treated glass

beads. Alternatively, the suspension can be transferred into a lysing matrix B 2 ml tube (MP Biomedicals) containing 500 μl of acid phenol.

5. Lyse the cells in a FastPrep instrument, with the following settings: Speed: 6.5, Time: 30 s.
6. Place tubes on ice for 1 min and repeat the FastPrep lysis using the same settings.
7. Centrifuge at 13,000 × *g* for 5 min at 4 °C.
8. Carefully transfer the aqueous phase (upper layer) into a 2 ml microfuge tube without disturbing the middle layer. Discard the tube with the remaining layers.
9. Add 1 ml TRI reagent to the recovered aqueous phase and mix gently by repetitive pipetting. Incubate for 5 min at room temperature.
10. Add 100 μl chloroform–isoamyl alcohol (24:1) and mix gently by repetitive pipetting.
11. Incubate for 3 min at room temperature.
12. Centrifuge at 13,000 × *g* for 5 min at 4 °C and carefully transfer the aqueous phase (upper layer) into a new 2 ml microfuge tube. Discard the tube with the remaining layers.
13. Add 200 μl chloroform–isoamyl alcohol (24:1) to the recovered aqueous phase and mix gently by repetitive pipetting. Incubate at room temperature for 5 min.
14. Centrifuge at 13,000 × *g* for 5 min at 4 °C and carefully transfer the aqueous phase (upper layer) into a 1.5 ml microfuge tube.
15. Add 500 μl isopropanol and mix by inverting the tube. Incubate for at least 15 min on ice.
16. Centrifuge at 13,000 × *g* for 15 min at 4 °C to precipitate RNAs.
17. Carefully remove the supernatant without disturbing the pellet and wash the pellet by adding 1 ml 70% ethanol.
18. Centrifuge at 13,000 × *g* for 5 min at 4 °C, pipette off ethanol and allow the pellet to dry (about 5 min at room temperature under vacuum conditions or 15 min on the bench, avoid to overdry the pellet).
19. Resuspend each pellet in 100 μl of RNA buffer preheated to 37 °C.
20. Incubate at 37 °C for 15 min to help for pellet dissolution. Place samples on ice.
21. Add 10 μl of TURBO DNase buffer and 2 μl of TURBO DNase to your RNA samples (*see* **Note 3**).

22. Incubate at 37 °C for 60 min.
23. Add 10 μl of DNase inactivation reagent.
24. Incubate at room temperature for 2 min. Mix the samples three times during the incubation time.
25. Centrifuge at 10,000 × *g* for 1 min at room temperature.
26. Carefully transfer the supernatant into a new centrifuge tube. Make sure to not transfer any of the DNase inactivation reagent-containing pellet as it would interfere with further applications. A second round of centrifugation and supernatant transfer can be performed if any of the pellet has been pipetted.
27. Measure the RNA concentrations using a NanoDrop ND-1000 spectrophotometer (*see* **Note 4**).
28. After quantification, aliquot the RNA samples (to avoid repetitive freeze–thaw cycles) and either store at −80 °C or use immediately. At this step the RNA quality could be checked by agarose gel electrophoresis or on Agilent 2100 Bioanalyzer if available.

3.2 Reverse Transcription and Quantitative Real-Time PCR (qRT-PCR)

Prior to qRT-PCR reaction, one or two reference (housekeeping) gene(s) need to be selected to serve as internal standard for normalization of data. Design of primers for target sequences is also a key step and use of dedicated programs such as Primer3web (bioinfo.ut.ee/primer3/) is strongly recommended. General recommendations for primer design include a primer melting temperature (Tm) of around 60 °C, a primer size close to 20 nucleotides and a product size ranging between 100 and 150 bp.

1. Mix 1 μg of total RNA with 1 μl of $p(dN)_6$ random primer and DEPC-treated water up to 25 μl in a microcentrifuge tube.
2. Incubate at 70 °C for 10 min to remove any secondary structure and place immediately on ice to prevent the refolding.
3. Add the following components to the microcentrifuge tube:

AMV RT 5× buffer	10 μl
dNTPs (25 mM)	4 μl
RNasin	1 μl
AMV RT enzyme	1 μl
DEPC-treated water	9 μl

4. Incubate at 37 °C for 2 h and then at 85 °C for 5 min.
5. Add 50 μl DEPC-treated water to the mixture to dilute the cDNAs to 10 ng/μl. The cDNA is ready for qRT-PCR (*see* **Note 5**). Store cDNAs at −20 °C or use immediately.

6. For a typical qRT-PCR reaction, mix the reaction components below in a PCR strip or a 96 well plate:

FastStart Universal SYBR Green Master (Rox)	10 μl
Water	6 μl
Forward primer (10 μM)	1 μl
Reverse primer (10 μM)	1 μl
cDNA (10 ng/μl) (from 3.2 step 5)	2 μl

7. Run samples using a qRT-PCR machine. A standard qRT-PCR program is as follows:

PCR step	Temperature (°C)	Time	Number of cycles
Enzyme activation	95	10 min	1
Denaturation	95	15 s	40
Annealing/extension	60	1 min	40

8. The number of PCR cycles required to reach the threshold cycle (C_T) will be used for comparison of gene expression levels. For each sample, normalize the quantity of cDNAs of the target gene to the quantity of cDNAs of the reference gene (s). Use the comparative C_T method ($2^{-\Delta\Delta CT}$) to get the expression fold change [16].

3.3 Denaturing Urea Polyacrylamide Gel Electrophoresis for Northern Blot

Using this method, RNAs are denatured in urea, which disrupts the secondary structure, and are separated on the basis of their size as they travel through a polyacrylamide gel toward the anode. This method results in excellent separation of small size RNAs ranging between 5 and 1000 nucleotides. For detection of larger transcripts denaturing formaldehyde-containing agarose gel electrophoresis would be more suitable followed by capillary transfer procedure. The protocol described here employs presealed plastic cassettes (Criterion empty cassettes, Bio-Rad), but it is readily adaptable for use with any other type of gel casting. Ready-to-use RNAse-free precast gels can be also used.

1. Prepare 200 ml acrylamide gel solution with the desired acrylamide percentage depending on the separation range (Table 1) (*see* **Note 6**). The gel solution can be used for a series of experiments.
2. Heat the solution with stirring to dissolve urea until a homogenous solution is obtained (*see* **Note 7**).
3. Filter the solution through 0.45 μm membrane filter. The solution can be kept at room temperature for up to 2 months if protected from light (e.g., wrapped in aluminum foil).

Table 1
Recipe for 8 M urea-denaturing acrylamide gel

% Final acrylamide	3.5	5	8	10	15	20
Separation range (nucleotides)	>500	150–500	90–350	60–150	30–60	5–30
Urea (g)	100	100	100	100	100	100
TBE 10× (ml)	20	20	20	20	20	20
40% acrylamide–bis-acrylamide (ml)	17.5	25	40	50	75	100
DEPC-treated water (ml)	87.5	80	65	55	30	5

Adapted from [21]

4. To prepare two midi gels, transfer 30 ml of the gel solution prepared in **step 3** into a 50 ml conical tube. Add 60 μl of 25% APS and 30 μl of TEMED to the gel solution and mix gently.
5. Pour the gel into the cassette and gently insert the corresponding comb (*see* **Notes 8** and **9**). Allow to polymerize for approximately 30 to 40 min at room temperature. The remainder of the gel solution in the 50 ml conical tube can be used to check the polymerization state of the gel.
6. When the gel is polymerized, gently remove the comb and the rubber seal from the bottom edge.
7. Assemble the gel apparatus and fill the tank (top and bottom chambers) with 1× TBE buffer.
8. Prerun the gel at 25 mA for 20 min (*see* **Note 10**).
9. While prerunning the gel, prepare all samples to be analyzed as follows:
   (a) Mix 10 μl of each RNA sample adjusted at a concentration of 0.5 μg/μl (5 μg RNA) with an equal volume of gel loading buffer II. Note that the sample volume will depend on the well capacity of the gel.
   (b) Mix 7 μl of 0.1–2 kb RNA ladder with an equal volume of gel loading buffer II.
   (c) Denature samples and RNA ladder by incubation 5 min at 95 °C and place immediately on ice.
10. Just before loading the samples into the wells of the gel, rinse wells thoroughly with a pipette to flush urea out of the wells using 1× TBE buffer. Load 20 μl of the samples and 14 μl of the RNA ladder into the wells without introducing any air bubbles (*see* **Note 11**).
11. Run the gel at 50 mA for approximately one hour. Gel loading Buffer II (Ambion) contains two tracking dyes, xylene cyanol and bromophenol blue that are used to monitor the progress of

Table 2
Dye migration in denaturing polyacrylamide gels

% Acrylamide	Xylene cyanol (nucleotides)	Bromophenol blue (nucleotides)
3.5	210	55
5	130	35
8	75	19
10	55	12
15	28	8
20	28	8

Adapted from [21]

migration. The migration of the tracking dye depends on the acrylamide gel composition (Table 2). In routine use, the gels are run until the bromophenol blue gets out and the xylene cyanol is about 2 cm from the bottom of the gel.

12. Disassemble the gel-running apparatus and separate the gel plates with the help of a spatula. Let the gel set on one of the plastic plates. Cut off a corner of the gel for orientation.
13. Lay clear Saran wrap over the top of the gel, flip the gel over and detach the gel from the plate. The gel is very fragile and needs to be handled with adequate care. Saran wrapping of the gel is helpful for further handling.
14. Stain the gel by soaking in 1× TBE buffer with 0.5 μg/ml ethidium bromide for 10 min (*see* **Note 12**).
15. Eliminate the excessive dye by soaking in 1× TBE buffer for 5 min.
16. Capture the image with a gel documentation system for quality and loading controls of the RNAs.

3.4 Electrotransfer of RNAs for Northern Blot

This protocol has been optimized for use with a Criterion blotter apparatus but should be readily applicable for use with any wet electrotransfer system.

Before starting the protocol, sufficient amount of transfer TBE buffer (about 2 l) should be chilled to 4 °C to avoid excessive heating during transfer.

1. Separate RNA sample on a urea-denaturing gel as described above (Subheading 3.3).
2. Cut a Hybond N+ membrane and six pieces of Whatman 3MM filter paper to the size of the gel. Alternatively, two pieces of commercial precut blot filter paper can be used.

3. Wet the two fiber pads from the electrotransfer apparatus, the membrane and the two Whatman filter pads in chilled 1× TBE buffer.
4. Assemble the transfer stack in an electrotransfer cassette as follows:
   (a) Lay one fiber pad on the anode side (black grid) followed by three pieces of Whatman filter paper (or one piece of precut blot filter paper).
   (b) Carefully place the gel on top of the filter papers followed by the membrane (*see* **Note 13**).
   (c) Top with three more pieces of filter paper (or one piece of precut blot filter paper).
   (d) Squeeze out any air bubble by rolling a roller or a pipette over the filters.
   (e) Finally place the second fiber pad on the top of the filter papers on the cathode side (red grid).
   (f) Clamp the cassette closed.
5. Finish assembling the wet transfer apparatus and fill the tank with cold 1× TBE buffer.
6. Place the ice block provided for Criterion Bio-Rad transfer system in the back of the cell, drop a stir bar into the tank and run at constant voltage (50 V) for 30 min on a magnetic stirring device.
7. Disassemble the apparatus and remove the membrane from the transfer stack. Cut off a corner of the membrane for orientation. Keep the RNA surface facing up throughout all subsequent steps.
8. UV-cross-link the transferred RNAs to the membrane using a UV cross-linker set at 0.51 J/cm^2.
9. Lay the membrane on Saran and place onto a UV transilluminator to visualize RNAs. Use a pencil to mark the location of the ladder bands. Alternatively, prelabeled ladder can be used for determination of transcript size. Some available RNA ladders include transcripts that can be radioactively labeled or probed with ladder-specific probe for size determination directly on autoradiographs.
10. Wash the membrane for 5 min in DEPC-treated water and dry by incubation at room temperature for at least 30 min (*see* **Note 14**).

3.5 Radioactive Labeling of Probes for Northern Blot

A nucleic acid probe, specifically designed for the sequence of interest, is used for the detection of the RNA of interest. The probe and the target nucleotide sequence need to be complementary to each other. Alternatively, the probe can also be nonisotopically labeled (*see* **Note 15**). The protocol below, however, focuses only on the radioactive labeling of the probes as radiolabeled probes are still probably today the most sensitive and they are the ones that are routinely used in our lab (*see* **Note 16**). Oligonucleotides, double-stranded DNA and in vitro-transcribed RNA can all be used as hybridization probes. Oligonucleotide probes (>20 nucleotides) are the probes of choice to target a short, specific sequence within target genes or riboswitch regions (Fig. 4). Double-stranded DNA probes are classically generated by PCR and can comprise all or most of the target gene or riboswitch region (Fig. 4). They generally provide greater specificity than oligonucleotides because of their longer sequence. However, the probe is double stranded, and a denaturation step is therefore necessary prior to hybridization in order for the specific strand to hybridize with the target RNA. PCR probes are often less sensitive than oligonucleotide probes due to the tendency of the DNA strands to rehybridize to each other. An RNA probe, produced by in vitro transcription reaction, is the best probe in tereporterrms of sensitivity. This is because RNA–RNA hybrids are more stable than DNA–RNA hybrids. Thus, they can be hybridized and washed under more stringent conditions, i.e., higher temperature, which results in lower background. However, these probes can be difficult to work with as they are easily degraded by RNases.

According to your choice of the probes proceed to Subheadings 3.5.1, 3.5.2 or 3.5.3 followed by the membrane hybridization procedure described in Subheading 3.6.

3.5.1 Radioactive Labeling of Oligonucleotide Probes

The oligonucleotide probes need to be designed in antisense orientation to the RNA sequence to be detected (Fig. 4). In order to use the protocol below, the oligonucleotide probes need to be larger than 20 nucleotides to ensure the specificity of hybridization and efficient recovery when using the Micro Bio-Spin P-30 gel columns for labeled oligonucleotide purification.

1. To radioactively label oligonucleotides, mix the reaction components below in a RNAse-free microcentrifuge tube:

Reverse primer (10 μM)	2 μl
[γ-^{32} P]ATP (10 mCi/ml)	3 μl
Polynucleotide kinase buffer A 10×	1 μl
DEPC-treated water	3 μl
Polynucleotide kinase	1 μl

2. Incubate at 37 °C for 30 min and heat-inactivate at 70 °C for 10 min. Readjust the final volume at 10 μl if evaporation occurred.

3. Purify the radioactive oligonucleotides with Micro Bio-Spin P-30 gel columns.
4. Check for the radioactive signal using available counter (Geiger counter or liquid scintillation counter).
5. The purified radioactively labeled oligonucleotide probe can be stored at −20 °C until use but only for a few days, since ^{32}P isotope has a half-life of 14 days.

3.5.2 Radioactive Labeling of Double-Stranded DNA Probes

1. Before amplification of the probe by PCR, label the 3′ primer with [γ-^{32}P]ATP in a microcentrifuge tube as follows:

Reverse primer (10 μM)	2 μl
[γ-^{32} P]ATP (10 mCi/ml)	3 μl
Polynucleotide kinase buffer A 10×	1 μl
DEPC-treated water	3 μl
Polynucleotide kinase	1 μl

2. Incubate at 37 °C for 30 min and heat-inactivate at 70 °C for 10 min.
3. Perform in parallel a radioactive and a control nonradioactive PCR reaction. Any one of the many PCR systems available is suitable for this amplification. The following recipe describes the use of Taq DNA polymerase supplied by Thermo Scientific:

10× buffer	5 μl
$MgCl_2$ (25 mM)	5 μl
dNTPs (10 mM)	1 μl
Forward primer (10 μM)	1 μl
Nonradioactive reverse primer (10 μM) or Radioactive reverse primer	2 μl or 10 μl
Taq polymerase	0.5 μl
Genomic DNA (100 ng/μl)	1 μl
DEPC-treated water	Up to 50 μl

The PCR cycle parameters should be determined empirically for each combination of primers and DNA template. However, a typical PCR cycling program for the amplification of short fragments (< 500 bp) is as follows:

PCR step	Temperature (°C)	Time	Number of cycles
Initial denaturation	94	5 min	1
Denaturation	94	30 s	35
Annealing	55	30 s	35
Extension	72	30 s	35
Final extension	72	5 min	1

4. Load and resolve the control (nonradioactive) PCR product on a freshly prepared 2% agarose gel. Visualization of a band at the expected size will serve as a control for the quality of the PCR amplification.
5. Purify the radioactive PCR product using the Nucleospin extract II kit (Macherey-Nagel) or equivalent PCR cleanup system. Elute in 50 μl DEPC-treated water.
6. Check for the radioactive signal using the available counter (Geiger counter or liquid scintillation counter).
7. The purified radioactively labeled double-stranded DNA probe can be stored at −20 °C until use but only for a few days, since ^{32}P isotope has a half-life of 14 days.

3.5.3 Radioactive Labeling of RNA Probes

This method can be used to increase the sensitivity of the detection in the case transcripts with low expression levels are used. A sequence corresponding to the probe is first amplified by PCR using a classic forward primer and a reverse primer containing a floating tail harboring a promoter sequence recognized by T7, T3, or SP6 polymerase (*see* Table 3 and **Note 2**).

1. For the PCR reaction, a polymerase having proofreading function and generating PCR products with blunt ends (to prevent unspecific transcription initiation) should be used. The following recipe describes the use of Phusion polymerase supplied by Thermo Scientific that fulfills the requirements mentioned above:

Table 3
Promoter sequence recognized by T7, T3, or SP6 RNA polymerase

T7 promoter	TAATACGACTCACTATA**G**GGAGA
T3 promoter	AATTAACCCTCACTAAA**G**GGAGA
SP6 promoter	ATTTAGGTGACACTATA**G**AAGNG

Transcription start sites are indicated in *bold*

10× HF buffer	5 μl
dNTPs (10 mM)	1 μl
Forward primer (10 μM)	2.5 μl
Reverse primer (10 μM)	2.5 μl
Taq polymerase Phusion	0.5 μl
Genomic DNA (100 ng/μl)	1 μl
DEPC-treated water	Up to 50 μl

The PCR cycle parameters should be determined empirically for each combination of primers and DNA template. However, a typical PCR cycling program using a high-fidelity Taq polymerase for the amplification of short fragments (< 500 bp) is as follows:

PCR step	Temperature (°C)	Time	Number of cycles
Initial denaturation	98	1 min	1
Denaturation	98	10 s	35
Annealing	55	20 s	35
Extension	72	15 s	35
Final extension	72	5 min	1

2. Purify the PCR product using the Nucleospin extract II kit (Macherey-Nagel) or equivalent PCR cleanup system. Elute in 20 μl DEPC-treated water.
3. Measure the DNA concentration of the purified PCR sample using a NanoDrop ND-1000 spectrophotometer.
4. For the in vitro transcription reaction, mix the reaction components of MAXIscript T7 kit (Ambion) below in a microcentrifuge tube:

10× buffer	2 μl
ATP (10 mM)	1 μl
CTP (10 mM)	1 μl
GTP (10 mM)	1 μl
[α-^{32} P]UTP (10 mCi/ml)	5 μl
T7 polymerase	2 μl
Purified PCR product (from 3.5.3 **step 2**)	1 μg
DEPC-treated water	Up to 20 μl

5. Incubate at 37 °C for 20 min.
6. Add 1 μl of DNaseI and incubate at 37 °C for 15 min to degrade the matrix DNA.
7. Add 1 μl of 0.5 M EDTA to stop the enzymatic reaction.
8. Purify the radioactive RNA probes with Micro Bio-Spin P-30 gel columns or equivalent.
9. Check for the radioactive signal using the available counter (Geiger counter or liquid scintillation counter).
10. The purified radioactively labeled RNA probes can be stored at −20 °C until use but no longer than a few days, since ^{32}P isotope has a half-life of 14 days.

3.6 Membrane Hybridization for Northern Blot

Before starting preheat the ULTRAhyb buffer at 50 °C, and preheat the hybridization tubes and hybridization oven at hybridization temperature (42 °C or 68 °C according to the probe used).

1. Place the hybridization tube on a rotation scaffold in a hybridization oven. Incubate with gentle rotation at 42 °C (if using a radioactive oligonucleotide or PCR probe) or 68 °C (if using a radioactive RNA probe) for 2 h.
2. If using a radioactive oligonucleotide or PCR probe, denature the probe by incubation at 95 °C for 10 min and place immediately on ice for at least 2 min.
3. Add the radioactive probe into the hybridization tube. Incubate with gentle rotation at 42 °C (if using a radioactive oligonucleotide or PCR probe) or 68 °C (if using a radioactive RNA probe) for 14–24 h.
4. Preheat the washing buffer I and II to the same temperature as the hybridization temperature.
5. Wash twice with 50 ml of washing buffer I for 5 min each with gentle rotation at the hybridization temperature. Drain the used washing buffer into a radioactive liquid waste container.
6. Wash twice with 50 ml of washing buffer II for 15 min each with gentle rotation at the hybridization temperature. Drain the used washing buffer into a radioactive liquid waste container.
7. Lay the membrane on a Whatman piece and allow the membrane to dry for at least 10 min.
8. Wrap the membrane into saran and expose the membrane to a storage PhosphorImager screen. The required exposure time will depend on the intensity of the signal and will range between 2 and 15 days.
9. Scan the screen with a Typhoon imager (GE Healthcare Life Sciences) or equivalent to acquire high-quality images.

3.7 Membrane Stripping

After hybridization, the membrane can be stripped of the radioactive probe and reprobed up to three times.

1. Soak the membrane in 100 ml of boiling DEPC water containing 0.5% SDS and incubate for 10 min in hot solution with gentle agitation.
2. Wash the membrane once in DEPC water at room temperature for 5 min with gentle agitation.
3. Drain the membrane briefly on absorbent paper. Place the membrane between two Whatman paper pieces and wrap with aluminum foil. Wrapped membrane can be kept at −20 °C.

3.8 Experimental Data Examples

To provide a principle of how Northern blot works and to validate the functionality of c-di-GMP-responsive riboswitch candidates, we have included examples derived from our work on *Clostridium difficile* c-di-GMP-responsive riboswitches.

C-di-GMP might be an important signal in *C. difficile*. In contrast to most Gram-positive bacteria, this pathogen encodes a large number (up to 37) of c-di-GMP turnover enzymes [17]. Moreover, the presence of 16 c-di-GMP-responsive riboswitches has been described in this bacterium. *C. difficile* is one of few bacteria containing both c-di-GMP-responsive riboswitches: 12 type I (c-di-GMP-I) and 4 type II (c-di-GMP-II). We have detected the expression of all 16 riboswitches in our RNA-seq analysis and confirmed the expression of nine of them by Northern blot [14]. To analyze the relevance of c-di-GMP in *C. difficile*, we modulated the intracellular level of this signaling molecule by overexpression of the diguanylate cyclase-encoding gene *dccA* on a plasmid under the control of the anhydrotetracycline (ATc)-inducible P*tet* promoter. This strain was used for further analysis of associated phenotypes and of the effects on the target gene and regulatory RNA expression. Coordinated control of motility and biofilm formation by c-di-GMP through the action of specific riboswitches was demonstrated in *C. difficile* [9, 14, 18, 19].

To assess the functionality of c-di-GMP-responsive riboswitches in *C. difficile*, we analyzed the effect of c-di-GMP level on the transcription of corresponding genes (Fig. 5). Positive and negative effects on premature termination of transcription of target gene were observed for "OFF" and "ON" switches of type I, respectively. In *C. difficile*, for the c-di-GMP-riboswitches of type I having "OFF" switch action, the repression of targeted genes in the presence of high c-di-GMP could be observed [14]. Under these conditions, c-di-GMP binds to the mRNA at the riboswitch motif, leading to premature termination of transcription and mRNA is hardly detectable (Fig. 5a). In the presence of low c-di-GMP, conformational changes within riboswitch region allow the

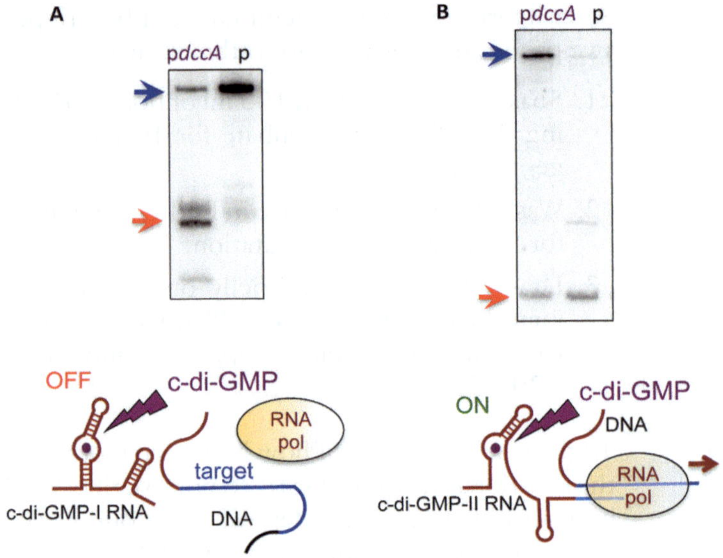

Fig. 5 Northern blot profiles for c-di-GMP-I "OFF" (**a**) and c-di-GMP-II "ON" (**b**) riboswitches. For Northern blot analysis with riboswitch-specific probes the RNA samples were extracted from a control strain carrying empty vector (p) and diguanylate cyclase-overexpressing strain (p*dccA*) at late exponential growth phase in the presence of ATc inducer. Under *dccA*-inducing conditions, c-di-GMP binds to the mRNA at the riboswitch motif of c-di-GMP-I (**a**), leading to premature termination of transcription (terminated transcript of small size indicated by *red arrow*) and mRNA (*blue arrow*) is barely detected. In the presence of low c-di-GMP, conformational changes within riboswitch region allow the synthesis of read-through transcript corresponding to full-length mRNA (*blue arrow*). Inverse regulation is observed in the case of c-di-GMP-II-targeted genes (**b**)

synthesis of read-through transcript corresponding to full-length mRNA. On the contrary, the expression of genes regulated by the c-di-GMP riboswitches of type II was induced in the presence of high c-di-GMP concentrations. Accordingly, Northern blot profiles showed anti-termination of transcription when diguanylate cyclase expression was induced (Fig. 5b).

4 Notes

1. The RNases are ubiquitous, very stable, and robust RNA-degrading enzymes. Creating a RNase-free environment prior to starting this protocol is strongly recommended, and this can be achieved by following the special precautions below:
   - Clean all bench surfaces and equipment with commercially available RNase decontamination reagents such as RNase AWAY (Ambion) or RNaseZap (Ambion).
   - Always wear gloves when working with RNA and change them frequently to avoid contamination.

- Reserve separate reagents and equipment for RNA work only.
- Use RNase-free plasticware and filter tips.
- Use pipettes dedicated to RNA work.
- Always keep the RNA samples on ice when preparing an experiment.

2. An example, derived from our work on *Clostridium difficile* c-di-GMP-responsive riboswitches [14], is provided here to help design riboswitch specific oligonucleotides. The DNA sequence of the type II c-di-GMP-responsive riboswitch Cdi2_3 is as follows:

   5′-ATATAATTGTATTTTTAATATAAATAAAGCAATTTTAA TTGTTTTTATGAAATATTATAGAGATGTT**TGAAGTATAT TCTATTATTGGGCACCT**TATGGATATACTGAGTCAGT GGTGCAACCGGCTATGAATATAAATTTATTTATTTTCAT AGCTTTTTTTGTATAATTTTT-3′.

   - The sequence of the reverse riboswitch-specific oligonucleotide (Subheading 2.5.1 **item 3** and Subheading 2.5.3 **item 3**), annealing with the sequence highlighted in grey, is:
   - 5′-TTATATTCATAGCCGGTTGCAC-3′.
   - The sequence of the forward riboswitch-specific oligonucleotide (Subheading 2.5.2 **item 6** and Subheading 2.5.3 **item 3**), annealing with the sequence complementary to that in bold, is:

     5′-TGAAGTATATTCTATTATTGGGCACCT-3′
   - The sequence of the reverse riboswitch-specific oligonucleotide, annealing with the sequence highlighted in grey and containing a floating tail harboring a promoter sequence recognized by T7 polymerase (Subheading 2.5.3 **item 4**), is:

     5′-TAATACGACTCACTATAGGGAGATTATATTCATA GCCGGTTGCAC-3′ with the underlined sequence corresponding to the floating tail.

3. The total RNA prepared by standard RNA extraction methods is not completely free from DNA contamination. In order to remove any residual genomic DNA in your samples, a DNase treatment is strongly recommended (*see* **Note 5**).

4. Nucleic acids have their absorption maximum at 260 nm and this optical density is therefore used by the NanoDrop to determine the RNA concentration in your sample. Quality of the purified total RNA should also be assessed using the NanoDrop. The 260/280 nm ratio is a good estimate of how pure the total RNA is. A 260/280 nm ratio of 1.9–2.1 is indicative of pure RNA. A lower value suggests contamination by proteins, phenol or other contaminants that absorb at 280 nm. A second measure for the purity of the RNA sample is the

260/230 nm ratio which should be >1.8. A lower value suggests contamination by phenol, ethanol, EDTA or other contaminants that absorb at 230 nm. Electrophoresis-based methods could be also used to assess the RNA quality.

5. A control lacking reverse transcription needs to be set up to ensure that contaminant DNA has been efficiently eliminated during the TURBO DNase treatment. A convenient way to carry out this control is to make a small aliquot of your purified RNA diluted at a concentration of 10 ng/μl and to use this diluted RNA instead of the cDNA in a qRT-PCR run (as described in Subheadings 3.2, **steps 6** and 7). Specific primers for one of the reference gene are used in this assay. The C_T value obtained for the RNA sample needs to be at least five cycles above the C_T value obtained for the corresponding cDNA to consider that the RNA sample is not significantly contaminated with gDNA. If the RNA samples fail the test, it needs to be retreated with TURBO DNase. The control reaction lacking reverse transcriptase could be also included during reverse transcription step.
6. Acrylamide is a neurotoxin in its unpolymerized form and is also a carcinogen. Wear gloves and safety glasses when handling acrylamide-containing solutions.
7. If the gel solution is to be used the same day as it is prepared, allow it to cool down before using. The gel solution needs to be at room temperature for the polymerization step.
8. When APS and TEMED have been added to the acrylamide gel solution, the gel mixture should be poured immediately and the comb should be inserted as fast as possible because the gel will start solidifying after a few minutes depending on the ambient temperature.
9. Ensure that the wells are covered with the polyacrylamide solution. If any bubble can be seen around the wells, remove the comb and try inserting it again.
10. The prerun of denaturing gel is an important step. It removes ammonium persulfate and others free ions that may interact with RNA. The prerun also heats up the gel to help denaturing the RNA.
11. Excess of urea in the bottom of the wells of the gel interfere with sample loading and cause uneven RNA bands. Introduction of air bubbles when loading the samples also leads to uneven RNA bands.
12. Ethidium bromide is a highly toxic carcinogen. Wear gloves and safety glasses when handling ethidium bromide-containing solutions.

13. The gel is very fragile, and it can be quite difficult to transfer it to the Whatman without breaking it. A useful tip is to place a dry Whatman piece on top of the gel and to flip over. Then, the Whatman piece can be wet together with the gel in cold 1× TBE buffer before being assembled in the electrotransfer cassette.
14. The UV cross-linked membrane can be allowed to dry overnight at room temperature. For longer storage, the membrane should be placed between two Whatman filter paper pieces, wrapped with aluminum foil and kept at −20 °C.
15. Nonisotopically labeled probes are becoming increasingly popular today because of safety considerations as well as cost and disposal of radioactive waste products. In addition, nonradioactive probe can be stored for at least a year and the time needed to detect the signal is strongly reduced. Many nonisotopic labels are available (fluorophores, haptens, biotin, and digoxigenin) and some of them seem to achieve the same sensitivity as radioactive label.
16. All steps involving radioactivity handling should be done with extreme care in a dedicated room and behind a radioactive protective shield. Always wear gloves and safety glasses when working with radioisotopes. Any radioactive contamination should be cleaned up using appropriate procedures.

References

1. Nudler E, Mironov AS (2004) The riboswitch control of bacterial metabolism. Trends Biochem Sci 29(1):11–17. doi:10.1016/j.tibs.2003.11.004
2. Serganov A, Nudler E (2013) A decade of riboswitches. Cell 152(1):17–24. doi:10.1016/j.cell.2012.12.024
3. Romling U (2012) Cyclic di-GMP, an established secondary messenger still speeding up. Environ Microbiol 14(8):1817–1829. doi:10.1111/j.1462-2920.2011.02617.x
4. Hengge R (2010) Cyclic-di-GMP reaches out into the bacterial RNA world. Sci Signal 3 (149):pe44. doi:10.1126/scisignal.3149pe44
5. Lee ER, Baker JL, Weinberg Z, Sudarsan N, Breaker RR (2010) An allosteric self-splicing ribozyme triggered by a bacterial second messenger. Science 329(5993):845–848. doi:10.1126/science.1190713
6. Smith KD, Shanahan CA, Moore EL, Simon AC, Strobel SA (2011) Structural basis of differential ligand recognition by two classes of bis-(3′-5′)-cyclic dimeric guanosine monophosphate-binding riboswitches. Proc Natl Acad Sci U S A 108(19):7757–7762. doi:10.1073/pnas.1018857108
7. Smith KD, Strobel SA (2011) Interactions of the c-di-GMP riboswitch with its second messenger ligand. Biochem Soc Trans 39 (2):647–651. doi:10.1042/BST0390647
8. Sudarsan N, Lee ER, Weinberg Z, Moy RH, Kim JN, Link KH, Breaker RR (2008) Riboswitches in eubacteria sense the second messenger cyclic di-GMP. Science 321 (5887):411–413. doi:10.1126/science.1159519
9. Bordeleau E, Purcell EB, Lafontaine DA, Fortier L-C, Tamayo R, Burrus V (2015) Cyclic Di-GMP riboswitch-regulated type IV pili contribute to aggregation of *Clostridium difficile*. J Bacteriol 197(5):819–832. doi:10.1128/JB.02340-14
10. Chen AG, Sudarsan N, Breaker RR (2011) Mechanism for gene control by a natural allosteric group I ribozyme. RNA 17 (11):1967–1972. doi:10.1261/rna.2757311
11. Rosinski-Chupin I, Soutourina O, Martin-Verstraete I (2014) Riboswitch discovery by

combining RNA-seq and genome-wide identification of transcriptional start sites. Methods Enzymol 549:3–27. doi:10.1016/B978-0-12-801122-5.00001-5

12. Dar D, Shamir M, Mellin JR, Koutero M, Stern-Ginossar N, Cossart P, Sorek R (2016) Term-seq reveals abundant ribo-regulation of antibiotics resistance in bacteria. Science 352(6282): aad9822. doi:10.1126/science.aad9822
13. Burge SW, Daub J, Eberhardt R, Tate J, Barquist L, Nawrocki EP, Eddy SR, Gardner PP, Bateman A (2013) Rfam 11.0: 10 years of RNA families. Nucleic Acids Res 41(Database issue):D226–D232. doi:10.1093/nar/gks1005
14. Soutourina OA, Monot M, Boudry P, Saujet L, Pichon C, Sismeiro O, Semenova E, Severinov K, Le Bouguenec C, Coppée J-Y (2013) Genome-wide identification of regulatory RNAs in the human pathogen *Clostridium difficile*. PLoS Genet 9(5):e1003493. doi:10.1371/journal.pgen.1003493
15. Burhenne H, Kaever V (2013) Quantification of cyclic dinucleotides by reversed-phase LC-MS/MS. Methods Mol Biol 1016:27–37. doi:10.1007/978-1-62703-441-8_3
16. Livak KJ, Schmittgen TD (2001) Analysis of relative gene expression data using real-time quantitative PCR and the 2(−delta delta C (T)) method. Methods 25(4):402–408. doi:10.1006/meth.2001.1262
17. Bordeleau E, Fortier LC, Malouin F, Burrus V (2011) c-di-GMP turn-over in *Clostridium difficile* is controlled by a plethora of diguanylate cyclases and phosphodiesterases. PLoS Genet 7 (3):e1002039. doi:10.1371/journal.pgen.1002039
18. Purcell EB, McKee RW, Bordeleau E, Burrus V, Tamayo R (2016) Regulation of type IV pili contributes to surface behaviors of historical and epidemic strains of *Clostridium difficile*. J Bacteriol 198(3):565–577. doi:10.1128/JB.00816-15
19. Purcell EB, McKee RW, McBride SM, Waters CM, Tamayo R (2012) Cyclic diguanylate inversely regulates motility and aggregation in *Clostridium difficile*. J Bacteriol 194 (13):3307–3316. doi:10.1128/JB.00100-12
20. Nelson JW, Sudarsan N, Phillips GE, Stav S, Lunse CE, McCown PJ, Breaker RR (2015) Control of bacterial exoelectrogenesis by c-AMP-GMP. Proc Natl Acad Sci U S A 112 (17):5389–5394. doi:10.1073/pnas.1419264112
21. Sambrook J, Fritsch EF, Maniatis T (1989) Molecular cloning: a laboratory manual. Cold Spring Harbor Laboratory, Cold Spring Harbor, NY

Chapter 30

Isothermal Titration Calorimetry to Determine Apparent Dissociation Constants (K_d) and Stoichiometry of Interaction (n) of C-di-GMP Binding Proteins

Bruno Y. Matsuyama, Petya V. Krasteva, and Marcos V.A.S. Navarro

Abstract

Isothermal titration calorimetry (ITC) is a commonly used biophysical technique that enables the quantitative characterization of intermolecular interactions in solution. Based on enthalpy changes (ΔH) upon titration of the binding partner (e.g., a small-molecule ligand such as c-di-GMP) to the molecule of interest (e.g., a receptor protein), the resulting binding isotherms provide information on the equilibrium association/dissociation constants (K_a, K_d) and stoichiometry of binding (n), as well as on changes in the Gibbs free energy (ΔG) and entropy (ΔS) along the interaction. Here we present ITC experiments used for the characterization of c-di-GMP binding proteins and discuss advantages and potential caveats in the interpretation of results.

Key words C-di-GMP, C-di-GMP sensor proteins, Intermolecular interactions, Receptor–ligand interactions, Isothermal titration calorimetry (ITC), Dissociation constant (K_d), Binding stoichiometry

1 Introduction

Cyclic diguanylate monophosphate (c-di-GMP) is an intracellular second messenger prevalent among bacteria. While it was discovered as an allosteric regulator of a biofilm-promoting cellulose synthase in *Gluconacetobacter xylinus* about three decades ago [1], it was not until the dawn of the twenty-first century that its central role in bacterial physiology began to emerge. Today, the cyclic RNA-based dinucleotide has gained appreciation as a key mediator of a plethora of adaptational and virulence strategies, including biofilm formation, extracellular matrix and toxin secretion, phenotypic variation, and host immune response modulation [2, 3].

C-di-GMP is synthesized and degraded by dedicated families of enzymes, including GGDEF domain-containing diguanylate cyclases for its synthesis from GTP and EAL or HD-GYP domain-containing phosphodiesterases for its degradation to linear di-GMP (pGpG) or

Karin Sauer (ed.), *c-di-GMP Signaling: Methods and Protocols*, Methods in Molecular Biology, vol. 1657,
DOI 10.1007/978-1-4939-7240-1_30, © Springer Science+Business Media LLC 2017

GMP, respectively [4, 5]. These protein domains feature significant conservation and take their names after consensus motifs found in their primary sequence. Bioinformatics studies largely enabled by the recent wealth of genome sequencing data have identified, often in multiple copies, c-di-GMP turnover enzymes in the majority of sequenced species, while structural and mechanistic studies have helped pinpoint catalysis-essential features to distinguish between catalytically active and divergent enzymes [5, 6].

Contrary to these c-di-GMP turnover modules, receptors for the dinucleotide demonstrate remarkable diversity in conserved domain architecture, c-di-GMP binding motifs, and downstream effector function [2–4]. Based on phyletic distribution similar to that of c-di-GMP turnover enzymes and a likely role in c-di-GMP-triggered physiological responses, PilZ domains were the first to be predicted and subsequently confirmed as sensory modules for the dinucleotide [7, 8]. The so-called inhibitory sites, or I-sites, on both catalytically active and incompetent GGDEF domains (e.g., WspR and PelD of *Pseudomonas aeruginosa* [9, 10]), as well as divergent EAL domain active sites (e.g., FimX and LapD of *Pseudomonas* spp. [11, 12]), have also been reported to serve as c-di-GMP-binding regulatory motifs that are somewhat amenable to predictive approaches.

Many identified c-di-GMP sensors, however, harbor either novel folds (e.g., eukaryotic STING, BldD of streptomycetes [13–16]) or unexpected c-di-GMP binding modules typically associated with different regulatory inputs (e.g., response receiver domains and ATPase domains [17–19]). Thus, the identification of many c-di-GMP sensors has been either the result of "educated guesswork" based on the proteins' physiological effects (e.g., VpsT of *Vibrio cholera*, FleQ of *Pseudomonas aeruginosa)* or of unbiased screening approaches such as capture compound-based purification (e.g., BldD of *Streptomyces* spp.) [2].

Isothermal titration calorimetry (ITC) has proven a valuable technique in the mechanistic studies of c-di-GMP recognition by the dinucleotide's targets. Based on measuring precise aliquots of heat (i.e., enthalpy changes) released or taken up along the titration of a ligand into its interacting partner, ITC represents a relatively uncomplicated yet remarkable method in that it provides a complete thermodynamic profile of the binding reaction [20]. The experimental raw data consist of a series of peaks representing titration-associated thermal power variations necessary to maintain equal the temperatures of a water-filled reference cell and the sample cell where the interaction occurs (Fig. 1). The areas of these peaks are subsequently integrated, plotted against the molar ratio of the binding partners, and the resultant isotherm is fitted into a binding model from which the stoichiometry of binding, the dissociation constant, as well as changes in the Gibbs free energy and entropy can be quantitatively determined (Fig. 1). Additional

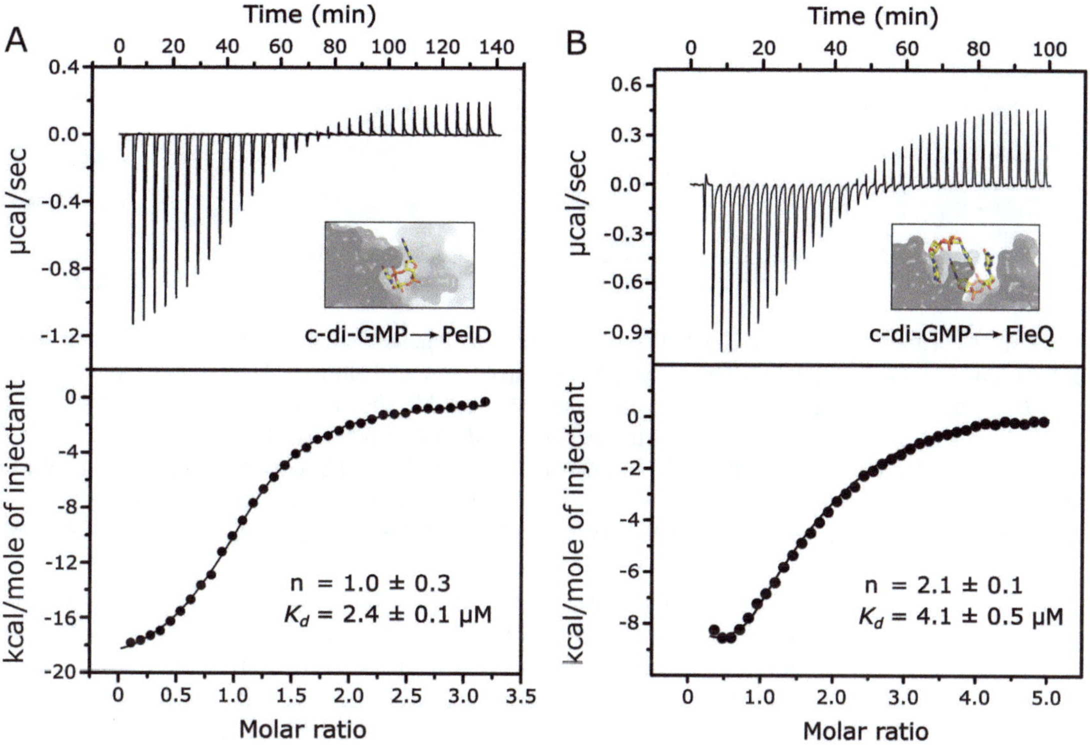

Fig. 1 Representative examples of c-di-GMP titration to *P. aeruginosa* PelD (**a**) and FleQ (**b**). In both panels, *top graphs* show the heat variations recorded over the course of the experiment (raw thermogram) and *bottom panels* show the integrated thermogram plotted against the molar c-di-GMP–protein receptor ratio (binding isotherm). Fit using the one site-model in the Origin software yielded the stoichiometry (*n*) and affinity constant (K_d). *Insets* show a single c-di-GMP interacting with the I-site of the degenerate-GGDEF domain of PelD, PDB code 4EU0 [28], (**a**) and an intercalated c-di-GMP dimer bound at the site in the AAA+ domain of FleQ, PDB code 5EXX [18] (**b**)

advantages of the method are that no sample labeling or immobilization are required and that the interacting molecules, although required at significant concentrations, can be subsequently recovered for other applications (e.g., crystallization).

A remarkable trait of c-di-GMP as a ligand is that it can adopt various conformations and oligomeric states [2, 4, 21]. As a result, the binding stoichiometry varies significantly among studied receptors, and ITC has proven a powerful method to provide an accurate assessment of this parameter. It should be noted, however, that albeit its advantages, ITC is not a "magic bullet" and orthogonal approaches are to be applied for accurate ligand binding characterization [22]. For example, a binding stoichiometry of $n = 1$ can reflect monomeric c-di-GMP binding a monomer of its target molecule (as in PelD or FimX of *P. aeruginosa*, Figs. 1a and 2) or dimeric c-di-GMP binding to a dimer of its protein partner (as in VpsT of *V. cholerae*, Fig. 2). Similarly, a binding stoichiometry of $n = 2$ can mean either dimer-to-monomer binding (as in FleQ of

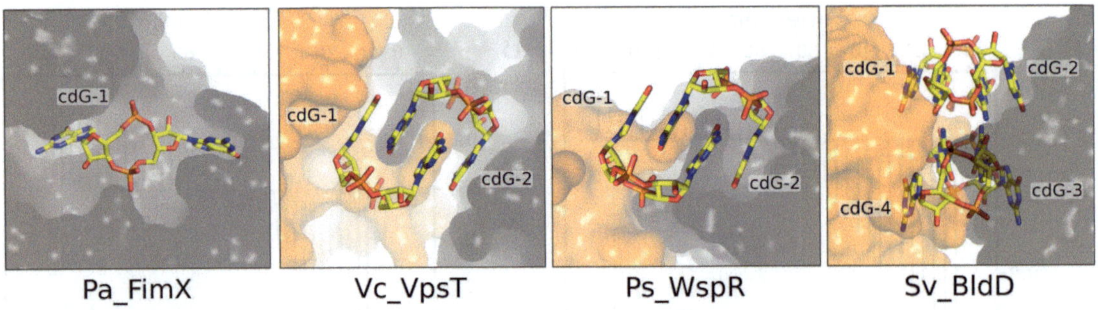

Fig. 2 Modes of interaction of c-di-GMP with protein receptors. Degenerate EAL c-di-GMP receptors, such as *P aeruginosa* FimX (Pa_FimX), interact with a single ligand in an extended form (Pa_FimX; PDB code 3HV8 [11]). Fully stacked intercalated c-di-GMP dimers are recurrently found in sites formed by oligomeric interfaces of protein receptors. For example, in crystal structures of *V. cholera* VpsT (Vc_VpsT; PDB code 3KLO [17]) and *P. syringae* WspR (Ps_WspR; PDB code 3I5A [25]) dimeric c-di-GMP binding bridges adjacent REC and GGEDF domains, respectively. Recently, a tetrameric form of c-di-GMP (two neighbor intercalated c-di-GMP dimers) was identified at the oligomeric interface of *Streptomyces venezuelae* BldD (Sv_BldD; PDB code 4OAY [16])

P. aeruginosa, Fig. 2) or tetramer-to-dimer interaction (as in BldD of *Streptomyces* spp., Fig. 2). Also, any conformational changes, such as oligomerization or intramolecular rearrangements, caused by the ligand-to-protein interaction would contribute to the measured heat effects and would skew the calculated affinities and thermodynamic parameters.

Here, we describe ITC experiments performed with *P. aeruginosa* FleQ and PelD and *V. cholera* VpsT as a guideline for a general-purpose protocol. Minor adjustments and modifications of this protocol can be employed to measure c-di-GMP affinity and stoichiometry of other protein receptors.

2 Materials

ITC is a very sensitive technique, implicating that the presence of impurities or aggregates in buffers and protein solutions could generate unwanted artifactual signals in the measurements. For best results, use HPLC grade water and chemicals, and filter all buffers through a 0.2 μm pore size polyvinylidene fluoride (PVDF) membrane.

2.1 C-di-GMP

1. C-di-GMP can be enzymatically synthesized using an active diguanylate cyclase and GTP as a substrate. Relatively straightforward, one-step reaction protocols utilizing different enzymes have been reported [23, 24]. We typically use the highly active I-site mutant variant $R^{242}A$ of *P. aeruginosa* WspR to synthesize c-di-GMP [25], which is then purified by reversed-phase high-performance liquid chromatography (HPLC) using a C18 column (*see* **Note 1**). A typical reaction

for synthesis of c-di-GMP contains 3 μM purified enzyme and 500 μM GTP.

2. Alternatively, chemically synthesized c-di-GMP can be purchased as a lyophilized sodium salt from vendors such as Biolog and Sigma.

2.2 Size Exclusion Chromatography (SEC)

1. Fast Protein Liquid Chromatography (FPLC) set-up equipped with an automated fraction collector and a UV absorbance detector at 280 nm (e.g., AKTA-Purifier).
2. Column: Superdex 200 10/300 GL (GE Healthcare Life Sciences) (*see* **Note 2**).
3. SEC buffer: 20 mM Tris–HCl pH 8.0, 300 mM NaCl, 5 mM $MgCl_2$ (*see* **Note 3**). To prepare the buffer, weigh the appropriate amounts of each component necessary for 1 L of buffer. Mix the components with about 900 mL of ultrapure water (sensitivity of 18 MΩ cm at 25 °C) and carefully adjust pH to 8.0 with HCl. Add water to complete the volume to 1 L. Alternatively, stock solutions of buffer components can be prepared first (1 M Tris–HCl pH 8.0, 5 M NaCl, 1 M $MgCl_2$, etc.).
4. Ultracentrifugal filters with a polyethersulfone (PES) membrane for protein concentration. Use devices with appropriate size and molecular weight cutoff (e.g., 10, 30, or 100 kDa).
5. PVDF membrane filters with 0.2 μm pore size.

2.3 ITC Data Acquisition and Analysis

1. Isothermal microcalorimeter equipped with two cells, a water-filled reference cell, and the sample cell. Several microcalorimeters, which differ in sample volumes usage and throughput, are manufactured by companies such as Microcal (now acquired by Malvern) and TA Instruments. Although this protocol is based on the MicroCal VP-ITC, any equipment suited for protein–ligand binding analysis can be used.
2. Origin 7.0 software including the Microcal ITC add-on routine.

3 Methods

3.1 Preparation of Protein and C-di-GMP Samples

To obtain high-quality ITC data, the protein sample should have a high degree of purity and homogeneity. Methods to optimize protein purification and stability are out of the scope of this chapter, but the reader could easily find them elsewhere. We recommend heterologous expression of the c-di-GMP receptor fused with a hexahistidine tag (6×His-tag), which allows for one-step protein purification through immobilized metal-affinity chromatography (IMAC) using Ni^{2+} or Co^{2+}-loaded nitrilotriacetic acid-agarose resins. Preferably use expression vectors that contain encoding sequences for protease cleavage sites, so that the 6× His-tag of the purified protein can be cleavaged after the IMAC. It is important to always perform a size exclusion chromatography as the last purification step prior to ITC to remove protein aggregates and obtain a homogenous protein sample, as described below.

1. Follow the preestablished purification protocol for the c-di-GMP receptor of interest (*see* **Note 4**).
2. Concentrate the protein solution with an appropriate PES centrifugal filter until the protein concentration is near the solubility limit. Monitor the protein concentration regularly using the theoretical protein extinction coefficient (calculated for example through the online tool ExPASy ProtParam [26]) and the recorded absorbance at 280 nm (*see* **Notes 5** and **6**).
3. Filter the concentrated sample through a 0.2 μm PVDF membrane. Alternatively, centrifuge the protein sample for 15-min at 15,000–20 000 × *g* to pellet aggregates.
4. Transfer the supernatant to a clean Eppendorf tube.
5. Dispense the concentrated protein solution in 20 μL aliquots in thin-wall PCR tubes and flash-freeze in liquid nitrogen (*see* **Note 7**). Store aliquots at –80 °C (*see* **Note 8**).
6. Before the ITC run, rapidly thaw protein aliquots necessary for one experiment and store in ice. One ITC experiment with a VP-ITC instrument requires about 2 mL of protein sample at the concentration determined for the ITC experiment (*see* ITC experimental design in Subheading 3.2).
7. Centrifuge samples again (15,000 × *g*, 15 min, 4 °C) to remove eventual particulate material. Transfer supernatant to a new tube.
8. Load protein sample into a Superdex 200 10/300 GL column preequilibrated with SEC buffer (*see* **Note 9**).
9. Collect 0.5 mL fractions using a flow rate in accordance with the recommended pressure limit. Monitor the absorbance at 280 nm and pool together the fractions containing the c-di-GMP receptor. Store in ice.
10. Collect approximately 20 mL of flow-through buffer and store on ice (*see* **Note 10**).
11. Measure the protein concentration as described above. According to the ITC experimental design (*see* Subheading 3.2), adjust the protein concentration by centrifugation with an appropriate PES centrifugal filter or by dilution using buffer collected from the previous step.
12. Dissolve lyophilized c-di-GMP to a concentration of 2 mM (approximately 1.4 mg/mL), with the flow-through SEC

buffer. As for the protein solution, prepare 500 μL of c-di-GMP solution at the desired concentration for the ITC experiment by diluting the 2 mM stock solution with the flow-through SEC buffer (*see* **Notes 11** and **12**).

13. Before starting the ITC experiment, degas the protein sample and the c-di-GMP solution to prevent bubble formation.

3.2 ITC Experimental Design

When designing an ITC experiment, several parameters should be tested and optimized to obtain good estimates of affinity constants, stoichiometry, and binding energies. This can be achieved by varying the concentrations of the protein receptor and the c-di-GMP solution, temperature of the experiment, and titration parameters (injection volume and spacing time), as described below.

1. Accurate fits on an ITC experiment require proper concentrations of the protein receptor and c-di-GMP. Set up the instrument as described below, considering the following parameters.
   (a) Cyclic di-GMP should be loaded into the injection syringe due to its high solubility.
   (b) The protein receptor goes into the sample cell of the ITC at a concentration inferred from the "c" value, $c = n \times M_{\text{cell}}/K_{\text{d}}$, where n is the number of ligands per protein molecule (i.e., the stoichiometry), and M_{cell} is the protein concentration, expressed in the same units as the affinity constant K_{d}. This value determines the shape of the binding isotherm, which is critical to obtain accurate measurements of thermodynamic and stoichiometric parameters.
   (c) Optimal sigmoidal binding shapes are achieved with $10 < c < 100$ [27], whereas higher and lower "c" values result in binding curves too steep or shallow, respectively (Fig. 3).
   (d) Known affinities of c-di-GMP toward protein receptors vary over two orders of magnitude (~0.1–10 μM), so a pilot experiment with a protein concentration ($[P_0]_{\text{cell}}$) of 20–30 μM should be used in the cell.
   (e) The concentration of c-di-GMP in the syringe ($[L_0]_{\text{syringe}}$) should be about $20 \times [P_0]_{\text{cell}}$, which is usually large enough to reach binding saturation before half of the titration.
   (f) Vary $[P_0]_{\text{cell}}$ and $[L_0]_{\text{syringe}}$ according to the binding curve. Increase or decrease the protein concentration if the curve is too shallow or steep, respectively. We usually perform ITC experiments changing the protein concentration by fivefold until obtaining an optimal binding isotherm.

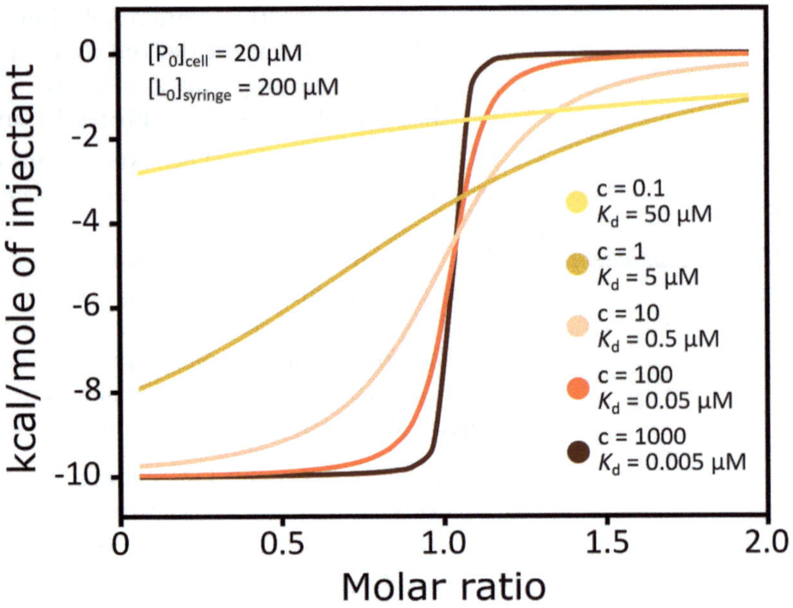

Fig. 3 Influence of the "c" value on the shape of ITC binding curves. The data were generated using the one-site model. Fixing $n = 1$ (stoichiometry of 1:1, protein–ligand) and $\Delta H = -10$ kcal/mol, curves were simulated for different affinity constants (K_d). In the simulations, the initial concentrations of protein in the cell ($[P_0]_{cell}$) and ligand in the syringe ($[L_0]_{syringe}$) were 20 and 200 μM, respectively

2. Initially set up the ITC experimental temperature close to the room temperature, so that equilibrium will be quickly attained. In our laboratory, we usually carry out experiments at 20–25 °C. If necessary, test experiments at different temperatures (*see* **Note 13**).
3. To optimize titration parameters, first set the injection volume for c-di-GMP to 8–10 μL.
   (a) About 30 injections will be performed, assuring a good number of points for fitting.
   (b) It is essential that after each injection the heat signal returns to the baseline before the next one.
   (c) An interval of 200–300 s between injections is usually sufficient.
   (d) Increase the spacing time if equilibration is not observed.

3.3 Data Acquisition and Analysis

After setting up the experimental parameters as described in Subheading 3.2, the ITC run can be initiated to obtain data for the thermodynamic parameters of the binding event.

1. Before loading samples, thoroughly clean injection syringe and sample cell by rinsing several times with HPLC grade water.

Follow the manufacturer's cleaning protocol to avoid damages to the ITC equipment.

2. Repeat the cleaning procedure with SEC flow-through buffer.
3. Load the protein sample and the c-di-GMP solution into the equipment (*see* also Subheading 3.2, **step 1**).
4. Carefully load the sample cell with at least 1.8 mL of protein solution using an adequate Hamilton syringe (*see* **Notes 14** and **15**). Since 1.45 mL completely fills the sample cell of MicroCal VP-ITC equipment, remove the excess volume in the chamber above the sample cell.
5. The injection syringe of the VP-ITP equipment has a volume of 300 μL, but to completely fill it we recommend the preparation of about 500 μL of titration solution.
6. Connect a plastic syringe to the injection syringe port with a silicone tubing.
7. Insert the needle of the injection syringe into c-di-GMP solution and slowly draw it until a little excess volume is flowing through the silicone tubing attached to the injection syringe fill port.
8. Once excess volume is detected in the silicone tubing, close the fill port immediately and detach the silicone tubing (*see* **Note 16**).
9. Clean the injection syringe needle with a kimwipe and carefully insert it into de ITC sample cell (*see* **Note 17**).
10. Input the desired experimental parameters into the controlling software (temperature, the number of injections, injection volume, the time between injections, etc. as determined in Subheading 3.2) and start the experiment.
11. After the experiment is done, thoroughly clean the equipment again.
12. Once the instrument has been cleaned, perform control runs.
    (a) Perform a blank control titration of c-di-GMP into the sample cell filled with flow-through SEC buffer to determine the heat of c-di-GMP dilution.
    (b) If necessary, also titrate buffer into the protein solution to determine heat changes promoted by protein dilution (*see* **Notes 18** and **19**).
    (c) The concentrations of the c-di-GMP solution and the protein sample in the control experiments should be the same used for the binding experiment.
13. To determine *n*, K_d and the thermodynamic parameters of the binding event, fit the acquired data using the Microcal ITC macro implemented in the Origin 7.0 software.
    (a) First, load the data file and check the quality of raw data for signs of artifacts in the raw thermogram.

(b) Repeat the step above by loading the c-di-GMP control data and subtract the dilution heat from the binding isotherm (*see* **Note 20**).

(c) Select the single-site binding model to fit the data (Fig. 1). Despite the diversity of interaction modes displayed by c-di-GMP-receptor systems, there is no report of protein receptor bearing more than one distinct binding site.

4 Notes

1. The reaction uses commercially obtained GTP as a substrate for an in-house purified enzyme and can require the addition of bivalent ions (e.g., Mg^{2+}). Following empirically determined incubation for full product conversion, the enzyme can be removed by heat denaturation and filtration and the c-di-GMP can be purified by HPLC on a C18 column using ammonium acetate-based buffers and methanol gradient. The eluted c-di-GMP can then be lyophilized and stored. Once optimized, milligram amounts of c-di-GMP can be synthesized within 2 days, excluding the time necessary for expression and purification of the selected diguanylate cyclase. However, by our experience, optimization of the entire process, especially the HPLC procedures for product purification, could be complicated and time-consuming. Over the last decade prices for the dinucleotide have decreased dramatically making the direct purchasing option relatively cost-effective.
2. Superdex 200 separates molecules with molecular weights between 10 and 600 kDa. The type of size-exclusion chromatography resin can be changed according to the protein of interest. Superose 6 and Superdex 75 provide appropriate alternatives for larger and smaller protein targets, respectively.
3. The buffer components should be optimized for maximum stability and monodispersity of the samples. Although commonly used for biological purposes, Tris and HEPES buffers have high heats of ionization (ΔH_{ion}) and thus are not recommend as first options for ITC experiments. Ideally, buffers with $\Delta H_{ion} \sim 0$, such as phosphate, acetate, and citrate, should be used. If the c-di-GMP receptor is optimally stable in a Tris- or HEPES-based buffer, perform the ITC experiment in triplicate to ensure a more accurate measurement of binding enthalpy.
4. To obtain a well-shaped sigmoidal binding curve, several ITC runs will likely need to be performed in order to optimize the experimental setup. We usually perform large-scale protein preparations that yield enough sample for multiple experiments

and freeze in small voluble aliquots. Using the same purified protein and performing a SEC after thawing protein sample, as suggested in this protocol, has produced consistently reproducible ITC results.

5. The behavior of the protein sample during concentration should be empirically tested. The sample should be gradually concentrated via short, 1–2 min long runs, between which the concentrate should be gently homogenized with a micropipette. The centrifugation should be stopped if visible aggregates start to appear, which could indicate that the concentrated protein has reached its solubility limit. For example, purified *P. aeruginosa* FleQ usually concentrate up to 30–40 mg/mL.
6. The protein's extinction coefficient can be estimated from its primary sequence using Web-based tools, such as Expasy ProtParam [26]. Because this method considers the extinction coefficient of aromatic residues in solution, which could be affected by their local environment in a folded protein, the absorbance measurement should be carried out under denaturing conditions (e.g., 6 M guanidinium chloride). The absence of tryptophan residues yield misestimated extinction coefficient, and a different method to measure protein concentration, such as Bradford or BCA assays, should be employed.
7. Test the protein stability under freezing–thawing cycles. If necessary, add glycerol up to 20% v/v to the protein sample prior to freezing.
8. Storing the protein of interest in small aliquots eliminates the need for repeated freeze–thaw cycles and thus minimizes potential freezing damages to the protein sample.
9. This step is used to remove eventual protein aggregates and exchange the protein sample buffer. Although not optimal for elution peak resolution, up to 500 μL protein sample can be loaded onto the Superdex 200 10/300 GL column. The flow rate (~0.5 mL/min) can vary with the type of SEC column used and the particular experimental setup. The new generation "Increase" columns from GE Healthcare allow for increased pressure limit and flow rate and feature improved elution peak resolution.
10. Make sure that the absorbance at 280 nm, used to monitor protein elution, has returned to the baseline before start collecting buffer. Use this buffer for necessary protein and c-di-GMP dilutions, rinsing, and ITC baseline determination.
11. C-di-GMP is highly soluble (~50 mg/mL in aqueous solution). To prepare the c-di-GMP stock solution, initially, add the necessary buffer to roughly obtain a solution with a concentration of 2 mg/mL. A precise estimation of your stock

solution can then be carried about by measuring its absorbance at 260 nM, using the c-di-GMP extinction coefficient of 26,600 OD/M/cm.

12. Buffer mismatch between ligand and protein solutions lead to large heats of dilution that could mask the binding signal. Although several ITC protocols recommend extensive dialysis of the macromolecule and ligand against buffer in the same container, often the dialysis process negatively affects the quality of protein sample, especially if the c-di-GMP receptor is not very stable over extended periods in solution. By our experience, dilution of thoroughly desalted c-di-GMP with the flow-through buffer collected from SEC assures buffer matching without compromising data quality. If c-di-GMP is purified in-house, ammonium acetate-based buffer should be used for the HPLC purification as it is essentially volatile.
13. Temperature is an important factor that affects the binding enthalpy for a given titration due to heat capacity changes ($\Delta C_p = \Delta H/T$). If there is solid evidence that the protein receptor under study binds c-di-GMP but the binding isotherm presents poor signal-to-noise ratio, it is possible that the heat of binding at the experimental temperature (e.g., 20°C) is close to zero. Before concluding that ITC is not suitable to study your system, perform test experiments at 10–15 and 30-35 °C.
14. Before loading solutions, make sure that the injection syringe and sample cell are completely dry to avoid dilutions errors.
15. Make sure no air has entered the Hamilton syringe when loading it with about 1.8 mL of protein solution. With utmost care insert the needle of the Hamilton syringe until it gently touches the bottom of the ITC protein cell. Then, lift the needle just enough so there is no more contact with the reservoir. Avoiding movement of the syringe, initially inject about 1.5 mL very slowly and then quickly inject the remaining volume. This procedure assures that any air bubbles eventually attached to the cell during the slow 1.5 mL injection will be displaced.
16. To avoid poor baselines and artifactual signals, the titration solution into the injection syringe should also be air-free. Although the procedure to fill the injection syringe is simple, we recommend the assistance of a second person to perform this task, since it requires manual operation of the plastic syringe, manipulation of sensitive equipment and controlling of the injection syringe through a software interface (fill, purge, close commands). If visible bubbles are noticed after the filling procedure, place the injection syringe needle back into remaining c-di-GMP solution and perform two cycles of purge/refill using the controlling software.

17. Clean the needle thoroughly but never directly touch the vane of the syringe tip. Make sure that the injection syringe is properly inserted into the ITC sample cell so that the needle is perpendicular to the bench. Bent injection syringe needle is also a potential cause of poor data baseline.
18. Even taking utmost care to match ligand and protein solution buffers, several systems present considerable heat released in dilution control experiments, often due to ligand oligomerization or aggregation. This is exactly the case for the c-di-GMP molecule, which forms concentration-dependent dimers and higher oligomers in solution [21]. Due to the high concentrations usually required for the c-di-GMP titrant solution, most probably a considerable endothermic thermogram will be observed in a control dilution experiment. Oligomerization of c-di-GMP is enhanced by the presence of divalent metals in solutions, such as Mg^{2+} and Mn^{2+}. Accordingly, dilution effects are more pronounced when the system buffer contains these ions [21], as observed in the thermograms of Fig. 2. Experiments with *P. aeruginosa* PelD (Fig. 1a) and FleQ (Fig. 1b) were similarly designed, using a c-di-GMP solution concentration of 700 μM and the same buffer composition, except for the presence of 5 mM $MgCl_2$ in the FleQ buffer. Although both titrations reached binding saturation, the endothermic heat recorded in the final titrations is higher in the FleQ ITC experiment.
19. Dilution effects of the protein solution in the sample cell could also occur due to concentration-dependent oligomeric changes. Some c-di-GMP receptors not only present concentration-dependent oligomeric equilibrium even in the absence of c-di-GMP, but also undergo significant c-di-GMP-induced oligomeric changes (e.g., *V. cholera* VpsT and *P. aeruginosa* FleQ [17, 18]). Such effects should be considered when designing and analyzing ITC results of a new receptor.
20. If there are any artifacts, such as baseline spikes and unexpected peaks, the corresponding integrated data points should be removed before fitting. In addition, as the very first injection is typically of lower volume than the rest of the injection series, its data point should also be removed from the fitting.

Acknowledgments

Work in the Navarro laboratory is supported by Fundação de Amparo à Pesquisa do Estado de São Paulo under Grant 2009/13238-0. The Krasteva laboratory is supported by the Institute for Integrative Biology of the Cell (I2BC) and by a 2016 ATIP-Avenir grant from the *Centre National de la Recherche Scientifique*.

References

1. Ross P et al (1987) Regulation of cellulose synthesis in *Acetobacter xylinum* by cyclic diguanylic acid. Nature 325(6101):279–281
2. Krasteva PV, Sondermann H (2017) Versatile modes of cellular regulation via cyclic dinucleotides. Nat Chem Biol 13:350–359
3. Jenal U, Reinders A, Lori C (2017) Cyclic di-GMP: second messenger extraordinaire. Nat Rev Microbiol 15(5):271–284
4. Krasteva PV, Giglio KM, Sondermann H (2012) Sensing the messenger: the diverse ways that bacteria signal through c-di-GMP. Protein Sci 21(7):929–948
5. Schirmer T, Jenal U (2009) Structural and mechanistic determinants of c-di-GMP signalling. Nat Rev Microbiol 7(10):724–735
6. Galperin MY, Nikolskaya AN, Koonin EV (2001) Novel domains of the prokaryotic two-component signal transduction systems. FEMS Microbiol Lett 203(1):11–21
7. Amikam D, Galperin MY (2006) PilZ domain is part of the bacterial c-di-GMP binding protein. Bioinformatics 22(1):3–6
8. Ryjenkov DA et al (2006) The PilZ domain is a receptor for the second messenger c-di-GMP: the PilZ domain protein YcgR controls motility in enterobacteria. J Biol Chem 281 (41):30310–30314
9. De N et al (2008) Phosphorylation-independent regulation of the diguanylate cyclase WspR. PLoS Biol 6(3):e67
10. Lee VT et al (2007) A cyclic-di-GMP receptor required for bacterial exopolysaccharide production. Mol Microbiol 65(6):1474–1484
11. Navarro MV et al (2009) Structural analysis of the GGDEF-EAL domain-containing c-di-GMP receptor FimX. Structure 17 (8):1104–1116
12. Navarro MV et al (2011) Structural basis for c-di-GMP-mediated inside-out signaling controlling periplasmic proteolysis. PLoS Biol 9(2):e1000588
13. Huang YH et al (2012) The structural basis for the sensing and binding of cyclic di-GMP by STING. Nat Struct Mol Biol 19(7):728–730
14. Shang G et al (2012) Crystal structures of STING protein reveal basis for recognition of cyclic di-GMP. Nat Struct Mol Biol 19 (7):725–727
15. Shu C et al (2012) Structure of STING bound to cyclic di-GMP reveals the mechanism of cyclic dinucleotide recognition by the immune system. Nat Struct Mol Biol 19(7):722–724
16. Tschowri N et al (2014) Tetrameric c-di-GMP mediates effective transcription factor dimerization to control Streptomyces development. Cell 158(5):1136–1147
17. Krasteva PV et al (2010) Vibrio cholerae VpsT regulates matrix production and motility by directly sensing cyclic di-GMP. Science 327 (5967):866–868
18. Matsuyama BY et al (2016) Mechanistic insights into c-di-GMP-dependent control of the biofilm regulator FleQ from Pseudomonas aeruginosa. Proc Natl Acad Sci U S A 113(2): E209–E218
19. Wang YC et al (2016) Nucleotide binding by the widespread high-affinity cyclic di-GMP receptor MshEN domain. Nat Commun 7:12481
20. Freyer MW, Lewis EA (2008) Isothermal titration calorimetry: experimental design, data analysis, and probing macromolecule/ligand binding and kinetic interactions. Methods Cell Biol 84:79–113
21. Stelitano V et al (2013) Probing the activity of diguanylate cyclases and c-di-GMP phosphodiesterases in real-time by CD spectroscopy. Nucleic Acids Res 41(7):e79
22. De N et al (2010) Biophysical assays for protein interactions in the Wsp sensory system and biofilm formation. Methods Enzymol 471:161–184
23. Korovashkina AS et al (2012) Enzymatic synthesis of c-di-GMP using inclusion bodies of Thermotoga maritima full-length diguanylate cyclase. J Biotechnol 164(2):276–280
24. Zahringer F, Massa C, Schirmer T (2011) Efficient enzymatic production of the bacterial second messenger c-di-GMP by the diguanylate cyclase YdeH from *E. coli*. Appl Biochem Biotechnol 163(1):71–79
25. De N et al (2009) Determinants for the activation and autoinhibition of the diguanylate cyclase response regulator WspR. J Mol Biol 393(3):619–633
26. Wilkins MR et al (1999) Protein identification and analysis tools in the ExPASy server. Methods Mol Biol 112:531–552
27. Turnbull WB, Daranas AH (2003) On the value of c: can low affinity systems be studied by isothermal titration calorimetry? J Am Chem Soc 125(48):14859–14866
28. Li Z et al (2012) Structures of the PelD cyclic diguanylate effector involved in pellicle formation in Pseudomonas aeruginosa PAO1. J Biol Chem 287(36):30191–30204

Part VIII

Targeting c-di-GMP Signaling

Chapter 31

Targeting c-di-GMP Signaling, Biofilm Formation, and Bacterial Motility with Small Molecules

Clement Opoku-Temeng and Herman O. Sintim

Abstract

Bacteria possess several signaling molecules that regulate distinct phenotypes. Cyclic di-GMP (c-di-GMP) has emerged as a ubiquitous second messenger that regulates bacterial virulence, cell cycle, motility, and biofilm formation. The link between c-di-GMP signaling and biofilm formation affords novel strategies for treatment of biofilm-associated infections, which is a major public health problem. The complex c-di-GMP signaling pathway creates a hurdle in the development of small molecule modulators. Nonetheless, some progress has been made in this regard and inhibitors of c-di-GMP metabolizing enzymes that affect biofilm formation and motility have been documented. Herein we discuss the components of c-di-GMP signaling, their correlation with biofilm formation as well as motility and reported small molecule inhibitors of c-di-GMP signaling.

Key words c-di-GMP, Diguanylate cyclase, Phosphodiesterase, PilZ, STING, Inhibitor

Nucleotide-based second messenger signaling systems are prevalent in both prokaryotic and eukaryotic systems. Cyclic dinucleotides have been found to regulate various phenotypes in bacteria and archaea [1]. In 1987, cyclic di-GMP was discovered as the first cyclic dinucleotide and found to regulate cellulose synthesis in *Gluconoacetobacter xylinus* (previously *Acetobacter xylinum*) [2]. It took the scientific community over 20 years to fully appreciate the importance of c-di-GMP signaling in regulating myriad processes in bacteria. Over the last decade, works from various groups have demonstrated that c-di-GMP regulates an astounding array of processes in bacteria, including biofilm formation, motility, cell cycle, and virulence factor production, among others (Fig. 1) [1, 3].

c-di-GMP signaling plays critical roles during the infection stages of several human pathogens, including *Pseudomonas aeruginosa*, *Clostridium difficile*, *Salmonella typhimurium*, and *Vibrio cholerae*. Flagellar motility, secretion systems and other virulence factors, which are partly regulated by c-di-GMP signaling, facilitate infection of mammalian cells and tissues by these pathogens [1, 4].

Karin Sauer (ed.), *c-di-GMP Signaling: Methods and Protocols*, Methods in Molecular Biology, vol. 1657, DOI 10.1007/978-1-4939-7240-1_31,

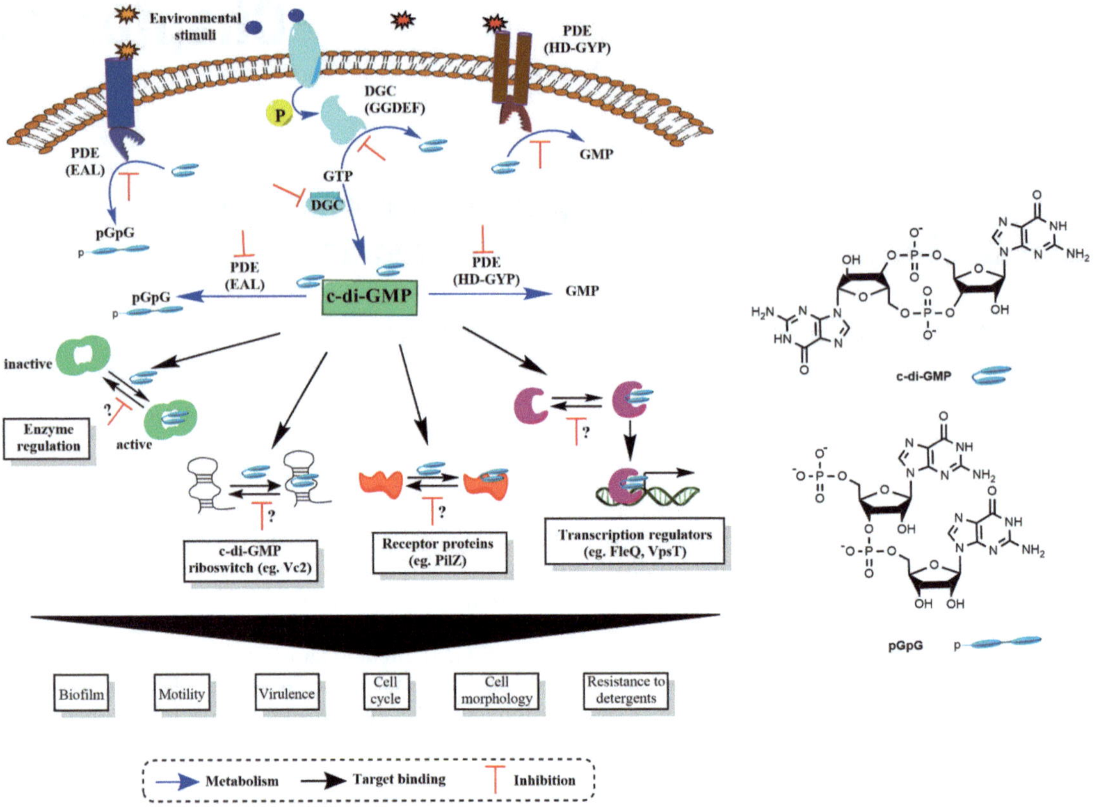

Fig. 1 A schematic of c-di-GMP signaling

Chronic infection of *P. aeruginosa* in the lungs of cystic fibrosis patients has in part been associated with biofilm formation, a process regulated by c-di-GMP signaling pathway [5] (Fig. 1). Biofilms have traditionally been notoriously difficult to clear using small molecule drugs, so the identification of c-di-GMP-mediated processes that lead to biofilm formation has ushered a renewed hope that new generation of anti-biofilm agents, which target c-di-GMP signaling, would emerge [6]. In this chapter, we discuss c-di-GMP signaling and the new class of small molecules that inhibit c-di-GMP metabolism proteins (synthases and phosphodiesterases). These cell permeable molecules can affect the intracellular concentrations of c-di-GMP, leading to the inhibition of biofilm formation or reduced motility.

1 Synthesis of c-di-GMP: GGDEF Domain Proteins

Bacteria that utilize c-di-GMP signaling possess enzymes known as diguanylate cyclases (DGCs), which catalyze the synthesis of the second messenger by condensing two molecules of guanosine triphosphate to first form the linear 5'pppGpG and then cyclizing it to c-di-GMP in the presence of Mg^{2+} ions, Fig. 1 [2]. The active site (A site) of these enzymes contains the GGDEF (Gly-Gly-Asp-Glu-Phe) or GGEEF (Gly-Gly-Glu-Glu-Phe) motif, which is essential for DGC activity [1, 7, 8]. The *Caulobacter crescentus* response regulator PleD was the first protein to be described to contain a GGDEF domain. Some DGC proteins, such as PleD (*C. crescentus*), PelD and WspR from *P. aeruginosa* as well as YdaM from *Escherichia coli* and DgcK and DgcL from *Vibro cholerae* possess an inhibitory site (I-site), characterized by an RxxD motif (where x is any amino acid), where the product c-di-GMP can bind to allosterically inhibit its own synthesis [1, 4, 9, 10]. Some bacteria harbor many types of c-di-GMP synthases or phosphodiesterases and it is still unclear why different versions of enzymes that make or degrade the same product are needed in the same cell. For example *P. aeruginosa* has 33 GGDEF containing proteins localized either in the cytoplasm or on the cytoplasmic membrane [1, 4, 11]. Perhaps these DGCs play different roles or respond to different cues to raise the intracellular levels of c-di-GMP. Recently, O'Toole and colleagues suggested that DGCs may require direct interaction with their targets to effect signaling. They demonstrated that the I-site of GcbC, a *P. fluorescens* DGC interacted with the target protein LapD [12, 13].

2 Degradation of c-di-GMP: EAL or HD-GYP Domain Proteins

To control the intracellular concentration of c-di-GMP bacteria express c-di-GMP specific phosphodiesterases (PDEs), which degrade the signaling molecule. Thus far, two main types of PDEs have been characterized, EAL and HD-GYP domain proteins [1, 8]. The EAL (Glu-Ala-Leu) domain-containing PDEs primarily linearize c-di-GMP into 5′-phosphoguanylyl-guanosine (5′–pGpG) (Fig. 1) and only slowly hydrolyze 5′-pGpG to GMP [14]. EAL-domain proteins require Mg^{2+} or Mn^{2+} ions for catalysis and are inhibited by Ca^{2+} [14]. The Glutamate residue in the EAL motif coordinates one metal ion in the active site. Some examples of EAL domain-containing PDEs include YahA and DosP from *E. coli*, RocR and PvrR from *P. aeruginosa* [1, 4]. HD-GYP (His-Asp and Gly-Tyr-Pro) domain-containing PDEs are the second group of c-di-GMP specific PDEs. These enzymes are capable of hydrolyzing c-di-GMP directly into two GMP molecules. Like the EAL domain proteins, HD-GYP proteins have a binuclear metal center with either Fe^{2+} or Mn^{2+} [1]. Examples of HD-GYP domain-containing PDEs include RpfG from *Xanthomonas campestris* pv. Campestris, Bd1817 from *Bdellovibrio bacteriovorus*, PmGH from *Persephonella marina* and PA4781 from *Pseudomonas aeruginosa*, which was shown to bind an array of transition metals including Ni^{2+}, Zn^{2+}, Co^{2+}, and Mn^{2+} [1, 15].

3 Receptors of c-di-GMP

c-di-GMP binds to several effector molecules, such as riboswitches, transcription factors, or enzymes. The PilZ domain was one of the earliest described c-di-GMP specific binding motifs [16] and is also one of the most extensively studied c-di-GMP effector molecules [1]. Examples of PilZ-containing proteins are Alg44 (*P. aeruginosa*), which is involved in alginate synthesis; DgrA (*C. crescentus*), which affects motility; and BcsA (*G. xylinus*), which regulates cellulose synthesis [17–19]. The RxxD domain is another well-studied c-di-GMP binding motif. In *C. crescentus* and *P. aeruginosa*, binding of c-di-GMP to the RxxD motifs of PopA and PelD effector proteins affect cell cycle progression and exopolysaccharide production respectively [20, 21]. Other c-di-GMP binding proteins, for example FimX, contain degenerate (catalytically dead) GGDEF or EALPhosphodiesterase (PDE):EAL domains [22, 23]. The regulation of gene expression is achieved by c-di-GMP binding to transcription factors or riboswitches, which are regulatory noncoding RNA elements found at the 5′ UTR of genes. In *X. campestris*, binding of c-di-GMP to Clp affects its DNA binding activity [24, 25]. The *P. aeruginosa* FleQ, an enhancer binding protein, serves to activate genes involved in flagella biosynthesis and a repressor of other genes like those involved in exopolysaccharide (EPS) biosynthesis [26, 27]. Hickman and Harwood showed that binding of c-di-GMP to FleQ relieves the repression of genes involved in EPS biosynthesis [26]. *V. cholerae* transcription regulator VpsT oligomerizes upon binding c-di-GMP [28]. In *Clostridium difficile* and *V. cholerae*, c-di-GMP has been shown to bind to the Vc2 riboswitch class I and 84Cd riboswitch class II respectively [29–32].

In *C. crescentus*, the cell cycle kinase CckA, a histidine kinase regulates cell cycle progression. Jenal and colleagues demonstrated that binding of CckA to c-di-GMP causes a shift in activity of CckA from kinase to phosphatase, which facilitates the initiation of replication [33, 34]. Another example of c-di-GMP affecting enzyme function is found in *P. aeruginosa* where the phosphorelay activity of the histidine kinase, SagS is inhibited by a protein–protein interaction with the PilZ protein HapZ at high concentrations of c-di-GMP [35]. Histidine kinases play important roles in biofilm formation. For example SagS regulates *P. aeruginosa* biofilm by controlling surface attachment [36] and has also been shown to be responsible for the extremely high tolerance of *P. aeruginosa* biofilms for antimicrobial agents [37].

c-di-GMP has been shown to bind to and activate STING (Stimulator of Interferon Genes, also known as MITA, MPYS and ERIS) in immune cells. This interaction stimulates type I interferon response serving as a means by which the immune system senses the presence of bacterial pathogens [38]. Recently, siderocalin, a component of the innate immune system with antibacterial activities was demonstrated to bind c-di-GMP. The binding was shown to inhibit the antibacterial activity of siderocalin, providing a new insight into how bacteria evade the immune system [39].

4 Intracellular c-di-GMP Concentration and Biofilm Formation

Perhaps the most established role of c-di-GMP signaling is regulating bacterial lifestyle by controlling the transition of bacteria from a planktonic to a sessile lifeform. Generally, high c-di-GMP concentration is thought to increase biofilm formation whereas low c-di-GMP concentration results in decreased biofilm formation.

The first stage of biofilm formation is surface attachment, which is initially reversible but then becomes irreversible. In *P. fluorescens* the intracellular concentration of c-di-GMP affects this process. The outer membrane adhesin, LapA plays an important role in the transition from reversible to irreversible surface attachment and hence biofilm formation [40]. This protein is regulated by the protease activity of LapG via cleavage of the N-terminal of LapA. At high cellular c-di-GMP concentration, the degenerate EAL domain of LapD binds c-di-GMP, thereby activating LapD to interact with and sequester LapG. Without the protease activity of LapG on LapA, *P. fluorescens* biofilm formation is enhanced. The reverse is true where low c-di-GMP inhibits the interaction between LapG and LapD allowing the former to cleave LapA [4, 41–43].

The PilZ domain protein YcgR regulates biofilm formation of *E. coli* upon binding to c-di-GMP. It has been shown that the YcgR–c-di-GMP complex interacts with components of the flagellar motor. This interaction has been noted to impede the proper flagellar motor function, thereby facilitating the transition from motile to sessile/biofilm lifestyle [44–46]. Fazli et al. observed that in *Burkholderia cenocepacia*, elevated levels of c-di-GMP enhanced biofilm formation. In that study, the authors showed that deletion of the c-di-GMP effector protein Bcam1349 resulted in decreased biofilm formation [47].

Transcriptional activation of the *mrk* operon in *Klebsiella pneumoniae* is mediated by the c-di-GMP effector protein MrkH, leading to the expression of type 3 fimbriae genes. The type 3 fimbriae of *K. pneumoniae* is involved in the attachment step of biofilm formation. Wilksch et al. showed that overexpression of the *K. pneumoniae* PDE MrkJ, which decreases intracellular c-di-GMP, abolished the transcriptional activation ability of MrkH. However, in a Δ*mrkJ* strain (contains higher intracellular c-di-GMP), the activity of MrkH was increased, leading to enhanced biofilm formation [48].

5 Inhibiting c-di-GMP Synthesis or Degradation

Based on the many phenotypes regulated by c-di-GMP and their relevance in bacterial survival, a few groups have embarked on the development of small molecules to inhibit c-di-GMP metabolism enzymes. This endeavor is however nontrivial because the genomes of most Gram-negative bacteria encode multiples of GGDEF, EAL, and HD-GYP domain-containing proteins, presumably with different active site geometries so inhibiting these myriad enzymes with a single compound is a high order. For example *P. aeruginosa* has 18 GGDEF-domain proteins, 16 GGDEF-EAL domain proteins, 5 EAL domain proteins, and 3 HD-GYP domain proteins [49]. Lory and colleagues observed that in *P. aeruginosa*, multiple DGCs and PDEs affect distinct phenotypes. For example, overexpressing the DGC PA2870, and other GGDEF domain proteins like SiaD and PA0575, had no effect on biofilm formation, while the DGCs WspR, RoeA, and PA3702 increased biofilm formation [50]. Clearly inhibiting biofilm formation in *P. aeruginosa* via single targeting of one DGC is not going to be an effective strategy. In some bacteria, although there might be a few DGCs that control the global c-di-GMP concentrations, other DGCs with lower catalytic activity might still play important roles. For example, the genome of *Salmonella enterica serovar* Typhimurium encodes 12 GGDEF domain proteins and 14 EAL domain proteins [7]. But the DGC AdrA contributes significantly to the total c-di-GMP pool (about 1000x more than two other active DGCs STM2123 and STM3388) [51]. On the other hand, STM2123 and STM3388 (but not AgrA) affect the expression of extracellular matrix components, including exopolysaccharide cellulose and curli fimbriae via the transcriptional regulator CsgD [51]. Perhaps a promiscuous DGC inhibitor is what is needed to control c-di-GMP-mediated biofilm formation.

Benziman and colleagues identified the earliest inhibitors of c-di-GMP synthase. A **glycosylated triterpenoid saponin** (Fig. 2a) was purified from extracts of garden pea (*Pisum sativum*) and was shown to inhibit *A. xylinus* DGC in a noncompetitive manner with an IC_{50} value of 5 μM. They also demonstrated that the antibiotic **Papulacandin B** (Fig. 2a) also had DGC inhibitory activity with IC_{50} value of 70 μM [52, 53].

Waters and coworkers used a c-di-GMP inducible transcriptional reporter in *V. cholerae* to identify a number of compounds that inhibited the DGC activity of *V. cholerae* VC2370. Some of these compounds also inhibited the DGC activity of WspR from *P. aeruginosa* WspR. These compounds, such as **DI-3** (N-(4-anilinophenyl)benzamide) and **DI-10** (N-{[(2-phenylethyl)amino]carbonothioyl} benzamide) (Fig. 2a) could significantly inhibit *V. cholerae* biofilms (Fig. 2b). Compound **DI-3** also inhibited *P. aeruginosa* biofilms, and so perhaps it is possible to identify and develop broad-spectrum DGC inhibitors [54].

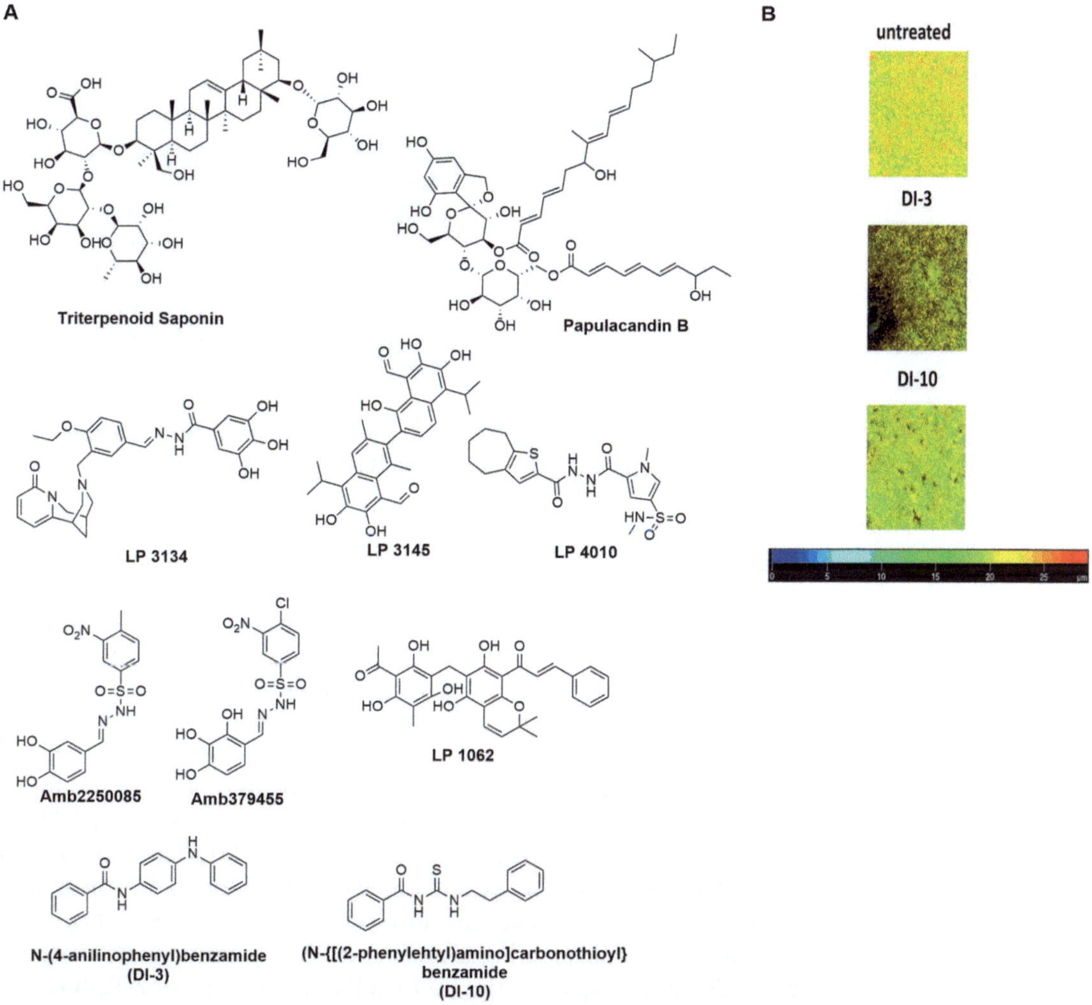

Fig. 2 (**a**) Structures of c-di-GMP DGC inhibitors. (**b**) Inhibition of *V. cholerae* biofilm by the DGC inhibitors DI-3 and DI-10. Reproduced with permission from [54] Copyright © 2012 American Society for Microbiology

Palys and coworkers used an *in silico* virtual screen to identify *Thermatoga maritima* DGC (tDGC) inhibitors, **LP 3134**, **LP 3145**, **LP 4010**, and **LP 1062**, Fig. 2a, [55]. These inhibitors were shown to inhibit *P. aeruginosa* and *Acenitobacter baumanii* biofilm formation. Interestingly, the authors noted that while all four inhibitors dispersed preformed *P. aeruginosa* biofilms, only **LP 3134** could disperse preformed *A. baumanii* biofilms [55]. Following this discovery, Rinaldo and colleagues also used a 3D pharmacophore model of the active site of PleD, a *C. crescentus* DGC to identify **Amb2250085** and **Amb37945** (Fig. 2a) as PleD inhibitors with half maximal inhibitory concentration (IC_{50}) values of approximately 11 μM. The authors also noted that **Amb37945** could inhibit the DGC activities of WspR and YfiN from *P. aeruginosa* [56].

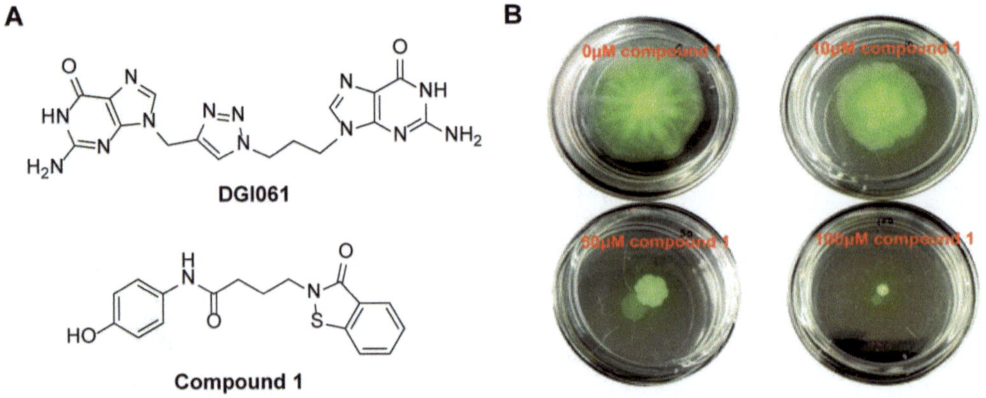

Fig. 3 (**a**) Structures of c-di-GMP PDE inhibitors. (**b**) Inhibition of *P. aeruginosa* swarming by **Compound 1**. Reproduced with permission from [58]. Copyright © 2016, The Royal Society of Chemistry

Some c-di-GMP PDEs regulate virulence factors production. Just like DGCs, not all PDEs regulate the global concentrations of c-di-GMP so some of these virulence-associated PDEs (as long as they do not regulate biofilm dispersal) could be targeted with small molecules. Unlike the situation with DGC inhibition, there is a paucity of inhibitors against c-di-GMP PDEs. An important observation from the study by Lory and colleagues was that some PDEs in *P. aeruginosa* had no effect on biofilm formation. For example, mutating genes that encode PA3947 (*rocR*), PA3825, PA2123, and PA2200, all of which showed *in vitro* PDE activity, did not affect biofilm formation relative to wildtype. However, a mutation of *pvrR* resulted in a decrease in biofilm formation [50]. Lory and colleagues also observed that *rocR* and *pvrR* mutants were avirulent towards mice [50]. Also, the HD-GYP domain regulator, RpfG of *X. campestris* was shown to be involved in virulence of the plant pathogen [57]. PDE inhibitors could therefore serve as anti-virulence agents [58–62]. Recently, Sintim and coworkers identified a benzoisothiazolinone derivative **Compound 1** (Fig. 3a), as an inhibitor of *P. aeruginosa* RocR PDE activity [58]. It was shown that the inhibitor was specific for RocR as the compound did not inhibit the activities of other PDEs tested, including PA4108, PvrR and DipA from *P. aeruginosa* as well as YahA from *E. coli* and snake venom phosphodiesterase (SVPD) from *Crotalus atrox*. The RocR inhibitor did not inhibit biofilm formation but inhibited swarming motility (Fig. 3b) [58]. Cutruzzolà and colleagues have also reported a PDE inhibitor DGI061 (Fig. 3a). The authors however did not report on the effect of this inhibitor on bacterial phenotype [62].

6 Conclusions

The essential roles of c-di-GMP signaling in bacteria cannot be overemphasized. A significant amount of work has gone into characterizing the molecular mechanisms underlying c-di-GMP signal pathway including metabolizing enzymes, receptors/effectors and phenotypic outputs. This has enhanced our understanding of the c-di-GMP signaling network. However, the rate of discovery of inhibitors has lagged behind. Potent inhibitors of c-di-GMP signaling may have anti-biofilms and anti-virulence applications. Due to the complexities of c-di-GMP signaling, it will be prudent to consider the potential implications of targeting any given component. Nonetheless, inhibitors that affect the phenotypic outputs such as biofilm formation or virulence are beneficial in that they could serve as tools for unraveling the complex network of c-di-GMP signaling. We are optimistic that more potent inhibitors of c-di-GMP signaling will be found and that some will proceed into the clinic, analogous to inhibitors of cGMP/cAMP signaling that have found applications in modern medicine.

Acknowledgment

We acknowledge funding from the NSF: CHE-1307218 and CHE-1636752.

References

1. Römling U, Galperin MY, Gomelsky M (2013) Cyclic di-GMP: the first 25 years of a universal bacterial second messenger. Microbiol Mol Biol Rev 77(1):1–52
2. Ross P, Weinhouse H, Aloni Y, Michaeli D, Weinbergerohana P, Mayer R, Braun S, Devroom E, Vandermarel GA, Vanboom JH, Benziman M (1987) Regulation of cellulose synthesis in *Acetobacter xylinum* by cyclic diguanylic acid. Nature 325(6101):279–281
3. Opoku-Temeng C, Zhou J, Zheng Y, Su J, Sintim HO (2016) Cyclic dinucleotide (c-di-GMP, c-di-AMP, and cGAMP) signalings have come of age to be inhibited by small molecules. Chem Commun (Camb) 52(60):9327–9342
4. Kalia D, Merey G, Nakayama S, Zheng Y, Zhou J, Luo Y, Guo M, Roembke BT, Sintim HO (2013) Nucleotide, c-di-GMP, c-di-AMP, cGMP, cAMP, (p)ppGpp signaling in bacteria and implications in pathogenesis. Chem Soc Rev 42(1):305–341
5. Costerton JW, Stewart PS, Greenberg EP (1999) Bacterial biofilms: a common cause of persistent infections. Science 284 (5418):1318–1322
6. Rabin N, Zheng Y, Opoku-Temeng C, Du Y, Bonsu E, Sintim HO (2015) Biofilm formation mechanisms and targets for developing antibiofilm agents. Future Med Chem 7(4):493–512
7. Römling U, Gomelsky M, Galperin MY (2005) C-di-GMP: the dawning of a novel bacterial signalling system. Mol Microbiol 57 (3):629–639
8. Schirmer T, Jenal U (2009) Structural and mechanistic determinants of c-di-GMP signalling. Nat Rev Microbiol 7(10):724–735
9. Paul R, Weiser S, Amiot NC, Chan C, Schirmer T, Giese B, Jenal U (2004) Cell cycle-dependent dynamic localization of a bacterial response regulator with a novel di-guanylate cyclase output domain. Genes Dev 18(6):715–727
10. Chan C, Paul R, Samoray D, Amiot NC, Giese B, Jenal U, Schirmer T (2004) Structural basis of activity and allosteric control of diguanylate cyclase. Proc Natl Acad Sci U S A 101 (49):17084–17089

11. Galperin MY, Nikolskaya AN, Koonin EV (2001) Novel domains of the prokaryotic two-component signal transduction systems. FEMS Microbiol Lett 203(1):11–21
12. Dahlstrom KM, Giglio KM, Sondermann H, O'Toole GA (2016) The inhibitory site of a diguanylate cyclase is a necessary element for interaction and signaling with an effector protein. J Bacteriol 198(11):1595–1603
13. Dahlstrom KM, Giglio KM, Collins AJ, Sondermann H, O'Toole GA (2015) Contribution of physical interactions to signaling specificity between a diguanylate cyclase and its effector. MBio 6(6):e01978–e01915
14. Schmidt AJ, Ryjenkov DA, Gomelsky M (2005) The ubiquitous protein domain EAL is a cyclic diguanylate-specific phosphodiesterase: enzymatically active and inactive EAL domains. J Bacteriol 187(14):4774–4781
15. Rinaldo S, Paiardini A, Stelitano V, Brunotti P, Cervoni L, Fernicola S, Protano C, Vitali M, Cutruzzolà F, Giardina G (2015) Structural basis of functional diversification of the HD-GYP domain revealed by the *Pseudomonas aeruginosa* PA4781 protein, which displays an unselective bimetallic binding site. J Bacteriol 197(8):1525–1535
16. Amikam D, Galperin MY (2006) PilZ domain is part of the bacterial c-di-GMP binding protein. Bioinformatics 22(1):3–6
17. Merighi M, Lee VT, Hyodo M, Hayakawa Y, Lory S (2007) The second messenger bis-(3′-5′)-cyclic-GMP and its PilZ domain-containing receptor Alg44 are required for alginate biosynthesis in *Pseudomonas aeruginosa*. Mol Microbiol 65(4):876–895
18. Ryjenkov DA, Simm R, Römling U, Gomelsky M (2006) The PilZ domain is a receptor for the second messenger c-di-GMP: the PilZ domain protein YcgR controls motility in enterobacteria. J Biol Chem 281(41):30310–30314
19. Christen M, Christen B, Allan MG, Folcher M, Jenö P, Grzesiek S, Jenal U (2007) DgrA is a member of a new family of cyclic diguanosine monophosphate receptors and controls flagellar motor function in *Caulobacter crescentus*. Proc Natl Acad Sci U S A 104(10):4112–4117
20. Lee VT, Matewish JM, Kessler JL, Hyodo M, Hayakawa Y, Lory S (2007) A cyclic-di-GMP receptor required for bacterial exopolysaccharide production. Mol Microbiol 65 (6):1474–1484
21. Duerig A, Abel S, Folcher M, Nicollier M, Schwede T, Amiot N, Giese B, Jenal U (2009) Second messenger-mediated spatiotemporal control of protein degradation regulates bacterial cell cycle progression. Genes Dev 23(1):93–104
22. Navarro MV, De N, Bae N, Wang Q, Sondermann H (2009) Structural analysis of the GGDEF-EAL domain-containing c-di-GMP receptor FimX. Structure 17(8):1104–1116
23. Qi Y, Chuah ML, Dong X, Xie K, Luo Z, Tang K, Liang ZX (2011) Binding of cyclic diguanylate in the non-catalytic EAL domain of FimX induces a long-range conformational change. J Biol Chem 286(4):2910–2917
24. Tao F, He YW, Wu DH, Swarup S, Zhang LH (2010) The cyclic nucleotide monophosphate domain of *Xanthomonas campestris* global regulator Clp defines a new class of cyclic di-GMP effectors. J Bacteriol 192(4):1020–1029
25. Chin KH, Lee YC, Tu ZL, Chen CH, Tseng YH, Yang JM, Ryan RP, McCarthy Y, Dow JM, Wang AH, Chou SH (2010) The cAMP receptor-like protein CLP is a novel c-di-GMP receptor linking cell-cell signaling to virulence gene expression in *Xanthomonas campestris*. J Mol Biol 396(3):646–662
26. Hickman JW, Harwood CS (2008) Identification of FleQ from *Pseudomonas aeruginosa* as a c-di-GMP-responsive transcription factor. Mol Microbiol 69(2):376–389
27. Arora SK, Ritchings BW, Almira EC, Lory S, Ramphal R (1997) A transcriptional activator, FleQ, regulates mucin adhesion and flagellar gene expression in *Pseudomonas aeruginosa* in a cascade manner. J Bacteriol 179 (17):5574–5581
28. Krasteva PV, Fong JCN, Shikuma NJ, Beyhan S, Navarro M, Yildiz FH, Sondermann H (2010) Vibrio cholerae VpsT regulates matrix production and motility by directly sensing cyclic di-GMP. Science 327(5967):866–868
29. Smith KD, Lipchock SV, Ames TD, Wang J, Breaker RR, Strobel SA (2009) Structural basis of ligand binding by a c-di-GMP riboswitch. Nat Struct Mol Biol 16(12):1218–1223
30. Sudarsan N, Lee ER, Weinberg Z, Moy RH, Kim JN, Link KH, Breaker RR (2008) Riboswitches in eubacteria sense the second messenger cyclic di-GMP. Science 321(5887):411–413
31. Lee ER, Baker JL, Weinberg Z, Sudarsan N, Breaker RR (2010) An allosteric self-splicing ribozyme triggered by a bacterial second messenger. Science 329(5993):845–848
32. Smith KD, Shanahan CA, Moore EL, Simon AC, Strobel SA (2011) Structural basis of differential ligand recognition by two classes of bis-(3′-5′)-cyclic dimeric guanosine monophosphate-binding riboswitches. Proc Natl Acad Sci U S A 108(19):7757–7762

33. Lori C, Ozaki S, Steiner S, Böhm R, Abel S, Dubey BN, Schirmer T, Hiller S, Jenal U (2015) Cyclic di-GMP acts as a cell cycle oscillator to drive chromosome replication. Nature 523(7559):236–239
34. Dubey BN, Lori C, Ozaki S, Fucile G, Plaza-Menacho I, Jenal U, Schirmer T (2016) Cyclic di-GMP mediates a histidine kinase/phosphatase switch by noncovalent domain cross-linking. Sci Adv 2(9):e1600823
35. Xu LH, Venkataramani P, Ding YC, Liu Y, Deng YY, Yong GL, Xin LY, Ye RJ, Zhang LH, Yang L, Liang ZX (2016) A cyclic di-GMP-binding adaptor protein interacts with histidine kinase to regulate two-component signaling. J Biol Chem 291(31):16112–16123
36. Petrova OE, Sauer K (2011) SagS contributes to the motile-sessile switch and acts in concert with BfiSR to enable *Pseudomonas aeruginosa* biofilm formation. J Bacteriol 193 (23):6614–6628
37. Gupta K, Marques CN, Petrova OE, Sauer K (2013) Antimicrobial tolerance of Pseudomonas aeruginosa biofilms is activated during an early developmental stage and requires the two-component hybrid SagS. J Bacteriol 195 (21):4975–4987
38. Burdette DL, Monroe KM, Sotelo-Troha K, Iwig JS, Eckert B, Hyodo M, Hayakawa Y, Vance RE (2011) STING is a direct innate immune sensor of cyclic di-GMP. Nature 478 (7370):515–518
39. Li W, Cui T, Hu L, Wang Z, Li Z, He ZG (2015) Cyclic diguanylate monophosphate directly binds to human siderocalin and inhibits its antibacterial activity. Nat Commun 6:8330
40. Hinsa SM, Espinosa-Urgel M, Ramos JL, O'Toole GA (2003) Transition from reversible to irreversible attachment during biofilm formation by *Pseudomonas fluorescens* WCS365 requires an ABC transporter and a large secreted protein. Mol Microbiol 49(4):905–918
41. Monds RD, Newell PD, Gross RH, O'Toole GA (2007) Phosphate-dependent modulation of c-di-GMP levels regulates *Pseudomonas fluorescens* Pf0-1 biofilm formation by controlling secretion of the adhesin LapA. Mol Microbiol 63(3):656–679
42. Newell PD, Monds RD, O'Toole GA (2009) LapD is a bis-(3′,5′)-cyclic dimeric GMP-binding protein that regulates surface attachment by *Pseudomonas fluorescens* Pf0-1. Proc Natl Acad Sci U S A 106(9):3461–3466
43. Newell PD, Boyd CD, Sondermann H, O'Toole GA (2011) A c-di-GMP effector system controls cell adhesion by inside-out signaling and surface protein cleavage. PLoS Biol 9(2):e1000587
44. Fang X, Gomelsky M (2010) A post-translational, c-di-GMP-dependent mechanism regulating flagellar motility. Mol Microbiol 76(5):1295–1305
45. Paul K, Nieto V, Carlquist WC, Blair DF, Harshey RM (2010) The c-di-GMP binding protein YcgR controls flagellar motor direction and speed to affect chemotaxis by a "backstop brake" mechanism. Mol Cell 38(1):128–139
46. Boehm A, Kaiser M, Li H, Spangler C, Kasper CA, Ackermann M, Kaever V, Sourjik V, Roth V, Jenal U (2010) Second messenger-mediated adjustment of bacterial swimming velocity. Cell 141(1):107–116
47. Fazli M, O'Connell A, Nilsson M, Niehaus K, Dow JM, Givskov M, Ryan RP, Tolker-Nielsen T (2011) The CRP/FNR family protein Bcam1349 is a c-di-GMP effector that regulates biofilm formation in the respiratory pathogen *Burkholderia cenocepacia*. Mol Microbiol 82(2):327–341
48. Wilksch JJ, Yang J, Clements A, Gabbe JL, Short KR, Cao H, Cavaliere R, James CE, Whitchurch CB, Schembri MA, Chuah ML, Liang ZX, Wijburg OL, Jenney AW, Lithgow T, Strugnell RA (2011) MrkH, a novel c-di-GMP-dependent transcriptional activator, controls *Klebsiella pneumoniae* biofilm formation by regulating type 3 fimbriae expression. PLoS Pathog 7(8):e1002204
49. Valentini M, Filloux A (2016) Biofilms and cyclic di-GMP (c-di-GMP) signaling: lessons from *Pseudomonas aeruginosa* and other bacteria. J Biol Chem 291(24):12547–12555
50. Kulasakara H, Lee V, Brencic A, Liberati N, Urbach J, Miyata S, Lee DG, Neely AN, Hyodo M, Hayakawa Y, Ausubel FM, Lory S (2006) Analysis of *Pseudomonas aeruginosa* diguanylate cyclases and phosphodiesterases reveals a role for bis-(3′-5′)-cyclic-GMP in virulence. Proc Natl Acad Sci U S A 103 (8):2839–2844
51. Kader A, Simm R, Gerstel U, Morr M, Römling U (2006) Hierarchical involvement of various GGDEF domain proteins in rdar morphotype development of *Salmonella enterica serovar Typhimurium*. Mol Microbiol 60 (3):602–616
52. Ohana P, Delmer DP, Carlson RW, Glushka J, Azadi P, Bacic T, Benziman M (1998) Identification of a novel triterpenoid saponin from *Pisum sativum* as a specific inhibitor of the diguanylate cyclase of *Acetobacter xylinum*. Plant Cell Physiol 39(2):144–152

53. Ohana P, Delmer DP, Volman G, Benziman M (1998) Glycosylated triterpenoid saponin: a specific inhibitor of diguanylate cyclase from *Acetobacter xylinum*. Biological activity and distribution. Plant Cell Physiol 39(2):153–159

54. Sambanthamoorthy K, Sloup RE, Parashar V, Smith JM, Kim EE, Semmelhack MF, Neiditch MB, Waters CM (2012) Identification of small molecules that antagonize diguanylate cyclase enzymes to inhibit biofilm formation. Antimicrob Agents Chemother 56(10):5202–5211

55. Sambanthamoorthy K, Luo C, Pattabiraman N, Feng X, Koestler B, Waters CM, Palys TJ (2014) Identification of small molecules inhibiting diguanylate cyclases to control bacterial biofilm development. Biofouling 30(1):17–28

56. Fernicola S, Paiardini A, Giardina G, Rampioni G, Leoni L, Cutruzzola F, Rinaldo S (2016) *In silico* discovery and *in vitro* validation of catechol-containing sulfonohydrazide compounds as potent inhibitors of the diguanylate cyclase PleD. J Bacteriol 198(1):147–156

57. Ryan RP, Fouhy Y, Lucey JF, Jiang BL, He YQ, Feng JX, Tang JL, Dow JM (2007) Cyclic di-GMP signalling in the virulence and environmental adaptation of Xanthomonas campestris. Mol Microbiol 63(2):429–442

58. Zheng Y, Tsuji G, Opoku-Temeng C, Sintim HO (2016) Inhibition of *P. aeruginosa* c-di-GMP phosphodiesterase RocR and swarming motility by a benzoisothiazolinone derivative. Chem Sci 7(9):6238–6244

59. Zhou J, Watt S, Wang J, Nakayama S, Sayre DA, Lam YF, Lee VT, Sintim HO (2013) Potent suppression of c-di-GMP synthesis via I-site allosteric inhibition of diguanylate cyclases with 2′-F-c-di-GMP. Bioorg Med Chem 21(14):4396–4404

60. Wang J, Zhou J, Donaldson GP, Nakayama S, Yan L, Lam YF, Lee VT, Sintim HO (2011) Conservative change to the phosphate moiety of cyclic diguanylic monophosphate remarkably affects its polymorphism and ability to bind DGC, PDE, and PilZ proteins. J Am Chem Soc 133(24):9320–9330

61. Shanahan CA, Gaffney BL, Jones RA, Strobel SA (2013) Identification of c-di-GMP derivatives resistant to an EAL domain phosphodiesterase. Biochemist 52(2):365–377

62. Fernicola S, Torquati I, Paiardini A, Giardina G, Rampioni G, Messina M, Leoni L, Del Bello F, Petrelli R, Rinaldo S, Cappellacci L, Cutruzzolà F (2015) Synthesis of triazole-linked analogues of c-di-GMP and their interactions with diguanylate cyclase. J Med Chem 58 (20):8269–8284

Chapter 32

Discovering Selective Diguanylate Cyclase Inhibitors: From PleD to Discrimination of the Active Site of Cyclic-di-GMP Phosphodiesterases

S. Rinaldo, G. Giardina, F. Mantoni, A. Paiardini, Alessio Paone, and Francesca Cutruzzolà

Abstract

One of the most important signals involved in controlling biofilm formation is represented by the intracellular second messenger 3′,5′-cyclic diguanylic acid (c-di-GMP). Since the pathways involved in c-di-GMP biosynthesis and breakdown are found only in bacteria, targeting c-di-GMP metabolism represents an attractive strategy for the development of biofilm-disrupting drugs. Here, we present the workflow required to perform a structure-based design of inhibitors of diguanylate cyclases, the enzymes responsible for c-di-GMP biosynthesis. Downstream of the virtual screening process, detailed in the first part of the chapter, we report the step-by-step protocols required to test the positive hits in vitro and to validate their selectivity, thus minimizing possible off-target effects.

Key words Diguanylate cyclases, Inhibitors, Virtual screening, Phosphodiesterases, PleD, RocR

1 Introduction

C-di-GMP signaling systems represent promising targets to hamper biofilm formation and virulence in many bacterial pathogens; this can be achieved mainly through the inhibition of the activity of diguanylate cyclases (DGCs), the enzymes yielding c-di-GMP from two molecules of GTP. The rational approach aimed at targeting DGCs necessarily needs to deal with enzymes that are well characterized from a structural and functional point of view.

PleD from *Caulobacter crescentus* and WspR form *Pseudomonas aeruginosa* were the first diguanylate cyclases (DGC) to have their structure solved [1–4], and are among of the best characterized [5]. The pioneering work on these enzymes unveiled the key residues involved in substrate binding and allosteric regulation. The

S. Rinaldo and G. Giardina contributed equally to this work.

Karin Sauer (ed.), *c-di-GMP Signaling: Methods and Protocols*, Methods in Molecular Biology, vol. 1657,
DOI 10.1007/978-1-4939-7240-1_32, © Springer Science+Business Media LLC 2017

existence of a general product inhibition mechanism for DGCs was shown to be related to the presence of a conserved noncompetitive c-di-GMP binding site, named I-site, to be distinguished from the GTP binding site (A-site), where catalysis occurs (Fig. 1a, b).

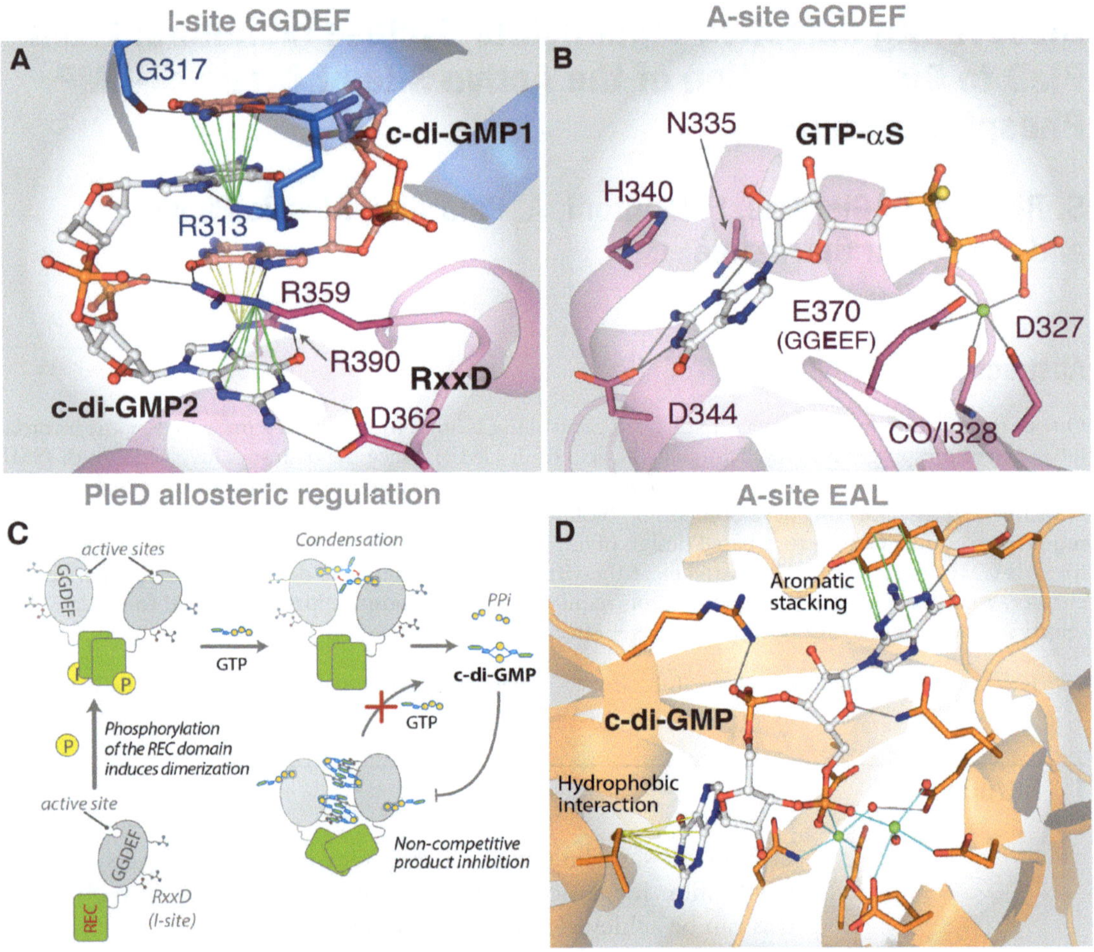

Fig. 1 PleD structure: I-site and A-site. (**a**) I-site of PleD in complex with a c-di-GMP dimer (*white* and *salmon*). The dimer cross-links two different domains (*pink* and *blue*). Residues from the RxxD motif (R359 and D362) and the other interacting residues are shown as sticks. H-bonds are shown as *thin grey lines*, while π-cation interactions between arginine residues and the guanines are shown in *green* and *yellow* (PDB id 2WB4 [unpublished]). (**b**) GTP-αS (*white*) bound to the A-site of PleD (*pink*). The guanine base interacts only with N335 and D344 through H-bonds, while the oxygen atoms from the β and γ phosphates coordinate the magnesium ion together with E370 (from the GGEEF motif), D327 and the main chain carbonyl of I328 (CO/I328). One of the two axial positions of the octahedral coordination is unoccupied (PDB id 2V0N chain B [2]). (**c**) Scheme of the allosteric regulation of PleD. The inactive protein is monomeric and phosphorylation of D53 in the REC domain induces dimerization, allowing catalysis to occur. C-di-GMP binding to the I-site cross-links the two GGDEF domains blocking the enzyme in a nonproductive conformation (product feedback inhibition mechanism). (**d**) The c-di-GMP binding mode to the active site of EAL PDEs is very conserved. C-di-GMP binds as a monomer in a trans-conformation with the two guanine bases interacting with conserved residues by hydrophobic, H-bonds and π-stacking interactions. Only one phosphate, the one undergoing hydrolysis by an activated water molecule, is coordinated to the metal centre (PDB id 4Y9Q unpublished)

C-di-GMP is able to bind to the I-site as an intercalated dimer (*cis*-dimer), interacting with an arginine and an aspartate residue belonging to a conserved motif RxxD. Binding of c-di-GMP to the I-site blocks the PleD enzyme in an inactive conformation by a homo or hetero domain cross-linking (Fig. 1c). However, having a conserved RxxD region is a necessary, but not sufficient condition for a DGC to display product feedback inhibition, as recently demonstrated for $\mathrm{YfiN_{HAMP\text{-}GGDEF}}$ from *P. aeruginosa* [6]. Therefore, in principle, DGCs may be inhibited by targeting either the I-site, when present, or the A-site. In this regard, it must be noted that the structural data available so far shows that, in the GGDEF domain, a single A-site is present, bound to nonhydrolyzable GTP analogs [2, 7]. However, it has been established that condensation of two GTP molecules in DGCs requires a dynamic change of the quaternary structure of the domains (monomer → dimer transition) to enter catalysis. The dynamics of the quaternary state assembly, as already evidenced in other prokaryotic and eukaryotic molecular systems [8–10], might be modulated by yet unidentified parameters, and may have relevance for drug design purposes.

Literature data report many studies focused on the rational synthesis of c-di-GMP analogs aimed at targeting the I-site [11–13], employing different strategies of chemical synthesis; however, these molecules are still far from being used as anti-biofilm agents. One potential limitation is that c-di-GMP (and likely its analogs) does not cross the bacterial membrane and, as reminded above, many DGCs lacks the I-site to control their activity [6]. Moreover, given that c-di-GMP is also the substrate of phosphodiesterases (PDEs) (Fig. 1d), c-di-GMP analogs could also act as potent inhibitors of these enzymes [12, 13], an effect that should be in principle avoided to improve the efficacy of DGC inhibition in the bacterial cell. All these considerations should be taken into account in the design of c-di-GMP analogs and therefore a validation test on a reference PDE such as RocR from *P. aeruginosa* (see below) and, possibly, on a DGC lacking the I-site such as $\mathrm{YfiN_{HAMP\text{-}GGDEF}}$ from *P. aeruginosa* should be part of the experimental pipeline.

Studies targeting the I-site to identify compounds structurally different from c-di-GMP analogs have not been reported yet; however, in principle, such an approach might be feasible, by means of virtual screening (VS) studies. VS makes use of computer-based methods to identify new compounds showing some activity against a protein target.

On the other hand, all the successful structure-based studies carried out so far to target the A-site [14, 15] made use of a VS approach, which implied the prior identification of the key interactions responsible for GTP binding at the A-site. Among the identified compounds, only one actually showed anti-biofilm activity, possibly *via* off-target effects [14]. In principle, for the VS approach, the ideal choice is to perform in vitro validation of identified compounds on the VS target (such as PleD for the protocols described herein), given that in vitro validation does not necessarily require a DGC harboring a canonical I-site. Control experiments to validate the specificity of those compounds targeting the A-site are difficult to design, since in principle all proteins coping with GTP could represent possible off-targets; inhibitor selectivity is to be probed by assays on bacterial cells together with toxicity tests on eukaryotic cells. For this reason, the rational approach for targeting the A-site based on genuine GTP analogs is expected to be very challenging: indeed the broad spectrum of biological activities of GTP increases the chance of off-target effects also in eukaryotes. In the latter case, it is therefore imperative to identify compounds with completely different scaffolds, compared to GTP, but still able to bind the A-site and selectively inhibit the activity of DGCs.

In light of the above-mentioned considerations, we propose a working plan (Fig. 2), showing the choice of the target and all the controls required to target alternatively the A- or the I-site of DGCs (Fig. 2). Given the abundance of biochemical data available, PleD or WspR represent ideal in silico and in vitro models for validation studies on both A- and I-site inhibition. It should be mentioned that PleD requires to be preactivated with BeF_3, which mimics the phosphorylation of its accessory regulatory REC domain, responsible for dimerization and consequent activation of the enzyme (Fig. 1c) [2]. Therefore, the use of less complex DGC targets should also be considered for future novel screening campaigns [6, 16].

In this chapter we report all the methodological procedures required to perform in silico and in vitro inhibition studies with PleD [14, 15]. We also report the detailed protocols to perform PDE assays on RocR from *P. aeruginosa*, a well-known reference system for PDE studies, whose structural and functional details are known [17, 18]. This step is crucial for the identification of suitable candidates to be taken forward for in vivo studies (where both DGCs and PDEs are present).

For kinetic studies, we have chosen to focus on circular dichroism (CD) spectroscopy and reverse phase HPLC (RP-HPLC), the only two available approaches allowing quantitative measurements on both DGCs and PDEs, even though other techniques specific for the sole DGC or PDE have been published (including the Enzcheck® phosphate assay, only for DGCs [11, 14, 15, 19, 20]).

Out of the scope of this chapter, we recall that targeting DGCs may also include other, not structure-based, approaches such as the drug repurposing strategy and the high throughput screening (reviewed in [21]), which benefit from the diverse c-di-GMP biosensors developed [22–27] and the novel in vitro DRaCALA approach [28]. In all cases, with the exception of the DRaCALA screening, these methods evaluate the overall decrease in the c-di-GMP levels and the phenotypic effect on biofilm, being the

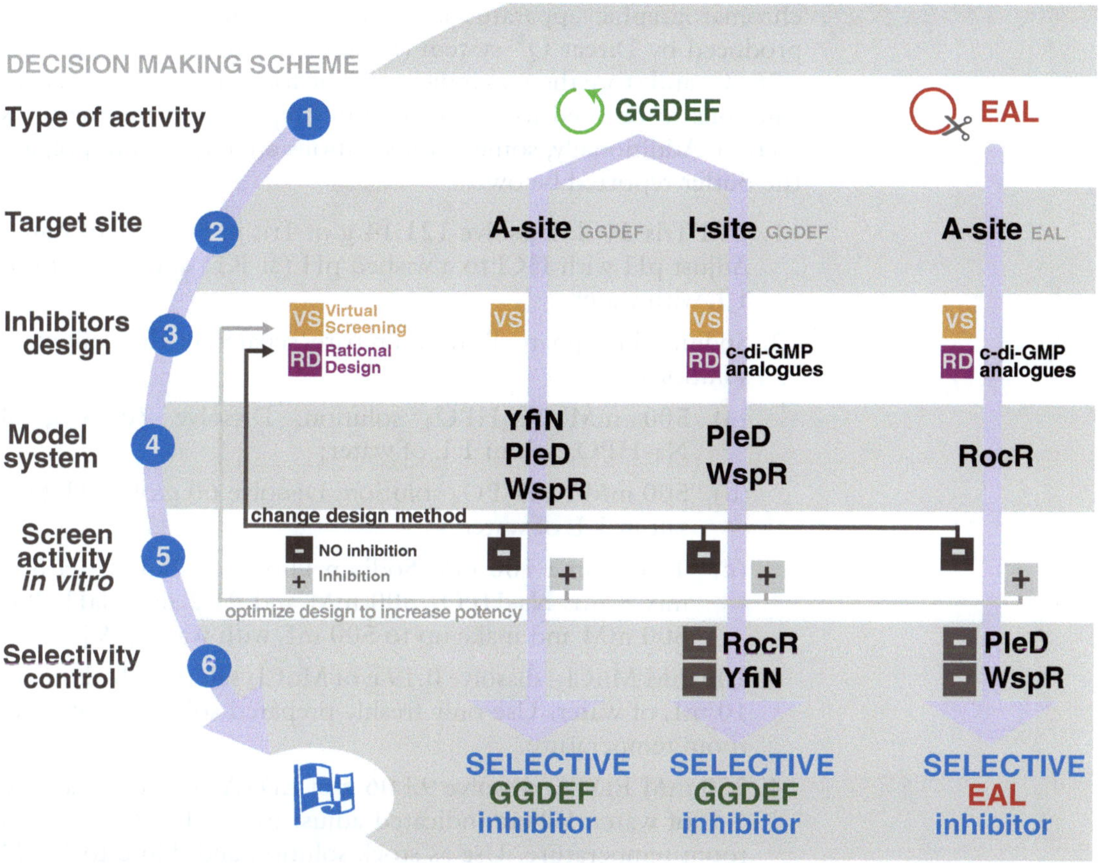

Fig. 2 Decision-making scheme to obtain in vitro inhibitors of DGCs. Flowchart showing the relevant steps (numbered *1–6*) for targeting both DGCs (first and second columns) and PDEs (third column). As mentioned in the text, the emerging role of selected PDEs in promoting biofilm makes this class of enzymes also possible targets for inhibition studies. *Minus* indicates lack of inhibition, while *plus* represents positive hits, and therefore inhibition

individual target difficult to assess. However, the recent evidence showing that lowering c-di-GMP intracellular levels can lead to increase of biofilm formation, makes the scenario, if possible, even more complicated [25], extending the possible strategies to interfere with biofilm to those targeting PDEs (*see* Fig. 2, third column, for a possible workflow for PDE inhibitor design).

Structure-based design of selective inhibitors of DGCs will undoubtedly greatly benefit from more biochemical and structural data to refine and implement the design process, and novel ideas to allow the positive hits to be effective in vivo.

2 Materials

2.1 Solutions

All reagents used are of high purity grade (≥99%). Prepare all solutions required for experiments and for cleaning the chromatographic apparatus and columns using ultrapure water produced by Direct-Q® system (or similar systems, 18.2 MΩ/cm a 25 °C) and store them as indicated. The following is a list of stock solutions to be used for the procedures reported in the methods section. Additionally, some stock solutions are required to prepare the buffer reported below:

1. 1 M Tris–HCl: Dissolve 121.14 g of Tris in 800 mL of water. Adjust pH with HCl to a wished pH (at RT) and make up to 1 L with water.
2. Sodium Phosphate Buffer: Prepare separately the following solutions:
   (a) 500 mM Na_2HPO_4 solution: Dissolve 70.98 g of Na_2HPO_4 salt in 1 L of water;
   (b) 500 mM NaH_2PO_4 solution: Dissolve 60 g of NaH_2PO_4 salt in 1 L of water;
   (c) To obtain a 100 mM Sodium Phosphate buffer pH 5.8, mix 8 mL Na_2HPO_4 500 mM and 91.2 mL NaH_2PO_4 500 mM and make up to 500 mL with water [29].
3. 100 mM $MnCl_2$: dissolve 0.19 g of $MnCl_2$ salt (tetrahydrate) in 10 mL of water. Use only freshly prepared solutions. Store at room temperature.
4. 250 mM EDTA: dissolve 93.06 g of EDTA salt (dihydrate) to 1 L of water. Where indicated adjust pH with HCl. Store at room temperature. Use as stock solution and dilute to 1 mM (1:250), if indicated in the buffer listed below.
5. 1 M dithiotreitol (DTT): dissolve 1.54 g of DTT in 10 mL of water; make aliquots of 500 μL and store at −20 °C. Use a freshly thawed aliquot for the experiments and dilute to 1 mM (1:1000), when indicated in the buffer listed below.
6. 5 M NaCl: dissolve 292.2 g of NaCl salt in 1 L of water. Store at room temperature.
7. 1 M $CaCl_2$: dissolve 147.02 g of $CaCl_2$ salt (dihydrate) in 1 L of water. Store at room temperature.
8. 100 mM $NiSO_4$: Dissolve 26.28 g of $NiSO_4$ salt (hexahydrate) in 1 L of water. Store at room temperature.
9. 100 mM phenylmethylsulfonyl fluoride (PMSF): Dissolve 1.74 g of PMSF in 100 mL of isopropanol. The solution must be prepared under chemical hood. Store at −20 °C.
10. 1 M Isopropyl β-D-thiogalactoside (IPTG): Weight 2.38 g IPTG (Inalco) and dissolve with 10 mL of water. Filter-sterilize the solution using Minisart® syringe 0.2 μm filter (Sartorius) under laminar flow before storing at −20 °C in small aliquots.

11. 100 mM $BeCl_2$: add about 450 mL of water to a 2 L Erlenmeyer flask and gradually add 4 g of $BeCl_2$ powder (all the content of the ampoule; do not weight the powder) in 3–4 portions to avoid a too vigorous reaction, due to HCl gas formation, and gently shake the flask. After the solution has cooled adjust the pH with ca. 3.5 mL of NaOH 10 M (be careful that at pH 5.5 BeOH precipitates), adjust the final volume to 500 mL with water. Store the solution in a cabinet designed for toxic substances. Before handling beryllium-containing compounds, please refer to the proper Personal Protective Equipment reported on the safety sheet. All the procedures must take place under chemical hood, as beryllium is highly toxic. Moreover, dissolving of $BeCl_2$-powder in water is highly exergonic, yielding BeOH and HCl. $BeCl_2$ powder comes in an ampoule under inert atmosphere. Store the solution at room temperature.
12. 500 mM NaF: dissolve 2.09 g of NaF salt in 100 mL of water. Store at room temperature.
13. Nucleotides:
    (a) C-di-GMP (1 μmol) and pGpG (0.1 μmol) powders (as provided by the vendor), once dissolved in 1 mL of water, are stored at −20 °C as 1 and 0.1 mM stock solutions, respectively.
    (b) For GTP, once dissolved in water, adjust the pH to 7.0 with concentrated NaOH, and make aliquots of 20 μL to be stored at −20 °C as 4 mM stock solution.
    (c) The preparation of all nucleotides (listed in a and b) must be carried out in ice.
14. 1 mg/mL Proteinase K from *Engyodontium album*: flash freeze aliquots in liquid nitrogen once dissolved in water, store at −20 °C.
15. Lysis buffer: 20 mM Tris–HCl, pH 8.0, 50 mM NaCl, 1 mM PMSF. Add about 50 mL of water, 2 mL Tris–HCl 1 M, 1 mL NaCl 5 M and 1 mL PMSF 100 mM. Adjust pH with HCl to pH 8.0 and make up to 100 mL with water.
16. Reaction buffer: 20 mM Tris, pH 8.0, 100 mM NaCl, 10 mM $MgCl_2$, 2.5 mM $MnCl_2$.
17. Buffer A: 20 mM Tris–HCl, pH 8.0, 500 mM NaCl. Add about 200 mL of water, 10 mL 1 M Tris–HCl and 50 mL 5 M NaCl. Adjust pH with HCl to pH 8.0 and make up to 0.5 L with water. Store at room temperature.
18. Buffer B: 20 mM Tris–HCl, pH 8.0, 500 mM NaCl, 500 mM imidazole. Add about 200 mL water, 17.02 g of imidazole, 10 mL 1 M Tris–HCl and 50 mL 5 M NaCl. Adjust pH with HCl to pH 8.0 and make up to 0.5 L with water. Store at room temperature.

19. Buffer C: 25 mM Tris–HCl, pH 8.2, 500 mM NaCl, 20 mM imidazole. Add about 200 mL water, 0.68 g of imidazole, 12.5 mL 1 M Tris–HCl and 50 mL 5 M NaCl. Adjust pH with HCl to pH 8.2 and make up to 0.5 L with water. Store at room temperature.
20. Buffer D: 25 mM Tris–HCl, pH 8.2, 500 mM NaCl, 80 mM imidazole. Add about 200 mL water, 2.72 g of imidazole, 12.5 mL 1 M Tris–HCl and 50 mL 5 M NaCl. Adjust pH with HCl to pH 8.2 and make up to 0.5 L with water. Store at room temperature.

2.2 Cell Cultures

1. LB broth: Weight 7.5 g tryptone, 7.5 g NaCl, 3.75 g yeast extract, and add water to a final volume of 750 mL. Cover with silver paper and sterilize the LB.
2. LB broth containing 100 μg/mL ampicillin.
3. RPMI 1640 medium supplemented with 10% fetal bovine serum (FBS), 100 U/mL penicillin and streptomycin. For the complexity of the medium, ready-to-use RPMI 1640, commercially available such as Gibco or Sigma-Aldrich (or similar), is required to ensure reproducibility. Addition of 10% of FBS and 100 IU/mL penicillin and streptomycin to the original medium bottle to obtain a 1× ready-to-use medium should be done under sterile hood.
4. Trypsin: use a 1× ready commercial trypsin–EDTA solution for cell culture (such as from Gibco or Sigma-Aldrich).
5. Dulbecco's Modified Phosphate buffered saline (DPBS): Salt Concentration for a 1× solution (g/L): 8.0 g NaCl, 0.2 g KCl, 1.15 g Na_2HPO_4 (anhydrous), 0.2 g KH_2PO_4.

2.3 Chromatography

All solutions used for size exclusion chromatography (SEC) FPLC and reverse phase (RP) HPLC have to be filtered using 0.45 μm membrane filters before use.

1. Methanol, HPLC-grade.
2. Ethanol, HPLC-grade.
3. Size exclusion chromatography with HiLoad Column 26/60 Superdex 200. Column should be stored in 20% ethanol. Wash the with 2-3 column volumes of water before buffer equilibration.
4. RP-HPLC using a 150 × 4.6 mm reverse phase column (Prevail C8, Grace Davison Discovery Science, particle size of 5 μm). The column should be stored in 20% ethanol, washed in H_2O–methanol (98:2, v/v) and equilibrated with at least three volumes of 100 mM Phosphate buffer pH 5.8–methanol (98:2, v/v). Column final washing is performed in H_2O–methanol (80:20, v/v, at least five volumes).

5. 0.2 μm filters (Bilk GHP Acrodisc 13 mm) to remove protein from boiled samples.
6. High-Performance Liquid Chromatography (HPLC) system (such as AZURA® HPLC system, with UV detector and pumps integrated in a unique apparatus).
7. Fast Performance Liquid Chromatography (FPLC) system (such as ÄKTAprime plus system).
8. PD-10 (GE Healthcare) desalting column, for desalting or buffer exchange steps. Follow the protocol reported by the vendor.

2.4 Spectroscopy

1. UV-Vis spectra acquisition must be done in quartz cuvette or analogs, such as UVette (Eppendorf), which allows UV absorbance quantification, using a JASCO V-650 spectrophotometer (or similar).
2. For CD spectra, use a quartz cuvette with reduced volume, such as Hellma cuvette series 108B, 10 mm light path, which allows keeping the sample volume at 800 μL. Real-time analysis were carried out setting the realtime acquisition mode of a JASCO J-710 CD spectropolarimeter as follows: (a) wavelength: 282 nm; (b) data pitch: 0.5 s (c) scanning speed: 100 nm/min; (d) scanning mode: continuous (e) response: 4 s; (f) band width: 5 nm; (g) accumulation: 2 acquisitions.
3. C-di-GMP spectra were obtained setting the spectrum acquisition mode of CD instrument as reported: (a) spectra from 340 to 240 nm; (b) data pitch: 1 nm; (c) scanning speed: 100 nm/min; (d) scanning mode: continuous; (e) response: 4 s; (f) band width: 5 nm; (g) accumulation: 4 acquisitions.

3 Methods

3.1 Virtual Screening (VS) of A-Site Inhibitors

The basic goal of VS is to decrease the size of the huge chemical space of available small molecules, to a manageable number of prioritized hits that can be easily assayed in vitro against a specific target. A broad range of computational techniques can be applied to tackle this issue. In our work, we integrate two complementary approaches, i.e., pharmacophore searches and molecular docking, to screen the ZINC public library of commercially available compounds [30].

3.1.1 Ligands Library and Target Preparation

1. Download the drug-like and lead-like subsets of the ZINC database (http://zinc.docking.org/browse/subsets) [30] to generate low-energy conformations.
2. Apply the "Conformation import" function of MOE (The Molecular Operating Environment; Chemical Computing

Group®, Montreal, Canada; https://www.chemcomp.com). For each compound, apply the following settings: maximum 250 conformers for each compound; no input filters; constraints options kept at their default values; retain only the conformers with strain energy less than 4 kcal/mol.

3. Save the obtained conformations as a Moe Database Format (.mdb file).
4. Obtain the three-dimensional (3D) structure of PleD in complex with GTP-α-S from Protein Data Bank (PDB; accession number 2V0N [2]). Remove all other ligands except GTP-α-S and Mg ions by manually editing the PDB file (remove lines 8193–8289 and lines 8324–8333 with a text editor).
5. Automatically assign the protonation state and geometry of residues of PleD using the "Protonate 3D" function of MOE. Protonate 3D solves the macromolecular protonation state assignment problem by selecting a protonation state for each chemical group that minimizes the total free energy of the system (taking titration into account).
6. Visually inspect the final complex in order to verify the absence of steric clashes between GTP-α-S and the residues at the active site.

3.1.2 Pharmacophore Model Generation and Search

1. Pharmacophore modeling calculations are performed using MOE. Use the MOE "Pharmacophore Query" (PQ) function in order to build the initial Pharmacophore Hypothesis (PH), starting from the "PleD – GTP -α-S complex".
2. Through the "Unified Annotation Scheme", assign the pharmacophore annotation points (such as H-bond donor, H-bond acceptor, and hydrophobic). Insert the chemical features (coordinates, type of chemical feature, and sphere radius) corresponding to atoms or groups of atoms of GTP-α-S in the PH, as depicted in Fig. 3. MOE's default values for the radii of atom-based chemical features (1 Å) and projections (1.4 Å) are used.
3. Insert also an excluded volume, comprising residues at the active site of PleD located within a radius of 15 Å from any atom of PleD, in the final PH.
4. Use the final PH to screen the library of compounds obtained in 1.1, using the "Pharmacophore Search" option of MOE. Set all of the features as essential and filter all compounds not having at least five features of the PH.

3.1.3 Docking of Compounds

1. The compounds retrieved from the pharmacophore searches are docked into the active site of PleD, by means of Molegro Virtual Docker (MVD) software (CLCbio).

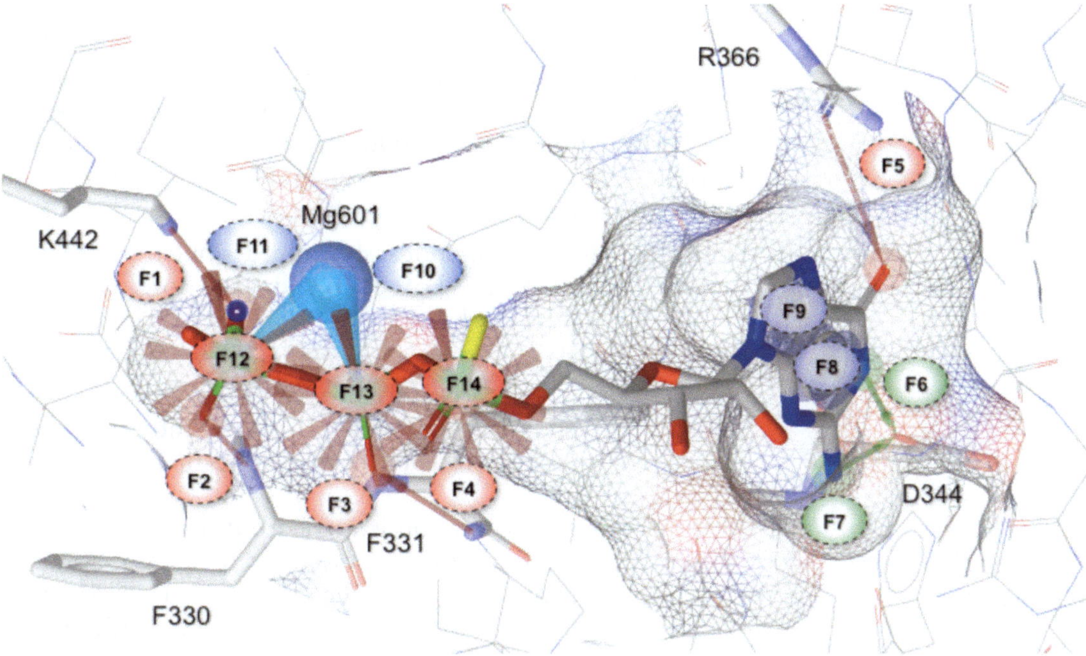

Fig. 3 Pharmacophore hypothesis of a competitive inhibitor of GTP bound to PleD. Protein is shown as *grey lines*, with the binding site shown as a Van der Waals surface and coloured according to polarity. GTP-α-S is also shown as sticks. Residues interacting with the pharmacophore map are represented as sticks and colored by atom type. F1–F5, hydrogen bond acceptors (*red arrows*); F6 and F7, hydrogen bond donors (*green arrows*); F8 and F9, aromatic centroids (*blue tori*); F10 and F11, metal binding groups (*cyan cones*); F12–F14, negative charges (*red cones*). Exclusion volume spheres are not shown for clarity

2. Prepare the clean three-dimensional (3D) structure of PleD by automatically assigning bond orders and hybridization, charges, and Tripos atom types. Keep the previously obtained explicit hydrogens.
3. Use the “Detect Cavities” function of MVD to identify the binding cleft of GTP-α-S. Keep parameters at their default values.
4. Select a search space with a 15-Å radius, centered on the GTP binding cavity, for docking purposes. Use grid-based MolDock score with a grid resolution of 0.30 Å as a scoring function, and MolDock SE was docking algorithm [31]. Keep the rest of parameters at their default values.
5. For each ligand, carry out a minimum of 10 runs. Rank the retrieved compounds according to their score (obtained with the “Rerank Energy” function implemented in MVD).
6. Compute the mean and standard deviation of the scores, and keep only those compounds scoring at >2.0 standard deviations from mean.

7. Redock the obtained compounds with AutoDock [32]. To this end, make use of the Lamarckian genetic algorithm (LGA) implemented in AutoDock, using the following values: number of individuals in population of 150, maximum number of energy evaluations of 2,500,000, maximum number of generations of 27×10^3, and rate of gene mutation of 0.02. Keep all other parameters at their default values.
8. Keep only those compounds showing a similar (root mean square deviation [RMSD] of <2.0 Å) docked pose, as assessed by MVD and AutoDock.
9. Rank and prioritize the obtained compounds based on mean predicted affinity of MVD and Autodock and commercial availability from vendors (listed in ZINC), and choose the top-ranking compounds (10–100) for in vitro testing.

3.2 *In Vitro* Screening of Selective DGC Inhibitors

3.2.1 Protein Preparation: Expression and Purification of PleD

PleD expression and purification is carried out as reported in [1, 33], with minor modifications. Expression vector used is a pET24 based plasmid containing the *pleD* gene in frame with a C-terminal His-tag for expression in *Escherichia coli* strain BL21 (DE3) pLysS (hereinafter referred to as BL21PleD strain).

1. Inoculate one colony of BL21PleD strain into 15 mL of LB medium containing 100 μg/mL ampicillin.
2. Grow overnight (15 h) by shaking at 180 rpm at 37 °C.
3. Dilute 1:100 into 2 × 750 mL of LB containing 100 μg/mL ampicillin and grow by shaking at 180 rpm at 30 ° C until an $OD_{600} = 0.5$ is reached.
4. Once an $OD_{600} = 0.5$ has been reached, induce PleD expression by 0.8 mM IPTG (final concentration) and allow to grow for an additional 3 h at 30 °C.
5. Collect cells by centrifugation (10 min at 5700 × *g* at 4 °C) and, if necessary, store at −20 °C or proceed with protein purification.
6. Resuspend cells with 80 mL of lysis buffer.
7. Lyse cells by sonication (carried out in ice for 15 min, amplitude 65%, 3″ of pulse and 9″ of pause);
8. Centrifuge the cell lysate 10 min at 22000 × *g* at 4 °C; collect the supernatant and discard the pellet.
9. Load the soluble fraction (i.e., the supernatant) on a Ni-HiTrap™ Chelating HP column (GE Healthcare) equilibrated with buffer A.
10. Wash the column in 10 mM imidazole and elute with two steps (i.e., 80 and 300 mM imidazole); to obtain the three buffers, mix buffer A and buffer B, by diluting buffer B 1:50, 1:6,25, 1:1,66, respectively, with buffer A. Pure protein elutes in the

300 mM imidazole pool; late fractions of the 80 mM imidazole pool may contain PleD, which is discarded for further purification steps.

11. Load the purified protein on a FPLC column for SEC separation (HiLoad 26/60 Superdex 200 prep grade GE Healthcare) equilibrated in the running buffer (buffer A with NaCl concentration reduced to 100 mM). Keep the input protein solution below a concentration of 2 mg/mL and 2 mL of total loading volume (if necessary dilute with running buffer) to allow c-di-GMP bound to the I-site to dissociate and also to remove excess of imidazole; ligand-free PleD is necessary to perform kinetic experiments and should peak as a monomer. Given the low concentration of the input, multiple SEC runs will be necessary to purify all the available protein. Run at 2 mL/min flow rate. The elution time for the nucleotide-free species is about 100 min (eluition volume = 200 mL). Note that a dialysis step does not improve significantly the amount of c-di-GMP-free protein and therefore this step has been removed from the purification procedure.
12. Check the UV spectrum of each fraction from the SEC and collect only those fractions whose 280/260 nm ratio is >1.25; this ratio accounts for the protein contribution (280 nm) vs. the nucleotide one (260 nm) and values above 1.25 refers to nucleotide free protein (*see* also **Note 1**). Flash-freeze in liquid nitrogen and store at −20 °C.

3.2.2 Protein Preparation: Expression and Purification of Other Systems

As previously mentioned, inhibition studies targeting the I-site rather than the A-site of PleD, greatly benefit from a cross-validation of the positive hits with a DGC lacking the feedback inhibition via canonical I-site, such as the cytoplasmic portion of $\text{YfiN}_{\text{HAMP-GGDEF}}$ from *P. aeruginosa* (Fig. 2). A detailed protocol for the purification of $\text{YfiN}_{\text{HAMP-GGDEF}}$ is reported in [6].

Moreover, putative I-site targeting compounds should be validated on a EAL PDE, such as RocR from *P. aeruginosa*, a mandatory step in particular for those compounds based on the c-di-GMP scaffold, which are in principle also substrate analogs of PDEs.

A detailed protocol of RocR expression and purification as a His-tag recombinant protein is reported in [3]. RocR activity is deeply affected by the presence of the His-tag (particularly after thawing) and therefore it is necessary to remove it, before using the enzyme in activity assays.

For those constructs containing a PreScission (GE Healthcare) cutting site [3], after purification of protein, proceed as follows:

1. Incubate RocR with 10 U/mg of PreScission (GE Healthcare) protease overnight at RT, in the presence of 1 mM DTT and 1 mM EDTA (final concentrations).

2. Remove EDTA and DTT with a PD10 Desalting (GE Healthcare) step (or analogous desalting columns).
3. Load the sample on an NiNTA Column (GE Healthcare) equilibrated in buffer C [3].
4. Elute the cleaved protein using buffer D and collect eluates (*see* **Note 1**).
5. Repeat the desalting process to remove imidazole (as in Subheading 3.2.2, **step 2**). Make aliquots and freeze in liquid nitrogen.

3.2.3 Protein Preparation: Thawing and Quantification

1. Thaw proteins (i.e., PleD and/or RocR) on ice.
2. Centrifuge at 9600 x *g* for 10 min at 4 °C and place the supernatant in a new vial.
3. Collect the UV spectrum to determine protein concentration. PleD and RocR have extinction coefficients at 280 nm of 0.3 and 0.54 $(mg/mL)^{-1}cm^{-1}$, respectively.

3.2.4 Activation of PleD

1. Incubate the enzyme solution for 10 min at room temperature (RT) in the reaction buffer (*see* **Note 2**).
2. Activate by adding 1 mM $BeCl_2$ and 10 mM NaF (final concentrations); incubate for 30 min at RT. During kinetic assays, keep the final concentration of activated PleD, hereinafter PleD*, at 0.5 μM.

3.3 In Vitro Activity Assays

3.3.1 Circular Dichroism Assay

1. Obtain a calibration curve of c-di-GMP as reported in [19] by collecting the spectra of different c-di-GMP solutions ranging from 3 to 50 μM (by diluting the 1 mM stock solution into the buffer reported in Subheading 3.2.4 and used for kinetics); collect spectra from 340 to 240 nm at the same T that will be used for the reaction and adjust instrument parameters to minimize signal to noise ratio. Plot the CD signal at 282 nm against c-di-GMP concentration at fit the data linearly.
2. To exclude that the CD spectrum of c-di-GMP is affected by the presence of the putative inhibitor, prepare a 100 μM inhibitor solution (without protein) plus 10 μM c-di-GMP. For hydrophobic compounds, keep the final DMSO concentration below 5%. As a control, prepare a solvent control by acquiring CD spectra of 10 μM c-di-GMP in the same % of DMSO (if required for compounds solubility, otherwise in the reaction buffer) in the absence of the inhibitor using conditions described in step Subheading 3.3.1, **step 1**. Collect the spectrum of the mixture and compare it with that of the sole c-di-GMP (Fig. 4a). For those molecules whose spectra at 282 nm affect that of c-di-GMP, either by overlapping or by changing the shape of the spectrum of the nucleotide, the method

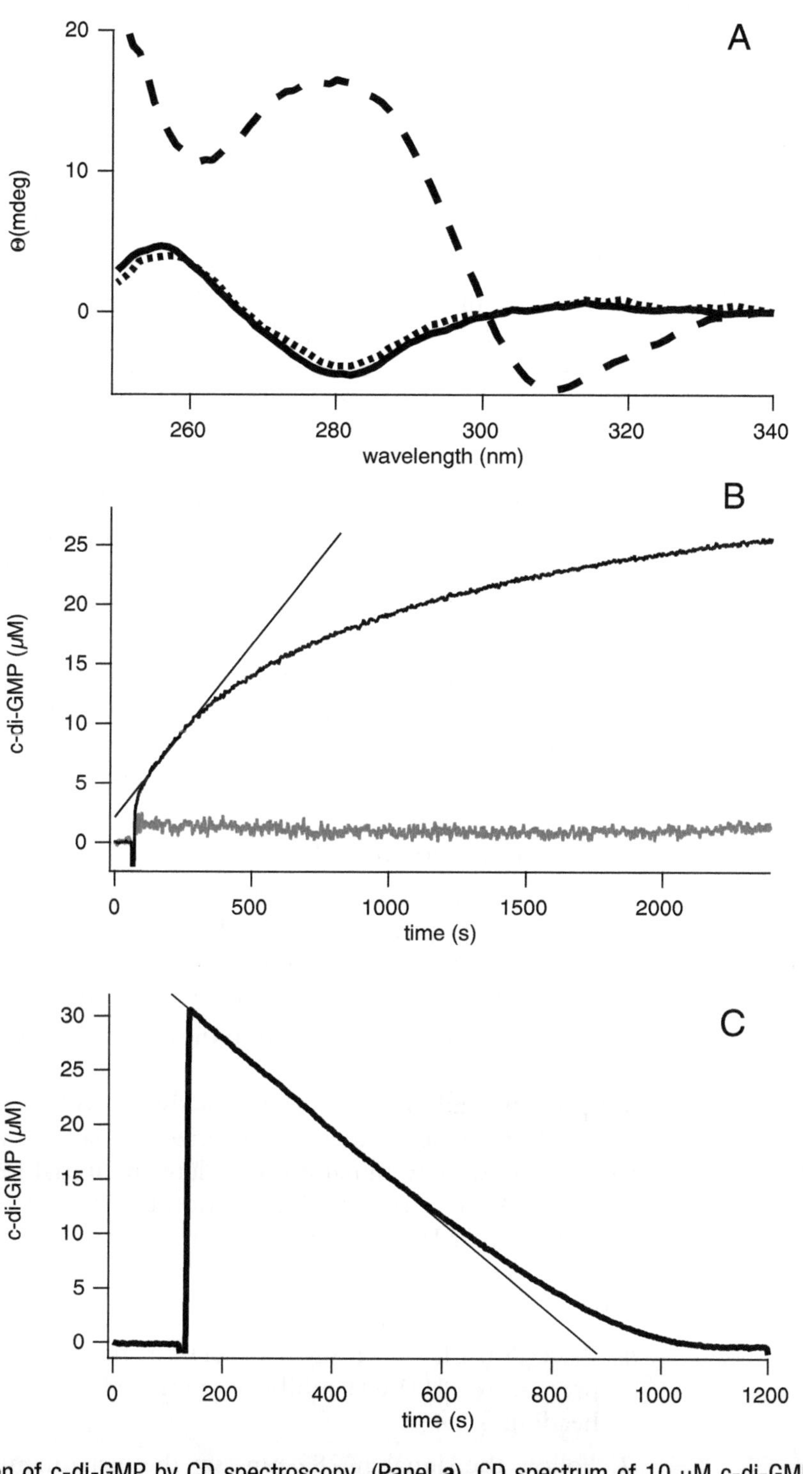

Fig. 4 Detection of c-di-GMP by CD spectroscopy. (Panel **a**), CD spectrum of 10 μM c-di-GMP (*continuous line*), used as a reference in comparison with spectra of a mixture of 10 μM c-di-GMP + 100 μM inhibitor (1) not interfering (*dotted line*) or (2) altering the c-di-GMP signal (*dashed line*). (Panels **b** and **c**) kinetics in real-time acquisition mode. Time course of the reaction catalyzed by the DGC PleD (Panel **b**) in the presence of 5% DMSO (*black curve*) or 100 μM LP1062, a known inhibitor of DGCs [14] (*grey curve*). (Panel **c**) time course of the reaction catalyzed by the PDE RocR as monitored in real-time by following the CD signal at 282 nm at 20 °C. In both panels (**b**) and (**c**) the raw data were converted into c-di-GMP concentration using a calibration curve (see text). The initial rate is obtained by the slope of the linear fit of the early points in the kinetics (*continuous thin line*)

cannot be applied. If this is the case, proceed with RP-HPLC analysis (*see* Subheading 3.3.3).

3. Incubate 0.5 μM PleD* with 100 μM of each compound (i.e., putative inhibitor) for 15 min. Keep final DMSO concentration below 5%; as a control incubate the protein with the same % of DMSO without inhibitor. Collect the spectrum.
4. After 15 min, add 100 μM GTP to start the reaction and incubate for an additional 30 min (*see* **Note 3**). Add 100 mM $CaCl_2$ to stop the reaction after 30 min. Collect the spectrum; keep the sample at 25 °C and final volume ≈ 900 μL. For medium-throughput screening, aliquot a solution of PleD* or c-di-GMP into two different 24-well plates and then add to each plate DMSO or the inhibitor solution, prior to GTP addition. This allows to speed up the procedure and to minimize variability; then, analyze 800 μL of each sample at 25 °C, in a 1-cm quartz cuvette. Proceed as described in Subheading 3.3.1, **steps 2–4**.
5. Baseline-correct the spectra by subtracting the buffer from the raw data, and by adjusting the signal at 340 nm to zero, as no optical activity is expected at this wavelength [19].
6. Extrapolate the concentration of c-di-GMP that has been produced by PleD* over the course of 30 min by using the calibration curve reported in 1.
7. Compare the amount of c-di-GMP produced by PleD* in the presence of each inhibitor with that obtained in the control experiment (no inhibitor added, with the same % of DMSO), to identify possible DGC inhibitors (*see* also **Note 4**).
8. Repeat the screening at least in duplicates (**steps 2–7**).

3.3.2 Determine the Initial Velocity of the Reaction by CD

Compounds leading to a residual activity ≤30% are considered positive hits to be further characterized, first by evaluating the IC_{50}. For IC_{50} determination, calculate the initial velocity of the reaction (v_0) at different inhibitor concentrations (at fixed substrate concentration of 100 μM). To obtain the initial rate of catalysis, the enzymatic assay should be performed in a real-time acquisition mode:

1. Place 800 μL of 0.5 μM of PleD* in the CD cuvette, in the presence of 100 μM inhibitor (prepared as reported in Subheading 3.2.4).
2. Follow the signal at 282 nm (set the instrument in real time acquisition mode at fixed wavelength) at 25 °C; collect a baseline for 5 min, then rapidly add 100 μM GTP (*see* **Note 5**).
3. Acquire the CD signal at 282 nm for 10 min (the acquisition time could be set in the instrument parameters panel); once acquired, discard the sample.

4. Graph the data acquired at 282 nm using a data processing software such as Igor Pro (Wavemetrics), KaleidaGraph (Synergy), or similar.
5. Analyze the data by linear fitting using the first few minutes of the kinetics (considering that for PleD the linearity is lost within the first 3–4 min, due to product inhibition) (Fig. 4b).
6. Convert the initial rate observed at each inhibitor concentration into % of residual activity (consider the initial rate observed in the absence of inhibitor to be 100%).
7. To obtain the IC_{50} value, plot log [inhibitor] μM vs. % of residual activity and fit data using the log-dose vs. response equation (using nonlinear fit software such as Prism).
   (a) Include points falling on *Y*-axis, which represents the 100% residual activity obtained in the absence of inhibitor.
   (b) Each point of the concentration–response curve should represent the mean value and standard deviation of three independent experiments by CD spectroscopy.

3.3.3 Cross-Validation of Inhibitors Targeting the I-Site of DGCs

As detailed above, positive hits could also be tested using YfiN-HAMP-GGDEF and RocR from *P. aeruginosa*, depending on the rationale of inhibitor design.

1. For $YfiN_{HAMP\text{-}GGDEF}$ kinetics, incubate 10 μM of the protein solution (150 mM NaCl, 20 mM Tris–HCl pH 7.5, and 1% glycerol) with 10 mM $MgCl_2$, 2.5 mM $MnCl_2$ (10 min at RT).
2. To study the possible inhibition of the PDE activity of *P. aeruginosa* RocR solution, incubate 0.5 μM of the enzyme solution in the reaction buffer for 10 min prior to start the experiments.
3. For both enzymes, add 100 μM of inhibitor and use the same % of DMSO in the positive control experiment.
4. Place the solution in the CD cuvette, follow the signal at 282 nm for 5 min and start kinetics by adding 100 μM GTP or 30 μM c-di-GMP, for $YfiN_{HAMP\text{-}GGDEF}$ and RocR, respectively (Fig. 4c);
5. Extrapolate the initial rate as detailed in Subheading 3.3.2 for real time measurements with PleD*.

3.3.4 RP-HPLC Assay

For those compounds whose CD spectra interfere with the signal of c-di-GMP at 282 nm (*see* Subheading 3.3.1, **step 2**), the nucleotide content of the reaction mixture (both DGC or PDE) must be evaluated by RP-HPLC.

1. To evaluate the yield of the reaction products (both c-di-GMP and/or pGpG), calibrate with different concentrations of the nucleotide(s) to calculate their concentration in the reaction

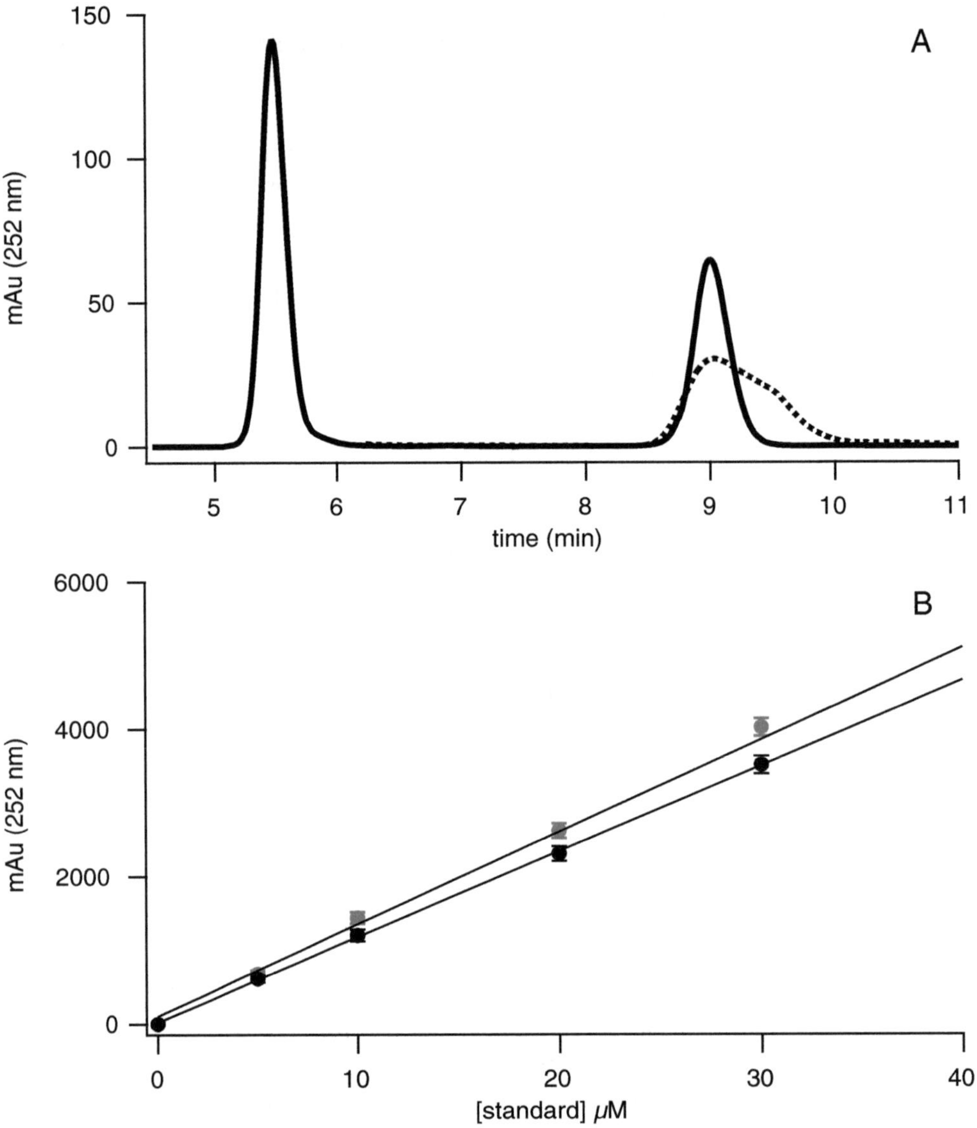

Fig. 5 Detection of c-di-GMP in the presence of putative inhibitors by RP-HPLC. (Panel **a**) RP-HPLC chromatogram of standard solutions of 10 μM pGpG and c-di-GMP (*continuous line*), superimposed with the chromatogram of a mixture of 10 μM c-di-GMP + 100 μM inhibitor, the latter interfering with c-di-GMP migration (see text for experimental details). The latter peak cannot be used for c-di-GMP quantification and therefore to evaluate c-di-GMP production/consumption. (**b**) Linear dependence of the peak area of c-di-GMP (in *grey*) or pGpG (in *black*) on the corresponding nucleotide's concentration; this calibration curves can be used to extrapolate c-di-GMP production/consumption in the presence of putative inhibitors, whose migration does not superimpose with that of c-di-GMP

mixture (Fig. 5b), by preparing different nucleotides solutions in the reaction buffer used in the downstream **step 3** (ranging from 3 to 30 μM, 300 μL of each solution).

2. Separate the nucleotide content of the samples (100 μL) by RP-HPLC using a C8 column as detailed in materials, at RT; use as

mobile phase 100 mM phosphate buffer pH 5.8/ methanol (98/2, v/v, flow rate 1 mL/min) and set the UV detector at 252 nm [19]. For an estimate of possible retention times *see* Fig. 5a; although the order of elution of the various species is not affected, slight changes in the retention times might occur depending on the HPLC apparatus used to run the chromatography.

3. Prepare 300 μL of the reaction mixture as reported above both for PleD* or RocR and start kinetics accordingly (*see* Subheading 3.2.4, 3.3.1, **steps 3** and **4** and Subheading 3.3.3, **step 2** and **3**, respectively).
4. In parallel, prepare a standard mixture (under the same experimental conditions reported in the previous step) by using 30 μM of synthetic c-di-GMP, pGpG (Biolog) and 100 μM GTP (GE Healthcare) in the presence of 100 μM inhibitor; this is crucial also to exclude comigration of the standard with the inhibitor. An example is shown in Fig. 5a.
5. Stop the reaction (both the enzymatic reaction and the standard mixture) after 30 min first by adding 50 μg/mL of proteinase K (by diluting 1:20 the stock solution) and then keep the sample for 10 min at 95 °C;
6. Remove protein precipitate by filtering the solution using 0.2 μm filters (Bilk GHP Acrodisc 13 mm). *See* also **Note 6** for more details.
7. Separate the nucleotide content of the samples as detailed in **step 2** of this paragraph.

3.4 Validation of Compounds on Bacterial Cells and Toxicity on Eukaryotes Cells

For positive hits, an assay should be performed to validate the compound(s) as potential anti-biofilm drugs. As mentioned in the introduction, many assays and c-di-GMP biosensors are available, which are detailed in refs. [22–27]. Given the wide range of bacterial cell-based validation assays, the choice of a specific technique will greatly depend on which type of biofilm and/or species should be targeted and on the rationale of the inhibition study. A detailed description of these techniques is beyond the scope of this chapter; please refer to specific literature (see above). However, a detailed protocol for performing a preliminary toxicity assay on eukaryotic cells is given below, given that this is a necessary and common step for any potential inhibitor to be effective, independently on the rationale of the design. We suggest to performe this step in particular for virtual screening compounds targeting the (GTP binding) A-site, to exclude the toxicity on eukaryotes, which can be due to off-target effects. Proceed as follows:

1. Seed lung adenocarcinoma A549 cells; 80.000 cells per well in 12 multiwell using RPMI medium containing 10% FBS and

penicillin/streptomycin and incubated at 37 °C and a 5% CO_2 atmosphere.

2. Following 24 h of incubation, add the putative inhibitor to the medium at a final concentration of 10, 100, or 300 μM; the control sample is obtained with the same % of DMSO. Incubate for 24 h at 37 °C in a 5% CO_2 atmosphere.
3. Remove the medium after 24 h of incubation with the inhibitor, wash the cells with 1 mL of PBS.
4. Add 200 μL of 1× trypsin–EDTA to each well. After 2 min or when the cells are visibly detached, add 1 mL of complete RPMI in each well, collect the cells by pipetting and transfer the cells to a 2 mL tube.
5. Centrifuge the cells at 0.2 × *g* for 10 min.
6. Remove the supernatant, and add 50 μL of trypan blue to the cell pellet.
7. Add 10 μL of the cells in duplicate into a Burker counting chamber.
8. Examine the cells using an inverted microscope.
9. Count the blue cells as dead cells (count the uncolored cells as vital).
10. Plot the ratio as a function of inhibitor concentration and compare it with the control sample.

4 Notes

1. It is recommended to validate the purification procedures by running SDS-PAGE of the different fractions.
2. The PleD activation step has been found to be less effective when done in the dark.
3. Thaw and keep the GTP stock solution in ice to minimize autohydrolysis; do not refreeze the GTP aliquot or reuse it for other experiments.
4. To ensure that the experiments have been properly set up, it is helpful to use a known DGCs inhibitor, such as LP1062 [14], which completely abolishes PleD* activity (Fig. 4b, grey trace and previously reported inhibitors [14, 15]).
5. For real time kinetics, the addition and the correct mixing of the substrate is crucial to obtain reproducible traces. Mix immediately the solution by pipetting 3–4 times from the bottom of the solution in the cuvette; for RocR assay, where the high increase of the signal due to c-di-GMP addition competes with c-di-GMP consumption due to RocR activity, it could be useful to invert the order of the addition (i.e., c-di-GMP as baseline and addition of protein to start kinetics).

6. During the RP-HPLC assay, the step in which the reaction is stopped is crucial to avoid misleading results. The sole boiling of aliquots of the reaction to stop catalysis leads to an overestimation of the observed reaction products and hampers reproducibility. For CD analysis, addition of $CaCl_2$ at certain time is a reliable and fast method. On the other hand, for RP-HPLC analysis, this method should be replaced with other strategies, if phosphate buffer is used, i.e., (a) addition of 50 μg/mL of proteinase K as described above or (b) addition of 25 mM EDTA pH 6.0 and then incubation of the sample for 10 min at 95 ° C; the latter method cannot be applied to CD analysis, given that the c-di-GMP signal depends on the presence of free Mn^{2+} ions.

References

1. Chan C, Paul R, Samoray D, Amiot NC, Giese B, Jenal U, Schirmer T (2004) Structural basis of activity and allosteric control of diguanylate cyclase. Proc Natl Acad Sci U S A 101 (49):17084–17089. doi:10.1073/pnas.0406134101
2. Wassmann P, Chan C, Paul R, Beck A, Heerklotz H, Jenal U, Schirmer T (2007) Structure of BeF3-modified response regulator PleD: implications for diguanylate cyclase activation, catalysis, and feedback inhibition. Structure 15 (8):915–927. doi:10.1016/j.str.2007.06.016
3. De N, Pirruccello M, Krasteva PV, Bae N, Raghavan RV, Sondermann H (2008) Phosphorylation-independent regulation of the diguanylate cyclase WspR. PLoS Biol 6 (3):e67. doi:10.1371/journal.pbio.0060067
4. De N, Navarro MV, Raghavan RV, Sondermann H (2009) Determinants for the activation and autoinhibition of the diguanylate cyclase response regulator WspR. J Mol Biol 393(3):619–633. doi: 10.1016/j.jmb.2009.08.030
5. Schirmer T (2016) C-di-GMP synthesis: structural aspects of evolution, catalysis and regulation. J Mol Biol 428(19):3683–3701. doi:10.1016/j.jmb.2016.07.023
6. Giardina G, Paiardini A, Fernicola S, Franceschini S, Rinaldo S, Stelitano V, Cutruzzolà F (2013) Investigating the allosteric regulation of YfiN from *Pseudomonas aeruginosa*: clues from the structure of the catalytic domain. PLoS One 8(11):e81324. doi:10.1371/journal.pone.0081324
7. Tarnawski M, Barends TR, Schlichting I (2015) Structural analysis of an oxygen-regulated diguanylate cyclase. Acta Crystallogr D Biol Crystallogr 71(Pt 11):2158–2177. doi:10.1107/s139900471501545x
8. An DR, Im HN, Jang JY, Kim HS, Kim J, Yoon HJ, Hesek D, Lee M, Mobashery S, Kim SJ, Suh SW (2016) Structural basis of the heterodimer formation between cell shape-determining proteins Csd1 and Csd2 from *Helicobacter pylori*. PLoS One 11(10):e0164243. doi:10.1371/journal.pone.0164243
9. Astegno A, Capitani G, Dominici P (2015) Functional roles of the hexamer organization of plant glutamate decarboxylase. Biochim Biophys Acta 1854(9):1229–1237. doi:10.1016/j.bbapap.2015.01.001
10. Patel D, Kopec J, Fitzpatrick F, McCorvie TJ, Yue WW (2016) Structural basis for ligand-dependent dimerization of phenylalanine hydroxylase regulatory domain. Sci Rep 6:23748. doi:10.1038/srep23748
11. Fernicola S, Torquati I, Paiardini A, Giardina G, Rampioni G, Messina M, Leoni L, Del Bello F, Petrelli R, Rinaldo S, Cappellacci L, Cutruzzolà F (2015) Synthesis of triazole-linked analogues of c-di-GMP and their interactions with diguanylate cyclase. J Med Chem 58 (20):8269–8284. doi:10.1021/acs.jmedchem.5b01184
12. Zhou J, Watt S, Wang J, Nakayama S, Sayre DA, Lam YF, Lee VT, Sintim HO (2013) Potent suppression of c-di-GMP synthesis via I-site allosteric inhibition of diguanylate cyclases with 2′-F-c-di-GMP. Bioorg Med Chem 21(14):4396–4404. doi:10.1016/j.bmc.2013.04.050
13. Wang J, Zhou J, Donaldson GP, Nakayama S, Yan L, Lam YF, Lee VT, Sintim HO (2011) Conservative change to the phosphate moiety

of cyclic diguanylic monophosphate remarkably affects its polymorphism and ability to bind DGC, PDE, and PilZ proteins. J Am Chem Soc 133(24):9320–9330. doi:10.1021/ja1112029
14. Sambanthamoorthy K, Luo C, Pattabiraman N, Feng X, Koestler B, Waters CM, Palys TJ (2014) Identification of small molecules inhibiting diguanylate cyclases to control bacterial biofilm development. Biofouling 30(1):17–28. doi:10.1080/08927014.2013.832224
15. Fernicola S, Paiardini A, Giardina G, Rampioni G, Leoni L, Cutruzzolà F, Rinaldo S (2016) In silico discovery and in vitro validation of catechol-containing sulfonohydrazide compounds as potent inhibitors of the diguanylate cyclase PleD. J Bacteriol 198(1):147–156. doi:10.1128/jb.00742-15
16. Deepthi A, Liew CW, Liang ZX, Swaminathan K, Lescar J (2014) Structure of a diguanylate cyclase from *Thermotoga maritima*: insights into activation, feedback inhibition and thermostability. PLoS One 9(10):e110912. doi:10.1371/journal.pone.0110912
17. Chen MW, Kotaka M, Vonrhein C, Bricogne G, Rao F, Chuah ML, Svergun D, Schneider G, Liang ZX, Lescar J (2012) Structural insights into the regulatory mechanism of the response regulator RocR from *Pseudomonas aeruginosa* in cyclic Di-GMP signaling. J Bacteriol 194 (18):4837–4846. doi:10.1128/JB.00560-12
18. Rao F, Yang Y, Qi Y, Liang ZX (2008) Catalytic mechanism of cyclic di-GMP-specific phosphodiesterase: a study of the EAL domain-containing RocR from *Pseudomonas aeruginosa*. J Bacteriol 190(10):3622–3631. doi:10.1128/JB.00165-08
19. Stelitano V, Brandt A, Fernicola S, Franceschini S, Giardina G, Pica A, Rinaldo S, Sica F, Cutruzzolà F (2013) Probing the activity of diguanylate cyclases and c-di-GMP phosphodiesterases in real-time by CD spectroscopy. Nucleic Acids Res 41:e79. doi:10.1093/nar/gkt028
20. Antoniani D, Rossi E, Rinaldo S, Bocci P, Lolicato M, Paiardini A, Raffaelli N, Cutruzzolà F, Landini P (2013) The immunosuppressive drug azathioprine inhibits biosynthesis of the bacterial signal molecule cyclic-di-GMP by interfering with intracellular nucleotide pool availability. Appl Microbiol Biotechnol 97 (16):7325–7336. doi:10.1007/s00253-013-4875-0
21. Opoku-Temeng C, Zhou J, Zheng Y, Su J, Sintim HO (2016) Cyclic dinucleotide (c-di-GMP, c-di-AMP, and cGAMP) signalings have come of age to be inhibited by small molecules. Chem Commun (Camb) 52(60):9327–9342. doi:10.1039/c6cc03439j
22. Sambanthamoorthy K, Sloup RE, Parashar V, Smith JM, Kim EE, Semmelhack MF, Neiditch MB, Waters CM (2012) Identification of small molecules that antagonize diguanylate cyclase enzymes to inhibit biofilm formation. Antimicrob Agents Chemother 56(10):5202–5211. doi:10.1128/aac.01396-12
23. Antoniani D, Bocci P, Maciag A, Raffaelli N, Landini P (2010) Monitoring of diguanylate cyclase activity and of cyclic-di-GMP biosynthesis by whole-cell assays suitable for high-throughput screening of biofilm inhibitors. Appl Microbiol Biotechnol 85(4):1095–1104. doi:10.1007/s00253-009-2199-x
24. Pawar SV, Messina M, Rinaldo S, Cutruzzolà F, Kaever V, Rampioni G, Leoni L (2016) Novel genetic tools to tackle c-di-GMP-dependent signalling in Pseudomonas aeruginosa. J Appl Microbiol 120(1):205–217. doi:10.1111/jam.12984
25. Groizeleau J, Rybtke M, Andersen JB, Berthelsen J, Liu Y, Yang L, Nielsen TE, Kaever V, Givskov M, Tolker-Nielsen T (2016) The anticancerous drug doxorubicin decreases the c-di-GMP content in Pseudomonas aeruginosa but promotes biofilm formation. Microbiology 162 (10):1797–1807. doi:10.1099/mic.0.000354
26. Kellenberger CA, Sales-Lee J, Pan Y, Gassaway MM, Herr AE, Hammond MC (2015) A minimalist biosensor: quantitation of cyclic di-GMP using the conformational change of a riboswitch aptamer. RNA Biol 12 (11):1189–1197. doi:10.1080/15476286.2015.1062970
27. Xie Q, Zhao F, Liu H, Shan Y, Liu F (2015) A label-free and self-assembled electrochemical biosensor for highly sensitive detection of cyclic diguanylate monophosphate (c-di-GMP) based on RNA riboswitch. Anal Chim Acta 882:22–26. doi:10.1016/j.aca.2015.04.061
28. Lieberman OJ, Orr MW, Wang Y, Lee VT (2014) High-throughput screening using the differential radial capillary action of ligand assay identifies ebselen as an inhibitor of diguanylate cyclases. ACS Chem Biol 9(1):183–192. doi:10.1021/cb400485k
29. Sambrook J (1989) Molecular cloning: a laboratory manual. Cold Spring Harbor Laboratory Press, Cold Spring Harbor, NY
30. Irwin JJ, Shoichet BK (2005) ZINC—a free database of commercially available compounds for virtual screening. J Chem Inf Model 45 (1):177–182. doi:10.1021/ci049714+
31. Thomsen R, Christensen MH (2006) MolDock: a new technique for high-accuracy

molecular docking. J Med Chem 49 (11):3315–3321. doi:10.1021/jm051197e

32. Morris GM, Huey R, Olson AJ (2008) Using AutoDock for ligand-receptor docking. Curr Protoc Bioinformatics Chapter 8:Unit 8 14. doi:10.1002/0471250953.bi0814s24

33. Paul R, Weiser S, Amiot NC, Chan C, Schirmer T, Giese B, Jenal U (2004) Cell cycle-dependent dynamic localization of a bacterial response regulator with a novel di-guanylate cyclase output domain. Genes Dev 18(6):715–727. doi:10.1101/gad.28950/418/6/715

Chapter 33

High-Throughput Screening for Compounds that Modulate the Cellular c-di-GMP Level in Bacteria

Julie Groizeleau, Jens Bo Andersen, Michael Givskov, Jens Berthelsen, and Tim Tolker-Nielsen

Abstract

Bacteria in the biofilm mode of growth cause numerous problematic infections due to their resistance to antimicrobials and the immune system. Because conventional antimicrobial compounds cannot efficiently eradicate biofilm infections, we urgently need new efficient anti-biofilm drugs. The secondary messenger c-di-GMP is a positive regulator of biofilm formation in many clinically relevant bacteria, and it is assumed that drugs that lower the intracellular level of c-di-GMP will force biofilm bacteria into a more treatable planktonic lifestyle. We describe a protocol for high-throughput screening of chemical libraries for compounds that lower the c-di-GMP level in bacteria, and potentially can serve as lead compounds in the development of novel biofilm dismantling drugs.

Key words Biofilm, c-di-GMP, High-throughput screening, Anti-biofilm drugs

1 Introduction

Biofilms formed by opportunistic pathogenic bacteria attain the highest levels of antibiotic tolerance and an almost unlimited capacity to survive in the infected host [1]. Thus conventional antimicrobial compounds often fail to eradicate biofilms. Consequently there is an urgent need to develop alternative measures to combat biofilm infections. A promising strategy, referred to as biofilm dismantling, is to force the bacteria out of the protective biofilm-state to a free-living mode where they are susceptible to the action of the immune system and antimicrobials.

Current research indicates that the secondary messenger c-di-GMP is a general controller of the biofilm lifestyle in various bacterial species [2]. High internal levels of c-di-GMP induce production of extracellular matrix components, which promotes biofilm formation, whereas low c-di-GMP levels downregulate the production of extracellular matrix components and lead bacteria to the planktonic mode of life [2, 3]. C-di-GMP is synthetized by

Karin Sauer (ed.), *c-di-GMP Signaling: Methods and Protocols*, Methods in Molecular Biology, vol. 1657,
DOI 10.1007/978-1-4939-7240-1_33,

diguanylate cyclases containing a GGDEF domain and is degraded by phosphodiesterases containing either an EAL or HD-GYP domain [2, 3]. Bacterial genomes often encode many different putative diguanylate cyclases and phosphodiesterases (e.g., 41 in the case of *Pseudomonas aeruginosa* PAO1), which frequently contain sensory domains that are thought to enable translation of diverse environmental cues into c-di-GMP levels, promoting either a biofilm lifestyle or a planktonic lifestyle. The GGDEF, EAL, and HD-GYP motifs of the diguanylate cyclases and phosphodiesterases are conserved in various bacterial species, making the enzymes interesting new targets for biofilm control.

Evidence has been provided that a reduction in the intracellular level of c-di-GMP through induction of a phosphodiesterase results in dispersal of already established biofilms both in in vitro systems and in a mouse biofilm infection model [3, 4]. Accordingly, it is anticipated that drugs that can lower the intrabacterial c-di-GMP level will force bacteria away from the biofilm lifestyle and into the much more susceptible planktonic lifestyle.

We have previously generated a collection of cell-based GFP reporters that can gauge the level of c-di-GMP in *P. aeruginosa* [5], and potentially can be used for high-throughput screens for compounds that lower the c-di-GMP content in bacteria [6]. The reporters are based on fusions of the c-di-GMP/FleQ-regulated *cdrA* gene to *gfp*, and might work in other bacterial species than *P. aeruginosa* if they produce FleQ homologs.

Here, we describe a high-throughput screening assay for identification of small molecules that can modulate the c-di-GMP level in *P. aeruginosa* (Table 1 and Fig. 1). The assay is based on assessing cellular c-di-GMP levels using a *P. aeruginosa* monitor strain containing one of our plasmid-based *cdrA-gfp* reporter constructs. The c-di-GMP level in the *P. aeruginosa* wild type is below the detection limit of our c-di-GMP reporters during planktonic growth.

Table 1
Workflow of the cell-based screening

Day	Subheading	Description
1	3.1 3.2	Preparation for data handling Preparation of the c-di-GMP monitor strain
2	3.3	Preparation of overnight cultures
3	3.4	Preparation of assay for *Z*'-factor determination
4	3.5, **steps 1–12** 3.5, **step 13**	Analysis of the Z'-factor of the screening process (validation of the method for doing HTS) Preparation of the screening plates and perform screening
5	3.6	Analysis of screening results and identification of hits

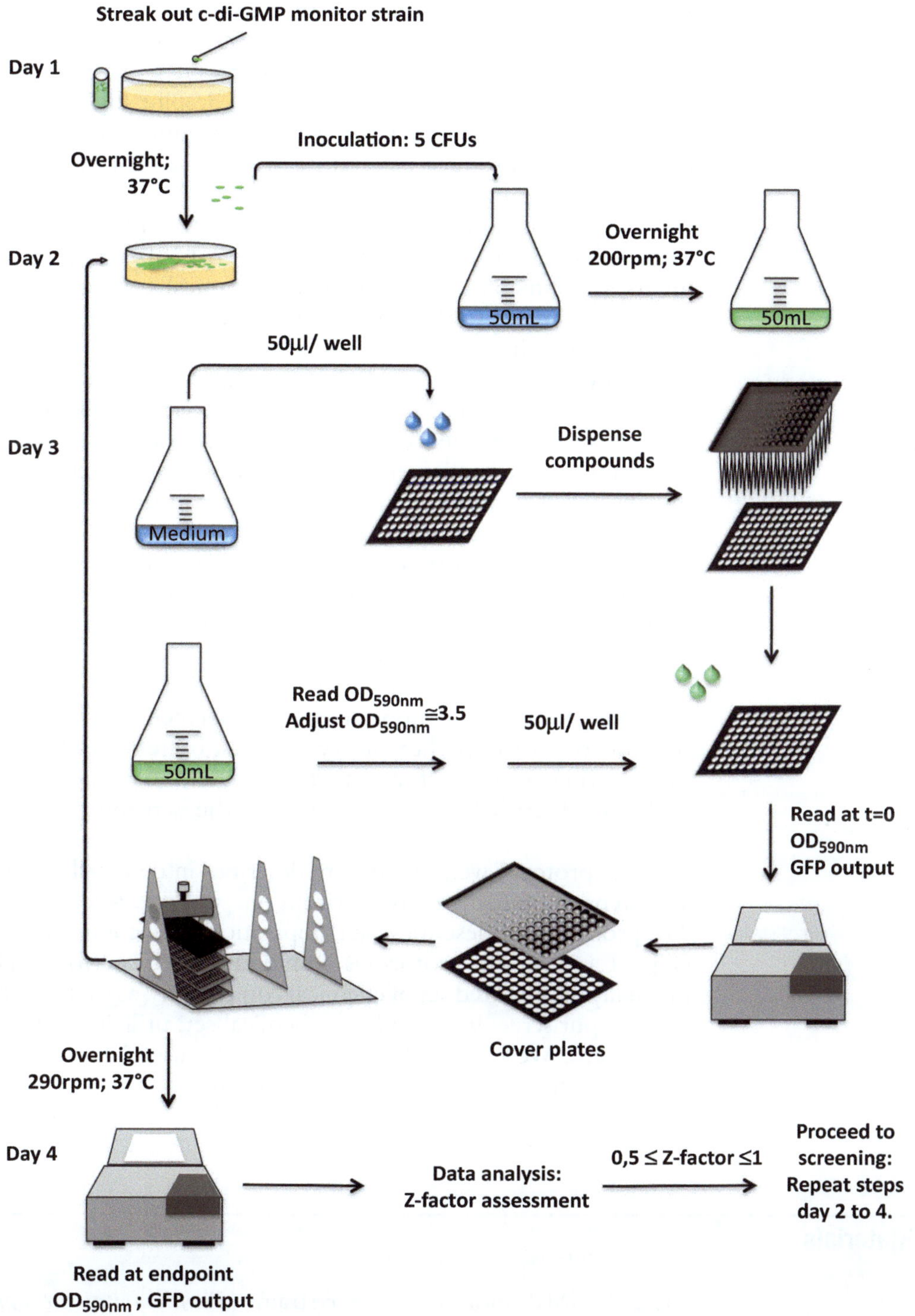

Fig. 1 Overview of the screening procedure

However, the monitor strain is engineered with a *wspF* mutation, which results in activation of the WspR diguanylate cyclase to produce high levels of c-di-GMP. Moreover, to ensure that the monitor strain remains fully suspended despite high c-di-GMP levels, it also harbors *pelA* and *pslBCD* mutations rendering the cells unable to produce Psl and Pel exopolysaccharides thus preventing cell aggregation.

High-throughput screening (HTS) has traditionally belonged to the pharma industry due to the high costs of screening equipments, consumables, and proprietary chemical compound collections not available to the public. But in recent years, the costs have lowered, and chemical collections are becoming more available to the academic community. High-throughput screenings comprise biochemical screening for modifiers of an isolated protein activity, as well as screening on live cells, both prokaryotic and eukaryotic. Screening on live bacteria has been performed extensively throughout the last decades in the pharma industry in the search for new antibiotics, utilizing a reduction in cell viability as readout [7]. In contrast, live eukaryotic cells are often screened for compounds that modify cellular function, using high-content screening analysis in which several cellular markers are followed during the screen. The live bacterial screen we describe here has been designed to score a modification of the level of c-di-GMP without affecting bacterial growth. One important aspect of screening live bacteria is that, in contrast to higher eukaryotic cells, bacteria multiply many times during even a brief screening period, and an endpoint control of bacterial growth is fundamental in the screening procedure (Fig. 1).

The protocol we present here describes in detail all the necessary steps to perform a screen within one working week (Table 1). The protocol also describes data preparation, collection and analysis (Figs. 1 and 2). The protocol is simple, and can be performed both manually on a limited set of chemical compounds, as well as in high-throughput screening on a large chemical set, utilizing automated equipment. We believe this method will be useful for identifying chemicals that affect bacterial c-di-GMP metabolism and signaling pathways, and can serve as lead compounds in the development of novel anti-biofilm drugs.

2 Materials

1. c-di-GMP monitor *reporter*strain (*Pseudomonas aeruginosa ΔwspF pelA pslBCD* containing PcdrA::gfp (mut3)^{C} (*see* ref. 5).
2. Ultrapure water (e.g., MilliQ water). TOC $\leq$ 5 ppb, 18.2 MΩ cm, sterile and pyrogen free.

a.

1_20160927	1	2	3	4	5	6	7	8	9	10	11	12
A	BIO356	BIO357	BIO358	BIO359	BIO360	BIO361	BIO362	BIO363	BIO364	BIO365	BIO366	SNP
B	BIO367	BIO368	BIO369	BIO370	BIO371	BIO372	BIO373	BIO374	BIO375	BIO376	BIO377	SNP
C	BIO378	BIO357	BIO358	BIO359	BIO360	BIO361	BIO362	BIO363	BIO364	BIO365	BIO366	SNP
D	BIO389	BIO390	BIO391	BIO392	BIO393	BIO394	BIO395	BIO396	BIO397	BIO398	BIO399	SNP
E	BIO400	BIO401	BIO402	BIO403	BIO404	BIO405	BIO406	BIO407	BIO408	BIO409	BIO410	DMSO
F	BIO411	BIO412	BIO413	BIO414	BIO415	BIO416	BIO417	BIO418	BIO419	BIO420	BIO421	DMSO
G	BIO422	BIO423	BIO424	BIO425	BIO426	BIO427	BIO428	BIO429	BIO430	BIO431	BIO432	DMSO
H	BIO433	BIO434	BIO435	BIO436	BIO437	BIO438	BIO439	BIO440	BIO441	BIO442	BIO443	DMSO

b.

Abs590 1_20160927	1	2	3	4	5	6	7	8	9	10	11	12
A	0,623	0,625	0,623	0,615	0,614	0,615	0,613	0,612	0,629	0,615	0,619	0,801
B	0,644	0,620	0,622	0,621	0,614	0,610	0,619	0,622	0,623	0,630	0,621	0,804
C	0,622	0,634	0,642	0,626	0,619	0,617	0,628	0,621	0,625	0,624	0,630	0,809
D	0,621	0,617	0,622	0,619	0,610	0,612	0,607	0,612	0,609	0,618	0,616	0,804
E	0,627	0,619	0,625	0,625	0,619	0,612	0,619	0,610	0,603	0,610	0,643	0,634
F	0,632	0,618	0,634	0,622	0,618	0,611	0,620	0,607	0,625	0,618	0,609	0,614
G	0,645	0,625	0,622	0,625	0,624	0,611	0,600	0,610	0,613	0,602	0,609	0,612
H	0,615	0,632	0,632	0,631	0,621	0,632	0,616	0,634	0,632	0,633	0,641	0,617

Fig. 2 (**a**) Example of the layout of a 96-well microtiter screening plate. The plate ID is written in the *top left corner* of the screening plate layout. BIOxxx refers to wells containing test compounds in culture with a final DMSO concentration of 1%. SNP refers to wells containing 50 μM of SNP in culture with a final DMSO concentration of 1%, which serve as a positive control for c-di-GMP reduction. DMSO refers to wells containing culture with a final DMSO concentration of 1%, which serve as a negative control. (**b**) Examples of screening data readouts. Cell density (*Abs590*) and fluorescence intensity (*Gfp*) and subsequent data analysis (fluorescence intensity/cell density = RFU) are displayed as heat maps. In the heat map of cell densities (*grid Abs590*), *red* represents the highest cell density measured, *white* represents the cell density average of DMSO treated cultures, while *green* represents a cell density of 0. In the heat map of fluorescence intensity (*grid Gfp*), *red* represents the highest fluorescence intensity measured, *white* represents the fluorescence intensity average of the DMSO treated cultures, while *green* represents the lowest fluorescence intensity measured. In the heat map of fluorescence/cell density (*grid RFU*) *red* represents the highest RFU value calculated, *white* represents the RFU value average of the DMSO treated cultures, while *green* represents the lowest RFU value calculated. (**c**) An example of a plate displaying a hit which lowers the RFU output of the c-di-GMP monitor strain by more than 20%

Gfp 1_20160927	1	2	3	4	5	6	7	8	9	10	11	12
A	51107	54645	57529	50135	53716	53038	53229	50340	52085	58126	55495	23663
B	50710	53409	49064	49202	49396	50654	48835	55705	48026	52535	54900	24261
C	54327	51009	49693	53774	52812	50312	53673	52291	53201	53569	57382	24700
D	43773	57114	59571	51359	54144	53333	47695	50085	52996	60282	52189	23247
E	50943	54247	52405	52808	49995	53331	51656	56126	57186	51373	51478	60697
F	55767	55086	63543	53303	66692	52928	57322	53137	66569	56330	67094	56487
G	55168	58160	54325	52485	51149	54856	50354	53740	52974	55248	54244	56332
H	49885	55657	52950	56390	53750	52778	52968	48461	55298	74157	55617	57477

RFU 1_20160927	1	2	3	4	5	6	7	8	9	10	11	12
A	81992	87432	92304	81583	87498	86235	86770	82316	82788	94482	89689	29532
B	78725	86089	78907	79218	80420	82976	78866	89546	77133	83364	88404	30185
C	87296	80514	77428	85937	85386	81497	85435	84170	85071	85891	91129	30541
D	70452	92508	95846	82933	88792	87125	78616	81838	86987	97547	84679	28913
E	81287	87673	83794	84549	80743	87154	83472	92022	94840	84203	80030	95768
F	88263	89068	100163	85764	107980	86578	92443	87529	106497	91206	110096	92010
G	85492	93006	87274	84012	81973	89792	83884	88141	86354	91792	89133	92093
H	81175	88058	83734	89320	86613	83551	86000	76469	87553	117229	86734	93222

Fig. 2 (continued)

3. LB medium: Add 5 g of yeast extract, 10 g of tryptone, 10 g of NaCl to 1 L of ultrapure water. Adjust the pH to 7.4 with NaOH. Autoclave 20 min at 120 °C.
4. LB 1.5% agar plate with 100 μg/mL Gm: add 15 g of agar to 1 L of nonautoclaved LB medium. Autoclave 20 min at

C.

1_20160927	1	2	3	4	5	6	7	8	9	10	11	12
A	-	-	-	-	-	-	-	-	-	-	-	SNP
B	-	-	-	-	-	-	-	-	-	-	-	SNP
C	-	-	-	-	-	-	-	-	-	-	-	SNP
D	BIO389	-	-	-	-	-	-	-	-	-	-	SNP
E	-	-	-	-	-	-	-	-	-	-	-	-
F	-	-	-	-	-	-	-	-	-	-	-	-
G	-	-	-	-	-	-	-	-	-	-	-	-
H	-	-	-	-	-	-	-	-	-	-	-	-

Fig. 2 (continued)

120 °C. Cool down the solution until the bottle can be handled with bare hands. Add 100 mg of gentamicin sulfate salt to 1 L of LB 1.5% agar medium.

5. 1 M $MgCl_2$ stock solution: add 203.3 g of $MgCl_2 \cdot 6H_2O$ to 1 L of ultrapure water, filter-sterilize through a 0.22 μm filter, keep at 4 °C.
6. 1 M $CaCl_2$ stock solution: add 147.02 g of $CaCl_2 \cdot 2H_2O$ to 1 L of ultrapure water, filter-sterilize through a 0.22 μm filter, keep at 4 °C.
7. A-10 buffer: 20 g/l $(NH_4)_2SO_4$, 60 g/l $Na_2HPO_4 \cdot 2H_2O$, 30 g/l KH_2PO_4, 30 g/L NaCl, pH of final solution should be 6.4 ($\pm$ 0.1), autoclave 20 min at 120 °C. Store at 4 °C.
8. 10% glucose: add 10 g of glucose to 100 mL of ultrapure water, autoclave 20 min at 120 °C, keep at 4 °C.
9. 20% casamino acids: add 20 g of casamino acids to 100 mL of ultrapure water, autoclave 20 min at 120 °C, keep at 4 °C.
10. Trace metals stock solution: add 200 mg of $CaSO_4 \cdot 2H_2O$, 20 mg of $MnSO_4 \cdot H_2O$, 20 mg of $CuSO_4 \cdot 5H_2O$, 20 mg of $ZnSO_4 \cdot 7H_2O$, 10 mg of $CoSO_4 \cdot 7H_2O$, 12 mg of $NaMoO_4 \cdot H_2O$, 5 mg of H_3BO_3, to 1 L of ultrapure water, filter-sterilize through a 0.22 μm filter. Prevent from light and store at 4 °C.

11. Stock solution of 200 mg/l $FeSO_4$: add 10 mg of $FeSO_4 \cdot 7H_2O$ to 50 mL of ultrapure water, filter-sterilize through a 0.22 μm filter. Protect from light, keep at 4 °C (*see* **Note 1**).
12. Solution of 100 mg/mL gentamicin sulfate: add 100 mg of gentamicin sulfate salt to 1 mL of ultrapure water, filter-sterilize through a 0.22 μm filter, store at 4 °C.
13. Screening medium (250 mL final volume): In this order, to 0.25 mL 1 M of $MgCl_2$, add 0.025 mL of 1 M $CaCl_2$, complete to 225 mL with ultrapure water. Add 25 mL of A-10 buffer, 5 mL of 10% glucose (0.2% final), and 6.25 mL of 20% casamino acids (0.5% final). Add 0.250 mL of the stock trace metals solution, 0.250 mL of Stock solution of 200 mg/L $FeSO_4$, and 0.250 mL of the solution of 100 mg/mL gentamicin sulfate (approx. 100 μg/mL final) (*see* **Notes 2** and **3**).
14. 100% DMSO, which meets EP and USP specifications, keep at room temperature, prevent from light (*see* **Note 4**).
15. Sodium nitroprusside (SNP) stock solution: Dissolve 7.45 mg of SNP in 5 mL of 100% DMSO. Aliquot the stock solution and protect it from light. Store at −20 °C (*see* **Note 5**).
16. 0.9% salt-water solution: add 9 g of NaCl to 1 L of ultrapure water.
17. Compound library: dissolve library compounds in 100% DMSO. Store at −20 °C.
18. Incubator, set to 37 °C.
19. Orbital table shaker suitable for plate holder. We recommend using a table shaker type KS501 digital, from IKA. Its orbital amplitude of 30 mm diameter enables optimal oxygenation for the volume of culture used in our screening protocol. (*see* **Note 6**).
20. Erlenmeyer flask 50 mL.
21. Multidispenser to dispense fresh medium or culture, type FluidX Xrd-384, eight channels (*see* **Note 7**).
22. Liquid handling robot with 96 head tips and single use 25 μL tips, type Caliper Life Sciences Zephyr (PerkinElmer), to dispense compounds.
23. Plate reader: Viktor™ x4 PerkinElmer 2030 Multilabel Reader (*see* **Note 8**).
24. Air-permeable membrane lid preventing evaporation of growth media, suitable for 96-well plates. We recommend using the air-permeable sandwich cover for polystyrene 96 low round

well multiplates from Kuhner, duetz-system. Some pins enable the lid to be placed easily; they limit evaporation and can easily be cleaned between screens (*see* **Note 9**).

25. Plate holder which enables the simultaneous shaking of sixteen 96-well plates with their air-permeable membrane lid on. We recommend using the Universal Clamp for 12–16 low well multiplates from Kuhner, duetz-system. This plate holder fits the recommended table shaker. Plates are easily placed with their lid on and kept in place under high-shaking conditions.
26. Screening plate: black 96-well plate with flat optical bottom (Nunc).

3 Methods

The following is a description for screening 1408 compounds/day (single dose) or 704 compounds/day in duplicate (*see* **Note 10**). The workflow is organized by days to better reflect timing. Subheading 3.4 and 3.5, **steps 1–12** are designed for testing the suitability of the reporter system by determining the ***Z*-factor** while Subheadings 3.5, **step 13** and 3.6 describe the procedure for actually screening library compounds.

3.1 Preparation for the Data Handling of the Screening: Day 1 (Morning)

1. Before embarking on any screening campaign, we recommend to create one file per compound library or screening project. This can easily be done using Excel (*see* **Note 11**).
2. Generate the layout of each screening plate. Depending on if you want to work with single measurement or technical duplicate, attribute one to two wells per compound, remember to have control wells in each plate. Fifty micromolar SNP is used as a positive control for c-di-GMP lowering, and 1% v/v DMSO is used as a negative control for treatment as compounds are dissolved in DMSO (e.g., Fig. 2a) (*see* **Note 12**).
3. Give each plate layout an identity (ID) (e.g., plate #_Date; plate # should be unique to each plate) (e.g., Fig. 2a).
4. In the eventuality you did create an Excel file for your screening campaign, we recommend to paste the layout of your plates screened within a day into one Excel sheet named by the date of screening (*see* **Notes 13** and **14**).
5. Save and close the file.

3.2 Preparation of the c-di-GMP Monitor Strain: Day 1 (Afternoon)

1. Streak the c-di-GMP monitor strain (*see* ref. 5) from your freezer stock culture at −80 °C, onto LB-agar plates (1.5% agar) supplemented with 100 μg/mL of gentamicin.
2. Incubate overnight (approx. 18 h) at 37 °C.

3.3 Preparation of Overnight Cultures: Day 2 (Afternoon)

1. Transfer 15 mL of freshly prepared screening medium (described **step 13** in Subheading 2) into a sterile 50 mL Erlenmeyer flask.
2. Inoculate the 50 mL Erlenmeyer flask with five single colonies of the c-di-GMP monitor strain (*see* Subheading 3.2).
3. Repeat Subheading 3.3, **steps 1** and **2** above to generate a technical duplicate.
4. Place the cultures at 37 °C and 200 rpm for 18 h. The OD_{600nm} of the overnight culture should be around 3.5 (*see* **Note 6**).

3.4 Preparation of the Assay for Z′-Factor Determination: Day 3

A Z′-factor analysis of the method will estimate if the activity range of your assay is broad and sufficient to be used as a screening procedure. To do so, SNP (50 μM final) is used as a positive control for c-di-GMP lowering, and DMSO (1% v/v final) is used as a negative control. We highly suggest to verify the Z′-factor of your screening every time new medium is prepared.

1. Allow the 5 mM SNP solution (dissolved in 100% DMSO) and the 100% DMSO stock solutions to thaw at room temperature in the dark (*see* **Notes 4** and **5**).
2. Prepare two microtiter plate layouts: we recommend using half of the plate for each control. For example, write "SNP" in all wells of column 1–6 and "DMSO" in all wells of column 7–12. Write the ID of the two plates layout generated on the 96-well plates used for the Z′-factor determination assay (*see* **Note 13**).
3. Using the dispenser, distribute 50 μL of sterile medium into the all the wells (*see* **Note 15**).
4. Using a liquid handling robot or multi-pipette, dispense as indicated in your layouts 1 μL of SNP 5 mM (dissolved in 100% DMSO; SNP 50 μM final) and 1 μL of DMSO 100% per well (DMSO 1% v/v final).
5. Prepare the bacterial inoculum by diluting the overnight culture of the c-di-GMP monitor strain from Subheading 3.3 50-fold into 10 mL of screening medium (*see* **Notes 6**, **16**, and **17**).
6. Dispense 50 μL of the diluted bacterial culture into all wells of the plate prepared **step 4** (*see* **Note 18**).
7. Repeat for the technical replicate of the overnight culture.
8. Vortex the plates to make sure compounds are correctly resuspended into the cultures (*see* **Note 19**).
9. Place the plate into the plate reader. Make sure the plate is placed in the right orientation in the plate reader.
10. Start the program to read the absorbance at 590 nm (Abs590nm) and the fluorescent output of the monitor strain (e.g., for GFP: $\lambda_{EX} = 485$ nm; $\lambda_{EM} = 535$ nm) (*see* **Note 20**).

11. For each plate, name the output data file from the plate reader as follow: "plate ID_t = 0 h". This data point will be used to remove the background for the Abs590nm and fluorescence, which could bias the final data analysis.
12. Place the membrane lid onto the 96-well plates (*see* **Note 21**).
13. Place and lock the plates onto the plate holder in the shaker, and incubate at 37 °C at 290 rpm overnight (approx. 18 h) (*see* **Note 22**).
14. In order to be able to start your screening right after the validation of the Z'-factor of your screening method (*see* Subheading 3.5, **steps 1 to 12**), we recommend preparing an overnight culture as described in Subheading 3.3. Thus, you will be able to use this overnight culture to start screening your compounds as soon as Day 4 afternoon (*see* Subheading 3.5, **step 13**).

3.5 Evaluation of the Screening Setups, Determination of the Z′-Factor: Day 4 (Morning), and Start of the Chemical Library Screening: Day 4 (Afternoon)

1. Using a dispenser, add 100 μL of sterile salt-water 0.9% in all wells and vortex to homogenize the cultures of the 96-well plate (*see* **Note 23**).
2. Place the 96-well plate into the plate reader. Make sure the plate is placed in the right orientation in the plate reader.
3. Start the program to read the absorbance at 590 nm (Abs590nm) and the fluorescent output of the monitor strain (e.g., for GFP: $\lambda_{EX} = 485$ nm; $\lambda_{EM} = 535$ nm) (*see* **Note 20**).
4. For each plate name the output data file from the plate reader as follow: "plate ID_t = 18 h".
5. To ease the data analysis, we recommend copying the data from the file "plate ID_t = 18 h" into the file "plate ID_0h".
6. For each well, correct the Abs590nm and GFP output at the endpoint of the assay for their background. Use the following formula: $Abs590nm_{corrected} = Abs590nm_{t\,=\,18h} - Abs590nm_{t\,=\,0h}$;$GFP_{corrected} = GFP_{t\,=\,18h}$-$GFP_{t\,=\,0h}$.
7. Calculate the ratio $GFP_{corrected}/Abs590nm_{corrected}$, which provides a measure of a relative fluorescent unit (RFU). In this screening procedure RFUs are used to discriminate hit compounds. Thus, RFUs are used to validate the Z'-factor of the assay.
8. For all SNP treated cultures, calculate the mean of the RFUs ($Mean_{Low}$) and RFUs standard deviation ($Stdev_{Low}$).
9. For all DMSO treated cultures, calculate the mean of the RFUs ($Mean_{High}$) and RFUs standard deviation ($Stdev_{High}$).
10. Calculate the absolute value of $Mean_{High} - Mean_{Low}$ (Abs ($Mean_{High} - Mean_{Low}$)).

11. Calculate the Z-factor using the following formula:
$Z' = 1 - ((3 \times Stdev_{High} + 3 \times Stdev_{Low})/Abs(Mean_{High} - Mean_{Low}))$.
12. If your Z-factor has a value in the range 0.5 and 1, then your assay is suitable for high-throughput screening and you can start to prepare your screening plates (*see* following step) (*see* **Note 24**).
13. Prepare the plates for screening the chemical library (Day 4, afternoon):
Repeat Subheading 3.4 but following the layout of the plates prepared in Subheading 3.1, **step 2** (*see* **Notes 16** and **17**).

3.6 Results and Analysis: Day 5

1. Follow Subheadings 3.5, **steps 1–7**.
2. Calculate the percentage of reduction of the RFUs between treated ($RFU_{treated}$) and untreated cultures (RFU_{DMSO}). The following formula can be used: Variation (%) = ($RFU_{treated} - RFU_{DMSO})/RFU_{DMSO} \times 100$ (*see* Fig. 2b).
3. Calculate the percentage of reduction of the OD590nm between treated ($Abs590nm_{treated}$) and untreated cultures ($Abs590nm_{DMSO}$). The following formula can be used: Variation (%) = ($Abs590nm_{treated} - Abs590nm_{DMSO})/Abs590nm_{DMSO} \times 100$ (*see* Fig. 2b).
4. Calculate the percentage of reduction of the GFP output between treated ($GFP_{treated}$) and untreated cultures (GFP_{DMSO}). The following formula can be used: Variation (%) = ($GFP_{treated} - GFP_{DMSO})/GFP_{DMSO} \times 100$ (*see* Fig. 2b).
5. For our purpose, we selected compounds lowering the RFU of the c-di-GMP monitor strain by more than 30% without impacting the growth (Abs590nm) more than 20% (*see* Fig. 2c).

4 Notes

1. $FeSO_4$ solution should be kept at 4 °C for one month maximum. If the yellow color appears to intensify or the solution gets turbid, discard and prepare a fresh one.
2. If extensive screening is planned, it is possible to prepare a large batch of the screening medium and store it at 4 °C. This way, the Z-factor for the batch does not need to be repeated before each screening. However, we do not recommend keeping the batch of screening medium for more than a week to avoid any metal precipitation.

3. We assume that any other medium could be used with this screening procedure as long as the Z'-factor of the assay (*see* Subheading 3.4 and 3.5, **steps 1–12**) lies between 0.5 and 1.
4. DMSO is light sensitive and hygroscopic and should be kept away from light and kept closed in a dry place.
5. Solution should be kept at −20 °C. Because of its light sensitivity, the SNP can get degraded. We recommend to aliquot the stock solution of SNP 5 mM to optimize the freshness of the SNP during the screening process. The SNP stock solution can be thawed at room temperature in the dark more than one time. However, any change in the coloration of the SNP stock might indicate a degradation of the SNP. Thus any solution showing any bluish or greenish coloration should be discarded.
6. Using our growth conditions, the Abs590nm value of the overnight culture will be approximately 3.5. Depending on your shaker the Abs590nm of your overnight culture might differ. If the Abs590nm is below 3.5, we recommend increasing the shaking intensity of the overnight culture. The use of a bigger Erlenmeyer flask will likely increase the aeration of the culture and might increase its yield. Note that, if your shaker does not allow growing an overnight culture with an Abs590nm around 3.5, this method may still be suitable for screening purpose. Follow the protocol using your overnight culture and calculate the Z'-factor of your assay (*see* Subheading 3.4, **steps 1–12** and Subheading 3.5).
7. We assume that any dispenser could be used that is capable of distributing a volume of 50 μL in 96-well plates under sterile conditions.
8. We assume that any plate reader could be used that supports a 96-well microtiter plate format, capable of reading absorbance at 590 nm and performing excitation/emission read of GFP derivatives (e.g., $\lambda_{EX} = 485$ nm; $\lambda_{EM} = 535$ nm).
9. The recommended membrane lids will only work on plates that have empty, and not solid, space between each well.
10. This procedure could easily be scaled up by increasing the number of plate holders and membrane lids.
11. For our screening purpose we used Excel, but any software that supports the layout and file format of the data file generated by the plate reader can be used. If intensive library screening is planned, we highly recommend using software controlling the plate reader that is capable of performing the data analysis; this greatly helps the screening process. Note that it is possible to create macro program in Excel, which enable to automatize the data analysis, therefore increasing the ease and efficiency of your screening campaign.

12. When we screen our chemical library, we use four wells for each control. However, we postulate that two wells for DMSO control and two wells for SNP 50 μM controls would be sufficient. For consistency, we recommend to always keep the controls at the same position.
13. In order to ease the tracking of data and plates, we recommend identifying the plates used to determine the *Z*-factor (*described* Subheading 3.4) such as: "letter_date (X_YEARMMDD)" of the *Z*-factor assay and write the ID in the top left corner cell of the layout. For example, name one of your plate used to determine the *Z*-factor of the assay the 29th of September 2017 "A_20170929" and the other "B_20170929". For the plates used to screen compounds, we recommend identifying them such as: "number-date (N_YEARMMDD)" of the screening and write it in the top left corner cell of the layout.
14. Screening an entire library might require multiple days. For a single library of compounds, we recommend having one Excel file containing different sheets each identified by date. Each sheet should contain the layout of the plates screened on a single day. We recommend naming each sheet by the date of the screening day. Remember to give an ID to each plate as well, e.g. "1_20170929," "2_20170929"; these IDs would correspond to the layout of plates screened the 29th of September 2017. This makes the tracking of the compound analysis easier. When you need to find the data file corresponding to your compound "X", search for "X" in your Excel file. Then by looking at the number of the plate and the date, you can recreate the name of the data file from the plate reader.
15. To prevent contamination of the liquid handling robot, we chose to dispense the library compounds into sterile medium. This also ensures enhanced mixing of the compounds. An alternative to this procedure is to prepare a culture by diluting an overnight culture 1:100 into screening medium, then fill the wells with 100 μL of this culture and add 1 μL of compound per well, followed by mixing. Note that if you choose this procedure, you can proceed directly to Subheading 3.4, **step 8**.
16. Depending of the length of the tubing of your dispenser, you might need to prepare a larger volume of the bacterial culture. We recommend evaluating this volume before starting to prepare the medium and culture for the plate assay.
17. If you chose to dispense the compound in 100 μL of culture as suggested in **Note 15**, the volume of cultures need to be increased accordingly.
18. After dispensing the culture, the dispenser should be rinsed with water, and then a solution of 0.5% chlorine or 70% ethanol should be run through and left for some hours to disinfect the

machine. The dispenser should be rinsed with water and culture medium before use.

19. We recommend using a plate vortex to mix the contents present in the microtiter plates. An alternative is to place the plates with membrane-lid on, into the shaker at 290 rpm for 1 min as described Subheading 3.4, **steps 12** and **13**, and then read $t = 0$.

20. For our screening procedure, we used the Viktor™x4 2030 Multilabel Reader (PerkinElmer). We created a program to read both the Abs590nm and the GFP output ($\lambda_{EX} = 485$ nm; $\lambda_{EM} = 535$ nm) of our c-di-GMP monitor strain. This way, for each plate, Abs590nm and GFP data were exported into the same file, which eases their analysis and tracking. If you used the plate annotation system described in Subheadings 3.1, **steps 2** and **3 Note 13**, remember to make sure that the name of the data file output of the plate reader corresponds to "ID plate_t = Xh", X depending if your read at $t = 0$ h or at the end point $t = 18$ h.

21. If the membrane of the membrane lid has not been in contact with the culture, the membrane lid can be reused between the assay without being autoclaved but we recommend to wipe the silicone surface with 70% ethanol prior to use.

22. Shaking should provide maximal aeration as it has been observed that low oxygenation drastically decreases the output of the c-di-GMP monitor strain, reducing the dynamic range of the assay.

23. We recommend using 0.9% NaCl solution or any isotonic solution in order to prevent cell lysis and growth. By doing so, plates can be kept at 4 °C after reading until analysis of the data and in case a reading error is suspected plate can be read again. Use an isotonic solution which does not impact the fluorescence output neither the Abs590nm.

24. If your Z'-factor is lower than 0.5, it could be due to either a lack of oxygenation and/or poor growth. You can try to increase the shaking condition and/or to slightly increase the trace metal and ferrous iron concentrations. However, we observed that at concentrations above 1 μM of ferric iron the output of the c-di-GMP monitor strain decreases, lowering the dynamic range of the assay.

Acknowledgments

This work was supported by grants from and the Danish Council for Independent Research (DFF–1323-00177) to TTN, and from the Danish Strategic Research Council and the Lundbeck Foundation (R198-2015-486) to MG.

References

1. Costerton JW, Stewart PS, Greenberg EP (1999) Bacterial biofilms: a common cause of persistent infections. Science 284:1318–1322
2. Hengge R (2009) Principles of c-di-GMP signalling in bacteria. Nat Rev Microbiol 7:263–273
3. Gjermansen M, Ragas P, Tolker-Nielsen T (2006) Proteins with GGDEF and EAL domains regulate *Pseudomonas putida* biofilm formation and dispersal. FEMS Microbiol Lett 265:215–224
4. Christensen LD, Van Gennip M, Rybtke MT et al (2013) Clearance of *Pseudomonas aeruginosa* foreign-body biofilm infections through reduction of the cyclic Di-GMP level in the bacteria. Infect Immun 81:2705–2713
5. Rybtke MT, Borlee BR, Murakami K, Irie Y et al (2012) Fluorescence-based reporter for gauging cyclic di-GMP levels in *Pseudomonas aeruginosa*. Appl Environ Microbiol 78:5060–5069
6. Groizeleau J, Rybtke M, Andersen JB et al (2016) The anti-cancerous drug doxorubicin decreases the c-di-GMP content in *Pseudomonas aeruginosa* but promotes biofilm formation. Microbiology 162:1797–1807
7. Silver LL (2011) Challenges of antibacterial discovery. Clin Microbiol Rev 24:71–109

Chapter 34

Genetic Tools to Study c-di-GMP-Dependent Signaling in *Pseudomonas aeruginosa*

Livia Leoni, Sarika Vishnu Pawar, and Giordano Rampioni

Abstract

Pseudomonas aeruginosa infections are often difficult or impossible to treat, mainly due to its ability to form antibiotic-resistant biofilms. Since c-di-GMP signaling strongly influences *P. aeruginosa* biofilm development and sensitivity to antibiotics, it is considered a promising target for the development of anti-biofilm drugs and it is under intensive investigation. However, studying c-di-GMP signaling in *P. aeruginosa* is challenging, mainly due to (1) the multiplicity of enzymes involved in c-di-GMP metabolism, (2) the difficulty to extract and measure c-di-GMP intracellular levels by chemical methods, and (3) the lack of genetic tools specifically dedicated to this purpose.

Here, a bioluminescence-based reporter system convenient for studying cellular processes or compounds expected to cause an increase or a decrease in intracellular c-di-GMP levels produced by *P. aeruginosa* cultures is described. Bioluminescence is particularly appropriate in *P. aeruginosa* research, due to the high intensity of the signal and total lack of background noise. In addition, the use of genetic cassettes allowing the fine control of *P. aeruginosa* c-di-GMP intracellular levels via arabinose induction is described.

Overall, the genetic tools described here could facilitate investigations tackling the c-di-GMP signaling process on different fields, from cellular physiology to drug-discovery research.

Key words *Pseudomonas aeruginosa*, Cyclic-di-GMP, Whole-cell biosensors, Genetic tools, Screening, Biofilm inhibition

1 Introduction

The human pathogen *Pseudomonas aeruginosa* is considered a major model organism for studies concerning the second messenger 3′,5′-cyclic diguanylic acid (c-di-GMP). As in many other bacteria, the c-di-GMP intracellular levels in *P. aeruginosa* rely on the opposite enzymatic activity of diguanylate cyclases (DGCs) and phosphodiesterases (PDEs), involved in the synthesis and degradation of this molecule, respectively. A decrease in c-di-GMP intracellular levels hampers *P. aeruginosa* ability to form biofilm and increases its sensitivity to antibiotics [1, 2]. Consequently, *P. aeruginosa* c-di-GMP signaling is considered a promising target for the development of anti-biofilm drugs [1, 2].

Karin Sauer (ed.), *c-di-GMP Signaling: Methods and Protocols*, Methods in Molecular Biology, vol. 1657, DOI 10.1007/978-1-4939-7240-1_34, © Springer Science+Business Media LLC 2017

Several studies investigating the physiological role of c-di-GMP signaling exploited inducible genetic systems for the overexpression of genes encoding DGCs or PDEs, in order to artificially alter the intracellular levels of this second messenger in *P. aeruginosa* [1, 2]. The underlying rationale of this approach is that the overexpression of a DGC or a PDE gene would increase or decrease the intracellular concentration of c-di-GMP, respectively. Here, the use of improved genetic devices allowing for fine-tuning of intracellular c-di-GMP levels upon arabinose-dependent control is described. Three integrative plasmids carrying genetic cassettes for the arabinose-dependent expression of the *P. aeruginosa* DGC YfiN (also known as PA1120 or TpbB) or the PDEs RbdA (also known as PA0861) or PA2133 have been developed to this purpose (Fig. 1). These plasmids were named $pP_{BAD}YfiN^{Ind}$, $pP_{BAD}RbdA^{Ind}$, and $pP_{BAD}2133^{Ind}$, respectively [3]. Briefly, an AraC/P_{BAD} promoter–regulator couple, optimized to specifically avoid basal expression level in *P. aeruginosa*, was employed for the arabinose-dependent expression of the genes encoding YfiN, RbdA and PA2133 [3, 4], and the mini-CTX1 plasmid [5] was selected as delivery vector for the chromosomal integration of these expression devices. These genetic devices allow stable integration of the expression cassettes in the *attB* neutral site of the *P. aeruginosa* chromosome, without affecting the overall cell physiology and the need of antibiotics for positive selection. Since in *P. aeruginosa* high levels of c-di-GMP cause pellicle formation (cells clumping) in liquid cultures [6–8], this phenotype can be conveniently used to rapidly assess the ability of the expression cassettes to alter intracellular c-di-GMP levels (Fig. 2) [3].

c-di-GMP levels can be precisely measured only after cell disruption, solvent extraction and subsequent analysis by radioactive thin layer chromatography, mass spectrometry, or circular dichroism [9, 10]. These methods, although accurate and sensitive, could be difficult/inconvenient to use for research groups with limited expertise in analytical chemistry. To this end, a convenient plasmid for studying cellular processes or compounds expected to cause an increase or a decrease in intracellular c-di-GMP levels produced by *P. aeruginosa* cultures has been developed [3]. This plasmid, named pP*cdrA*::*lux*, carries a transcriptional fusion between a c-di-GMP responsive promoter (*i.e.*, the promoter of the *cdrA* gene) and the *luxCDABE* operon for bioluminescence emission. Bioluminescence is a high-performance signal not requiring substrate addition for detection and endowed with robust signal emission. Moreover, since *P. aeruginosa* does not emit light, light emission from engineered *P. aeruginosa* strains carrying the pP*cdrA*::*lux* plasmid is not affected by background noise [11, 12]. The ability to detect intracellular c-di-GMP levels lower than the basal levels normally produced by the wild type *P. aeruginosa* strain represents an innovative aspect of the pP*cdrA*::*lux* reporter device. Indeed, previously

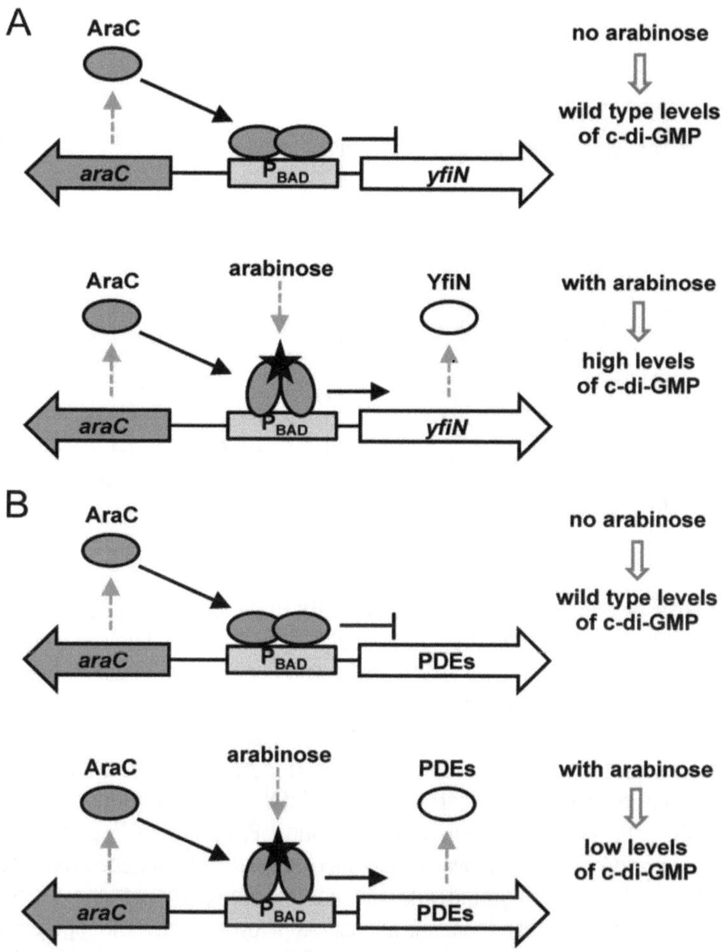

Fig. 1 Schematic representation of the genetic devices for the arabinose-dependent expression of the YfiN DGC (**a**) and of the RbdA or PA2133 PDEs (**b**). When arabinose is present, it binds to the AraC protein, leading to activation of the P_{BAD} promoter, hence allowing expression of the downstream genes. The arabinose-dependent induction of YfiN results in high levels of intracellular c-di-GMP (**a**), while the arabinose-dependent induction of the PDEs RbdA or PA2133 leads to a decrease in c-di-GMP levels (**b**). *Black arrows* indicate positive regulation; *black T-lines* indicate negative regulation. Modified from [3], with permission

described reporter systems based on *gfp* as reporter gene were able to detect a decrease in the c-di-GMP intracellular levels only in engineered bacterial strains overproducing c-di-GMP [13]. The reduced performance of the fluorescence-based reporter device with respect to pP*cdrA*::*lux* could be due to the lower intensity of the fluorescent signal when compared to light emission, and to the high fluorescent background of *P. aeruginosa* cultures.

The pP*cdrA*::*lux* plasmid could be used in any application in which an indirect and convenient method to detect variations in the intracellular levels of c-di-GMP in *P. aeruginosa* is required. As an example, Fig. 3 shows the response of *P. aeruginosa* PAO1

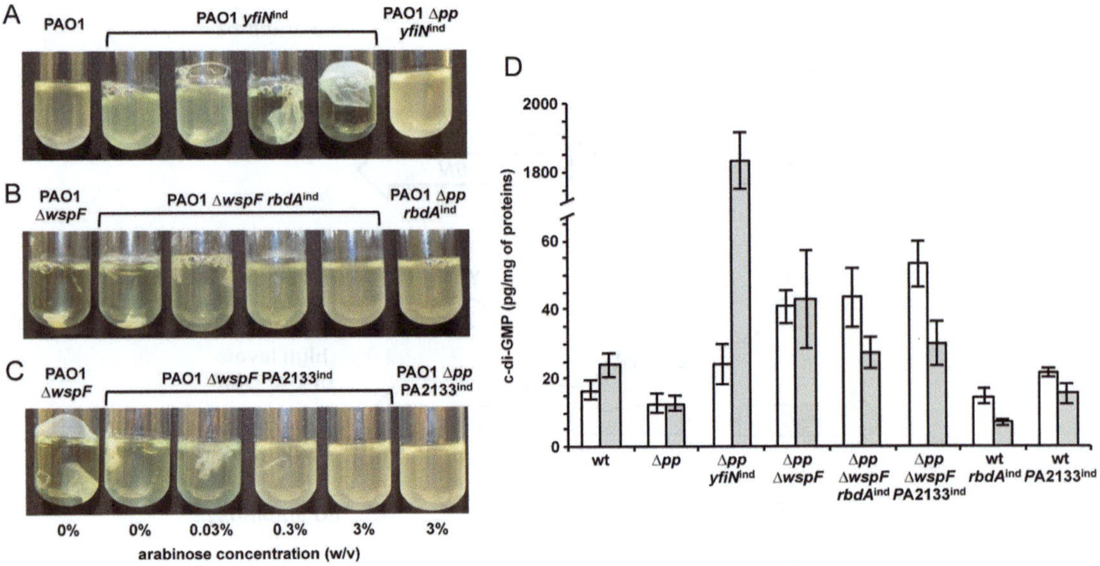

Fig. 2 Effect of arabinose-dependent induction of (**a**) YfiN, (**b**) RbdA, or (**c**) PA2133 on pellicle formation in the following *P. aeruginosa* strains: PA01 wild type; PA01 *yfiN*^{Ind}, PA01 derivative carrying pP$_{BAD}$YfiN^{Ind} plasmid integrated into the chromosome; PA01 Δ*pp yfiN*^{Ind}, PA01 *yfiN*^{Ind} derivative in which the *pel* and *psl* genes for exopolisaccarides biosynthesis and pellicle formation have been inactivated; PA01 Δ*wspF*, PA01 pellicle forming mutant carrying a deletion in the *wspF* gene; PA01 Δ*wspF rbdA*^{Ind}, PA01 Δ*wspF* derivative carrying pP$_{BAD}$RbdA^{Ind} plasmid integrated into the chromosome; PA01 Δ*pp rbdA*^{Ind}, PA01 derivative in which the *pel* and *psl* genes have been inactivated and carrying pP$_{BAD}$RbdA^{Ind} plasmid integrated into the chromosome; PA01 Δ*wspF* PA2133^{Ind}, PA01 Δ*wspF* derivative carrying pP$_{BAD}$PA2133^{Ind} plasmid integrated into the chromosome; PA01 Δ*pp rbdA*^{Ind}, PA01 derivative in which the *pel* and *psl* genes have been inactivated and carrying pP$_{BAD}$PA2133^{Ind} plasmid integrated into the chromosome. The strains were grown in M9-CAA supplemented with different arabinose concentrations, as shown below the figures. Photographs were taken after 14 h of growth at 37 °C in static conditions. (**d**) Intracellular levels of c-di-GMP measured by liquid chromatography tandem mass spectrometry in the indicated strains grown in the absence (*white bars*) and in the presence (*grey bars*) of 0.3% arabinose. Results are the average of three independent assays. Standard deviations are reported as error bars. Reproduced from [3], with permission

(pP*cdrA*::*lux*) to the well-known c-di-GMP inhibitor sodium nitroprusside (SNP) [14, 15]. Other possible applications of the pP*cdrA*::*lux* plasmid could be (1) to investigate possible alteration of intracellular c-di-GMP levels in a *P. aeruginosa* mutant with respect to the wild type strain, and (2) to identify new compounds affecting c-di-GMP intracellular levels in a screening campaign.

Overall, the genetic tools described here could facilitate investigations tackling c-di-GMP-dependent biofilm formation at different levels, from cellular physiology to drug-discovery research.

2 Materials

All solutions should be sterilized and prepared using double-distilled water.

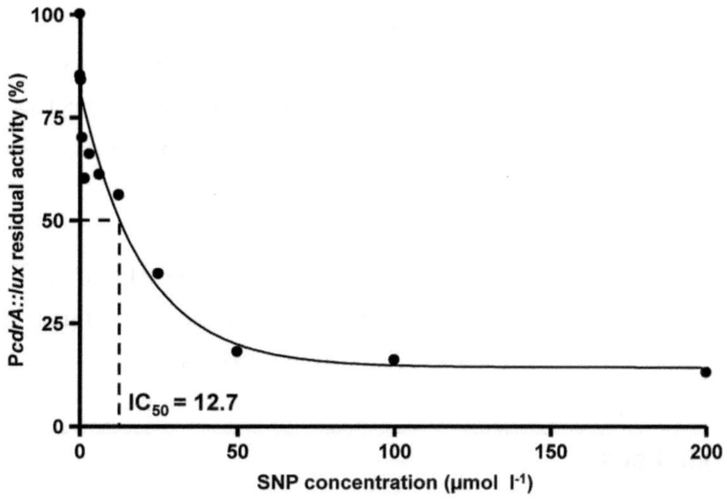

Fig. 3 Graph reporting the dose-dependent response of the biosensor *P. aeruginosa* PA01 (P*cdrA::lux*) to different SNP concentrations (indicated below the graph). The *line* is the best-fit curve generated by the software Prism, using the log-dose vs. response eq. *Y*-axis intercept represents the 100% residual activity obtained in the absence of SNP. Data fit allowed to extrapolate the IC_{50} being 12.7 μM. Reproduced from [3], with permission

1. Bacterial strains: *Escherichia coli* S17.1 λ*pir* carrying one of the following plasmids: $pP_{BAD}YfiN^{Ind}$; $pP_{BAD}RbdA^{Ind}$; pP_{BAD}-$PA2133^{Ind}$; pP*cdrA::lux* [3]; *P. aeruginosa* strain(s) of choice. The strains are available from the authors, upon request.
2. Luria–Bertani broth (LB): 10 g/L NaCl, 10 g/L tryptone, 5 g/L yeast extract.
3. M9-CAA: 82 mM Na_2HPO_4, 24 mM KH_2PO_4, 130 mM NaCl, 1 mM $MgSO_4$, 0.1 mM $CaCl_2$, 0.5% casamino acids.
4. Kanamycin (Km): prepare a 25 mg/mL stock solution in bi-distilled sterile water, and store at −20 °C.
5. Tetracycline (Tc): prepare a 20 mg/mL stock solution in bi-distilled sterile water containing 50% (v/v) ethanol, and store at −20 °C.
6. Nalidixic acid (Nal): prepare a 5 mg/mL stock solution in bi-distilled sterile water, add one or more drops of 1 M NaOH to allow dissolution of the powder, and store at −20 °C.
7. Luria-Bertani agar (LA): LB plus 15 g/L agar.
8. LA plates supplemented with 200 μg/mL Tc and 15 μg/mL Nal.
9. Sodium nitroprusside (SNP): freshly prepare a 50 mM stock solutions in bi-distilled sterile water before use.
10. Arabinose: prepare a 25% stock solutions (25 g/100 mL) in bi-distilled sterile water.

11. Saline solution: 9 g/L NaCl.
12. Black clear-flat-bottom 96-wells microtiter plates.
13. Automated luminometer-spectrometer plate reader.
14. Conical glass flasks, 50 mL.
15. Glass tubes, 10 mL.
16. Microcentrifuge.
17. Mcrocentrifuge tubes, 1.5 mL.
18. Disposable sterile loops, 10 μL.
19. Disposable petri dishes.

3 Methods

3.1 Generation of *P. aeruginosa* Strains with Governable c-di-GMP Levels

1. Introduce $pP_{BAD}YfiN^{Ind}$, $pP_{BAD}RbdA^{Ind}$ or $pP_{BAD}PA2133^{Ind}$ into *P. aeruginosa* strains by conjugation. To do so, grow overnight at 37 °C in 50 mL conical glass flasks:
   (a) The recipient *P. aeruginosa* strain of choice (*e.g.*, wild type PAO1 or PAO1 Δ*wspF*, a c-di-GMP overproducing mutant) in 5 mL of LB.
   (b) The donor strain *E. coli* S17.1 λ*pir* carrying one of the following plasmids $pP_{BAD}YfiN^{Ind}$, $pP_{BAD}RbdA^{Ind}$, or $pP_{BAD}PA2133^{Ind}$, in 5 mL of LB supplemented with 10 μg/mL Tc.
2. Centrifuge 1 mL of each of the cultures at 3000 × *g* for 5 min at room temperature in a 1.5 mL microcentrifuge tube.
3. Wash each of the bacterial pellets with 500 μL of saline solution twice.
4. Mix the bacterial suspensions derived from 1 mL of the donor and of the recipient strains in a 1.5 mL microcentrifuge tube, and centrifuge at 3000 × *g* for 5 min at room temperature.
5. Resuspend the resulting bacterial pellets in 20 μL of saline solution.
6. Spot the resuspended mixtures onto the center of an LA plate.
7. As a control, repeat **steps 2–3** for the recipient *P. aeruginosa* and the donor *E. coli* S17.1 λ*pir* strains.
8. Resuspend each resulting bacterial pellet in 20 μL of saline solution.
9. Spot 20 μL of the bacterial suspensions derived from 1 mL of the donor or the recipient strains alone onto the center of LA plates (*see* **Note 1**).
10. Incubate the plates for 6 h at 37 °C.

11. After 6 h incubation at 37 °C, recover the bacterial spots from LA plates with 10 μL sterile disposable loops, and resuspend them in 1 mL saline solution in 1.5 mL microcentrifuge tubes by vigorously rotating the loops.
12. Repeat the step above for both controls and the mixed samples.
13. Plate 100 μL of each of the resulting bacterial suspensions on LA plates supplemented with 200 μg/mL Tc and 15 μg/mL Nal (*see* **Note 1**).
14. Centrifuge each of the remaining 900 μL of the bacterial suspensions at 3000 × *g* for 5 min at room temperature.
15. Resuspend the pellets in 100 μL saline solution and plate on LA plates supplemented with 200 μg/mL Tc and 15 μg/mL Nal.
16. Incubate the plates overnight at 37 °C.
17. Select single colonies and restreak them on LA plates supplemented with 200 μg/mL Tc and 15 μg/mL Nal (*see* **Note 1**).
18. Incubate the plates overnight at 37 °C (*see* **Note 2**).
19. To visualize pellicle formation and cell clumping, inoculate the bacteria from the LA plates obtained in **step 18** in 3 mL M9-CAA in 50 mL conical glass flasks.
20. Incubate the cultures overnight (*i.e.*, 16–18 h) at 37 °C with shaking at 200 rpm.
21. Dilute the overnight cultures to an absorbance at 600 nm wavelength (A_{600}) of 0.02 in 10 mL glass tubes containing 2 mL M9-CAA supplemented with increasing concentrations of arabinose (*e.g.*, from 0 to 3%, as in the example shown in Fig. 2).
22. Incubate at 37 °C in static conditions for 14 h and check pellicle formation, as in the example shown in Fig. 2 (*see* **Note 3**).

3.2 Use of Bioluminescence for Monitoring c-di-GMP Levels

A *P. aeruginosa* strain containing the reporter plasmid pP*cdrA::lux* will emit light as a function of the intracellular c-di-GMP levels. Examples of utilization of this reporter system are: (a) to study the effect of a drug on c-di-GMP intracellular levels; (b) to investigate possible alteration of intracellular c-di-GMP levels in a *P. aeruginosa* mutant with respect to the wild type strain. Introduce the pP*cdrA::lux* plasmid into the desired *P. aeruginosa* strains by conjugation, as described in Subheading 3.1, to generate the reporter strain PAO1 (pP*cdrA::lux*).

1. Select the *P. aeruginosa* strains containing the pP*cdrA::lux* plasmid on LA plates supplemented with 300 μg/mL Km and 15 μg/mL Nal.
2. Grow the reporter strain [*e.g.*, PAO1 (pP*cdrA::lux*)] overnight in 50 mL conical flasks at 37 °C with 200 rpm shaking in 5 mL of M9-CAA supplemented with 300 μg/mL Km.

3. Dilute the culture to an A_{600} of 0.5 into M9-CAA medium without antibiotic.
4. Dispense 200 μL aliquots of the OD-adjusted reporter strain in at least three wells (replicates) of a black 96-well microtiter plate with flat and clear bottom.
5. Incubate the microtiter plate at 37 °C in static conditions.
6. Measure A_{600} and light counts per second (LCPS) with an automated luminometer–spectrometer plate reader, simultaneously (*see* **Note 4**). If required, repeat the measurements along time (*e.g.*, 4, 6, and 8 h).
7. Average the A_{600} and LCPS measurements of the replicates. Normalize the averaged LCPS to the averaged A_{600} to obtain P*cdrA* activity.

3.3 Use of the Bioluminescent Reporter Strain P. aeruginosa PAO1 (pPcdrA::lux) for the Screening of Molecules Affecting c-di-GMP Levels

1. Grow the reporter strain [*e.g.*, PAO1 (pP*cdrA::lux*)] overnight at 37 °C with 200 rpm shaking in 5 mL of M9-CAA supplemented with 300 μg/mL Km (*see* **Note 5**).
2. Dilute the culture to an A_{600} of 1.0 into M9-CAA medium and dispense 100 μL aliquots in a black 96-well microtiter plate with flat and clear bottom.
3. As untreated control, add 100 μL of M9-CAA in six wells containing the reporter strain.
4. Set up serial dilutions in M9-CAA of the compound to be tested to 2× the final required concentrations. For example, if the final concentration of the compound should be tested at 200, 20 and 2 μM, prepare serial dilutions of the compound of interest in M9-CAA at a concentration of 400, 40, and 4 μM.
5. Aliquot 100 μL of each compound in three microtiter wells (replicates) containing the reporter strain.
6. Incubate the microtiter plate at 37 °C in static conditions.
7. Measure A_{600} and light counts per second (LCPS) with an automated luminometer-spectrometer plate reader, simultaneously (*see* **Note 4**). If required, repeat the measurements along time (*e.g.*, 4, 6, and 8 h).
8. Average the A_{600} and LCPS measurements of the replicates. Normalize the averaged LCPS to the averaged A_{600} to obtain P*cdrA* activity.
9. Compare P*cdrA* activity of the treated samples to the one of the untreated control.
10. Plot the P*cdrA* activity of the treated samples relative to the concentration of the compound of interest. If the compound reduces c-di-GMP levels, P*cdrA* activity should be reduced with increasing concentration of the compound. An example of a dose-dependent response of *P. aeruginosa* PAO1 (pP*cdrA*::*lux*) to the well-known c-di-GMP inhibitor SNP [14, 15] is shown in Fig. 3.

4 Notes

1. Before starting, isolate your recipient strain on LA plates supplemented with 200 μg/mL Tc and 15 μg/mL Nal. There should be no growth of isolated colonies. Otherwise, identify the minimal inhibitory concentrations of Tc in LA plates supplemented with 15 μg/mL Nal. The control samples with the donor and recipient strains alone should not grow on the selective plates, thus ensuring that colonies grown on the selective plates are not contaminants. If colonies of the control samples grow on the selective plates, this could be due to medium contamination or spoilage of the selective plates. In this case repeat the procedure with sterilized media and freshly prepared selective plates.
2. The *P. aeruginosa* strains carrying the $pP_{BAD}YfiN^{Ind}$, $pP_{BAD}RbdA^{Ind}$ or $pP_{BAD}2133^{Ind}$ plasmids can be used right after isolation on LA plates supplemented with 200 μg/mL Tc and 15 μg/mL Nal. If Tc resistance is undesired, the Tc resistance cassette can be deleted by Fpl2-mediate recombination with the pFpl2 plasmid [5]. Briefly, this can be achieved by introducing the pFpl2 plasmid in the *P. aeruginosa* strains by transformation*Plasmid transformation* or electroporation, and by selecting the recipient strains on LA plates supplemented with Carbenicillin 400 μg/mL. The pFpl2 plasmid can be then cured by growing the strains overnight at 37 °C in 2 mL of LB supplemented with 10% (w/v) sucrose counterselection, and by isolating the resulting culture on LA plates supplemented with 10% (w/v) sucrose [5].
3. Wild type *P. aeruginosa* strains containing plasmid $pP_{BAD}YfiN^{Ind}$ are expected to form pellicles in an arabinose-dependent way, indicating an increase in c-di-GMP intracellular levels. Conversely, *P. aeruginosa* Δ*wspF* (or other *P. aeruginosa* mutants overproducing c-di-GMP) are expected to show arabinose-dependent pellicle dispersal when engineered with the $pP_{BAD}RbdA^{Ind}$ or $pP_{BAD}2133^{Ind}$ genetic cassettes (as in the example shown in Fig. 2). A measurement of c-di-GMP levels by liquid chromatography coupled with mass spectrometry is recommended as additional control (as in the example shown in Fig. 2d). A mass spectrometry-based method by Severin and Waters Chapter 7 can be found elsewhere in this book. Likewise, a method based on HPLC by Petrova and Sauer Chapter 4 is likewise available elsewhere in this book.
4. We routinely use a Wallac 1420 Victor^{3V} multiplate reader (PerkinElmer) as automated luminometer-spectrometer plate reader. For the Wallac 1420 Victor^{3V} multiplate reader relevant parameters for bioluminescence measurement are: emission aperture, large; counting time, 1 s. Relevant parameters for absorbance measurement are: filter 595/60; excitation aperture, normal; and reading time, 0.1 s.

5. If using a reporter strain different from PAO1 (pP*cdrA::lux*), we recommend to validate the reporter system by testing the effect of a known c-di-GMP inhibitor (*e.g.*, SNP, as in the example of Fig. 3) prior to starting the screening.

Acknowledgments

This work was supported by: Italian Cystic Fibrosis Research Foundation (FFC 10/2013 to L.L.; www.fibrosicisticaricerca.it); MIUR of Italy (RBFR10LHD1 to G.R.). We wish to thank all the researchers involved in c-di-GMP research and in particular our colleagues Marco Messina, Serena Rinaldo, Francesca Cutruzzolà, and Volkard Kaever.

References

1. Römling U, Galperin M, Gomelsky M (2013) Cyclic di-GMP: the first 25 years of a universal bacterial second messenger. Microbiol Mol Biol Rev 77:1–52
2. Ryan RP (2013) cyclic di-GMP signaling and the regulation of bacterial virulence. Microbiology 159:1286–1297
3. Pawar SV, Messina M, Rinaldo S, Cutruzzolà F, Kaever V, Rampioni G et al (2016) Novel genetic tools to tackle c-di-GMP-dependent signalling in *Pseudomonas aeruginosa*. J Appl Microbiol 120:205–217
4. Lo Sciuto A, Fernández-Piñar R, Bertuccini L, Iosi F, Superti F, Imperi F (2014) The periplasmic protein TolB as a potential drug target in *Pseudomonas aeruginosa*. PLoS One 9(8): e103784
5. Hoang TT, Kutchma AJ, Becher A, Schweizer HP (2000) Integration-proficient plasmids for *Pseudomonas aeruginosa*: site-specific integration and use for engineering of reporter and expression strains. Plasmid 43:59–72
6. Hickman JW, Tifrea DF, Harwood CS (2005) A chemosensory system that regulates biofilm formation through modulation of cyclic diguanylate levels. Proc Natl Acad Sci U S A 102:14422–14427
7. Kulasakara H, Lee V, Brencic A, Liberati N, Urbach J, Miyata S et al (2006) Analysis of *Pseudomonas aeruginosa* diguanylate cyclases and phosphodiesterases reveals a role for bis-(3′-5′)-cyclic-GMP in virulence. Proc Natl Acad Sci U S A 103:2839–2844
8. Starkey M, Hickman JH, Ma L, Zhang N, De Long S, Hinz A et al (2009) *Pseudomonas aeruginosa* rugose small-colony variants have adaptations that likely promote persistence in the cystic fibrosis lung. J Bacteriol 191:3492–3503
9. Spangler C, Böhm A, Jenal U, Seifert R, Kaever V (2010) A liquid chromatography-coupled tandem mass spectrometry method for quantitation of cyclic di-guanosine monophosphate. J Microbiol Methods 81:226–231
10. Stelitano V, Brandt A, Fernicola S, Franceschini S, Giardina G, Pica A et al (2013) Probing the activity of diguanylate cyclases and c-di-GMP phosphodiesterases in real-time by CD spectroscopy. Nucleic Acids Res 41(7):e79
11. Meighen EA, Szittner RB (1992) Multiple repetitive elements and organization of the *lux* operons of luminescent terrestrial bacteria. J Bacteriol 174:5371–5381
12. Fan F, Wood KV (2007) Bioluminescent assays for high-throughput screening. Assay Drug Dev Technol 5:127–136
13. Rybtke MT, Borlee BR, Murakami K, Irie Y, Hentzer M, Nielsen TE et al (2012) Fluorescence-based reporter for gauging cyclic di-GMP levels in *Pseudomonas aeruginosa*. Appl Environ Microbiol 78:5060–5069
14. Barraud N, Hassett DJ, Hwang SH, Rice SA, Kjelleberg S, Webb JS (2006) Involvement of nitric oxide in biofilm dispersal of *Pseudomonas aeruginosa*. J Bacteriol 188:7344–7353
15. Barraud N, Schleheck D, Klebensberger J, Webb JS, Hassett DJ, Rice SA et al (2009) Nitric oxide signaling in *Pseudomonas aeruginosa* biofilms mediates phosphodiesterase activity, decreased cyclic di-GMP levels, and enhanced dispersal. J Bacteriol 191:7333–7342

Index

Karin Sauer (ed.), *c-di-GMP Signaling: Methods and Protocols*, Methods in Molecular Biology, vol. 1657,
DOI 10.1007/978-1-4939-7240-1, © Springer Science+Business Media LLC 2017

D

E

MIX
Papier aus verantwortungsvollen Quellen
Paper from responsible sources
FSC® C105338

If you have any concerns about our products,
you can contact us on
ProductSafety@springernature.com

In case Publisher is established outside the EU,
the EU authorized representative is:
Springer Nature Customer Service Center GmbH
Europaplatz 3, 69115 Heidelberg, Germany

Printed by Libri Plureos GmbH
in Hamburg, Germany